AIR POWER

AIR POWER

From Kitty Hawk to Gulf War II:
A History of the People, Ideas and
Machines that Transformed War
in the Century of Flight

STEPHEN BUDIANSKY

VIKING
an imprint of
PENGUIN BOOKS

VIKING

Published by the Penguin Group
Penguin Books Ltd, 80 Strand, London WC2R 0RL, England
Penguin Group (USA), Inc., 375 Hudson Street, New York, New York 10014, USA
Penguin Books Australia Ltd, 250 Camberwell Road, Camberwell, Victoria 3124, Australia
Penguin Books Canada Ltd, 10 Alcorn Avenue, Toronto, Ontario, Canada M4V 3B2
Penguin Books India (P) Ltd, 11 Community Centre, Panchsheel Park, New Delhi - 110 017, India
Penguin Books (NZ) Ltd, Cnr Rosedale and Airborne Roads, Albany, Auckland, New Zealand
Penguin Books (South Africa) (Pty) Ltd, 24 Sturdee Avenue, Rosebank 2196, South Africa

Penguin Books Ltd, Registered Offices: 80 Strand, London WC2R 0RL, England

www.penguin.com

Published in Great Britain with a new subtitle by Viking 2003
Published simultaneously in the USA by Viking Penguin
1

Copyright © Stephen Budiansky, 2003

Printed in Great Britain by Clays Ltd, St Ives plc

A CIP catalogue record for this book is available from the British Library

ISBN 0-670-91251-4

To Bill Cook

CONTENTS

AUTHOR'S NOTE

A writer contemplating a subject as full of dramatic action, flamboyant personalities, hallowed institutions, and brilliant inventions as this one faces a temptation he must bravely resist. My aim from the start was to tell the story of air *power*— of the revolutionary transformations that the airplane has brought to the conduct, consequences, and meaning of war in the hundred years since its invention. It is a story that brings together some of the greatest events and greatest minds of that century, and one of the fascinations in researching this subject has been tracing the intriguing and often unexpected interactions among personalities, institutions, and technology that conspired to foment this revolution in the way wars are fought and won, and indeed in the way we have come to think about war itself.

But telling the story of air power is not the same as offering up a complete history of aerial combat or a definitive account of the men, the institutions, or the machines that have waged war in the air. As I soon discovered, the only way I could stick to my chosen path was if I was prepared to be quite ruthless. There is, accordingly, much that is justifiably famous in the history of military aviation that I simply had to abandon by the wayside if I was to have a prayer of getting where I was going.

To those who would condemn me for failing to mention this famous airplane or that decisive battle, this legendary squadron or that heroic flyer, I plead completely guilty, and only hope that I may seek mitigation on the grounds that my intent has been to follow my tale where it led me and not (as so much military history so often does) to provide an exhaustive cataloging of all who undoubtedly deserve credit. I would also appeal to the wisdom of the French saying that Winston Churchill always said was his favorite: "L'art d'être ennuyeux, c'est de tout dire"—"The art of being boring is to tell all."

I am deeply indebted to the great scholars of air power and aviation history without whose works I could never have found my bearings in this vast field. Many were also extraordinarily generous in their personal assistance to me: answering questions, suggesting sources, and offering much-appreciated critiques of portions of this work. I would like to thank in particular James S. Corum, professor of comparative military studies at the School of Advanced Air and Space Studies, Maxwell Air Force Base; Richard P. Hallion, the former United States Air Force Historian; Herman Wolk, Roger Miller, and Wayne Thompson of the U.S. Air Force History Support Office; and John D. Anderson, Jr., professor of aerospace engineering at the University of Maryland. If I have managed to get above the trees and see the forest at all, it is because of the trails these and many other scholars of air power and aeronautical history have blazed to the vistas.

In recounting specific incidents and details that illustrate and substantiate this story, I have, whenever possible, tried to consult original sources, including memoirs and personal letters; official publications, reports, and memoranda; and contemporaneous views as expressed in newspapers, films, and other popular media. I am grateful to the archivists and staffs of the United Kingdom Public Record Office, the Library of Congress Manuscript Division, the U.S. National Archives and Records Administration, the Imperial War Museum, the Royal Air Force Museum, the German Bundesarchiv-Militärarchiv, and the U.S. Air Force History Support Office for their kind assistance.

My sincere thanks also go to Ralph Erskine, whose broad knowledge of military and naval history, not to mention his exceptional critical eye, generosity, and sound judgment, has made me just one of the many writers who are in his debt; Bill Cook, for valuable discussions and advice; Will O'Neil, Chief Scientist, Center for Naval Analyses, for valuable suggestions and for providing many copies of articles from his remarkable personal aviation library; Jean Roberts, for sharing original research and copies of documents on the life of S. F. Cody; David Mets and Sy Deitchman, for insights into the history of precision guided weapons; Ephraim Asculai, for helpful comments on early drafts; Maj. John Beaulieu, Office of the U.S. Air Force Historian, for kind assistance on many fronts; Yvonne Kinkaid, U.S. Air Force History Support Office, for helping me obtain copies of documents and answers to questions; Joseph Chambers and Bruce Holmes of NASA Langley Research Center and James Fallows of *The Atlantic Monthly,* for explaining principles of aerodynamics and flight; and Maj. Gen. Charles Metcalf, Ret., and Ron Hunt of the U.S. Air Force Museum, for allowing me an inside look at aircraft in the museum's collections. I also would like to give my special thanks to Peter and Celia David, whose kind hospitality made my research trips to London so pleasant and memorable.

I thank James Corum for lending the photographs from his extensive collection that appear here. The aircraft profile drawings were done by graphic artist Dave Merrill.

PART ONE

KITTY HAWK TO SAINT-MIHIEL, 1900–1918

1

VISIONS

It was an age of miracles.

The year 1900 began with an excited rush of newspaper articles, sermons, and speeches marveling over the transformations that had taken place in the century just past. "The nineteenth century," editorialized the *New York Times,* "has been marked by greater progress in all that pertains to the material well-being and enlightenment of mankind than all the previous history of the race." In "every department of science and intellectual activity," agreed the *Washington Post,* "we have gone beyond the wildest dreams of 1800."

People were not merely living in a miraculous age; they were keenly aware of living in a miraculous age, one in which there seemed no limit to what human ingenuity might do. Inventions were not merely providing new material comforts and easing burdens; they were breaking down the very certainties of centuries.

Change had come at a mind-spinning pace. The historian Mark Sullivan, born in 1874, wrote that as a boy he had carried a lantern "of a model as old, at least, as Shakespeare, a cylinder of tin with little jagged holes punched through it." Candles and candle molds were common household articles. Half of Americans still were farmers, and they still used tools that a farmer from a thousand years before would have had no trouble recognizing. Grain was mowed with handheld scythes and threshed on a barn floor using a flail made of two sticks joined together with a leather thong. As late as the 1880s, Sullivan recalled, a farmer who wanted a barn went out to the woods with an axe, chopped down oaks, trimmed them, and got his neighbors together for a barn raising. The blacksmith's shop and the gristmill were still fixtures of every rural hamlet, plying trades unaltered in their essentials since the Middle Ages.

The typical American or European of the mid-nineteenth century lived in a world that was not just medieval in its material and tangible dimensions; it was medieval in its cadences and habits of mind. The rhythms of life were set by the sun's rise and fall and the procession of the seasons. Men, and news, and knowledge, traveled at the speed a man or a horse could walk in a day—perhaps twenty-five miles, on a good day, on a good road, in good weather. Henry Adams, the historian and educator who struggled in his autobiography to fathom the world turned upside down that he now lived in, did not exaggerate when he observed that the "American boy of 1854 stood nearer the year 1, than the year 1900" in the education he was given.

In 1900 Henry Adams would stand in the Gallery of Machines at the Great Exposition in Paris and feel "his historical neck broken" as he contemplated the almost silently whirring dynamos that lit the fair's buildings and grounds. "He began to feel the forty-foot dynamos as a moral force, much as the early Christians felt the Cross," Adams wryly observed, his distant third-person voice perfectly echoing the disconnectedness from all things certain and familiar that the new century had ushered in. "The planet itself," he wrote, "seemed less impressive, in its old-fashioned, deliberate, annual or daily revolution, than this huge wheel, revolving within arm's length at some vertiginous speed." Even history, the laying out of an orderly sequence of events linked by cause and effect, history as Adams had practiced it as a professor at Harvard, had been stood on its head by this "sudden irruption of forces totally new." The year 1900, he conceded, "was not the first to upset schoolmasters. Copernicus and Galileo had broken many professorial necks about 1600; Columbus had stood the world on its head towards 1500; but the nearest approach to the revolution of 1900 was that of 310, when Constantine set up the Cross."

Modern social historians look back on the Victorians and belittle their naïve awe of science, technology, and progress, but Adams was no naïf, and what he expressed was what millions felt, and felt with perfect justice. Theirs was a world where the familiar bearings were simply gone, where religious belief, social conventions, even consciousness itself were being refashioned by the onrush of science; where even popes and kings might tremble before the impersonal forces of steam, steel, and electricity; where "everything was flexible, everything was possible," in the words of the historian Howard Mumford Jones. Inventions had annihilated distance and time. By 1900 there were a million and a half telephones in the United States; it was possible for a man in New York City to sit at a desk and carry on a conversation with a man in Omaha, 1,250 miles away, a feat that only a few decades earlier would have meant a journey of weeks. The network of railroad track had quadrupled since the Civil War, to 193,000 miles; a train now arrived and departed Chicago every four minutes. The railroads had freed travel from the weather, linked small towns with great cities, standardized time across the vast reaches of the continent. Every city and town used to keep its own local hours, resetting the clocks to 12:00 as the sun reached its zenith at noon each day; now even the smallest town was part of a rhythm and consciousness that pulsed to the tempo of the railroad and the metropolis. Electric lines and gas pipes tied individual houses to huge networks, pulling them out of their self-sufficient isolation into an unseen world beyond.

Distance and time were being annihilated by the way news now traveled, too. The decades after the Civil War had brought the Linotype machine and a new process for making white paper from cheap chemically digested wood pulp instead of expensive linen fibers, and daily newspapers sprang up everywhere. By 1900 there were 2,226 metropolitan dailies in the United States, many producing multiple editions each day as their high-speed rotary presses churned out the latest news, telegraphed from across the nation and around the world. The problems of even the most distant reaches of the world were now on people's minds and lips. In 1900 American farmers sent five thousand tons of wheat to help relieve a famine in India.

Invention was sweeping aside conventions and social distinctions as old as civilization itself. Where Marx failed, chemists and electrical engineers triumphed. The *San Francisco Examiner* observed that "in the span of a single life, the humblest artisan enjoys what kings could not purchase with their treasures a century ago." Costs of once unimaginable luxuries and conveniences plummeted. In 1900 Kodak introduced the Brownie camera; the camera cost one dollar and a roll of film that took six pictures cost ten cents, and suddenly everyone was a photographer; the stiff formal portraits of professional photographers were replaced in family albums with backyard scenes and youngsters mugging for the camera. Electric streetcars became so efficient they dropped their fares from a dime to a nickel. By 1901 there were 76,945 post offices in the United States, an all-time peak, and the recently introduced Rural Free Delivery system let loose a flood of mail-order retailing that freed customers from the tyranny of local merchants. The cornucopian fruits of the entire industrial and commercial energy of the nation were now directly available to farmers and housewives; everything from a suit to a book to a collie dog to a kitchen stove could be bought without leaving one's home.

Even death itself was unclenching the hold with which it had so untiringly and capriciously embraced humanity. As late as the 1870s people generally saw a doctor as a last resort, and with good reason, for the cure was often literally worse than the disease, and it usually wasn't much of a cure, either. By 1900 advances in microbiology and pathology were turning medicine, and public health in particular, into a science, and the results were nothing short of the miraculous. Pasteurization of milk, chlorination of water, and the drainage of swamps were eliminating diseases that had crippled and killed generations: brucellosis, yellow fever, malaria, typhoid, cholera.

And none of this happened without its being measured, and noted, and marveled over. It was the Age of Confidence, the Age of Optimism, the Age of Energy, the Age of Progress, but most of all it was the Age of Self-Consciousness, for people were filled with a palpable sense of living in a time of great consequence.

The transformations wrought by invention were nowhere more self-consciously on display than in the international expositions that Henry Adams and millions of others flocked to. Between 1876 and 1910 the United States staged a dozen of these pageants to progress; in all they drew a staggering one hundred million visitors. For the World's Columbian Exposition that opened in Chicago in 1893, the building of the fairground itself became an epitome of the limitless mutability of this amazing new world. On a boggy stretch of Chicago lakefront a million cubic yards of topsoil was shifted in three months, a million willows and ferns and other trees and shrubs trucked in and planted. Freight trains on newly laid track hauled in twenty thousand tons of iron and steel and seventy million board feet of lumber, and a shimmering white fantasy city of castles, temples, domes, Corinthian columns, and colossal sculptures arose on what had been a swamp just a few months before. The artist W. Hamilton Gibson hailed Chicago's "White City" as a "New Jerusalem," and he actually meant it.

The largest crowds were always to be found at the Electricity Building, a palace of forty thousand panes of glass whose centerpiece was a shaft seventy-eight feet high, covered with thousands of electric lights. The fair's organizers turned down a proposed plan to buy the Colosseum in Rome, dismantle it, ship it across the Atlantic, and reconstruct it "stone by stone" in Chicago, but the scheme would not actually have been out of keeping with the spirit the exposition sought to capture. Anything was possible.

The turn of the century brought not only an outpouring of reflections on how far mankind had come but also an irrepressible urge to project where it would go next. Newspapers sought out eminent persons to visualize the world a hundred years hence; the results, as Mark Sullivan recalled, "were usually grandiose." Some prognosticators, to be sure, seemed more concerned about the domestic and familial comforts that new inventions would bring; *Ladies' Home Journal* foresaw business travelers being able to phone their wives from aboard ship while crossing the Atlantic, while others were content to predict home ice-making machines ("everybody his own iceman").

Most, however, dwelt not upon the mundane material facts of new inventions that were likely to come, but on how these new machines would continue to transform life and society. In the view of most of these experts, there was almost nothing in the future that would not be touched by the tidal force of technological progress that the last century had unleashed. The Reverend Newell Dwight Hillis, a well-known clergyman and writer, looked into this crystal ball of progress and saw an all-encompassing vision of the world to come: "Laws are becoming more just, rulers humane; music is becoming sweeter and books wiser; homes are happier, and the individual heart becoming at once more just and more gentle."

If inventions could have great consequences, they could also have terrible consequences. In 1901, two years before the Wright brothers flew at Kitty Hawk,

H. G. Wells contributed a remarkable series of five articles to the *North American Review.* "Anticipations: An Experiment in Prophesy" was the title.

Among Wells's predictions was the perfection of the airplane, not in itself a terribly surprising prognostication from someone in 1901 setting out to be a prophet. But what Wells had to say next was rather more striking. "Directly that is accomplished," he wrote, "the new invention will be most assuredly applied to war."

Wars of the future, Wells continued, would be marked by a decisive struggle for the command of the air, and the bombs that would then rain down from aircraft would leave no spot on earth safe:

> The victor in that aerial struggle will tower with pitilessly watchful eyes over his adversary, will concentrate his guns and all his strength unobserved, will mark all his adversary's roads and communications and sweep them with sudden, incredible disasters of shot and shell. The moral effect of this predominance will be enormous. All over the losing country, not simply at his frontier, but everywhere, the victor will soar. Everybody, everywhere will be perpetually and constantly

An aerial battle in the world of the future, as depicted by the French cartoonist Albert Robida in his satirical *La guerre au vingtième siècle*, published in 1887.

looking up, with a sense of loss and insecurity, with a vague distress
of painful anticipations.

Wells's apocalyptic prophecies would become much more widely known a
few years later when he published a popular novel that elaborated this vision in
vivid detail. *The War in the Air* opens on a world of the not too distant future.
The English Channel has been spanned by a 150-foot-high bridge, monorails
and gyroscopically stabilized two-wheeled cars whisk people about their daily
business, shops are filled with produce from all over the world, and rumors are
buzzing that the armies of the world's great nations are conducting secret experiments with flying machines.

When war breaks out between Germany and America, a fleet of German airships suddenly appears over New York City, and this world of wonders becomes
a world of death. "She was the first of the great cities of the Scientific Age to suffer by the enormous powers and gross limitations of aerial warfare," explains
Wells's narrator. "She was wrecked as in the previous century endless barbaric
cities had been bombarded, because she was at once too strong to be occupied
and too undisciplined and proud to surrender in order to escape destruction." As
the line of German airships cruises the length of Broadway methodically dropping explosives, buildings and bridges collapse, and soon all of Manhattan is
engulfed in a sea of crimson flames, "one of the most cold-blooded slaughters in
the world's history, in which men who were neither excited nor, except for the
remotest chance of a bullet, in any danger, poured death and destruction upon
homes and crowds below."

The last chapter of the story takes places thirty years after the war. A young
boy and his uncle are walking through the deserted ruins of London, where a few
refugees and their derelict cows and pigs now wander the abandoned high
streets.

"But why did they start the War?" the boy asks.

"They couldn't stop theirselves," his uncle replies. "'Aving them airships
made 'em."

Wells was no pacifist. Shortly after *The War in the Air* appeared, he lent his
name to an influential body of British notables who were pressing the government
to recognize "the vital importance to the British Empire of aerial supremacy, upon
which . . . its very existence must largely depend." Wells meant his tale to be a
warning of a grim reality that must be realistically faced. As he explained much
later in his autobiography:

[even] before any practical flying had occurred, I reasoned that air
warfare, by making warfare three dimensional, would abolish the
war front and with that the possibility of distinguishing between
civilian and combatant or of bringing a war to a conclusive end. This
I argued, must not only intensify but must alter the ordinary man's

attitude to warfare. He can no longer regard it as we did the Boer War for example as a vivid spectacle in which his participation is that of a paying spectator at a cricket or base-ball match.

Wells wrote his science fiction with a serious purpose, and to a considerable extent that was how his readers took it. The lessons that Wells hoped to drive home in *The War in the Air* were in fact not far removed from the serious arguments about war and its nature then taking place in both the popular press and professional military journals. Since about 1895, all of the major mass-circulation periodicals in America—*The Atlantic Monthly, Harper's, Century, Scribner's, The Saturday Evening Post*—had been running a steady stream of articles discussing the future of war. War was the great popular intellectual issue of the day, much as religion and slavery had been a half century before. While some writers argued hopefully that civilization was moving beyond warfare as a means of settling disputes, a growing theme of many was that the twin forces of science and nationalism would make a modern conflict between industrialized powers far more destructive, and total, than anything the world had seen before. In 1901 Winston S. Churchill, then a twenty-six-year-old Member of Parliament, warned the House of Commons that the "small armies of professional soldiers" who fought decorous set-piece battles were a thing of the past; in the future, when "mighty populations are impelled on each other," winner and loser alike would suffer disaster when nations resorted to war. "The wars of peoples," he declared, "will be more terrible than those of kings."

Churchill would make the same point even more emphatically three decades later in an autobiographical account of his early days as a cavalry officer. (Among other extraordinary adventures, he took part in the last great cavalry charge in the British Army's history, the Battle of Omdurman in 1898.) "War, which used to be cruel and magnificent," Churchill wrote, "has now become cruel and squalid. . . . Instead of a small number of well-trained professionals championing their country's cause with ancient weapons and a beautiful intricacy of manoeuvre, sustained at every moment by the applause of their nation, we now have entire populations, including even women and children, pitted against one another in brutish mutual extermination."

There had been hints of what was to come in the industrialized slaughter of the American Civil War and the Franco-Prussian War of 1870–1871. European generals of the old school had at first haughtily denied that there were any lessons professional military men could learn from the American Civil War battles ("a contest in which huge armed rabbles chased each other around a vast wilderness," Helmuth von Moltke sniffed). But the three decades of relative peace had left a growing void of uncertainty over what would happen if "civilized" nations once again took up arms against one another, and that uncertainty was increasingly filled with foreboding.

Many writers now called attention to the series of innovations in the science of weaponry that had taken place in the 1880s and 1890s: smokeless powder, repeating rifles, long-range artillery, high-explosive shells. Even those who adamantly defended war as a necessity or indeed as an ennobling force for the cultivation of manly virtues and the advancement of civilization—as a surprisingly large number of great men still did—drew the future of warfare in generally apocalyptic terms. To "realists" like Hiram Maxim, whose contribution to the march of progress had been the invention of the machine gun, the apocalyptic face of a war in which "every science" had been pressed into its service was ultimately to the good. By making war "appalling to contemplate it makes nations pause," he explained. "It has led men who are good students of human nature to assert that the best way to preserve peace is to make war as terrible as possible—terrible in its toll of blood and money, terrible in its widespread ravages, and terrible in its uncertainty." Others argued that even if the prospect of appalling slaughter did not prevent war from breaking out, a fierce and terribly fought war would save lives in the long run, as it would inevitably be over far quicker than a war fought with less effective means.

A more chilling, and accurate, prediction of what this terrible new destructiveness on the battlefield would mean came from Ivan S. Bliokh, a Polish Jewish banker and railroad financier who wrote under the name of Jean de Bloch. After making his fortune, de Bloch spent fourteen years studying and thinking about the nature of war in the modern world, and in 1898 he published in Russian a six-volume treatise, *The Future of War in Its Technical, Economic, and Political Relations*. The next year an English translation appeared in both Britain and the United States, and his work quickly became widely known and much discussed. Military professionals still rank it among the greatest theoretical treatises of military strategy of the nineteenth century, certainly the greatest to be penned by an amateur. De Bloch argued that the increased firepower, rapidity, and range of artillery and the machine gun, coupled with the inherent lack of maneuverability of ever-larger armies, meant that the advantage in warfare had decisively shifted to the defensive. Stalemate on the battlefield was inevitable. Nations that went to war would be locked in a suicidal test of wills that would pit not just their armies but their entire reserves of industrial and economic power, and of civilian morale, against one another. When Wells and other futurists spun their visions of total war from the air, they were speaking to an already familiar idea: that scientific progress had become an unstoppable, transformative force in warfare.

Among those who were particularly impressed by de Bloch's arguments was Czar Nicholas II. Fearful that his nation's industrial and economic backwardness would place it at a terrible disadvantage in a world whose fate increasingly rested upon scientific and technological mastery, the Czar issued an appeal for an international conference that would seek the "lofty aim" of "general peace and a possible reduction of excessive armaments." On May 18, 1899, the Czar's birthday, representatives of the great powers assembled at The Hague.

Among the Russian proposals quickly approved by the conference was a "prohibition of the discharge or projectiles of any kind from balloons or by similar new methods." At the behest of the American representative, however, the issue was reopened a few days later and the delegates agreed to limit this prohibition to a five-year period only. The argument advanced by the American delegation for this proposal was a precocious foreshadowing of what would become *the* fundamental debate over air warfare for the century to come. And the position the Americans were taking was the one that would become the quintessential American position in all of those debates to come.

The rationale that had been advanced for banning this new weapon even before it existed, the American delegate Captain William Crozier noted in a lengthy speech, was that it was necessarily inaccurate and indiscriminate and would strike combatants and noncombatants alike. That was certainly the case for balloons that drifted at the whim of prevailing winds. But, Crozier insisted, future aircraft might not be so capricious: "Who can say that such an invention will not be of a kind to make its use possible at a critical point on the field of battle, at a critical moment of the conflict, under conditions so defined and concentrated that it would decide the victory . . . localizing at important points the destruction of life and property, and . . . sparing the sufferings of all who are not at the precise spot where the result is decided? Such use tends to diminish the evils of war."

Four years before a man first successfully piloted a heavier-than-air craft, a year before Count Ferdinand von Zeppelin had even demonstrated that an engine-driven lighter-than-air craft could be steered with any certainty, the threat of attack from the air was vivid in the minds not only of the man on the street but of statesmen and generals. So too was the seductively powerful thought that by precisely delivering an overwhelming strike from the sky at the very outset of war, armed aircraft could prove not just a revolutionary but a decisive force in conflict.

When the first aircraft began to appear in the military forces of the world's armies a few years later, people saw them with a strange sort of double vision. One image was of the primitive, fragile craft that actually stood before their eyes, dangerous and unreliable, capable of carrying only the daring or the foolhardy a few dozen miles and with no more than a few extra pounds to spare for carrying anything else, about as practical or fearsome a weapon of war as a pop gun. The other image, no less real in the minds of many, was the one they had come to know so well from the futuristic visions of technological prophets. This was the image of an apocalyptic instrument of total war, a weapon whose destructive power was different not only in magnitude but in kind from anything that admirals and generals and politicians had ever grappled with before, a weapon that would change not just the conduct of war but its very meaning.

"When airships and airplanes appeared," observed the historian Lee Kennett, it was accordingly inevitable that "extravagant and impossible things would sometimes be expected of them."

For all of the anticipation that had been built up around the airplane, the Wright brothers' first flight brought forth no thunderclaps or trumpet blasts; indeed, it was scarcely noticed at all. Those in the aeronautical world who did notice it did not all believe it.

Partly that was because the brothers were deliberately tight-lipped about the details of their invention so as not to jeopardize patents they had filed for. But mostly the Wrights were paying the familiar price of being pioneers: their rivals for the most part simply did not grasp what they had accomplished. The reason the Wrights flew was that they had solved a series of problems of absolutely fundamental importance in the science of flight, and they did it at a time when their rivals were not even aware that these *were* the fundamental problems.

Had the world understood the principles that made the Wright Flyer soar off the sands of Kitty Hawk where others had failed, the world would not have been able to doubt that it had in truth soared. Yet for years doubts would linger. The French, who saw themselves at the forefront of aeronautical science, from time to time suggested the Wrights were simply *bluffeurs.* As late as 1906 the *New York Herald,* in an editorial that appeared in its Paris edition, expressed an acid skepticism that reflected what many suspected: "The Wrights have flown or they have not flown. They possess a machine or they do not possess one. They are in fact either fliers or liars. It is difficult to fly. It is easy to say, 'We have flown.' "

Even after the legitimacy of the Wrights' claim as the first to fly was universally accepted, many who should have known better tended to belittle their accomplishment; the pair of bicycle mechanics from Dayton, who after all had never even graduated from high school, were at best inspired tinkerers or clever empiricists. Through a certain amount of practical persistence, trial and error, and no little daring, they had managed to get in just under the wire in the race against more scientifically able rivals.

There is no doubting that the Wrights were practical men, and persistent. As craftsmen they were superb. When the brothers decided to begin manufacturing their own line of bicycles in 1896, they fitted out the back room of their bicycle store as a small machine shop; without further ado they began cutting metal tubing for frames, turning cranks and hubs on a lathe, and constructing by hand their own wooden and metal wheel rims. When they needed a new power source to drive the line shaft for their machinery, they matter-of-factly, and with no previous experience, designed and built a one-cylinder internal combustion engine and ran it off the city gas line.

Once they began their flying experiments in 1899, they routinely shrugged off, or even laughed off, physical obstacles and logistical challenges any one of which might have defeated less self-reliant and practical-minded men. In late 1899 Wilbur obtained from the Weather Bureau in Washington tables of reported wind velocities at all weather stations in the country. The brothers sought a place with sustained winds of fifteen to twenty miles an hour so they could begin practicing with a man-lifting glider that would incorporate the features that they had so far tested only on kites. A place without too many hills or trees and with soft sand to cushion the inevitable hard landings was also an advantage. Kitty Hawk, North Carolina, seemed to best fit the bill, and so to Kitty Hawk they would go. The fact that it was a remote, inaccessible spot on the barrier islands that lay thirty-five miles off the mainland, inhabited only by a few fishermen and the staff of a coast guard station—and that indeed Wilbur would set off from Dayton without a clear idea of how to get there—seems not to have particularly troubled him. Wilbur departed Dayton on the 6:30 train on the evening of September 6, 1900. With him were trunks and crates containing the parts of the glider and all the tools needed to assemble it; the sixteen-foot spruce boards for the wing spars he planned to buy when he reached Norfolk, Virginia. Except for a visit to the Columbian Exposition in 1893, he hadn't been more than a day's bicycle ride from Dayton in a decade.

Many things went wrong. Arriving in Norfolk after twenty-four hours on trains and a steamer, Wilbur found there was no spruce to be had in any of the local lumberyards; after chasing around town in the 100-degree heat, he settled on less satisfactory white pine boards. Loading his growing pile of luggage on another train, he continued on to Elizabeth City, North Carolina; there he found it was impossible to get a boat to take him on the final leg of his journey. After three days, he finally found a disheveled local fisherman, Israel Perry, who assured him he could carry him across. The leaking skiff that was to take them to Israel's anchored boat three miles away sank to the gunwales when Wilbur's lumber and trunk were put aboard. On reaching the boat, Wilbur was dismayed to find that "it was in worse condition if possible than the skiff. The sails were rotten, the ropes badly worn and the rudder-post half rotted off, and the cabin so dirty and vermin-infested that I kept out of it first to last." They set out; the wind came up; the flat-bottomed boat began to take on water; the foresail tore loose with a roar, then the mainsail went, and they barely escaped being driven into a sandbar. Two days later Wilbur finally reached his destination. That was his introduction to Kitty Hawk.

Just trying to *live* at as godforsaken a spot as Kitty Hawk, let alone conduct dangerous and exacting experiments in the science of flight, would have been counted a major adventure for most ordinary people. Yet in their frequent letters home Wilbur and Orville always made humorous light of their difficulties, from the bad food to the sandstorms to the swarms of mosquitoes that chewed through their socks and underwear and left welts the size of "hen's eggs." In successive years they built wooden sheds, drilled their own deep well for water, and assembled, repaired, and modified a succession of "flying machines" on the spot with nothing but hand tools and the occasional assistance of the few locals.

All of this dogged ingenuity makes it easy to be distracted from the fact that the Wrights—or, to be more accurate, Wilbur Wright—possessed an extraordinary gift that went far beyond mere mechanical skill and physical adroitness. Though it was true that neither of the brothers finished high school, they had been encouraged by their family to pursue knowledge and to trust in their own judgment. Their calculations and papers show a gift for clear and logical thinking, a solid understanding of basic engineering concepts such as force and energy, and a meticulous attention to accurately recording the results of every test and experiment. Even more, they show a confidence in themselves that was truly exceptional. On May 30, 1899, Wilbur had written to the Smithsonian Institution requesting publications about flying experiments that had been carried out to date; this marked the beginning of his serious study of the problem. Only a few short months later he had identified the key failures of his predecessors and boldly conceived an original solution.

That Wilbur was so remarkably unintimidated by the failures of others, and equally uninfluenced by the approach then being actively pursued by such great scientific men as Samuel Langley, the Secretary of the Smithsonian, and Hiram Maxim, both of whom were conducting well-publicized flying experiments, owes much to the influence of his father, Milton Wright. A bishop of the United Brethren Church, Milton was uncompromising on matters of principle—and to him everything was a matter of principle. The bishop would spend years, often with Wilbur's help, fighting a lonely but ultimately triumphant battle against a fellow church official whom he had accused of embezzling funds; the church leaders, fearing a scandal, had instead sought to oust Bishop Wright and hush up the matter. He was not a popular man, but popularity counted nothing to him; only the truth mattered.

Aside from the cranks and crackpots that the pursuit of human flight had always attracted (among recent examples were Captain John W. Veiru, an old steamboatman who unveiled plans for a fish-shaped flying machine driven by a paddle wheel, and the Reverend Burrell Cannon, whose "Ezekiel Airship" was based on biblical descriptions), the Wrights' predecessors and competitors fell into two schools. One was the "brute force" approach, epitomized by Langley and Maxim. Working with ample financial backing ($50,000 from the U.S. Army in Langley's case, $150,000 out of his own pocket in Maxim's), these men conceived the problem of flight as fundamentally one of generating enough propulsive force to drive a craft into the air. Maxim built a four-ton machine driven by two 180-horsepower steam engines and actually succeeded in getting it to lift a few feet off the ground before crashing as it hurtled down a guided track. Langley invested most of his energies trying to develop as powerful and light an engine as possible for his "Aerodrome"; after driving the S. M. Balzer Company of New York into bankruptcy trying to fulfill his subcontract for an efficient engine, he hired his own engineer and produced a superb 52-horsepower five-cylinder radial engine that weighed a mere 124 pounds. It was more than a decade ahead of the state of the art, and remained unequaled until the eve of the First World War.

Yet neither Langley nor Maxim gave much thought to how to control an aircraft once in the air. To the extent they and other early experimenters did think about control, they mostly saw it as a matter of keeping the machine stable against the buffeting forces of sudden wind gusts; and that, in turn, was viewed mostly as a matter of ensuring that the machine itself had sufficient inherent stability, since it was assumed no human pilot could react quickly enough. Many designers adopted the technique of attaching the wings with an upward dihedral angle; if the aircraft dipped to the left, the left wing would present a more horizontal attitude to the air, generating more lift, while the right wing would present a higher angle, reducing lift, and these forces would tend to right the craft. Others developed "regulators," complex spring-loaded wings or tail surfaces, designed to effect the same kind of automatic corrections.

Significantly, neither Maxim nor Langley ever considered climbing aboard their crafts themselves. The designers of the brute-force school tended to refer to their pilots as "chauffeurs."

The other school was epitomized by the German mechanical engineer Otto Lilienthal. From 1889 until a fatal accident cut short his life in 1896, Lilienthal took to the air in a series of hang gliders that he had designed and built. He began with short downhill hops and gradually increased his glides, particularly after 1893 when he began to soar off of an artificial fifty-foot-high hill that he had had built for him outside of Berlin. Lilienthal's gliders could be controlled by the pilot shifting his weight from side to side, and although he had plans to eventually add a motor, he believed he first had to acquire the skill of adjusting to wind gusts in the air. His fatal crash occurred when a thermal eddy sent the glider into a stall and it nosed straight into the ground from fifty feet up. Lilienthal's spine was broken and he died the next day.

In the spring of 1900, Wilbur Wright, having determined to begin his own man-carrying glider experiments that summer, wrote to Octave Chanute, the "grand old man" of aeronautics. His letter surely ranks among the most extraordinary in the history of invention and engineering. Chanute, then sixty-eight years old, was a national figure. A civil engineer, he had designed and built the first bridge over the Missouri River and had served as chief engineer for the Erie Railroad. His interest in flight led him to correspond extensively with Lilienthal and other experimenters, and in 1894 he had published a compendium on the state of the art of heavier-than-air flight, *Progress in Flying Machines*. As a result, Chanute had become the unofficial but universally acknowledged clearinghouse for developments in aeronautics.

Wilbur's letter was the first of four hundred that he and Orville would eventually exchange with Chanute over the next decade. Chanute from the start offered enthusiastic and invaluable moral support to the brothers, and in the years ahead the Wrights never failed to show their gratitude to him for that support and friendship. But his technical advice was, truth be told, something considerably less than invaluable; indeed, he frequently seems to have missed the entire point of the theoretical and practical results the Wrights reported to him. In that very first letter Wilbur was deferential but forthright. He explained his plans for his upcoming experiments and put his finger unerringly on the failures of those who had tried to fly to date. "What is chiefly needed," he wrote, "is skill rather than machinery." Far more important than motors was, first of all, learning how to control a glider in actual flight. As Wilbur would later note, Lilienthal was the first to realize that "balancing was the *first* instead of the *last* of the great problems in connection with human flight."

But over the course of five years Lilienthal had managed to amass a total of only five hours in the air. And it was abundantly clear to Wilbur even at this early date that Lilienthal's method of control by weight shifting was simply inadequate to the task. Having watched buzzards soar, Wilbur explained to Chanute, he had concluded that birds regain lateral balance by actively rotating the tips of their wings to differentially alter the lifting force on each; "the bird becomes an animated windmill and instantly begins to turn, a line from its head to its tail being the axis."

Not only had Wilbur recognized that an effective means of lateral control—what modern aircraft designers call roll—was the essential missing ingredient in all flying experiments that had been carried out to date, but he had already hit on a practical solution to the problem. It was what would come to be known as "wing warping," and, as Orville Wright later recounted the story, Wilbur stumbled upon the idea one day in July 1899 when he was toying with a small cardboard box with its two ends removed. He realized that the whole box could be given a "helicoidal twist" that would make the top and bottom surfaces twist up on one side and down on the other. The Wrights quickly built a biplane kite incorporating the idea; control wires that could be worked from the ground applied the twist to the wings.

The slightly melodramatic story of Wilbur's chance discovery of the solution was inevitably portrayed by biographers in years to come as a crucial moment,

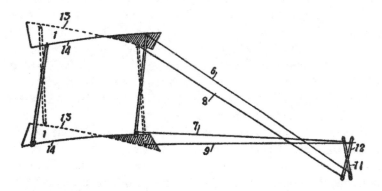

The Wrights quickly hit upon the idea of twisting the wings to provide lateral control, and built a kite to test the concept. Their ability to control all axes of motion in flight—roll, pitch, and yaw—put them years ahead of competitors.

the Wright brothers' "Eureka"; but, as Orville once wrote to complain to one writer, the episode was being made into something "dramatic beyond its importance." He explained: "It was one of our few discoveries made purely by accident of observation rather than as a result of study. It was not the revelation of a basic principle—it was merely a better mechanical embodiment of a basic principle which we had already discussed for several months."

Study was the essence of the Wrights' method. It was what allowed them at the very start of their experimental work to recognize the importance of gaining hours of actual experience in the air where others had slighted that; it had led them to grasp the essential importance of an effective system of active steering and control; and perhaps most remarkable of all, it led them to promptly conclude that the issue that had been the obsession and ruin of other aviation pioneers—building a powerful and light engine—was not an issue to worry about at all. This conclusion was the result of some simple but perfectly adequate back-of-the-envelope calculations. Using an ordinary grocer's pull scale, the brothers had measured the force their glider exerted on a tether when it was flown as a kite; the angle of the string relative to the ground and the laws of trigonometry allowed them to calculate the component of that force that acted in the horizontal direction—which was the machine's drag, basically its wind resistance. (The Wrights termed this "drift.") They then made a clever calculation of the energy expended by a bicyclist and used the result to figure the wind resistance added by having a pilot aboard their glider. Finally they calculated how powerful an engine had to be to overcome the total horizontal resistance offered by their glider.

The result was surprisingly low: a six-horsepower engine weighing two hundred pounds would be perfectly adequate. And, they noted, builders had already made engines that delivered twice as much horsepower per pound. When the time came to add an engine to their machines, an engine would be available for the asking.

The brothers returned to Kitty Hawk each year as their business permitted. They gained many hours of flying time, suffering a good many crashes—"well digging," they jocularly called it when they dove nose-first into the sand—but also learning how to deal with the caprices of the winds. They made a series of crucial modifications to their design that made the controls far more responsive, allowed their craft to make reliable turns, and decreased the danger of stalls of the kind that had sent Lilienthal plunging to his death.

The Wrights' systematic and studious approach would pay off once again as they confronted two other fundamental problems of flight: the correct calculation of the lift generated by wings of different shapes and the design of efficient propellers. Like the problem of control, these were problems that others were scarcely aware *were* problems. In their characteristic manner, the Wrights not only identified their crucial importance but attacked them with extraordinary self-assurance, even when their results led them to challenge the body of accepted wisdom.

Besides taking to the air in his daring gliding experiments, Lilienthal had, beginning in 1866, carried out an extensive series of measurements of the aerodynamic forces acting on airfoils. They were the first such data to be systematically collected, tabulated, plotted, and published. Chanute had reprinted some of Lilienthal's results in an article in 1897, which was how the Wright brothers came to learn of them.

Lilienthal's tables provided numerical coefficients needed to estimate lift and drag on a wing. Both forces varied with a wing's angle of attack—that is, its angle up or down relative to the oncoming wind—and Lilienthal had tabulated the lift and drag coefficients for many different angles of attack. At any given wind speed, the total lift or drag force acting on a wing could be calculated by multiplying the appropriate coefficient by the surface area of the wing, and then multiplying that result by a fundamental physical constant known as Smeaton's coefficient, which relates wind velocity to air pressure.

Until a wing begins to stall and lose lift altogether, lift increases with increasing angle of attack. But so, too, does drag. The Wrights recognized from the start that one key to a successful flier thus lay in designing wings that would generate enough lift at a low angle of attack in order to minimize drag. For their 1901 glider, they had nearly doubled the wing area, from 165 to 308 square feet, with the expectation, based on Lilienthal's coefficients, that enough lift would be generated to support the weight of the glider plus a man in a wind of as little as seventeen miles per hour, and with an angle of attack of at most three or four degrees.

At Kitty Hawk that summer they were sorely disappointed. In his diary entry for July 30, Wilbur recorded their mounting frustrations. Lift was no more than a *third* of the calculated value. The brothers had hoped to be able to soar for extended periods in an eighteen-mile-per-hour wind, but because of the glider's surprisingly poor performance they were lucky to get five minutes of gliding in a day.

It was frustrating, but it was also puzzling; indeed it was almost incomprehensible. Upon their return to Dayton the Wrights cast about for a possible explanation. They kept going back and forth in their view about whether Lilienthal's data was flawed. Then on October 6 Wilbur wrote to Octave Chanute to report that they had carried out a simple and eye-opening experiment. According to Lilienthal's tables, the drag force of an air stream striking perpendicularly against a flat plate eight inches by twelve inches should exactly equal the lift generated by an eight-by-eighteen-inch wing encountering that same air stream at an angle of attack of 5 degrees. The Wrights accordingly attached two metal plates of the appropriate shapes and sizes at right-angle locations on the rim of a bicycle wheel, mounted the wheel horizontally on a fork suspended over the front of a bicycle, and pedaled off to generate a good breeze. The wheel was free to rotate to a position where the two forces exactly balanced. To their surprise, the airfoil's lift force equaled the force on the flat plate only when it reached an eighteen-degree angle of attack to the wind.

Lilienthal's tables were, or so it seemed, flat wrong.

There were actually several different reasons for the discrepancy between calculation and reality. Most of the error, in fact, was not due to mistakes in Lilienthal's tables; modern measurements have shown them to be remarkably accurate. The problem was rather that the Wrights, and Chanute and others, had failed to realize that the coefficients Lilienthal had accurately obtained applied *only* to wings of the particular cross-section and planform shape he had used for his measurements. (It would also shortly become evident to the Wrights that the generally accepted value of Smeaton's coefficient was too large by about a third, further exaggerating the total predicted lift

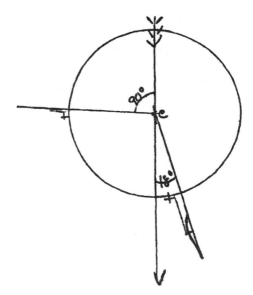

Wilbur Wright's sketch of the crucial bicycle-wheel experiment. Arrow indicates wind direction. Air striking the flat plate generates a counterclockwise force on the wheel; lift from the airfoil, a clockwise force. The forces balanced only when the airfoil was at an angle of 18 degrees, much greater than the 5 degrees predicted by Otto Lilienthal's tables.

for any given wing configuration.) But whatever the cause of the discrepancy—and the Wrights continued to wrangle over whether Lilienthal was right or wrong—the results of the bicycle-wheel experiment galvanized them to conceive and carry out experiments that would soon yield "the most valuable technical data in the history of applied aerodynamics" up to that time, in the words of the aviation historian John Anderson.

The Wrights did not invent the wind tunnel. But no one had collected much systematic data from wind-tunnel tests, and the Wrights now set out to fill that void. From mid-October to mid-December 1901 they tested more than two hundred different airfoil shapes in their tunnel: long skinny wings and short stubby wings; wings with a high arch and with a low arch; wings with the peak of the arch close to the leading edge and with the peak in the middle; wings with square ends and wings with pointed ends. The tunnel itself was a simple wooden box six feet long with a glass window on the top; a wind of up to thirty miles per hour was generated by a fan belted to the drive shaft in their machine shop.

By the time they had to stop their experiments to begin building the next season's bicycles, they had the answers they were looking for. Most important, they had discovered that a long wing that was relatively narrow in its fore-to-aft dimension (what in modern terminology is called a high aspect ratio) generated significantly more lift than a short, stubby wing of the same total surface area. For their 1902 glider they doubled the aspect ratio of their wings, stretching the span from twenty-two feet to thirty-two feet while shrinking the fore-to-aft width, or chord, from seven to five feet. The new wing configuration generated a lift-to-drag ratio of about 6, which would compare extremely favorably to that of fighter aircraft built in the World War a full decade later. (The famous Fokker E.III monoplane, for example, had a lift-to-drag ratio of 6.4.)

This dramatic improvement was at once apparent in photographs of the 1901 and 1902 gliders being flown as kites over the sands of Kitty Hawk. The 1901 glider pitches up like a fractious horse, struggling at a high angle of attack while the tether to the ground runs more horizontal than vertical, indicating a very poor lift-to-drag ratio. The 1902 glider seems to soar effortlessly and almost directly overhead, with the tether almost straight up and down.

True to their convictions about the relative unimportance of the propulsion problem, it was only in December 1902 that the Wrights first wrote to manufacturers seeking to purchase an engine. They sought a power plant that would deliver eight or nine horsepower and weigh no more than about 180 pounds. The replies were all unsatisfactory; undaunted, the Wrights decided they would simply build it themselves. The simple four-cylinder engine that they and their machinist Charlie Taylor produced actually generated twelve horsepower at the required weight. Although meticulously crafted, its design bordered on the crude. It had no carburetor, fuel pump, or spark plugs, and it certainly set no landmarks in the art of engine design. But it was good enough, and that was all that mattered. The Wrights from start to last knew that successfully flying a powered airplane depended least of all on the power plant. The patents they would apply for would not even mention the "comparatively unoriginal aspect of mating power to the airframe," as the aviation historian Peter Jakab put it.

Infinitely more significant was the final problem the Wrights now turned to. Next to the solution of the control problem, it would rank as the Wrights' greatest, though probably least widely known, achievement. It would certainly stand as the greatest testimony to their entire approach to invention. Langley and Maxim had fitted their aircraft with large, flat-bladed propellers modeled on a boat's screw. But Wilbur, once he began to think about the problem in late 1902, immediately saw that an airplane propeller was really just a wing traveling through a spiral course. The thrust it generated in the forward direction, was, in other words, exactly like the vertical lifting force generated by a wing. No one had thought of that before. The wind-tunnel tables the Wrights had already constructed would show them how to optimize the shape of the propeller to maximize its thrust.

But when they began to wrestle with the problem in earnest they kept going around in circles. Not only did the different parts of the propeller move at different speeds as it spun through the air—the outer tip faster and the center slower—but the angle of attack of different parts of the blade relative to the motion of the true air flow depended on the force generated by the propeller itself: the speed at which the airplane was moving through the air and the speed at which the propeller made the air move. "It seemed impossible to find a starting point," they wrote. Orville and Wilbur debated the matter, each proposing ideas that the other knocked down; "after long arguments we often found ourselves in the ludicrous position of each having been converted to the other side, with no more agreement than when the discussion began." But after months of thinking and further arguing, and after filling five notebooks with sketches, calculations, and equations, they had worked out a way to mathematically describe the flow of air over a propeller blade. It was a spectacular theoretical achievement, and so confident were they of the result that their propellers were in fact built solely on the basis of calculation. Later tests would show that their propellers had an efficiency of 70 percent in converting engine power to thrust (compared to 52 percent for Langley's and about 85 percent for an advanced variable-pitch modern propeller).

Their development of what became *the* modern theory of aircraft propeller design marked the apotheosis of their method of self-reliance and independent thought. The difference between the self-assurance of a crank or an egomaniac and the self-assurance and distrust for the opinion of others that Bishop Wright had instilled in his sons was that Wilbur and Orville were as merciless toward their own pet ideas as they were toward the ideas of anyone else. It was that combination of trusting one's ability to recognize the truth, without ever being carried away with one's monopoly on the truth, that was at the heart of their genius. In a letter he wrote in April 1903, Wilbur summarized his approach to the matter: "No truth is without some mixture of error, and no error so false but that it possesses some element of truth. If a man is in too big a hurry to give up an error he is liable to give up some truth with it, and in accepting the arguments of the other man he is sure to get some error with it. . . . After I get hold of a truth I hate to lose it again, and I like to sift all the truth out before I give up an error."

The autumn of 1903 was a particularly rainy, cold, and difficult one at Kitty Hawk. The brothers had arrived in September for their fourth sojourn at this spot they had come to love for its desolate beauty and quiet; this time their luggage included six hundred pounds of airplane parts, including the engine and propellers they had built over the winter and spring.

Many days the rain and wind halted work and drove them indoors. Some mornings were so cold that the water froze in their washbasin. Wilbur reported in one letter home that they had had to revise the system they had developed the previous year for classifying how cold it got at night; in addition to 1, 2, 3, and 4 blanket nights they now had 5 blanket nights, 5 blanket and 2 quilt nights, and once they had even had a 5 blanket, 2 quilt, fire, and hot water jug night. The ups and downs in the weather and in their work prompted Orville to joke in a post-card to Charlie Taylor that the "flying machine market has been very unsteady. . . . These fluctuations would have produced a panic, I think, in Wall Street, but in this quiet place it only put us to thinking and figuring a little." On November 28 the stock hit a new low: one of the propeller shafts cracked during a test run of the engine. There was nothing for it but to go back to Dayton and build a replacement, but with success this close they did not hesitate. Orville left for home two days later, fabricated two replacement shafts in record time, and was on his way back a week later. On the return trip to Kitty Hawk he read a newspaper account of the spectacular crash of Langley's powered Aerodrome on December 8. On its second attempt to take off with a pilot at the controls, the plane had trundled off its launch catapult atop a houseboat in the Potomac River and immediately plunged nose-first into the water. The pilot was rescued but the plane was a wreck, and Langley, who would quickly become the butt of criticism and ridicule for having wasted so much of the taxpayers' money, reluctantly abandoned his project forever.

Orville arrived back at the camp on December 11. Three days later the machine was ready. Wilbur won a coin toss and took the controls for the first attempt; the plane took to the air briefly, but Wilbur almost immediately over-controlled the elevator and sent it into a stall. The plane crashed to the sand, breaking the forward elevator and a landing skid. But by this point there were no longer any doubts in their minds. Wilbur cabled home the next day:

MISJUDGMENT AT START REDUCED FLIGHT [TO] HUNDRED AND TWELVE [FEET]. POWER AND CONTROL AMPLE. RUDDER ONLY INJURED. SUCCESS ASSURED. KEEP QUIET.

On December 17 the Flyer was ready to go again. Despite strong winds and freezing temperatures, at 10 a.m. they hoisted a signal flag at their camp to alert the crew at the nearby lifesaving station that they would make an attempt to fly that day. It was Orville's turn this time. Before mounting the machine he set up his camera on a tripod between their camp buildings and the launching rail that would guide the Flyer on its takeoff run, and he carefully trained the lens on a point a few feet short of the end of the track, just where he expected the Flyer to take to the air. John T. Daniels, a member of the lifesaving crew, had never oper-ated a camera before, but Orville placed him in charge of the shutter release. At 10:35 a.m., after letting the engine run for a few minutes to warm up, Orville released the rope that held the Flyer against the force of its turning propellers and the machine started down the track. As Wilbur ran alongside to steady the

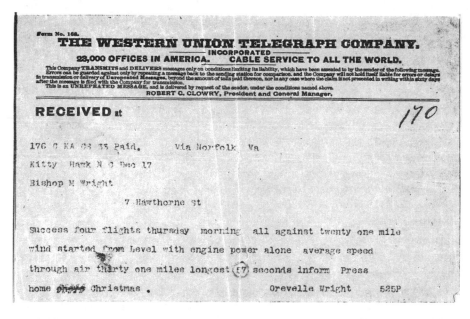

The Wrights' telegram to their father announcing the first successful flights at Kitty Hawk on December 17, 1903. *(Library of Congress)*

wing, Daniels, no doubt to his eternal subsequent relief, tripped the shutter perfectly on cue to capture an image that, in Peter Jakab's apt description, remains "one of the most famous photographs ever taken."

The reports of the Wright brothers' flights that appeared in the press were at a minimum garbled; most were ludicrous. The *New York Times* ran a sixteen-line item on its front page of December 26 revealing that certain unnamed inventors of an "airship" recently tested at Kitty Hawk were anxious to sell their device to the government for "scouting and signal work, and possibly torpedo warfare." The device was described as "an adaptation of the box kite idea, with a propeller working on a perpendicular shaft to raise or lower the craft, and another working on a horizontal shaft to send it forward." The story was sandwiched between longer items that the editors, and probably most readers, paid more attention to: There was a column devoted to the tragic tale of a British girl who had sailed across the ocean only to learn that her fiancé had been killed in a train wreck, and another story with the memorable headline CAKES KILLED A WOMAN: MRS. LEIBY ATE THREE DOZEN OF THE CHRISTMAS KIND AND DIED.

Even more distorted and embroidered accounts of the Wrights' airplane appeared abroad. After a few weeks of this, the Wrights tried to straighten things out by releasing their own brief account, but the damage had been done. The facts would remain tainted by fiction for years.

But their task was delicate, and in more ways than one. While they laughed off some of the more ridiculous stories, they were in fact deeply offended and troubled by others, especially a number that had suggested they were mere pupils or protégés of Chanute's. The brothers wanted to say enough to establish their claim in the court of public opinion while not jeopardizing it in patent court: their basic patent application, filed on March 23, 1903, would not issue for three years, and during that period of limbo a premature publication of the salient details of their invention could invalidate their claim. It was an impossible line to walk. In their statement to the press they had included a sentence that would sow the seeds of a later painful falling out with Chanute. "All of the experiments," they wrote, "have been conducted at our own expense, without assistance from *any* individual or institution." Chanute sent a slightly stiff reply asking just what they had meant by that, and the brothers smoothed things over by explaining that all they had meant was that since they "had paid the freight" they had the right to refuse to make their invention "public property."

But their secrecy had a more subtle purpose. Although talk of scouting and signaling and launching torpedoes was a reporter's invention, the fact was that the Wrights from the very first saw the only viable customer for their invention as the military forces of a "great Government." That was in part because there seemed to be no obvious commercial market for the airplane; it was also, as they would later explain to the British military attaché who had visited Dayton to inquire about their invention, because they were wary of being swindled by private companies or financial syndicates.

Selling their invention to a "great Government" also had its risks, however, risks that patents would not necessarily protect them against once the details of their work and its full significance were widely known: "Governments often appropriate inventions useful in warfare," Orville noted. Thus the brothers' plan was that when the time came they would sell their invention outright, together with their research and data, for a substantial sum up front, and that would be an end to it. They would not compete with "mountebanks for a chance to earn money in the mountebank business," as Wilbur once expressed it. They wanted to be compensated for their invention, period. For the scheme to work, though, they had to be the first—the unquestioned first—to offer a government a fully practical airplane of actual military value, and that was going to take some further experimentation. To maintain their lead over rivals in the meanwhile, absolute secrecy was essential.

Just a couple of weeks after their first flight, the Wrights had already had one brush with a mountebank. In deference to Chanute, though much against their wishes, they had allowed Chanute and his employee Augustus Herring to visit them at Kitty Hawk during the 1902 season and test two of Chanute's gliders. The gliders proved to be complete flops, Herring and Chanute left dejected, and that was the last they had heard from Herring. The Wrights were accordingly astonished to receive a letter from Herring dated December 26, 1903, claiming that he had independently solved the problem of flight and proposing that he be given a one-third interest in the Wrights' invention. Herring claimed that he had already been offered a substantial sum for his "rights" in interference suits against the Wrights. It was a none-too-subtle attempt at blackmail, and the Wrights disdained to reply. But it was an abrasive reminder of what they would be up against in trying to reap a profit from their work.

Unfortunately for the Wrights, their own government was at the moment about the last one on earth willing to entertain an offer to buy an airplane. Having been stung by the savage criticism in the press and in Congress that followed Langley's highly publicized debacle, the U.S. War Department decided to wash its hands of the business of heavier-than-air flight altogether. At the end of December 1903 the Cabot brothers of Boston, influential friends of Chanute's, urged their even more influential relation Senator Henry Cabot Lodge to take notice of the Wright brothers' successful flight and its potential value to the government; the senator passed the information on to the War Department, which discreetly ignored it.

So gun-shy were the American authorities that, a year later, when the Wrights were ready to make a more official approach to the government, they were sent a reply that, as one historian has put it, "came very near the border-line of official insanity." But part of the confusion lay in the fact that the Wrights were proposing what was apparently a wholly novel arrangement as far as the War Department's Board of Ordnance and Fortification was concerned. As they had planned all along, the Wrights were offering to enter into a contract under which, for a price agreed to in advance, they would deliver an airplane that performed to specifications agreed to in advance. They would receive nothing if they failed to deliver. But the Board apparently took that as a request for financial support of further experimentation, and explained that they were very sorry but they received a great many such requests and before they would consider any further expenditures on flying machines "the device must have been brought to the stage of practical operation without expense to the United States." The brothers concluded that this bafflingly illogical reply to their proposal was merely an odd bureaucratic way of slamming the door in their faces—"a flat turndown."

Meanwhile, the Wrights in spring 1904 had begun work on a series of improvements to their machine that would keep them at the forefront of aviation. Flying now from Huffman Prairie, a cow pasture eight miles from Dayton, they experimented with changes in balance and controls, and on September 20 Wilbur made the first-ever circular flight, staying aloft a minute and a half and covering almost a mile.

A few weeks after that, there arrived in Dayton a representative of a "great Government" who seemed far readier to do business than the Wrights' own countrymen. Lieutenant Colonel John Edward Capper was commandant of the

British Army's balloon units. He had come to America to see the aviation exhibits at the St. Louis World's Fair, but he had traveled to Dayton specially to see the Wrights, having heard about their work from Chanute, still the indispensable and universally known clearinghouse of information about all things aeronautical.

Though the Wrights would not let Capper see their airplane and told him they were not ready to talk business, they did show him photographs and their engine and discussed their work at length; all in all they succeeded in making a deep and favorable impression on their British visitor. Capper, on his return to England, filed a lengthy report. He had found the brothers to be courteous, well educated, and capable men: "I do not think are likely to claim more than they can perform. . . . [They] have satisfied me that they have at least made far greater strides in the evolution of the flying machine than any of their predecessors." Capper zeroed in on the "military importance" of the Wrights' work:

> If carried to a successful issue, we may shortly have as accessories of warfare, scouting machines which will go at a great pace, and be independent of obstacles on the ground, whilst offering from their elevated position unrivalled opportunities of ascertaining what is occurring in the heart of an enemy's country.

England, he warned, was "very backward" in the race to develop this new weapon. On departing from Dayton, he had asked the brothers "to give Great Britain the first chance" when their machine had reached a practical stage.

With those words the global aviation arms race began. But it was a slow-motion arms race, and it often seemed to move in reverse. Though Capper was keenly aware of the military significance of the Wrights' invention, the bureaucrats of the British War Office proved to be no less bureaucratic than those of the American War Department.

On March 1, 1905, the Wrights were at last ready to "talk business," and wrote to the War Office proposing to furnish an "aerial scouting machine" that could carry two men and supplies of fuel for a flight of at least fifty miles at thirty miles per hour. But the negotiations quickly hit the same impasse of misunderstanding that had hampered the Wrights' approach to the United States government. The Wrights insisted that their buyer had to sign a binding contract *before* they would reveal any of their secrets, including even a look at their machine. But in turn, the brothers undertook to assume all of the financial risk: the purchaser would not have to pay them a penny if the machine failed to live up to their promises. This was apparently more than the official mind could comprehend. The correspondence between the Wrights and the War Office throughout 1905 would be comical if it did not make such agonizing reading. Over and over, the Wrights explained their position; over and over, the British representatives replied, in apparent genuine puzzlement, that they only "wish to see your machine fly," and then they would be happy to enter into negotiations.

All sorts of bizarre interpretations over the years have been ascribed to the Wrights' position, with some even suggesting that Wilbur was possessed by a sort of Freudian inability to part with his creation. But in fact the deal they were proposing was rational and straightforward; their letters to officialdom were models of clarity; and their only fault was in deviating from the orthodox arrangements that usually governed the relations between governments and military contractors.

In the summer of 1906 Capper tried once again to break the logjam; citing a letter from the Wrights in which they mentioned that they were "offering to various governments our complete invention," he urged his superiors that he be allowed to ask the Wrights for a price. On July 31 the brothers stated their terms. For $100,000 they would furnish an airplane, train a British pilot to fly, and grant full rights to the British government to manufacture for its own use additional planes based on their patents. For an additional $100,000 they would supply all of their formulas and tables.

But by this point Capper himself had begun to waver. For one thing the price was breathtakingly high; it would be difficult to secure approval for such a sum. But more important was that Capper, recently promoted to the post of Superintendent of the Balloon Factory, in effect the head of all aeronautical research for the British Army, began to have the itch to make his own mark on aviation history. Working from fragmentary descriptions of the Wrights' airplane that had begun to appear in European aviation journals, inventors in both France and Britain were hastening to build their own "Wright-type" planes and catch up with the Americans. Capper laid his cards on the table in September when he recommended turning down the Wrights' offer, explaining to the Director of Fortifications and Works:

> I cannot advise any further action being taken on this matter. I consider the prices asked by the Wright Bros. are out of all proportion to the benefits to be gained. . . . As regards the Flying Machine itself: I have but little doubt that we shall soon be able . . . to turn out within a reasonable time a Flying Machine on much the same lines as that of the Wright Brothers, but we hope superior to it in several essentials, at an infinite fraction of the cost demanded by them.

Capper's actions were neither the first nor the last act of self-delusion on the part of military authorities when it came to the technological realities of aviation. But his conviction that Britain should attempt to surpass the Wrights on its own was given a further spur that fall of 1906 when from France came the stunning news that the aviator Alberto Santos-Dumont had begun to get his heavier-than-air craft off the ground. After making a few short hopping flights, Santos-Dumont flew for 60 meters on October 23, 1906; on November 12, before a crowd of deliriously cheering spectators at a field near Paris, he managed to stay aloft for a flight of 220 meters.

The Wrights at once understood that Santos-Dumont's airplane had more in common with a brick than a bird in its flying characteristics; it was minimally controllable at best, and what kept it in the air was mostly a matter of momentum and daring. The Wrights were not at all concerned that Santos-Dumont or others posed a serious technological challenge to them, but they were very concerned that people might think so. Even experts such as Octave Chanute had not really understood the crucial technological innovations that underlay the Wrights' machine; Chanute wrote to the Wrights that he fancied Santos-Dumont "is now very nearly where you were in 1904." Wilbur pointed out why that was emphatically not the case, but conceded that the "real disturbing element is the general *belief*" that other aviators were about to overtake them: "As a hindrance to business this is almost as bad as reality."

Capper might be forgiven for failing to grasp a distinction between impressions and reality that confused many. His faith in the man he had chosen to catapult Britain past the Wright brothers was, however, quite another matter. Lieutenant John Dunne's chief claim to fame was, apparently, that he was the well-connected son of a famous general. Invalided home after falling ill during the Boer War, he had begun to occupy himself designing and building gliders. Capper brought Dunne to the Balloon Factory in June 1906 and immediately began to put increasing stock in him. The dream of Britain's taking the world lead in military aviation apparently got the better of his judgment. The fact that Dunne's work was carried out in great secrecy did not help, either, for it would not have stood up to much scrutiny.

The press nonetheless caught hints of secret experiments being carried out at Blair Atholl, a ducal estate in the Scottish highlands, and the *Daily Express* correspondent managed to sneak a look at Dunne's machine through "a pair of powerful field glasses." On September 27, 1907, the paper ran a breathless account under large headlines:

BRITAIN'S FIRST AEROPLANE.
EXPERIMENTS WITH AERIAL
FIGHTING MACHINES.
SUCCESS ASSURED.
WORK IN SECRECY
IN THE HIGHLANDS

The truth was considerably more prosaic. On one glide, the plane crashed into a wall with Capper piloting, leaving him with a gash on his ear and blood pouring down his face. The British experimenters nonetheless proceeded to fit the glider with two engines they had optimistically brought with them to Scotland. With the engines opened full throttle, the plane was launched down a track laid on a hillside, whereupon it immediately plunged nose-first into the ground, and that was the end of that.

The Wrights meanwhile still lacked a definite buyer as the year 1907 came to a close. They had pursued possible contracts with France, Germany, even Russia. Several French aviators—Henri Farman, most notably—were now flying regularly, and the Wrights' first serious American competitor, Glenn Curtiss, backed by the famous inventor Alexander Graham Bell through a group called the Aerial Experiment Association, was about to take to the air, too. The Wrights had not flown since 1905. They realized they needed to do something dramatic to reestablish in public and official opinion the preeminent position they did in truth still command.

In spring 1908 the Wrights announced they would conduct a series of demonstration flights in France. Wilbur arrived in Paris on May 29; the crates containing the parts of the Flyer, thoroughly jumbled by the clumsy going-over they had been given by French customs officials, arrived two weeks later. The weeks dragged on as Wilbur methodically replaced broken parts, sewed the wing fabric sections together, and tested the engine. During one of those tests, on July 4, a radiator hose broke loose and Wilbur was struck on the arm and side with a jet of boiling water. He fainted from the pain; a foot-long blister welled up on his arm and took months to heal. The French were suspicious of all the delays. "Le bluff continue," declared one newspaper.

A month later, under cover of night, the fully assembled craft was finally moved to a shed on the Hunandières racecourse outside of Paris. Four days later, on August 8, Wilbur made a short test flight. It caused a sensation. Then came the real flying demonstrations: Wilbur circled, did figure eights, climbed to ninety feet, stayed in the air twenty minutes at a time. The French, who had been unable to contain their sneers, now could scarcely find words to express their admiration. "We are as children compared to the Wrights!" gasped René Gasnier, one of France's pioneer aviators. "We are beaten! We just do not exist!" agreed Léon Delagrange. "Mr. Wright has us all in his hands. What he does not know is not worth knowing," declared Paul Zens. "Veelbur Reet" caps became the rage all over France. The French Senate extended an invitation to Wilbur and greeted him with a standing ovation. The frying pan Wilbur had used for cooking at his hangar improbably wound up in the Louvre.

Wilbur's flights were indeed a total revelation. Here was a plane that flew gracefully, maneuverably, effortlessly. The pilot was no longer a daredevil, like a circus performer shot from a cannon; he was a sultan in majestic command of his flying carpet.

The European fliers, to a man, had still believed that the secret of flight lay in inherent stability. To the extent that they equipped their airplanes with a means for affecting their lateral balance at all, this was viewed only as a sort of corrective function that the pilot could call on to keep the wings level: the small and not very effective flaps of these early European airplanes were referred to by such terms as "balancing planes" or "righting planes" or "lateral equilibrators." None were able to make their machines bank into a turn the way the Wrights' airplane did; at best they could make rather spastic lurches to the left or right by use of the rudder. The Wrights' three-axis control system—giving the pilot full, independent control over the airplane's up-and-down pitch, left-and-right yaw, and lateral roll—was what gave the Flyer its masterful maneuverability, and it would henceforth be a feature of every successful airplane. The Wrights' highly efficient propeller design would be immediately and widely copied, too. "Mr. Wright," declared Edouard Surcouf, a French designer of airships, "has solved the problem of flight."

Other nations had finally begun to arrive at the same conclusion. The British, notably, were not among them: in the spring and summer of 1908 Capper had *again* spiked a deal between the War Office and the Wrights, tacking on a series of last-minute conditions to the official "Specifications for a Military Flying Machine" he had earlier drawn up, and which the Wrights were preparing to offer a bid on. The conditions were composed with malice aforethought; they were deliberately designed to disqualify the Wrights' machine from consideration. The new terms, six handwritten pages, arrived in Dayton in August 1908 while the brothers were both away; Bishop Wright forwarded them to Wilbur with a covering note that cut to the heart of the matter: "They only ask that an applicant should jump over the moon! through a hoop!! six times!!!" And so for a third time the British War Office lost its chance to acquire the Wrights' invention.

The United States government's inability to close a deal with the Wrights had always had more to do with lingering political embarrassment over the Langley debacle, compounded by bureaucratic inertia created by the War Department's contracting regulations, than with a lack of enthusiasm for military aviation in high places. With the passage of time and the increasing attention that aviation was manifestly arousing abroad, the hand of America's influential aviation enthusiasts was only strengthened at home. As the British government temporized and haggled, the Wright brothers' own country finally began to reclaim the initiative. It was President Theodore Roosevelt himself, persuaded by some influential lobbying, who had broken the bureaucratic logjam, and he directed the War Department to take more interest in flying machines. On May 11, 1907, the Board of Ordnance and Fortification had written to the Wright brothers inquiring about terms; the brothers had replied, offering a machine for $100,000, again on their standard terms of assuming all risk in return for a commitment up front. The negotiations dragged out; the brothers dropped their price to $25,000; but at last the wheels of bureaucracy began to turn and on December 23, 1907, the U.S. Army issued a request for bids for an airplane. The bidding was supposed to be a mere formality, the specifications having been negotiated with the

Wrights in advance. But, inevitably, all the cranks came out of the woodwork: the Army received forty-one proposals.

Among them was a bid from Augustus Herring, who claimed he could build an airplane for five thousand dollars less than the Wrights' price. The Wrights at once suspected that Herring was yet again up to no good. They were right. His earlier blackmail scheme having come to naught, Herring now approached the Wrights with a new shakedown. Presuming that the Army would have to award him the contract as the low bidder, he traveled to Dayton and offered to turn his winning bid over to the Wrights—for a cut. The Wrights were disgusted but not surprised. The Army, however, managed to thwart Herring's skullduggery by announcing it would accept both bids. Herring, after much bluster, turned up at Fort Myer outside of Washington on October 12, 1908, and announced he was making "technical delivery" of his flying machine—in the form of two small suitcases and an "innovation trunk" that, he claimed, contained all of the parts of an engine and airplane he had built in fulfillment of his contract. Nobody was impressed.

For by then Orville had put on his own flying spectacle at Fort Myer to rival what Wilbur was doing in Paris. Orville set record after record, staying in the air more than an hour at a stretch and climbing to two hundred feet. On September 17, 1908, he was carrying Lieutenant Thomas Selfridge as a passenger when one of the propellers cracked and struck the stay wire holding the tail. The plane plunged to the ground. Orville was seriously injured and was carried to the hospital at Fort Myer, where he remained for six weeks. Selfridge was killed. Had such a tragedy happened a few years earlier, it might well have brought the embryonic program of military aviation to a halt. Now nothing could stop it. Two weeks after Orville got out of the hospital he wrote to his brother that the Army officers had all hastened to assure him that "the U.S. is going to be our best customer." A year later the Army officially accepted delivery of the Wrights' airplane. In the official acceptance tests, Orville broke all previous duration records for a flight with a passenger, circling the field 79½ times in 1 hour 12 minutes 37⅘ seconds. On July 30, 1909, carrying Lieutenant Benjamin D. Foulois as passenger, the Flyer underwent its official speed trial, covering a ten-mile course from Fort Myer to Alexandria and back in 14 minutes 40 seconds. The average speed, 45.8 miles per hour, earned the brothers a bonus of $5,000 over the agreed price of $25,000, $1,000 for each mile per hour over forty. The U.S. Army so became the first military organization in the world to acquire an airplane.

The rest of the world was now close behind.

2

BOGEYMEN

Among the many prizes showered upon Orville and Wilbur Wright following their dramatic public flying demonstrations in 1909 was a medal presented to the brothers by the French Peace Society. Like so many inventors of new technologies of warfare before and since, the Wrights reassured themselves that they "were introducing into the world an invention which would make further wars practically impossible." Writing from the grim perspective of 1917, Orville recalled that he and Wilbur had "thought governments would realize the impossibility of winning by surprise attacks" once the airplane made it impossible to conceal troop movements on the ground: "No country would enter into war with another of equal size when it knew that it would have to win by simply wearing out its enemy."

The armies of the world that began to interest themselves in aviation in 1908 and 1909 were not, of course, aiming at so lofty a purpose as making war impossible. But they tended to agree with the Wrights' basic point: the military role of the airplane would be as a "scout" or observer, and nothing more. Aircraft could help keep a watchful eye over enemy movements; all other talk of battles in the sky and death raining from above was the product of the overworked imaginations of science-fiction writers. Certainly most professional military men pooh-poohed the idea that airplanes would ever revolutionize warfare. Some, indeed, wanted nothing to do with this new contraption at all.

In Britain the foremost proponent of that view was General Sir William Nicholson, Chief of the Imperial General Staff. In October 1908 the Committee of Imperial Defence had appointed an Aerial Navigation subcommittee to investigate "the dangers to which we would be exposed on sea or on land by any developments in aerial navigation reasonably probable in the near future" and to determine what, if any, "naval and military advantages" Britain herself might derive from the use of airships or airplanes. Nicholson spoke for the Army on the subcommittee and made it perfectly clear that as far as he was concerned the answer to both questions was "none at all." At one meeting he asserted that there was no point in Britain's spending any money on developing its own aircraft, since "as soon as we have got, at great expense, and made a very good aeroplane, other nations will make aeroplanes in the same way, and we shall be no better off than we were before." At another meeting he laid down the law (apparently a result of his own experiments in which he had himself driven around in his staff car) that it was impossible to use an airplane for observation since no one could reliably spot anything when traveling faster than thirty miles per hour. One of Nicholson's officers on the General Staff stood up after a lecture on aviation at the Royal United Service Institution in London and denounced the suggestion that the airplane would be a weapon of any significance in the next great war. The public, he warned, should not be "carried away with wild ideas." The idea of air-to-air combat was particularly absurd, he said, since it would mean "certain death" for both combatants.

Even the aerial enthusiasts in the world's militaries—and there were a good many of them, despite the caricatured image of military conservatism so ably represented by General Sir William Nicholson and his ilk—took a generally cautious view about aviation's military possibilities. Airmen were well aware of the limitations and temperamental natures of their primitive machines, and of how much practical work and experience it would take to learn to control them in the air and maintain them on the ground. Before any serious thought could be given to integrating airplanes into the organization and operation of a military field force, an array of new technical skills had to be mastered. On the theory that they were at least likely to know something about how to make complicated gadgets work, it was the engineering corps in the British and French armies and the Signal Corps in the U.S. Army that were handed the initial responsibility for airplanes.

Major George O. Squier of the U.S. Signal Corps was one of the most enthusiastic of the new breed of military airmen (he was also one of the very few officers in any of the world's armies at that time to have earned a Ph.D.), but in a review article he contributed to the new magazine *Flight* in 1909, he was careful to point out that a chief object of the Corps in purchasing a Wright airplane was "as a stimulus to develop practical aviation in the United States" and to "advance a new art of interest to the nation as a whole." The actual wartime applications of this technology were modest at best: "reconnoitering," delivering messages, and "carrying individuals of high rank and command to points where their personality is needed." The airplane was "not likely to become a burden bearing ship," Squier concluded.

Military airmen like Squier saw themselves as forward-looking and open to new ideas; they believed it was important for an army to keep up with all the latest scientific developments and even to conduct its own scientific research. But not every new scientific discovery was going to immediately appear in the hands of the troops. While many pioneering airmen had grand visions of what air power might someday do, they knew that that someday was not now. The important thing was to begin to learn about this new technology without delay and to keep up with developments in the field so that when its day finally came, the army would not be starting from scratch.

Richard Burdon Haldane, Britain's Secretary of State for War, saw things much the same way. Though he mistakenly dismissed the Wrights as mere "empiricists," in the larger sense he grasped the situation correctly. Most experimental work on airplanes *was* in the hands of amateurs and empiricists. Notable among them were the British Army's own heavier-than-air experimenters. After Lieutenant Dunne's fiasco in the Highlands, the baton had passed to another amateur member of Colonel Capper's Balloon Factory who, even in success, seemed to affirm all of Haldane's worst apprehensions about what was wrong with the way the British Army was running the show. On October 16, 1908, Samuel F. Cody lifted off from Farnborough Common in the slightly fantastic-looking biplane he had constructed to his own design. With the Union Jack fluttering from a wing strut, Cody flew 1,400 feet before crashing as he attempted to make a sharp left bank to avoid a clump of trees. It was Britain's first successful heavier-than-air flight.

How Cody came to be in the employ of the Balloon Factory was a tale in itself. An American, Cody had come to Europe as a performer in a touring Wild West act; among the stunts that he became famous for was racing on horseback against a champion bicycle rider. (Cody would win.) In 1898 he and his family began performing a melodrama entitled "The Klondyke Nugget"—Cody played the mustachioed villain—that proved a sensational hit on the British stage. The comfortable income he enjoyed from this success permitted him to devote more time to his lifelong hobby of kite flying. By 1904 he had persuaded the British authorities to conduct tests of the "man lifting" kites he had built, and by 1906 he had secured a two-year contract as "Chief Instructor in Kiting at the Balloon School," at £1,000 a year plus free fodder for his horse. Cody soon began to press for permission to experiment with adding a motor to his kites, and in June 1906 he was authorized to spend £550 to acquire a fifty-horsepower French engine.

Samuel Cody may have won the hearts of the British public; when he was killed in a crash in 1913 after his airplane broke up in flight, more than a hundred thousand people lined the route of his funeral procession. But he was hardly the kind of aeronautical researcher Haldane was looking for. Cody did have a good intuitive and empirical sense of aerodynamic principles, but he was first and last a showman. Major portions of his life story were total inventions; even after he left the stage and became famous to the British public for his aviation exploits, he kept alive the story of an adventurous boyhood in Texas, where, among other dramatic occurrences, his family was massacred by Indians and he, the sole survivor, crawled nine miles to safety with a gunshot wound in the thigh. In fact Samuel Cody had grown up in Davenport, Iowa; his real name was Cowdery; and he had once been sued by the real Buffalo Bill Cody for appropriating the name "Wild West Show" and claiming that he was the "son of Buffalo Bill." None of this quite fit with Haldane's determination that Britain needed to build a "real scientific Department of the State for the study of aerial navigation."

In 1909 the state of aviation art undoubtedly justified Haldane's view of things. Since the Wright brothers' 1902 wind-tunnel experiments and their 1903 derivation of propeller theory, the amount of aerodynamic science that had been brought to bear on the design of airplanes was close to nil. Steady advances were being made in aircraft performance, but almost entirely as a result of a return to the old "brute force" approach of pre-flight days: the assault on records was being propelled mostly by ever more powerful engines. The basic design of the airplane itself was remarkably conservative, and would remain so even through most of the First World War. Builders basically took what they knew worked and made small modifications, trusting their own intuition and experience about what "looked" right. To test the strength of their machines, they might pile up some sandbags on the wings, but no one knew for sure how an airplane would behave until it flew. It was routine for a builder to make countless adjustments to his design, some minor and some not so minor—shifting the wings, changing the rudder configuration, moving weight around to rebalance the craft   only after he had gingerly taken his plane for a series of gradually lengthening hops that revealed its quirks and defects. Wind-tunnel research at places like England's National Physical Laboratory or the laboratory the French engineer Gustave Eiffel had established in Paris in the shadow of the tower that bore his name had almost no influence on what aircraft designers did in the field.

In many ways it was a classic clash of cultures. Nineteenth-century scientists had disdained flight as the preoccupation of cranks and had done little to develop aerodynamic theories that could be applied to the practical problems of flight. Now that men *were* flying, and setting records and capturing the world's attention, it was the turn of the aviators to look upon the ivory-tower professors with contempt. The men who built and flew airplanes had developed their own tradition and culture, and it was one that placed great faith in practical experience and rules of thumb, not complex mathematical theories or laboratory experiments.

Yet none of these practical-minded airplane builders followed up even on the Wright's basic insight that reducing drag was the key to better performance. The strut-and-wire biplane had become the standard design mainly for structural reasons, not aerodynamic ones. Two wings joined by struts and diagonal bracing wires formed a very light and strong structure; it was basically a box girder of the kind that engineers used for bridges, the Pratt truss. The wings on any airplane are subjected to a variety of stresses: on the ground they are weighed down by their own gravity, while in flight they are subjected to the aerodynamic lift that must support the weight of the plane. The crisscrossing diagonal wires braced the wings against both forces. But aerodynamically the biplane had much going against it. The lifting forces of the two wings interfered with each other, generating less lift than a monoplane with the same total wing area would. The struts and wires generated enormous drag; even the thin piano-wire braces were major culprits, a counterintuitive fact that was seldom appreciated.

Aside from larger engines, the only other significant advances in aircraft design in this period were the replacement of the open-framework fuselage with a fabric-covered body and the adoption of ailerons in the place of the Wright's wing-warping system for roll control. Like others who saw Wilbur Wright fly in France in 1908, the French aviator Henri Farman immediately recognized the importance of roll control as the key to effective maneuvering, and just a few months later became the first to equip an airplane with truly effective ailerons of the modern type—flaps that moved up and down on the trailing edge of all four wing surfaces of his Voisin biplane.

The original Wright patent had covered both wing warping and ailerons, and the Wrights zealously enforced their claim, most notoriously by bringing suit against their fellow American aircraft builder Glenn Curtiss for infringement. The case would drag on for years and cost the Wrights much goodwill among the public at large and the aviation community in particular; many grumbled that the Wrights were "avaricious monopolists" who were thwarting progress. But when all was said and done, even after the Wrights' claims were ringingly upheld by the United States courts, Curtiss and others were never really hindered from fully exploiting the Wrights' three-axis control concept. Curtiss kept building and selling airplanes through it all. The matter became moot, finally, in 1917 when the United States entered the World War and the government worked out an arrangement whereby all of the country's aircraft manufacturers pooled their patents in a cross-licensing agreement; the Wright and Curtiss companies each received two million dollars in compensation for their respective rights.

The only inventors who were in truth set back by the Wright patent were the Wrights themselves. They had hoped to quickly sell their rights and return to the experimental work they loved; instead the drain that waging lawsuits placed on the Wrights' own energy and time ensured that they would fall inexorably behind the rapidly advancing state of the art. By 1911 they had already been surpassed by many of their competitors. The brothers would never again hold the commanding place in world aviation that they held for the first seven years of human flight.

In any case, the truth, as Haldane correctly diagnosed it, was that aviation progress in the long run was being hampered much more by a lack of new science and scientific minds than by any legal squabbling over the Wrights' original invention. And Haldane was determined to do something about it. As the Aerial Navigation subcommittee deliberated and took the testimony of experts through the winter of 1908–1909, Haldane played his cards shrewdly. He said almost nothing as General Nicholson calumniated the Army's aviation experiments in meeting after meeting, and he placidly went along with a final report that recommended that the Army abandon its work on airplanes altogether.

With the committee's recommendations formally approved, Haldane obediently terminated the airplane experiments at Farnborough and dismissed both Capper and Cody. He then turned around and immediately set up an Advisory Committee for Aeronautics, headed by the famous physicist Lord Rayleigh and aided by the National Physical Laboratory, to direct and carry out the nation's aeronautical research program. The committee, which would become the model for the similar organization in the United States that would in time become NASA, was given a broad mandate to identify the important problems in aeronautics and "to seek their solution by the application of both theoretical and experimental methods of research." The National Physical Laboratory, at Teddington near London, already had a small wind tunnel, and over the next few years it expanded its research facilities and began to recruit a growing cadre of scientifically trained researchers. It was one of several crucial events in 1909 that marked the beginning of a historic shift in the center of basic aeronautic science from America to Europe.

Haldane also at once set about recruiting a new man to succeed Capper as superintendent of the Balloon Factory at Farnborough. This time there would be no more dilettantes or military time-servers; he wanted "a civilian and an engineer" for the post, and he found such a man in Mervyn O'Gorman. A successful consulting engineer in the automobile business, O'Gorman was a scientifically versed expert who could be counted on to understand and put to use the basic scientific research of the Advisory Committee and the National Physical Laboratory. To make sure that there would be no interference from Nicholson or other obstructionist military men, Haldane arranged things so O'Gorman reported directly to the War Office, rather than through the military chain of command.

The crucial thing, Haldane perceived, was not merely to get one's hands on anything that could fly but to lay a solid foundation for future progress in aviation. By playing off the enthusiasts, who wanted the government to start buying airplanes at once, against Nicholson, who didn't want the government to have anything to do with airplanes at all, Haldane kept his own preferred middle course from attracting attention and criticism from both sides. It was a perfectly executed bureaucratic maneuver.

The Aerial Navigation subcommittee's report had denied the possibility that airplanes or airships posed any serious, immediate threat to Britain, but allowed that isolated bombing attacks on warships or dockyards from dirigibles—or possibly the landing of a small raiding party from such a craft—"could not be dismissed as an impossible operation of war." While acceding to the Navy's desire to purchase an airship of its own, the committee drew an important but subtle distinction that in many ways echoed Haldane's views about the difference between scientific research on this new class of weaponry and its actual deployment in the force. It was important to gain firsthand experience with airships, the report concluded, in order to be able to understand what, if any, dangers this new technology might pose in the hands of an enemy. But that was completely different from saying that they were practical weapons of war ready to be employed in the nation's military forces.

Such nuances largely escaped the enthusiasts in the civilian world. For one thing, flying was simply exciting: deliriously exciting.

In August 1909 the first international air meet was held in Reims, France. The weeklong event drew forty aircraft; it drew fifty thousand paying spectators each day. Glenn Curtiss brought his new airplane, powered by a fifty-horsepower V-8 engine, and set a new world speed record of 47.1 miles per hour; Henri Farman set a new duration record of more than three hours aloft. The rapturous French crowds forgot themselves in the emotion of the moment: patrons at an open-air café actually stood on their tables and shouted as the competitors flew by.

The men who were more than just excited spectators, the men who flew and owned airplanes, were a considerably smaller number, of course; but they were an unusually prominent, wealthy, and influential group whose ranks would soon grow to include Britain's First Lord of the Admiralty Winston Churchill and the Kaiser's brother, Prince Heinrich of Prussia. Many of the well-heeled new aviation enthusiasts were "sportsmen" who had been equally keen on motorcars. They had earlier formed automobile clubs that organized meets and rallies and offered prizes all over Europe. Now these same men began to form "aero" sport clubs in every country.

In a nationalistic age, it was inevitable that the race to achieve aviation firsts would increasingly be drawn along nationalistic lines. At first those who pressed their governments to spend money on aeronautics cast it as a matter more of national honor than of immediate military necessity. In any event, aviation proponents were quick to recognize that public enthusiasm for flight, especially when stoked by the winds of national pride, was a powerful new tool to advance their aims. When, in August 1908, the huge Zeppelin LZ 4 broke loose from her moorings during a squall and was destroyed, an appeal to the German public brought forth a flood of contributions; when it was over, six million marks had been raised to purchase a new Zeppelin for the government. Governments, aero clubs, and even newspapers in other countries followed with their own patriotic appeals. In Italy the Club Aviatori raised ten thousand dollars to buy a Wright Flyer in 1909 and bring Wilbur Wright to Rome to teach two officers, one from the navy, one from the army, to fly. Three years later, the Aero Club of Padua proposed to present the nation with a complete air force of one hundred airplanes. The target for its *Dati ali al'Italia* ("Give Wings to Italy") campaign was two million lire; three and a half million poured in. The French matched that outpouring with six million francs, which also bought a force of a hundred airplanes, plus training expenses for seventy-five pilots. A second German subscription raised seven million marks. Even Russia, Greece, and Switzerland launched aviation funds to which the public generously contributed.

Every new aviation milestone was noted and duly reported both in the popular press and in the raft of new specialty aviation magazines that had appeared: *Flight, Flugsport, The Aeroplane, Aero, L'Aérophile.* Aviation boosters became anxious at any signs that their own country was falling behind. It mattered less whether any of these milestones translated into an obvious military advantage—it was indeed hard to see how many of them did—than what they seemed to say about the prestige of a nation that could carry out such feats. Because aviation was so visible an activity, and one so widely followed by the public, it came to take on a psychological, almost mystical component as a symbol of a nation's might. It was the beginning of a disconnect between image and reality—what would sometimes be called the "shop window" phenomenon—that would, throughout the twentieth century, characterize how a nation's air force was perceived abroad.

It was not long before the campaigns for "air-mindedness" started to take on a much more overtly military tone. The aviation prizes being offered by industrialists and enthusiasts began to focus more explicitly on feats of direct military significance. In Germany the Prinzheinrichflug was awarded for proficiency in aerial reconnaissance; in France, André and Édouard Michelin, the tire manufacturers, sponsored a ten-thousand-dollar award for the greatest accuracy in bomb-dropping from an airplane. Alongside the "sport" clubs in each country there now began to appear national "leagues" urging military preparedness in the air.

Many of Britain's air enthusiasts had early on perceived a peculiarly British significance to flight that they now began to promote ever more stridently. When Santos-Dumont had flown in France in 1906, the press magnate Lord Northcliffe sent a famous and furious rejoinder to his editor at the *Daily Mail* for reporting the event in a straight factual account. The story was *not* that "Santos-Dumont flies 722 feet," Northcliffe fumed; the story was that "England is no longer an island. . . . It means the aerial chariots of a foe descending on British soil if war comes."

When Wilbur Wright made his spectacular flights in France two years later, there was no underplaying the story in England this time. If anything, the news had even a greater impact in England than in France; the British press was full of accounts of the wonders of the Wright airplane, and Major B. F. S. Baden-Powell, a leading British aviation enthusiast (and the brother of Robert Baden-Powell, who founded the Boy Scouts), declared to reporters: "That Wilbur Wright is in possession of a power which controls the fate of nations is beyond dispute."

Northcliffe was undoubtedly a genuine aviation enthusiast who wanted to help promote air-mindedness; he also was not averse to selling newspapers, and he knew that there was nothing like the fear of foreigners swarming over British soil to get the attention of the British reading public. For several decades "invasion novels" had been a popular literary genre in Britain. One wave had been set off in 1882 when the idea of a Channel tunnel linking England to the Continent was half-seriously bruited. Bookshops were soon filled with tales bearing titles such as *The Seizure of the Channel Tunnel, Surprise of the Channel Tunnel, Battle of the Channel Tunnel, The Battle of Boulogne,* and *How John Bull Lost London.* In *Surprise of the Channel Tunnel,* a large contingent of French restaurateurs, waiters, and bootmakers settle in Dover as a result of the prosperity brought by the tunnel traffic. They are in fact soldiers in disguise, waiting for the order to take over England. In *How John Bull Lost London,* the attack is spearheaded by a group of French tourists who leap from their excursion train to seize a beachhead for the ensuing invasion force.

The summer of 1908 had brought forth a similar outpouring of literary imagination, apparently triggered in part by British naval exercises that year whose "problem" was repelling an invading sea force. The 1908 contributions to the genre included *The Swoop, The Swoop of the Vulture, When England Slept,* and *The Great Raid.* This time it was usually the Germans who were doing the invading, though like their French predecessors they too usually took the form of waiter-soldiers who had taken up residence in England with sinister intent.

A standard theme in all of the invasion stories was that an apathetic populace had allowed itself to be caught fatally unprepared. The young P. G. Wodehouse saw in this the obvious stuff of parody and produced his own offering, *The Swoop! or How Clarence Saved England: A Tale of the Great Invasion.* In Wodehouse's version it is not only the Germans but seven other hostile forces—Russians, Swiss, Chinese, Young Turks, Moroccan Brigands, the Prince of Monaco, and the Mad Mullah—who invade England, all in competition with one another. Sharp-eyed Clarence, the story's young hero, discovers the invasion by reading the stop-press sports items in the local paper (SURREY DOING BADLY GERMAN ARMY LANDS IN ENGLAND) but cannot get anyone else to notice or care. The invaders advance across golf courses while indifferent members of the British public play on. On arriving outside the capital, the Germans decide to bombard London, on the grounds that if they don't, someone else probably will.

Unfortunately for Wodehouse, the book was a flop: in 1909 the threat of invasion was not something the British public was laughing about.

In such an atmosphere, the newly founded Aerial League of the British Empire had no difficulty in commanding press coverage when it sounded the alarm of Britain's dangerous vulnerability to a strike from the air. The League's appeal to Britons to "wake up" to the danger of aerial attack was picked up by all of the major newspapers and echoed in Parliament. Organized in January 1909 by a number of prominent air enthusiasts, the League made no bones about being an "agitating" body; its stated purpose was to advance the "vital importance to the British Empire of aerial supremacy, upon which its commerce, communications, defence, and its very existence must largely depend; and to urge these matters upon Parliament, public bodies, and public men." At a large and tumultuous public meeting that the League convened at Mansion House, the Lord Mayor of London's official residence, Lord Montagu of Beaulieu deplored the "backwardness and apathy shown by this country" in aviation matters and warned that "within a short period, say five years, the insularity of this country, as we understood it, might be destroyed." The league was flooded with letters and expressions of support from the powerful and prominent: Winston Churchill; Lord Curzon, the ex-viceroy of India; Prince Louis of Battenberg, commander of Britain's Atlantic fleet; and H. G. Wells, among many others.

A few weeks after the Mansion House gathering, Montagu addressed the National Defence Association and warned in even more vivid terms of what lay ahead for Britain if she neglected air supremacy: an attacking enemy could "come so swiftly and strike so directly" at the vital centers of the nation that Britain "would be almost paralysed before armies or navies could come to her aid." Among the vulnerable "nerve centers" that Montagu pointed to were the Houses of Parliament, railway stations, telephone and telegraph exchanges, and the stock markets. Northcliffe's *Daily Mail* eagerly picked up the theme, declaring that the loss of Britain's insularity was not five years in the future; insularity had in truth vanished with the nineteenth century. Now "a hostile airship, hovering over London, would be unassailable, and could inflict enormous damage."

Much of the anxiety in Britain was directed toward the German Zeppelins, which had begun making spectacular flights in 1908, staying aloft twelve hours at a time and covering 350 kilometers in one flight. German braggadocio fanned British fears. "What other people can produce an airship anywhere comparable to our Zeppelin cruisers?" boasted one German military journal. Illustrated postcards depicting a Zeppelin dropping bombs on a hapless British cruiser at sea enjoyed a great vogue in Germany.

Besides carrying bombs, British experts solemnly warned, the Zeppelins could carry men. Sir Hiram Maxim publicly and privately warned that even airplanes could soon ferry an invasion force across the Channel, delivering a hundred thousand troops between sunset and sunrise.

It was at about this time that people all over Britain started seeing what they were sure were German Zeppelins in the sky. A few scattered sightings had been reported in March 1909. On investigation, one had turned out to be a balloon carrying a couple of Chinese lanterns sent up by "two young men with the object of hoaxing the residents of the locality," according to a constable's report. But then on May 17 the *Daily Mail* reported that a "long sausage-shaped" object had been spotted in the air over the east coast of the country, and the story was off and running. A few days later all of the British papers carried a report of a "weird object" that had been seen in the skies of Cardiff; a traveler crossing Caerphilly Mountain, eight miles north of the city, even claimed to have bumped into two foreign men wearing heavy fur coats and caps who immediately jumped into their waiting airship and took off when he approached.

Northcliffe himself, who happened to be in Germany at the time, tried to damp down some of the hysteria, noting that the Germans were "beginning to believe that England is becoming the home of mere nervous degenerates." The

Germans had indeed learned a lesson that would shape their thinking in the Great War to come: If the British public had panicked at phantom airships, how would they react to a few real bombs dropped by real airships? (In late 1914, a German naval memorandum advocating Zeppelin attacks on London as a means of "forcing England to her knees" would refer to "the well-known nervousness" of the British public.)

Within a few weeks the sightings dwindled.

But Northcliffe was not averse to generating news himself to keep the fires burning. In June 1909 his *Daily Mail* pointedly reminded readers that the paper had the year before offered a prize for the first man to fly across the Channel; the offer still stood. By late July two contestants were in France preparing to make a try. In the early morning hours of July 25, while his rival stayed in bed having decided the weather was too bad for anyone to fly, Louis Blériot arose and saw that there was a lull in the storm. At 4:30 a.m. he took off into the dawn sky. Thirty-seven minutes later he set down behind Dover Castle.

This was a sensation to leave phantom airships in the dust. The next day Blériot was handed a check for £1,000 at a luncheon at the Savoy. The day after that, the *Daily Mail* ran a long article by H. G. Wells. "It is not the first warning we have had," Wells admonished his countrymen:

> It has been raining warnings upon us—never was a slacking, dull people so liberally served with warnings of what was in store for them. But this event—this foreign-invented, foreigner-built, foreigner-steered thing, taking out silver streak as a bird soars across a rivulet—puts the case dramatically. We have fallen behind in the quality of our manhood. . . . We are displayed a soft, rather backward people. Within a year we shall have—or rather they will have—aeroplanes capable of starting from Calais . . . circling over London, dropping a hundredweight or so of explosive upon the printing machines of *The Daily Mail* and returning securely to Calais for another similar parcel.

As this drumbeat for action grew louder, it was becoming difficult for Secretary Haldane to defend the calm, scientific approach to aviation. The Italians were buying a Wright Flyer; the French Army, following the Reims show, had bought two Wrights, two Farmans, and a Blériot; what did England have to show? In Parliament Haldane pointed out that airplanes were not yet practical weapons of war, Blériot's feat notwithstanding; the right thing to do, he insisted, was to follow the Germans' lead and methodically build up a solid foundation of scientific knowledge, rather than prematurely rushing to buy expensive machines that would immediately be made obsolete. Haldane probably made no friends among red-blooded Englishmen when he spoke affectionately of "my old university at Göttingen," where the German government had just established a chair in aeronautics and appointed an eminent engineer and mathematical physicist, Dr. Ludwig Prandtl, to fill the position. But even those who had supported the creation of the Advisory Committee for Aeronautics as a token of the government's seriousness were growing skeptical about Haldane's approach. Lord Montagu complained that the committee was composed of "theoretical and official people as opposed to practical men. . . . I do not recognise the name of any man on it of actual practical experience." In Parliament Haldane was relentlessly attacked. One member rose to declare furiously that the Secretary of State seemed to be "hypnotized by the blessed word *Science*" while "Frenchmen are landing like migratory birds on our shores!"

The proponents of aviation had found an argument that had legs. The professional military men remained virtually unanimous in their view that the only significant role of aircraft in war would be as scouts and observers for the ground forces, but that was not an argument that would either arouse the public or justify the dramatic infusions of cash that aviation proponents sought. So again and again aviation enthusiasts would emphasize the vulnerability of their nation to a strike from the sky. In Britain, it was not only the loss of "insularity" of their island nation but the large concentration of vulnerable targets so near its coasts, and thus the Continent, that was emphasized again and again in the press and the halls of power. Two years after Blériot's flight, during another debate in Parliament, Haldane explained that Britain did not require as many airplanes as France or Germany since Britain had a smaller army than either of those nations, and a ground force needed only so many eyes. C. G. Grey, the combative editor of the journal *The Aeroplane,* which had emerged as the leading—or at least the loudest—voice demanding more support for aviation, retorted at once. "We require an air service not for a very small army," he thundered, "but for an extremely rich, and an extremely helpless, and an extremely thickly populated country."

Every nation, though, could find weak and vulnerable points that lay dangerously exposed, at least in theory, to aerial attack by a hostile power; what had begun as Britain's native phobia, the special concern of no longer being an island, was soon a global epidemic. If any country was safe from air attack it surely was the United States. That did not stop the American inventor Riley E. Scott from pointing with concern to the fact that two of the nation's great cities, New York and San Francisco, lay right on the coasts: "A fleet of hostile aeroplanes . . . carrying high-explosive and incendiary bombs could soon produce havoc in the business district, probably starting a conflagration that could not be checked," he asserted. "The loss of life from fire, high-explosive bombs and panic would be appalling." And then there was America's crown jewel of engineering and the key to its command of the seas, the Panama Canal, now nearing completion; a few well-placed bombs, Scott declared, could paralyze the canal for weeks, possibly months.

If anyone noticed the incongruity of self-styled "practical men" embracing the fantasies of science-fiction writers as their chief debating point, they did not comment on it. The irony actually ran deeper. Imaginary aircraft with formidable capabilities that existed nowhere on earth were conjured up as the reason to equip one's army and navy with real aircraft, immediately—real aircraft that, with their extremely modest capabilities, would be useless in countering these formidable imaginary aircraft had they existed. What was most remarkable about the gulf between the perception and the reality of the airplane as a weapon of war was that it was easier to change the reality than to change the perception. Those who believed that aircraft would do great things were certain that they were seeing great things happen.

A few military airmen around the world had been tinkering with devices for dropping and aiming bombs or firing guns from aircraft, but these were mostly individual efforts and they had the spirit of one-off stunts directed more at scoring a conspicuous "first" than at responding to any clearly defined military needs. American aviators showed themselves especially adept at grabbing headlines for such feats. At a racetrack at Sheepshead Bay, New York, on August 20, 1910, a Lieutenant Jacob Fickel fired a pistol and a rifle at a target on the ground while flying a hundred feet overhead. Six months later, on January 15, 1911, a Lieutenant Myron Crissy, flying at a civilian air meet near San Francisco, had become the first aviator to drop a live bomb from an aircraft. On June 7 and 8 of the following year, Captain Charles De Forest Chandler managed to haul a Lewis machine gun up in one of the U.S. Army's Wright Model B airplanes at College Park, Maryland, and let off a few dozen rounds at a target on the ground, scoring a few hits from 250 feet and then 550 feet.

Demonstrating that such feats were at least not ruled out by the laws of physics had its value. But developing any of these ideas to the point that they were likely to be of consequence in a war was going to take a great deal more experimentation and refinement. Even Riley Scott, whose bomb-dropping apparatus had cinched the Michelin prize in 1912 and who often waxed eloquent about the looming possibilities of apocalyptic destruction from aerial bombardment, tried at times to balance this with an injection of realism about the vast technical and tactical problems yet to be solved. Not many people were listening. When Scott first tested his device at College Park in October 1911—it consisted of a simple sighting apparatus and a lever that released bombs from a rack mounted beneath the fuselage—the *Washington Post* ran a front-page story the following day quoting the Army pilot of the plane, who enthusiastically predicted that "it will not be long before the army will be able to completely destroy any large fortress or fort in the world" from the air.

Scott pointed out, however, that most bomb-dropping experiments were not being done under realistic conditions. "Bomb dropping has been a popular feature of recent aviation meets and enthusiastic reporters have given their imaginations full rein," he wrote in *Scientific American,* "with the result that cities have been destroyed, forts demolished, and battleships sunk. It must be observed, however, that all such tests have been made from heights of three hundred feet or less, at which range an aeroplane would not have a ghost of a show against machine guns, shrapnel rapid-fire guns or even rifle fire."

The gulf between perception and reality paradoxically only widened when airplanes actually appeared on the battlefield. At the start of 1909 the number of airplanes in the world's armies was zero; by 1910 it was fifty; by 1911 airplanes had been used in combat. The occasion was a colonial war between Italy and Turkey, and although the impact of air power during the conflict was somewhere between negligible and nonexistent, aviation enthusiasts took it as momentous affirmation of everything they had been saying. The aerial forces engaged in the war consisted of nine airplanes and two balloons that the Italians brought with them to Libya, the center of the fighting. On November 1 an Italian pilot took off from the desert, flew over the Turkish lines, and dropped four small Cipelli bombs, grenades, really, weighing about five pounds apiece, with a pin that the pilot had to pull with his teeth before lobbing them out of the cockpit.

The next day Italian newspapers trumpeted this first for Italian aviation. One carried the headline:

AVIATOR LT. GAVOTTI
THROWS BOMB ON ENEMY CAMP.
TERRORIZED TURKS SCATTER UPON
UNEXPECTED CELESTIAL ASSAULT.

It was a ridiculous exaggeration; it was also an uncannily perfect anticipation of a century of exaggerated claims to come for the effectiveness of bombs dropped from the air. The Turks' response to this first-ever use of an air-delivered explosive in combat would become equally iconic: They claimed the Italians had hit a hospital. (The Italians denied it but quietly checked out the story; they found that the hospital had indeed been hit, though probably earlier, by ordinary shelling. The Turks had not protested *that*; they knew there was no psychological advantage in claiming that something as prosaic as an artillery shell had struck their town.)

With both executor and victim of aerial bombardment united in their interest in overstating the effectiveness of this new weapon, the few voices of skepticism could scarcely make themselves heard. A few reasonably impartial foreign observers attempted to point out that many of the Italian bombs failed to detonate at all, or exploded harmlessly in the soft desert sand, and that in fact the Turks had not been scared in the least by the Italians' token raids. On the contrary, when one of the Italian dirigibles appeared overhead, the Turkish infantry invariably stood their ground and let loose a hail of rifle fire that often succeeded in riddling the balloons with holes; the dirigible crews had to keep a constant watch on the gas-pressure gauges to see if they had sprung a leak.

None of this dampened the enthusiasm of the Italians, or of the foreign press. The Italian army issued optimistic communiqués declaring that the results of the aerial attacks had exceeded all expectations. A correspondent for a British newspaper reported that this new weapon had revolutionized warfare, causing terrible destruction in the Turkish camp. Giulio Douhet, an Italian officer who would become the world's best-known proponent of the theory of strategic bombing in the decades following the First World War, declared that "a new weapon has come forth, the sky has become a new battlefield."

More triumphs followed. Guglielmo Marconi, the inventor of the radio, appeared on the scene to personally assist with an experiment in which a radio signal was successfully received by an airplane in flight. The Italian pilots conducted the first nighttime bombing raids. They took aerial photographs and motion pictures from the air.

The Italian general staff did eventually concede that the material damage caused by the aerial attacks may not have been quite as great as they had hoped, but it didn't matter: the strikes had had a powerful psychological effect. They had "dampened the enemy's ardor."

An uprising in French Morocco in 1912 and the Balkan wars of 1912–1913 provided further opportunities to demonstrate what airplanes could do; or, at least, opportunities for aviation proponents to issue bold claims about what airplanes could do. French pilots in Morocco had some success dropping bombs on such targets as markets, villages, and flocks of sheep that tended to put up little resistance. Small incendiary fléchettes, or darts, may or may not have been effective for their originally conceived purpose of setting fire to German Zeppelins in midair, but they worked like a charm in setting grain fields ablaze.

But in the Balkans, where the fighting was considerably less one-sided, the effect of aircraft was once again marked by a huge gulf between expectations and reality. The pilots were mostly foreign mercenaries; the more honest ones admitted that while their reconnaissance missions had been of some value, their attempts at bombing had not. "Absolutely worthless against fortifications," one pilot reported, "and scarcely dangerous for troops in dispersed order."

On the other hand, pilots found that they themselves were quite vulnerable to small-arms fire when flying at altitudes of three thousand feet or less. Yet the first question reporters who covered the fighting always asked was what effect the airplanes were having. They often seemed genuinely surprised when it was explained that the airplane had not instantly revolutionized warfare.

After every war, soldiers curse the generals and politicians who neglected to anticipate and prepare for war. Airmen, more than most, would draw an especially bitter pleasure in this pastime, forever repeating stories that demonstrated the obtuse shortsightedness of the great men who had dismissed the importance of the airplane. Britain had General Sir William Nicholson and General Sir Douglas Haig (before the First World War, Haig had once asserted that "flying can never be of any use to the Army" and suggested that an officer who was learning to fly was "wasting his time"); France had Marshal Ferdinand Foch, the future commander of Allied forces, who in early 1913 opined that "aviation is fine as a sport," but "as an instrument of war, *c'est zéro.*"

But in truth all of the major world powers had already begun to build up their air forces with remarkable rapidity by the time war began in Europe in August 1914. The buildup was all the more remarkable given that they still did not have a clear idea of what they would do with all of these airplanes.

As early as 1910, however, the French, the Germans, and the Americans began taking some first steps to try to answer that question on the basis of rigorous military principles. The French Army's fall maneuvers in 1911 proved that airplanes could consistently locate an enemy's position some sixty kilometers away. The French also concluded that a single officer acting as both pilot and observer had to stay in the air twice as long to gather the same amount of information as an observer flying as a passenger in a two-seat plane, who did not have his attention divided between watching the ground and controlling the plane. In the U.S. Army's Field Exercises held in Connecticut during August 1912, three Army airplanes flew reconnaissance missions and showed that even an outnumbered ground force obtained a "decided advantage" in maneuvering against an opposing force when the airplanes were spotting for it. In June and July the Signal Corps had carried out a series of experiments in sending radio signals from aircraft; the original idea had been for a passenger-observer to operate the transmitter, but it turned out the planes could not lift the combined weight of observer plus wireless equipment. Then one of the Army pilots, Lieutenant Benjamin Foulois, found that he was able to fly and operate a Morse code key with his thumb at the same time. It was that technique that he employed in the Connecticut maneuvers the following month.

In October of 1911, Captain Frederick Sykes was sent by the British War Office to observe the French maneuvers and to prepare a report on the role of aircraft in military operations. Sykes, who would be appointed the next year to command the Military Wing of Britain's newly created Royal Flying Corps, was an abrasive and ambitious man who was roundly disliked by many of his colleagues, but notably unlike many of those colleagues he was also a true intellectual. He read widely and he conscientiously kept up with military and political developments around the world. Convinced that a European conflict was soon inevitable and that aviation would play a decisive part in the coming war, he had enrolled in a course in aerodynamics at London University and learned how to fly in the summer of 1911. In his report, Sykes concluded that "there can no longer be any doubt as to the value of aeroplanes in locating an enemy on land and obtaining information." He estimated that an airplane could do in four hours what a conventional cavalry patrol would take four days to accomplish.

By 1912 the French, German, and Russian armies had already produced book-length manuals on aerial reconnaissance. Meanwhile, the first serious efforts at working out a more complete doctrine of air-power employment began to appear in military and political journals and in internal reports. In contrast to the broad-brush pictures of aerial devastation penned by professional agitators like the Aerial League, these treatises were often the work of professional military men and other strategic experts who tried to think through exactly how aircraft could best be integrated into military organizations and how they might most effectively be used to exert pressure against the enemy.

If their vision of the airplane as an offensive weapon was still futuristic, it was at least informed by sound military principles and a realistic understanding of how armies conducted operations in the field. Sykes, in his 1911 report, had suggested that aircraft could be used not just for reconnaissance but to make raids against the enemy's "vital points," such as bridges, junctions, and dirigible sheds, and also to control and adjust artillery fire by spotting the fall of shells fired by guns on the ground. Colonel Louis Jackson of the Royal Engineers, addressing the Royal United Service Institution in London in April 1914, laid out a careful analysis of installations that might effectively be targeted from the air using existing weapons or those reasonably expected to come into existence in the next few years. "The public mind has been disturbed for years by a vague menace," he noted, "and it is worth while to attempt to lay down in definite terms what the danger is." Jackson said that the nightmare scenarios of newspaper and magazine writers ("we are all familiar with that tiny phial of grey powder, the invention of some Russian or German chemist, which, if dropped on the dome of St. Paul's, would suffice to lay London in ruins") had actually made sensible people tend to go to the other extreme and reject "the whole danger [as] illusory." But he argued that fixed targets such as arsenals, dockyards, armament factories, and oil reservoirs could all come under aerial attack and would need to be defended in wartime.

Other experts suggested that aircraft might be especially effective if employed to attack important battlefield targets that lay beyond the range of conventional artillery, such as enemy reserve troops waiting in the rear or attempting to move up to the front on roads or rail lines. And nearly all had to concede that cities were extremely easy targets to hit and possibly quite vulnerable ones. Accuracy was not required; incendiaries could start large conflagrations; and if a strike came in the opening hours of a war, it could create panic and disarray and prevent mobilization.

Most of these analysts saw aviation not merely as an adjunct to the weaponry of existing army formations but as a fourth arm that would operate alongside infantry, cavalry, and artillery. Some argued, however, that the correct analogy was to be found in the relation of the army to the navy: the air, like the oceans, was a separate field of battle altogether, and it had to be patrolled and commanded not by a fourth arm of the army but by a third service of the armed forces. The danger that an enemy might launch a knockout blow from the air in the very first moments of a war required the creation of a "permanent, instantly available, and autonomous" force to counter it, a Lieutenant Poutrin of the French Army argued in a leading French military journal.

Many generals dismissed such theorizing as idle and premature speculation. Yet the fact was that to guide this emerging technology in a direction that was both realistic and militarily effective, the world's armies *had* to figure out a sound theory for its employment. In the absence of a clear directive of military needs that engineers could respond to, the technology tended to advance in response to its own internal evolutionary forces. And the sheer complexity of an airplane worked to steer innovation along a few narrow channels. The three basic forms that the airplane settled into in the years just before the war—the "tractor" biplane with a forward-facing propeller and the engine in the front, the "pusher" biplane with a rear-facing propeller and the engine in the back, and the tractor monoplane—were not conceived to be specialized types that favored particular performance characteristics, much less specialized military missions; they were just three tried and proven ways that all of the pieces of an airplane could be put together without disaster occurring.

Airplane designers still had only the vaguest understanding of how design choices translated into tradeoffs in performance characteristics such as speed, rate of climb, maneuverability, payload capacity, range, and maximum ceiling. The French aircraft builders now clearly led the world—they had quickly recovered their amour propre following the humiliation of Wilbur Wright's 1908 flights—but they did so from shops organized on the very Old World lines of the bourgeois small craftsman. Many of the French builders were independently wealthy and used their inherited fortunes to subsidize what were only marginally profitable enterprises. They were enthusiasts rather than businessmen or engineers. All were aviators themselves; only a few had scientific training.

The basic technological advances in aircraft performance that did occur at this time owed more to the French engine industry. The Gnome company's air-cooled rotary engine had dominated the Reims air show; by 1913 the firm employed some eight hundred workers at its factory near Paris, and though its manufacturing process still relied on considerable handwork it had automated much of the component fabrication with machine tools imported from the United States.

The rotary engine was lightweight, efficient, and reliable, if a bit odd to our modern eyes: the entire engine, cylinders and all, rotated about a fixed crankshaft. The propeller was simply attached to the engine and turned along with it. The considerable advantage was that even a large engine with multiple cylinders (the eighty-horsepower Gnome had nine) was efficiently cooled as the cylinders whizzed around through the air; the extra weight required for a water jacket and radiator could be shed altogether. Renault, France's next-largest engine maker, had meanwhile in 1911 introduced an air-turbine-cooled, seventy-horsepower V-8 engine. Propelled by these new power plants, French aircraft soon doubled the 1909 speed records to more than eighty miles per hour and pushed the altitude record from 1,500 feet to more than 12,000 feet.

Speed, duration, and altitude were obvious technological goals, but how these might relate to military effectiveness was still anybody's guess. Instead of military doctrine determining technology, technology was—by default—determining doctrine. On the other hand, doctrine that was not derived from realistic combat experience, or doctrine that demanded from technology what was scientifically impossible, could lead to a huge waste of effort. And unfortunately, in the years leading up to the Great War, both basic scientific research and realistic experience were in modest supply. In the years to come doctrine would sometimes be too far ahead of technology, and many futile efforts would be expended trying to meet unattainable goals; other times doctrine would be far too limited in its conception and retard technological progress that could have been had for the asking. Both failings bedeviled military aviation as it prepared for its first great test.

Beginning around 1912, military leaders made their first tentative efforts to impose some direction on what had been a largely undirected process in the design of aircraft and the development of air armaments. The French Army, correctly heeding the battlefield experience in the Balkans where aircraft had proved highly vulnerable to ground fire, went on a disastrous excursion attempting to build an airplane armored with thick metal plating. A few calculations might have shown the scheme to be utterly impractical; instead French aircraft designers wasted much valuable time and energy in 1913 and 1914 producing three worthless prototypes that turned out to be too heavy to fly.

The British and the Germans, having concluded that observation would be the key role of aircraft, fell into the other pitfall, of restraining technological development. Mervyn O'Gorman, the civilian engineer appointed to take charge of Britain's Balloon Factory (it had been renamed the Royal Aircraft Factory once the Royal Flying Corps was established), later acknowledged that the narrow conception of the military role of aircraft had almost strangled the development of British aircraft before the war:

> The ideas dominant prior to the war ascribed to aircraft the function of reconnaissance and scouting and for these alone had we made preparation. It is certain that outside of a few purely scientific and technical centres, we did not prior to the war attach a sufficiently vital importance to aircraft, nor believe in the extraordinary versatility of the instrument, even when demonstrations had been made and good reasons urged. This is explained partly because the aircraft user did not appreciate its potentialities himself, and partly because the somewhat exaggerated tone of some of its exponents had weakened their prestige.

The Royal Aircraft Factory, which had become the de facto center for designing, testing, and building most British military aircraft, had started out with the assumption that the *slowest* possible speed an airplane could be made to fly was best, given the requirements of reconnaissance missions. Attempts by certain French builders to set speed records, O'Gorman recalled, were viewed with disdain, as "astute efforts of commercial advertisement rather than as military achievement. . . . To call officially for construction to excell such speed was stated in England to be mere ostentation."

The BE2 airplane that was the staple machine of the Royal Flying Corps for the first two full years of the war was an inherently stable and thus relatively unmaneuverable airplane, and it was spectacularly slow, even by 1914 standards. Its top speed was about seventy-five miles per hour, some fifty miles per hour less than the 1913 world speed record, and it took more than half an hour to climb to ten thousand feet. (The BE2 was in fact so stable in the pitch and roll directions that a pilot could take off without touching the stick at all, just by opening up the engine and touching the rudder a bit to keep the nose straight as it taxied along the ground.)

BE2c
observation/light bomber
engine: 90-hp eight
    cylinder
top speed: 95 mph
armament: 1 machine gun

Larger airplane engines were disdained along with higher speeds. The Royal Aircraft Factory held an engine competition in 1914 and received entries ranging from 70 to 200 horsepower. But General Sir David Henderson, the commander of the RFC, decided that no engine above 100 horsepower would be needed, and the factory stopped all design work on large engines. At the outbreak of the war, the only engines available for British military aircraft were three French models ranging from 70 to 100 horsepower and an 80-horsepower engine made by the Royal Aircraft Factory.

The Germans took much the same view, favoring large, slow, and stable aircraft on the theory that these would furnish the most suitable platform from which to observe the ground below. In 1912 the Kaiser sponsored a prize for engine development that stipulated a maximum of 115 horsepower.

And as O'Gorman noted, no one—not the British, not the Germans, not the French—had "seriously entertained" the idea of "pursuing and catching an enemy" in the air. In July of 1914 a French officer, Captain Jean Faure, had tested a 37mm cannon on a Voisin biplane; he was told by his superiors that this showed a very commendable "spirit of research," but one that perhaps "smacks more of Jules Verne than of reality."

When war broke out that month, the major powers had about 1,000 operational front-line aircraft: Germany, 230; Russia, 190; France, 160; Britain, Italy, and Austria-Hungary each somewhere between 50 and 100. France's aeronautical appropriations in 1913 had been more than $7 million, Germany's and Russia's about $5 million, Britain's $3 million. Buying even the wrong airplanes was not cheap.

On the eve of war, Russia stood out as the most technologically backward of the great powers. Yet it was in St. Petersburg in the early morning hours of June 30, 1914, that an aircraft unlike anything that Russia's rivals had ever conceived lifted off into the dim glow that marked the end of the short summer night—the "white nights" of St. Petersburg, the Russians called them—and began a journey that would make aviation history. At one o'clock the still air around the Komendatsky Aerodrome was broken by the sound of four engines coming to life. A few minutes later a huge biplane began to bump across the field, pushed at first by twenty men lined up behind the wings, then slowly gaining speed on its own from the thrust of its four spinning propellers. To call the plane huge scarcely did it justice. Its wings spanned 100 feet, the same as a B-17 Flying Fortress of the Second World War. Its four automobile-type Argus engines, two delivering 125 horsepower and two 140 horsepower, were each more powerful than most of the single engines that were carried by nearly every airplane then in existence. The plane weighed more than three tons empty; for this flight it carried four men plus a ton of gasoline, oil, water, and spare parts including two complete replacement propellers.

At the controls for the takeoff was the plane's twenty-five-year-old inventor, Igor Sikorsky. The crew's destination for this flight was as audacious as the machine itself: Kiev, 800 miles due south. If all went well they would reach the city, Sikorsky's hometown, that night after a single refueling stop along the way at Orsha.

All did not go well. After their refueling stop they hit repeated downdrafts that knocked the plane to an altitude of less than 100 feet as Sikorsky struggled to keep it in the air; a gasoline line came loose from one engine, the spilled fuel igniting in a huge sheet of flame that the crew had to beat out with their overcoats while standing on the wing; they lost their way in low clouds, then dropped more than 1,200 feet as the craft went into a near-fatal spin. Yet, after an overnight repair stop, they found themselves nearing their destination the following morning. Breaking through the clouds at 900 feet, Sikorsky was greeted by the golden domes of Kiev's Lavra Cathedral directly ahead. A few minutes later they touched down at Kourencv Aerodrome, where the young inventor had made his first flight in an airplane just three years before.

It would have been an extraordinary journey for any airplane to undertake in 1914, but the Il'ya Muromets (Sikorsky had named the giant plane after a tenth-century Russian folk hero) was not just any airplane. When Sikorsky had proposed designing and building a four-engine plane, opinion in aviation circles ran strongly against him. He was told that such a large plane would be too heavy to get off the ground, or to safely control and land if it did get off the ground. There was nearly universal skepticism that a multiengine craft could work at all; if one engine failed, the asymmetric thrust of the remaining propellers would send the plane into a helpless spin.

But Sikorsky had arrived at the idea of building a multiengine plane precisely because of what he saw as the inherent danger of single-engine planes. It was a logical, if unique, outgrowth of his own remarkable—and brief—aviation career. The son of a medical doctor, a pioneering Russian psychologist, Sikorsky had enjoyed a comfortable upbringing and was enrolled as an engineering student at the Polytechnic Institute of Kiev when, in 1908, he read of the Wright brothers' spectacular flights that summer. He soon began to build and test models of a helicopter he designed; in January 1909 his older sister agreed to lend him the money he needed to travel to Paris, look over the flying machines there, and buy an engine and other parts to begin building his own full-scale aircraft. His family's friends and relatives, he later learned, tried to talk his father out of allowing such a trip: "They thought it almost outrageous to permit a boy not yet twenty years old to interrupt his studies to go to Paris with a huge sum of money which, they predicted, would be spent not on machinery but for totally different purposes." They needn't have worried: "At that time," Sikorsky explained, "nothing existed for me except the idea of the flying machine which was now so close to realization."

Over the next two years he made two more trips to Paris to keep in touch with the latest developments and purchase more powerful engines. He gave up on the helicopter, though he never forgot the idea, and decided to concentrate on making a more conventional airplane; on the third try he succeeded in building a machine that flew in a reasonably reliable manner.

By this time flying had become an obsession, and Sikorsky dropped out of the Polytechnic Institute so he could devote his entire efforts, often sixteen hours a day, to building a successful airplane. He drafted the neighborhood carpenter and plumber to help build components; Sikorsky himself was designer and test pilot, and he learned both roles as he went along. "At that time reliable information and aeronautical science were practically non-existent," Sikorsky later recalled, "and the pioneer designer and pilot had only his own meager experience, practical judgment and imagination to supply the necessary data on which to build his machine. When the plane was ready, and no one could tell whether or not it was any good, the designer had to become a pilot . . . having no instructor to explain or give advice." Crashes were frequent, and it was often difficult to figure out whether a problem was due to a bad design or unskilled piloting.

Through sheer doggedness Sikorsky succeeded by December 1911 in building a three-seater, the S-6-A, that set a world speed record for an airplane carrying a pilot and two passengers (about seventy miles per hour) and two months later won first prize in a Moscow competition. In September 1912 a modified version, the two-seater S-6-B, claimed the first prize of 30,000 rubles, about $15,000, in a competition sponsored by the Russian Army. With the prize money Sikorsky was able to pay back all that his family had advanced him.

By this time he had also attracted the attention and support of one of Russia's leading industrialists, Mikhail Shidlovsky, chairman of the Russo-Baltic Wagon Company. The firm built railroad cars and automobiles and was now interested in moving into the airplane business. One night during the Army competition,

Shidlovsky invited the young inventor to dinner. While sitting in his host's study afterward, Sikorsky began outlining his vision of the future of aviation. Sikorsky explained that he had always believed that to be truly practical airplanes needed to be much larger; they needed to have large, enclosed cabins so the crew could operate in comfort, especially in the difficult climate of Russia, and so pilots could relieve one another at the controls on long flights; they needed to have several engines as the only possible protection against the need to make a hazardous forced landing in case of an engine failure, a distressingly common occurrence. On one occasion, Sikorsky related, he had narrowly avoided killing himself squeezing between a freight train and a stone wall while landing in a small railroad yard after his engine quit; the culprit turned out to be a mosquito that had been sucked into the carburetor, blocking the air jet. Shidlovsky listened intently to all Sikorsky had to say. "Start the construction immediately," he replied.

Just half a year later, in May 1913, Sikorsky's first four-engine airplane, the Grand, began flying over St. Petersburg in a near-daily spectacle that stopped traffic on the streets below and drew crowds to the airfield. The Grand was a total departure from aviation practice of the day. Its enclosed cabin was fitted with a table and wicker chairs, a row of large windows for viewing the scenery below, even a washroom. Most pilots at the time insisted that feeling the speed and direction of the airstream against their face was necessary to sense the plane's movements and properly react to them. Sikorsky instead installed a series of instruments in his enclosed cabin to give the pilot the essential information he needed: altimeter, airspeed indicator, bank indicator. The Il'ya Muromets, which first flew in January 1914, was basically an enlarged and improved Grand; its cabin even included heat, a private cabin with a berth, and electric lights powered by a wind-driven generator. On one of its test flights it carried sixteen passengers—and a dog, the airfield's mascot.

In conceiving of such a giant bird, and designing and building it, Sikorsky was not responding to any stated military needs; he was simply following his intuition about what would work and what was technologically promising. But the upshot was that largely because of this one man's genius and persistence, Russia would enter the Great War with the only long-range airplane and the only heavy bomber in the world.

When Sikorsky and his crew landed at the Kourenev Aerodrome on July 1 after their epic flight, they were excitedly welcomed by the Secretary of the Aeronautical Society of Kiev. When he was through congratulating them, he mentioned the news that had just arrived: three days before, he told them, the heir to the Austro-Hungarian throne, Archduke Franz Ferdinand, had been shot and killed along with his wife by a Serbian assassin in Sarajevo.

3

REALITIES

For several weeks after the Archduke's assassination, war still seemed a distant prospect. "Why four great powers should fight over Serbia no fellow can understand," one member of the British government jotted in his diary in late July. Appealing for restraint and understanding, the Kaiser and the Czar, grandson and nephew of Britain's late Queen Victoria, exchanged cables signed "Willie" and "Nicky."

But once armies began to mobilize, the outcome was inevitable. Backed by its ally Germany, Austria delivered a humiliating ultimatum to Serbia demanding not only that the assassins, but anyone engaging in anti-Austrian propaganda in Serbia, be arrested. Serbia capitulated on all but one point: Austria's demand that its officials sit beside Serbian jurists in the investigation, trial, and sentencing of the conspirators. That matter, Serbia meekly demurred, should be submitted to the International Tribunal at The Hague to be decided. This was not what Austria had in mind. On July 28 Austria declared war. Serbia's ally Russia ordered a partial mobilization of its army; Germany responded in kind; France, Russia's ally, called up its three million citizen-soldiers.

In keeping with its long-standing plan for waging a two-front war, Germany prepared to launch a lightning assault through Belgium, skirting the French border defenses and seizing Paris in thirty-nine days. France would be out of the war before Russia could even finish mobilizing. On August 4 German troops set foot on Belgian soil, whose neutrality Britain had guaranteed in an 1839 treaty. All of Europe was now officially at war.

As the Germans wheeled north in the arc that would end at the French capital, the French dismissed the German thrust through Belgium as a feint and stolidly began to execute Plan XVII, a frontal assault on Alsace-Lorraine to the east. Thoroughly imbued with the offensive spirit, the French commander-in-chief, Joseph Jacques Césaire Joffre, ordered his troops into bayonet charges against the prepared German positions. In four days, from August 20 to 23, forty thousand French soldiers were killed and one hundred thousand wounded, mowed down by machine guns and heavy artillery. It was the first terrible lesson that this was an entirely new kind of war. The French First, Second, Third, and Fourth Armies fell back toward Paris. By the next day the French Fifth Army and the British Expeditionary Force, which had tried to hold the Germans' northern sweep at Mons across the border in Belgium, were in headlong retreat, too.

Joffre had insisted on a news blackout at the front; even the government in Paris had a difficult time learning of the disaster that had taken place, and in the chaos of retreat the situation became only more confused. The French Army's twenty-seven escadrilles, or "flights," of six aircraft apiece were all assigned to the five armies on the line, but as soon as the retreat began, the flow of spare parts, engines, and new and replacement aircraft—the infusions that the complex organism of military aviation demanded every day to remain alive—fell into immediate disarray. Nor was it a time for commanders to be much interested in something as new and untried as aviation. They had other things on their minds.

Fortunately for France, the military governor of Paris, General Joseph Simon Galliéni, was a notable exception to this rule. During the 1910 maneuvers he had taken a keen interest in aircraft and had quizzed the pilots at length about what airplanes could do. "He is one of the rare generals who believe in aviation," said Captain Georges Bellenger, one of the Army aviation commanders. Fortunate, too, was the fact that the aircraft builder Louis Breguet, now Corporal Breguet of the French Army, stationed in Paris, offered the services of one of his own proto-type airplanes to Galliéni. On September 2, as the garrison in Paris prepared for siege, Breguet took off with an observer to reconnoiter the approaching German forces to the north.

He returned with astonishing news: The German First Army under General Alexander von Kluck had turned. It was no longer heading toward Paris but had wheeled to the southeast, moving to bypass the city altogether. Von Kluck's supply train was enormous and increasingly stretched; his 84,000 horses alone required two million tons of fodder a day. And with the British now on his flank—having marched day and night from Mons, the retreating British force was now approaching the Marne River to the east of Paris—von Kluck decided he first had to eliminate that threat.

The next day observations by two of Captain Bellenger's aviators confirmed Breguet's report. Bellenger, unable to get his superiors to believe him, appealed directly to Galliéni's liaison officer; Galliéni urged Joffre to act at once. After some hesitation, risking all, the commander-in-chief ordered the forces defending Paris to move east and join the attack against von Kluck, catching him as his overextended lines of supply were exposed to the south of the Marne. Six thousand French troops were relayed to the battlefield in Paris taxicabs. The Germans were thrown back twelve miles north of the Marne, to the Aisne River; Paris was saved, and the German plan for a quick victory lay in ruins.

In his September 7 dispatch, Sir John French, commander of the British

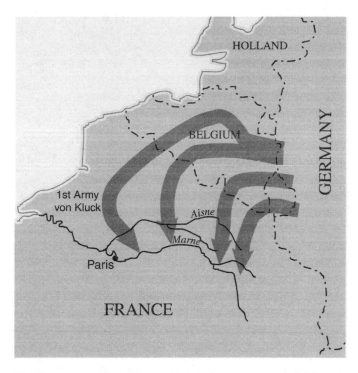

The German attack on France, August–September 1914. When von Kluck wheeled to the southeast to bypass Paris, French air patrols spotted the move in time for the Allies to rush troops to the Marne.

Expeditionary Force, singled out the airmen for praise: "They have furnished me with the most complete and accurate information which has been of incalculable value in the conduct of the operations."

On the eastern front, the Russians had mobilized more swiftly than the Germans had believed possible, but here it was the turn of German aviation to save the day. On August 17, two Russian armies, 370,000 strong in all, had begun the invasion of East Prussia, territory that is today part of Poland. Splitting up around the fifty-mile-long chain of the Masurian Lakes, the Russians' plan was to crush the German army beyond in a two-pronged attack. Both aerial reconnaissance and intercepted Russian radio messages alerted the Germans to the Russian intentions. The Germans concentrated their entire force against the first of the Russian pincers to arrive, General Alexander Samsonov's First Army, and routed the invaders, killing 30,000 and taking more than 100,000 prisoners, 500 field guns, and thousands of horses. The Battle of Tannenberg, the Germans proclaimed it, and it was to be their greatest victory of the war. General Samsonov's body was found in a pile of corpses; he had shot himself in the head with his revolver. "Without the airmen," declared Field Marshal Paul von Hindenburg, the German commander at the eastern front, "no Tannenberg."

That such relics of the old school as Hindenburg, Joffre, and French had each staked their fate in battle on information secured from aerial reconnaissance spoke volumes. Within a month the French Army had ordered a doubling of its air forces, to sixty-five escadrilles. A decision was quickly made to emphasize production of two-seater pusher-type airplanes, such as the Voisin and the Farman, which gave the forward-seated observer an unobstructed view of the ground; eight of the twelve different models of aircraft that the French had entered the war with were dropped.

Although the vital importance of aircraft for observation duty had been dramatically driven home, air operations remained ad hoc, hectic, and experimental throughout the autumn of 1914. Observers had had little realistic training in how to identify and interpret what they saw on the ground, and while Sir John French commended his airmen for furnishing the "most complete and accurate" information, a lot of it was anything but. In the opening days of the war one German observer excitedly reported that the British troops he had flown over were "thoroughly disorganized and running about their post in blind panic." In fact they were playing soccer.

The relative autonomy that aviation units enjoyed in the first weeks of the conflict—and the fact that their commanders often didn't yet know what to do with them—led to a burst of experimentation. Within a month virtually every role and mission that aircraft would come to play in every war of the future was tested in some form or another: strategic bombing, ground attack, forward artillery spotting, aerial photography, air-to-air combat, psychological warfare. On August 30, as German troops neared Paris, one German aviator took the initiative to fly over the city in his Taube monoplane and toss out five small bombs over the Gare de l'Est. Three failed to explode, but one that fell at the corner of rue Albouy and rue des Vinaigriers did, killing a woman as she stood in front of the shop of a baker and wine merchant. Along with the bombs he dropped a message: "The German Army is at the gates of Paris. You can do nothing but surrender. Lieut. Von Heldsen." Air crews returning from reconnaissance missions also experimented with dropping on the enemy troops below whatever they could get their hands on: steel darts, hand grenades, and 90mm artillery shells, sometimes with improvised fins attached.

Tales of enemy pilots chivalrously saluting one another as they passed in the air notwithstanding, some aviators wasted no time in trying to shoot down their foes. At first most pilots and observers were armed with nothing but pistols, intended only for self-defense if they were forced down behind enemy lines. The British, however, showed an eagerness to fight from the instant they left the ground. On August 22 the first German aircraft to be spotted by the British appeared over an RFC aerodrome at Maubeuge, close to the Belgian border. The crews of six British planes immediately "turned out armed with rifles" and gave chase; they also packed hand grenades that they optimistically thought would be useful for dropping on the German if they could maneuver into a favorable position directly above their foe. One of the British planes was a Farman that had been experimentally outfitted with a Lewis machine gun. None were any match for the German Albatross; the Lewis gun so weighed down the underpowered Farman that it was still struggling to get above 1,000 feet half an hour after the German had departed from the scene.

Undaunted, the British pilots went in for even more outré improvisations. One pilot tried to entangle the propeller of an enemy plane by dropping a weighted 150-foot-long cable in its path. Another scheme involved trailing an electrically detonated bomb fitted with grapnel hooks. By September 7 the British had actually managed to bring down five German airplanes without using any weapons at all—mainly by "maneuvering boldly" in what amounted to a game of chicken and forcing the enemy to land. On August 25 three British aircraft had thus chased an enemy monoplane; when the German landed, the pursuit continued on foot until the pilot disappeared into a nearby wood.

The popular aviation press expressed surprise that there had been so little aerial fighting. Army commanders expressed amazement that there had been any, and probably with more justification. Neither tactics nor equipment had anticipated air-to-air fighting. The general view was that trying to hit one airplane in flight with a bullet fired from another airplane in flight was an almost absurd proposition. Prewar experiments in aerial gunnery that had advanced beyond the stunt stage certainly were not encouraging, and little thought had gone into integrating guns to airplanes as part of an effective overall system. Pushers afforded the gunner a clear field of fire but had poor performance; tractor airplanes had only a narrow V-shaped gap between struts, wires, and propeller arc to fire through, forcing the gunner to shoot sideways. Correctly aiming a gun at an angle, taking into account the relative speeds of gunner and target, was in fact an extremely difficult problem. That most of the improvised air-to-air fighting in the first months of the war was done with carbines rather than machine guns only made it more miraculous to actually score a hit. The French captured a German order issued on October 2 that instructed pilots to avoid wasting time with aerial fighting at all; while it was true that French pilots sometimes "amused themselves" by taking potshots at German planes, the order noted, "there was nothing to worry about." Three days later, nonetheless, a French pilot, Joseph Frantz, and his mechanic-observer by the name of Quénault, shot down a German two-seater from their machine-gun-equipped Voisin pusher.

But such feats remained by far the exception. Airmen in the field could improvise aerial guns and bombs; they could not improvise a whole new airplane, and that was what it was going to take to make fighting and bombing something more than a matter of a few spirited ventures.

Innovations in observation techniques and equipment, by contrast, were swiftly incorporated into the operational air fleets during that first fall. By November the maneuver war on the ground was over. Thrown back to the Aisne River, the Germans dug in; then a race began as each side tried without success to outflank the other. When it was done, a line of opposing trenches stretched from the sea to the Alps. The terrible stalemate that would grip the Western Front for three years of brutal trench warfare had begun.

A few experiments had been made before the war using aircraft to adjust artillery fire; now as the opposing forces settled into the fixed positions, artillery "shoots" became the predominant mission for reconnaissance aircraft. At first spotters would have to land or drop messages in weighted tubes with streamers to communicate with the batteries, but quickly a system was developed for communicating by use of colored signal lights, smoke bursts, or movements of the wings. This in turn was quickly replaced by wireless telegraphy. The use of radios in airplanes had been the subject of much experiment, but none of the aircraft in service in the fall of 1914 had been designed to carry the heavy equipment, nor were they routinely supplied with it; in a matter of months, however, this experimental notion became an operational routine. In January 1915 British squadrons worked out a system called the "clock code": The observer would place over his map a celluloid disc centered on the enemy target; a series of concentric rings labeled with the letters A through F marked distances of 50, 100, 200, 300, 400, and 500 yards from the target; twelve pie slices were marked with the numerals I through XII of the clock, with XII pointing north. The observer in the aircraft would watch where the shells fell and report with a simple Morse code message: C9, B2, and so on. Because few aircraft were fitted with receivers (it was harder to pick up a signal from the ground in the air than vice versa, and the noise of the engines made it next to impossible to hear a message), troops on the ground communicated to the aircraft by laying out large white panels in prearranged patterns.

Panels were also used to mark the position of one's own forces so aircraft could keep commanders informed of the progress of an attack. A disastrous incident in September 1915, when French forces shelled their own infantry for three days despite reports from aviators that they were friendly troops, shocked the command into placing much more emphasis on such "contact patrols" and giving credence to their reports.

Experiments in aerial photography also advanced with unprecedented swiftness. During the early fall one British squadron took pictures of the German positions along the Aisne, but, as a General Staff report noted, "they were not regarded as of much interest." That was mainly because they were rather blurry. But by the end of the year high-quality photographs were routinely being taken, and by the end of the war the British alone would take more than 400,000 aerial pictures. The stagnation of the battlefront allowed aerial photographers to blanket every inch of the enemy position; detailed maps of trenchworks, fortifications, and rear areas were produced from the reconnaissance photos and distributed to infantry and artillery commanders.

SIGNALS MADE BY THE INFANTRY.

SIGNALS	PANELS USED WITH SIGNALLING TO INFANTRY AEROPLANES BATALLION	REGIMENT	BRIGADE	BY WIRELESS, SIGNAL LAMPS, SOUND, EARTH TELEGRAPHY	BY SIGNAL ROCKETS	OPTIONAL (USED) FOR USE OF BRIGADE OR DIVISION
Objective Attained.				—···——		The meaning of which must not be determined for once and all. These meanings must be explained in orders.
Request for Artillery Barrage.				————		
Request for Artillery fire in Preparation for Attack.				—·-———		I
Our field Artillery fire is short.				···		L
Our Heavy Artillery Fire is short.				···		III
We are ready for attack.				··		IV
We will not be ready for attack at designated hour.				————		V VI
We shall advance Lengthen Artillery fire.				····		L
Request supply of Cartridges				—·———		LI
Request supply of Hand Grenades.				————·—		IX
Understood ~ Message recd.				···—·—		

1. The signal "Understood" (given by panels) should be displayed after reception of all messages sent by wireless from infantry Aeroplane, for this is the only method of advising the Infantry observer that his radio apparatus is working properly.

2. The conventional signals sent by wireless, earth telegraphy, signal lamps, or sound should be repeated continuously in order to prevent their being taken for call letters.   C-917

One common method of signaling to aircraft entailed laying out large canvas panels on the ground in prescribed patterns. *(Author's photo/National Archives)*

Reconnaissance, artillery observation, and battlefield photography were hardly new ideas in warfare, but the speed and efficiency with which aircraft had shown they could carry out these functions resulted in a vast, practical shift of power in the military chain of command. Aerial observers were quite literally calling the shots. The RFC fliers attached to each army corps had the authority to direct the entire artillery of the corps onto a target if they spotted something that looked important and looked like it might not still be there if they waited for approval from the artillery commander. "At such times," noted the historian Lee Kennett, "a second lieutenant held the powers of a major general."

Aerial reconnaissance and observation was the mission from which all other aerial missions flowed. In a few short months it had become indispensable: by the time of the battle of Neuve Chapelle on March 10, 1915, British commanders were using photographic reconnaissance as the basis for their entire attack and artillery-fire plans. The instantaneous advantage conferred by having an airborne spotter on the scene while an artillery battle was under way had likewise become something no one could doubt. Troops on both sides noticed that whenever enemy aircraft appeared overhead the accuracy of incoming artillery fire instantly improved.

Air-to-air fighting had begun as something of a game; now it suddenly took on a far more serious military purpose. As a British General Staff report on aerial combat noted two years later, by 1915 it was already clear that aerial "fighting would be necessary on an ever-increasing scale to secure liberty of action for our artillery and photographic machines, and to interfere with similar work on the part of the enemy." The fighter was born of the counter-reconnaissance mission.

An effective fighter, as the improvisations of the fall of 1914 had already demonstrated, was more than just a plane with a machine gun thrown aboard. The two-seater reconnaissance planes that dominated the air forces of both sides were sorely wanting in every characteristic that was needed in a fighter. A fighter needed to be fast to pursue and catch up with a fleeing enemy; it needed a fast rate of climb to be able to get into the air quickly when the alarm was sounded that hostile aircraft were approaching the lines; it needed to be maneuverable to be able to get into a favorable position from which to fire; it needed a high maximum ceiling in order to gain the enormous tactical advantage that altitude offered. Altitude meant surprise, the ability to dive out of the sun on an unsuspecting foe. It also was akin to having what sailors in the age of sail called the "weather gauge": a ship that was to windward of the enemy always held the power to force a fight or break off the action. Similarly, a fighter at high altitude could command a large radius of action in the air below thanks to the speed that he would gain as he dived; he also had the choice of departing the scene before an enemy could climb to reach him.

Adding the weight of a machine gun to a two-seater only detracted from these characteristics. Some advocates of a heavy, two-seat fighter clutched at the hope that by freeing one man to concentrate all of his attention on shooting—and by giving him a flexible gun mounted on a pivot, which could sweep a large cone of the sky—the lack of maneuverability of the plane could be offset by the increased firepower of the gun. It was a vain hope, but one that died hard; not until such heavy "destroyers" were shot out of the sky in great numbers on both sides in the Second World War was it finally put to rest.

But other airmen were convinced immediately that the most deadly way to aim a gun in the air was by having the plane itself do the aiming. They also saw that among the existing types, the only one that came even close to offering the maneuverability and speed required for the job was the single-seat "scout."

The scouts had been designed for fast reconnaissance of a "strategic" kind, darting behind enemy lines to spot what the enemy was generally up to, not for the methodical collection of detailed tactical information that was the business of the stable and slow two-seaters. Scouts were small, light single-seaters with tractor propellers, often monoplanes in 1914, having top speeds typically twenty miles per hour faster than the larger reconnaissance planes. The trouble was there was no structurally safe place to mount a machine gun on a tractor mono-plane where it would still be in reach of the pilot—a necessity for reloading and for clearing the frequent jams that machine guns of that era were prone to.

Before the war the French aircraft engineer Raymond Saulnier had taken out a patent for a device that would allow a machine gun to fire through a rotating propeller; when the trigger was pulled, a mechanism synchronized the firing of the gun with the rotation of the engine so that it fired only at the instant when the propeller blades were not in the line of fire. He had not been able to get it to work in practice, however, largely because the moving parts of the Hotchkiss machine gun he was experimenting with were too heavy and sluggish to respond with the precise timing required.

In the fall of 1914 Roland Garros, a French flier who had been a test pilot for the Morane-Saulnier firm and knew of Saulnier's work, decided to try his own experiment: abandoning the synchronization con-cept, he simply fired a machine gun through a pro-peller protected with armored plating. The test (wisely carried out on the ground) showed that only one bullet in ten struck the blades. After obtaining leave to work with Saulnier over the winter to per-fect the idea, Garros returned to his escadrille on the northern front at Dunkirk with a Morane mono-plane fitted with the new armored-propeller firing system. From April 1 to 18, 1915, he used it to shoot down three German airplanes, a remarkable feat at a time when the total number of aerial engagements per month was about two dozen, with most encounters between German and French or British aircraft ending incon-clusively.

Morane-Saulnier N
scout
engine: 80-hp rotary
top speed: 90 mph
armament: 1 machine gun
   firing through armored
   propeller

On April 18 Garros was shot down by antiaircraft fire over German lines and was forced to land, whereupon he and his plane were captured. The plane was shipped at once to Berlin. Two months later the Germans had their answer: the first Fokker E.I monoplane, equipped with a synchronized machine gun and powered by an eighty-horsepower engine, went into service at the front. German inventors also had been tinkering with synchronized guns before the war, and apparently as soon as the military authorities saw the forward-firing gun that was the secret of Garros's success, they dug out these plans and set Anthony Fokker to work on them. On August 1, a Fokker flown by the soon-to-be-legendary ace Max Immelmann scored the first success with the new weapon, shooting down an unarmed RFC reconnaissance plane. The age of the true fighter had begun.

The Fokker E.I and its subsequent variants (the E.III, powered by a 100-

horsepower engine, would be the version produced in the greatest quantity) were in many ways unremarkable aircraft, but the impact of the forward-firing gun was enormous. Soon the Fokker had gained an almost mythic reputation. The British, hampered not only by the lack of a synchronized gun but also by a decision made in 1912 to halt all flights of monoplanes—they were believed to be structurally unsafe and prone to landing accidents as a result of their high landing speeds—fell back on two pusher models, the De Havilland DH2 and the Royal Aircraft Factory's FE2, to fill the gap in their fighter force. Though they were ungainly-looking planes and not great performers, the truth was they were more than a match for the Fokker in speed and rate of climb. The French Nieuport 11 "Bébé," a single-seat biplane with a Lewis gun mounted above the upper wing, proved even better, with a top speed of ninety-seven miles per hour versus the Fokker's eighty.

But the Fokker's reputation was, at least for a time, more important than any reality. "You were as good as dead if you as much as saw one," was the solemn belief of British fliers; rumor held that it could fly twice as fast as any Allied plane. The Germans helped to keep the myth alive by forbidding Fokker pilots to fly over Allied lines so that its secret would not fall into enemy hands. When the British finally did capture one in April 1916 (a second came into French hands in October 1916 when a German pilot landed by mistake in a fog at a French airfield), not only did the Allies discover the synchronization gear but they also found that the Fokker's speed and maneuverability were nothing exceptional. The captured plane was flown in mock combat against a Morane fighter and "a cheer went up" from a British squadron on the ground below as the French plane outpaced the German.

Fokker E.III
single-seat fighter
engine: 100-hp rotary
top speed: 87 mph
armament: 1 synchronized
7.92mm machine gun

Just as the British, and to some extent other nations, had before the war disdained speed as "mere ostentation," so they had disdained the tactics of air-to-air fighting that were now proving vital to survival. The distrust of the "sensational" in technical work was matched by an official discouragement of what was then disparagingly termed "stunting," Mervyn O'Gorman recalled. But pilots—at least the best pilots, who were always eager to bend the rules when it came to seeing what they could make their machines do—had little difficulty breaking free of such conservatism as soon as air-to-air fighting began in earnest.

The techniques required to maneuver onto the tail of an enemy aircraft, and to avoid ending up in the same uncomfortable position oneself, demanded that a pilot take advantage of every resource of three-dimensional motion an airplane was capable of. Besides circling and looping, a repertoire of deliberately erratic maneuvers was quickly developed to throw off or turn the tables on a pursuer.

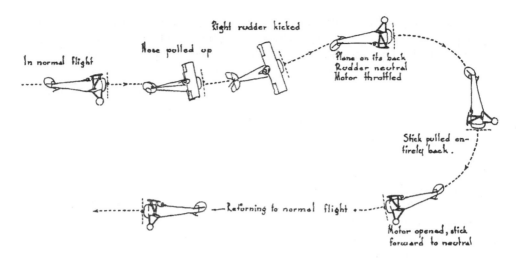

The British began the war with official disapproval of what was disparagingly termed "stunting." The advent of air-to-air combat led to a swift revision in attitude, and maneuvers such as this *renversement* became a standard part of pilot training.

Max Immelmann invented the maneuver that still bears his name, a half upward loop and half roll to quickly gain altitude and change direction. A similar maneuver, but with a downward half loop, was known in French as the *renversement*. A few particularly brave souls learned to throw their planes into a steep dive and pull out within feet of the ground, or even go into a spin (*vrille* in French) to fool their adversary into believing they had been hit and were out of control.

Aerial maneuvers were one thing; organizing groups of fighters and employing them to achieve a significant military effect was quite another. Such planning was going to take more than a cockpit-eye view. At first Allied commanders parceled their fighters out in ones and twos among existing reconnaissance units to provide protection for observation flights. The real or imagined threat of the Fokker worked to constrain the fighters' role even more narrowly; on January 14, 1916, RFC headquarters issued a "hard and fast" edict that "until the Royal Flying Corps has a machine as good as or better than the German Fokker," every reconnaissance aircraft "must be escorted by at least three other fighting machines" flying in "close formation." The French adopted much the same practice.

In effect this meant shrinking the fighter presence over the battlefront three-fold. It also handed all of the initiative to the Germans, who had now begun patrolling in pairs, and sometimes in groups of even three or four Fokkers. They could choose when and where to concentrate their numbers and pounce; it was impossible for the Allies to protect every single reconnaissance mission with a fighter escort that matched the potential threat that might come diving out of the sun.

The fighter had been born, but its relation to air power still lacked a clear focus, both for pilots and for the men who commanded them. That would begin to emerge only from the two momentous and terrible battles that were about to convulse the front in 1916.

Verdun was the logical outcome of the grinding stalemate of the trenches. At the start of the battle of Neuve Chapelle, the British had fired the 342 guns along their front for thirty-five minutes. They fired more shells in that barrage than in all of the Boer War. In one attack a thousand British troops surged forward in three waves; when none returned, those behind thought they had successfully taken the German trenches. In fact every single one had been killed. Attacks in which thousands died to advance a dozen yards were par for the course. Between attacks a constant hail of shells and sniper's bullets killed tens of thousands more month after month. In 1916 the explorer Ernest Shackleton, stranded in the Antarctic since shortly after the war had begun, finally reached the tiny British whaling station on the island of South Georgia. "Tell me, when was the war over?" he asked the station's manager. "The war is not over," the man replied. "Millions are being killed. Europe is mad. The world is mad."

At Verdun, General Erich von Falkenhayn, Germany's chief of the General Staff, planned to attack the French not to break through but to grind the enemy down in an orgiastic war of attrition. "France will bleed to death," he declared. Two fortresses guarded the road to Paris at Verdun. On February 21, 1916, Falkenhayn attacked with 850 guns, gas, flamethrowers, and a million men. In one day 7,000 horses were killed by the shelling. When it was all over ten months later, 300,000 men were dead.

To support the attack, Falkenhayn also ordered together an air force of unprecedented size. Concentrated in what the Germans termed a *Luftsperre*—an aerial blockade, or "barrage"—the German force of two-seater fighters maintained a constant patrol, cruising up and down the German side of the front to bar the French from penetrating it.

The French air forces were at first caught entirely by surprise, with only two escadrilles to counter the massed German fighters. Joffre at once summoned Major Tricornot de Rose of the Morane-Saulnier escadrille No. 12 to his headquarters and gave him carte blanche to do whatever was needed to counter the German air forces. De Rose quickly assembled fifteen escadrilles of Morane-Saulniers and Nieuport 11s; the order that went out to them on February 29 marked as clear a milestone in the history of air warfare as one is likely to find. "The mission of the escadrilles," it read, "is to seek out the enemy, fight him, and destroy him."

What followed was nothing less than a rout of the Germans, far less because of their ultimately outclassed airplanes or inferior numbers than because of a flawed operational concept. De Rose launched the first independent air campaign in history; it had the unambiguous aim to achieve air superiority, and it did so with dedicated fighter units that had no other mission than to sweep the skies of the enemy.

From Verdun on, noted the 1917 British General Staff report on air-to-air combat, the expansion of aviation forces was dictated almost entirely by the "expansion of aerial fighting." Both sides began to group more and more of their fighter aircraft into specialized units—"pursuit squadrons," or *groupes de chasse* or *Jagdstaffeln* ("hunting squadrons")—rather than parceling them out among all types of squadrons. The number of aircraft per fighter squadron was increased to twelve, to eighteen, and then to twenty-four. Fighters also began to operate in larger formations in the air to mass their firepower, with as many as a half dozen planes flying together, usually in a stepped V that afforded mutual protection:

1

2   3

4   5

Numbers 2 and 3 might typically fly fifty yards above and behind the Number 1 leader; Numbers 4 and 5 would be another fifty yards above and behind that; and sometimes a sixth plane would close the triangle, flying between Numbers 4 and 5. The rearward planes could keep an eye on the vulnerable flanks and tail of the leader, who could thus concentrate on scanning for targets in the skies ahead. In a tight situation, the members of the V could regroup into a defensive circle, each guarding the tail of the plane ahead. To direct maneuvers, the leader could fire colored flare pistols or dip his wings according to prearranged signals.

Above all, the lesson that Allied air commanders took away from Verdun was that to be thrown back on a defensive posture was to court disaster. The *Luftsperre* was fatally flawed, the British report concluded, because it was simply impossible to be everywhere at once:

> Owing to the unlimited space in the air, the difficulty one machine has seeing another and the accidents of wind and cloud, it is impossible for aeroplanes, however powerful and mobile, however numerous and skillful their pilots, to prevent determined opponents from reaching their objective. . . . In the air even more than on the ground, the true defence lies in attack.

The securing of air superiority was the mission that made all other missions possible in the air; and the only way to achieve air superiority was through mass and attack. The British attempted to apply that lesson in spades as they prepared for the Somme offensive, the Allies' great and dire attempt to break the deadlock on the Western Front in 1916. A policy of "relentless and incessant offensive" in the air was to guide the operations, and it paid off with astonishing effect. The Allies were able to carry out photographic reconnaissance flights with scant German interference and to shut out the Germans from doing the same for the first ten weeks of the battle; it also gave a free hand to RFC artillery spotters. Yet the enormous ground offensive, though it would drag on for five months, was doomed from the start. The plan was to neutralize German defenses in a week-long artillery bombardment preceding the attack; it would be a "walkover." A million and a half shells hurtled over the lines, but a million of those were shrapnel that had no effect on the dug-in Germans, and of the remaining half a million high-explosive shells, many were duds. On the morning of July 1, the guns fell silent and 100,000 British troops went over the top. By nightfall 20,000 were dead and 40,000 wounded. On the whole, the Allies' air superiority at least mitigated the disaster that did occur, but it could not rescue victory out of the first day's slaughter.

The imperative of gaining air superiority through relentless attack was a simple and powerful idea. But every simple and powerful idea can be abused; under its offensive-minded commander, Major General Hugh M. Trenchard, the RFC would abuse this one, to its great cost. Trenchard measured air superiority not merely in terms of winning freedom of operation for Allied air missions such as photoreconnaissance and artillery spotting, which was what really mattered; equally important, he insisted, was establishing moral ascendancy over the enemy. The troops in the trenches were exhorted to go out on patrols each night into no-man's-land; Trenchard saw forays over the line in the air much the same way. He also very traditionally, and absurdly, equated the depth that his fighter sweeps penetrated into enemy territory with occupying territory on the ground. (It was an idea that would die hard. A 1917 instructional pamphlet, "Fighting in the Air," told RFC pilots to follow this simple equation: "The further such patrols penetrate behind the hostile front the greater will be the moral effect of the success they gain.")

Such uncritical offensive-mindedness exacted a terrible, and needless, toll. The RFC began the Battle of the Somme on July 1, 1916, with 410 airplanes and 426 pilots; by the time the battle had staggered to its awful end in November, 782 British planes had been lost and 499 aircrew killed, an attrition rate of more than 100 percent. The relentless policy of offensive action and deep patrols ground up inexperienced pilots without mercy; one flyer, Cecil Lewis, recalled that pilots "were lasting, on an average, for three weeks." The Germans, meanwhile, had lost 359 aircrew in the same period.

A more soundly drawn lesson from Verdun and the Somme was the need for a unified command of all air units. Originally, all of the armies engaged on the Western Front had assigned the bulk of their aviation units to the command of army corps; the remainder, mainly for strategic reconnaissance, were assigned to higher levels of the echelon, armies and general headquarters. In reporting on the lessons of the French deployments, air commander Commandant Jean du Peuty emphasized that perpetuating this fragmented organization would inevitably risk the very kind of operational failure just exhibited by the Germans. "Fighting machines which are taking the offensive should be grouped together under a single command," he wrote. "Fighting squadrons which are dispersed among Army Corps sectors, and which come under the orders of the Commanders of such units, very rapidly become split up into small groups used for direct protection, barrage, or defensive patrols, and this entirely does away with their efficacy." In the early weeks of the Somme, the Germans erred not only in continuing to carry out barrage flights but also in having individual squadrons continue to pursue missions assigned to them by separate ground commands without an overall strategic focus. The German squadrons, du Peuty observed, operated in "watertight compartments" without any link between them.

By the end of 1916 the Germans, too, had begun to heed this lesson and were assembling a huge, autonomous fighter force. In July the Germans had sixteen single-seat fighters on the front; a month later the number had grown to sixty; by the end of the year the German fighter force consisted of thirty-three *Jagdstaffeln* of eighteen aircraft each. A single commanding general was placed in charge of all German Army aviation, and the organization was given the new name of Luftstreitkräfte ("air force").

Though they still tended to remain on their side of the line, the German fighters, having learned their lesson the hard way, abandoned the passive *Luftsperre* concept. Trenchard, however, failed to grasp the distinction and took the Germans' approach as a sign that they lacked courage and had suffered a fatal blow to morale. The British, Trenchard's fliers were informed, "as a nation like fighting," and like it all the more when they "are fighting against the odds." The Germans, by contrast, can easily be intimidated by a bold show of offensive spirit. "BRING DOWN YOUR HUN!" exhorted a manual for British fliers, and it went on to explain: "Only very occasionally does one meet a good enemy pilot. . . . When a Hun is attacked on our side of the lines, he is nervous and his chief idea is to get back to his own side where he can get aid."

The Germans, however, referred to the practice of sticking to their own side of the lines as "letting the customer come into the store." In September and October 1916, as commander of the newly formed Jagdstaffel 2, Oswald Boelcke alone shot down twenty British and French aircraft. The battle for air superiority was far from over, and it would require more than just a bold spirit to command the skies.

For one thing, the Germans in the fall of 1916 suddenly unleashed a potent new weapon. The Albatross D.I fighter that Boelcke and his protégés were given was powered by a new 160-horsepower, six-cylinder Mercedes engine; not only was its top speed of 105 miles per hour about 10 miles per hour faster than the best French and British fighters, but it mounted twin synchronized machine guns that gave it a decided advantage in firepower. Since at least the summer of 1915 the

British authorities had been aware that they needed a faster airplane powered by a bigger engine. Inspectors from the Royal Aircraft Factory visited the front on several occasions and talked to pilots and squadron leaders at length. In August

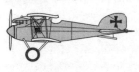

Albatross D.II
single-seat fighter
engine: 160-hp six
    cylinder
top speed: 105 mph
armament: 2 synchronized
    7.92mm machine guns

1915 one such report from the front had concluded that "the most important machine required at the present time is a single-seater capable of flying 110 miles per hour" and "very agile in the air." In December another inspector agreed: "Power is essential." In the same month, yet another visitor to the front lines concluded: "All of our home-designed machines appear over-weighted and under-powered."

From the outside, too, there had been a steady drumbeat of criticism of the Royal Aircraft Factory, and it grew louder with every German advance in fighter performance, real or imagined. But the larger context was that the fledgling British aircraft industry had never liked the government telling it what to do or, worse, competing directly with it in the design and manufacture of airplanes, and now it had a great club with which to beat that government. Mervyn O'Gorman, the superintendent of the Royal Aircraft Factory, had also apparently committed the unpardonable sin of brusquely turning back at the gates the powerful editor of *The Aeroplane,* Charles G. Grey, who was ever ready to champion the manufacturers' cause. Grey never forgave him. Week after week *The Aeroplane* poured out its bile in lead editorials signed "CGG." The factory was "a menace to British Aviation." The blueprints it sent to private firms contracted to build the BE2 were "so badly done that if the parts . . . were made as drawn the machine could not be induced to be got together." O'Gorman's factory was not only "utterly inefficient" as a manufacturing concern but had the blood of pilots on its hands: "The worst of the present designs of any independent aeroplane manufacturers are safer to fly than a B.E.2.C. built to R.A.F. [Royal Aircraft Factory] specifications would be. . . . The R.A.F. has always placed the opinion of its pseudo-scientists before the experience of practical men. The graves of R.F.C. officers are monuments to this grim fact."

By 1916 Grey's attacks were becoming even more vituperative and personal; one week he wrote: "As mentioned not long ago, Mr. O'Gorman was born an Irish gentleman. . . . One doubts whether Mr. O'Gorman's fellow countrymen of the Sinn Fein Society have actually done as much harm to the progress of Ireland as the staff of the R.A.F. has done to the progress of aircraft design and construction in England."

By this time the cry had been taken up in Parliament, too. Members echoed the charges that the BE2 was "murdering our pilots" and that the factory was undermining competition from private industry. An official government inquiry was launched. Though the investigation in the end became the pretext for removing O'Gorman, it largely exonerated the factory. O'Gorman's successor also strongly defended him against Grey's reckless charges.

What it came down to was that organizing an entire industrial system capable of designing and building airplanes and engines in response to a rapidly chang-ing set of military requirements that could not be anticipated in advance was a vastly more complex task than anything anyone had tried before. When the war broke out there was hardly an aircraft industry in England at all: In September of 1912 the factory had sent out an inspector to see how the Handley Page firm was getting on with its contract to build five airplanes. The inspector went to the company's address and found an old stable and riding school. The only sign of life was a few workmen who were fixing the roof; the box stalls for the horses were still in place, and the only evidence that anyone was planning to build any-thing were three carpenter's benches and a small treadle-driven fret saw, plus a few twist drills in the office next door. When the Royal Aircraft Factory held a military airplane competition in the summer of 1912, none of the entries from private firms met even the minimum standards; the factory's own BE2, designed by Geoffrey de Havilland, who was working for the factory at the time, was by far the best, and so the factory began to issue contracts to private firms to build them in quantity.

There was no denying that the Royal Aircraft Factory was huge. O'Gorman had built it up from a staff of about a hundred in 1909 to five thousand scientists, engineers, and workmen. But it was hard to see how it was really "undermining" private industry. The factory focused its efforts mainly on designing and building prototypes of new machines and conducting basic tests in metallurgy, chemistry, engines, and wireless. The factory did manufacture some airplanes itself, but by mid-1916 these had totaled only 77, versus 2,120 supplied by "the trade" since the start of the war. From August 1914 through March 1916, £13.3 million in orders for airplanes, engines, and spare parts had been placed with private firms, and £4.5 million of that had been for items completely designed by private man-ufacturers. O'Gorman also pointed out in his defense that, far from attempting to stifle competition, the factory had a file full of letters of thanks from manufactur-ers for the assistance they had received on technical problems. The factory tested components for manufacturers, assisted them in redesigning parts, and freely shared technical information on propeller design and stress calculations.

The government inquiry did however urge that "undesirable trade feeling" should be smoothed over, and as a sop ordered the factory to stop issuing designs directly to manufacturers and to stop building aircraft and engines itself. O'Gor-man's successor, Sir Henry Fowler, who had been chief mechanical engineer of the Midland Railway, looked into the matter of "trade feeling" and concluded that while it was true the manufacturers had their noses out of joint, the cause went to the heart of some much deeper problems.

One was the old rift between scientists and "practical men." Fowler con-cluded that the scientific staff of the factory, who "naturally . . . were largely drawn from Cambridge and ranged from the pure mathematician to those who had served several years in engineering shops," were often far ahead of the private firms in their knowledge of aerodynamics; they weren't always tactful about it, however, and "the tone of certain letters" had obviously been a bit condescending. (The staff of the factory were no doubt touchy themselves as a result of being such a large and visible group of men not in uniform; they were frequently suspected of being "shirkers" or deserters, and as one recalled, "It was extremely distasteful to be stopped in the street by police and made to produce papers, or to be presented with white feathers by crazed females." The problem was eventually solved by giving everyone on the staff RFC rank and uniform.)

But the deepest problem was that it was proving almost impossible to reconcile the quickly shifting demands of the battlefield with the realities of the manufacturing process. A typical First World War aircraft contained fifty thousand parts and took four thousand man-hours to build, and almost everything was made and fitted by hand, so it was easy to see how repeated design changes could throw a builder into despair. Yet the urgency of war required that design and production overlap from the start. "The necessities of war caused drawings to be got out at the first possible moment for production to commence," Fowler noted, "and alterations, often the result of experience in the field, had to be made."

While the manufacturers complained about being forced to make alterations in already-approved designs, the squadrons took matters into their own hands, which for a while only added to the chaos and the difficulty of matching design performance to military needs. Alterations extemporized in the field to accommodate gear such as wireless transmitters, bomb racks and bombsights, and cameras—none of which had been contemplated before the war—often had the effect of adding greatly to the wind resistance of the fuselage and dragging down the already substandard speed and rate of climb of the aircraft. Fliers themselves referred to their planes as "Christmas trees" after all of these ornaments had been attached.

By the second half of 1915, no normal peacetime system of design and manufacture could cope with the arms race that had emerged over the battlefields of France. Some of the confusion was avoidable, the result of overlapping lines of authority in which the Royal Aircraft Factory, the Directorate of Military Aeronautics in the War Office, the Directorate of Aircraft Equipment, and the Aircraft Inspection Department all were trying to tell the manufacturers what to do. The Admiralty and the War Office also worked at cross-purposes more than once, duplicating orders and, in one farcical episode at the beginning the war, actually outbidding each other: the Navy placed large orders for BE2 aircraft with the same firms that were building them for the Army, but at considerably higher prices than the Army had already negotiated, which resulted in the price going up from £600 to £900 per plane. The Cabinet appointed an Air Board that was supposed to coordinate overall strategic and procurement policy, but it had advisory powers only and ended up doing little more than providing an official forum for the Army and Navy to exchange insults that had previously been exchanged privately. Venting his frustration about the infighting and backbiting over aircraft production, Sefton Brancker, the Director of Military Aeronautics, wrote to General Trenchard in April 1916: "I get more and more impressed with the rottenness of our system and our institutions. . . . The Boches will beat us yet unless we can hang our politicians and burn our newspapers and have a dictatorship."

But much of the difficulty was inherent. Design and manufacturing were by their nature conservative businesses, especially with something so complex as an aircraft: every design change meant building new jigs and altering shop procedures, and at a time when aerodynamics was still a very inexact science, also introduced the risk of unanticipated structural or safety flaws. The Royal Aircraft Factory had begun the war with a standard practice of building all aircraft with a safety factor of six; weighted sandbags would be piled on the wings to test their strength and they had to be able to withstand a force six times that expected in normal flight. In a letter to manufacturers on August 28, 1915, the Directorate of Military Aeronautics informed manufacturers that in the case of airplanes "showing a marked advance over present types," a safety factor of four would be permitted. But the new reconnaissance machine designs coming out of the Royal Aircraft Factory were still heavy and underpowered. The RE7 and RE8, which appeared in 1916, immediately became sitting ducks for German fighters. The BE2s were still being made, too, and they were even worse. Breaking loose of the conservatism of "business as usual" required a new way of thinking in which change was as natural as standardization; it also required a critical mass of experienced designers and builders who could competently push the envelope past its accustomed boundaries.

This was true in engine design as well, on which so much now depended in the qualitative arms race against the Germans. The air-cooled rotary engines hit a practical upper limit at about 150 to 200 horsepower, since the centrifugal force of the engine not only produced a gyroscopic effect that tended to make the entire airplane rotate about its axis but also tended to tear the engine itself apart: valve rods, rocker arms, and even entire cylinders were known to come flying off. About 20 to 25 percent of the engine horsepower was wasted in the "windage" of the rotating cylinders. The rotary engines also required frequent maintenance; every forty hours, or even more frequently, they had to be completely torn down and rebuilt in a *révision générale,* as the French termed it.

The major problem with water-cooled engines, on the other hand, was their extra weight, about one pound per horsepower. Unlike automobile engines, aircraft engines had to be light and able to operate continuously at full power; there was only so much that the existing French and British automobile industrial base could contribute directly to the solution. Worse, many key components of high-performance engines, including the specialty steels and aluminum required for high temperature and strength, had been imported from Germany before the war.

But by 1916 the Allied aero-engine industry, especially in France, was finally getting on its feet and able to respond to the persistent calls from the front for larger engines. If there was one particular breakthrough, it had come when Marc Birkigt, a Swiss engineer living in Spain, offered his services to the French government. His "Hispano-Suiza" engine was a true revolution in engine design. The block was made of an aluminum alloy. The cylinders, in a V-8 arrangement, were formed of steel sleeves screwed into the block itself rather than protruding above. This "monobloc" design was lightweight; it protected the valves and related moving components by placing them inside the engine case; and it provided for excellent heat transfer and cooling. The first models were 150 horsepower, but it was clear the design could readily be scaled up to exceed 200 horsepower. The first Hispano-Suiza engines appeared on the front in late 1916 and early 1917 in the British SE5a and the Spad VII, which immediately set new speed records of 120 miles per hour or more. They were followed by the 220-horsepower Spad XIII. By the end of the war, more than twenty thousand 220-horsepower Hispano-Suiza engines had been manufactured in France, and many more in England and America under license. Though reliability problems plagued the Hispano-Suiza–equipped Spad VIIs and XIIIs throughout 1917 (the engines' tendency to seize was ultimately traced to an oil pipe inside the crankcase that would burst when excess pressure built up), the fundamental technological leap forward they represented gave the Allies a lead the Germans would never close.

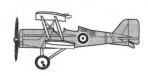

SE5a
single-seat fighter
engine: 200-hp V-8
top speed: 138 mph
armament: 1 synchronized,
  1 over-wing machine
  gun

A captured Hispano-Suiza fell into German hands in October 1916 and orders went out to duplicate it, but the German aero-engine industry was simply too far behind by then; having been firmly instructed by the authorities to make reliability and fuel economy the priorities in engine design, German industry had dutifully focused its efforts on perfecting the in-line six-cylinder engine, and when the orders finally came to widen the efforts it was too late. By the end of the war, the Allies were producing 300-, 400-, and even 600-horsepower V-8 and V-12 engines, which the Germans never came close to matching.

With the SE5a and the Spads, the Allies at last had their answer to the "scourge" of the Fokker and the Albatross. Not only were they fast, but they finally closed the lead in firepower the Germans had held ever since they introduced the synchronized gun on the Fokker E.I in 1915 and the twin synchronized guns on the Albatross D.I in 1916, both of which the Allies had been painfully slow to emulate. (The SE5a had one synchronized gun and one that fired over the top wing; the Spad XIII was the first Allied fighter to mount twin synchronized guns.)

Further boosting their advantage, the Allies finally had in place a system—the result of a lot of improvisation and many false starts—that was beginning to be able to respond swiftly to changes at the front. It was not a very rational or efficient system; indeed it was hardly a system at all in the strict sense of the term. But out of the chaos of multiple, and sometimes duplicative and competing, government departments and private firms, there had emerged a critical mass of innovative British and French talent. Now when the authorities recognized they needed a new capability, the odds were that there was someone out there who knew how to supply that need and could do it in reasonably short order. The Germans had a much more structured system of organizing industry, and that was proving both a strength and a weakness.

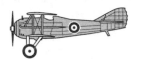

Spad XIII
single-seat fighter
engine: 220-hp V-8
top speed: 135 mph
armament: 2 synchronized
.303-in. machine guns

The rapid expansion of air operations in 1916 posed an unprecedented challenge not only to the supply of aircraft but also to the supply of men. From the start there had been something of a rough-and-ready attitude toward the training of military fliers. Throughout his career Benjamin Foulois would insist with a straight face that in 1910, stationed at Fort Sam Houston as the United States Army's only active pilot, he had actually been ordered to teach himself to fly. His "unique training method," he said, was to take the Army's sole Wright Flyer up and, whenever he crashed it, send a letter to the Wright brothers describing what had gone wrong and asking for advice. Foulois called himself the Air Force's first and only "correspondence school" pilot. It was a bit of a tall tale; Foulois had in fact received lessons from Orville Wright in a Wright Flyer at College Park, Maryland. But there was no doubt that a lot of his advanced training consisted of trial and error, with a lot of the latter. In the course of his four months of flying at Fort Sam Houston, he exhausted the $150 the Army had budgeted for repairs and was forced to contribute $300 from his own lieutenant's pay to cover all of the broken wings, rudders, propellers, and elevators.

The Wright airplanes, in which the pilot and passenger sat side by side, had introduced the dual-control system that would become the standard instruction method in the American and British air services during the World War. The Wrights also invented the first simulator, rigging up an old plane on sawhorses in a back room of their shop; the lateral control lever operated a clutch and moving belt that rocked the plane left or right as the lever was moved by a pilot trainee. Lieutenant Henry H. "Hap" Arnold, sent by the Army to Dayton in 1911 to learn to fly, recalled his initiation of days lurching about on this hobbyhorse plane.

The Wrights probably were partly responsible for setting things on the wrong track, however, by insisting that it was no harder to learn to fly than to learn to ride a bicycle. (They had been stung by criticism that "only acrobats" could fly their airplanes. In truth the 1903 Flyer was barely controllable and probably no one but the brothers could have managed their first flights. But their later airplanes were much more manageable, even if they retained an unintuitive control system in which left and right bank was controlled by a lever moved forward and back by the right hand; a second lever, operated by the left hand, controlled pitch; and the rudder was actuated by a short lateral-moving knob mounted atop the bank lever. It was Louis Blériot who devised the far more rational control system that soon became standard, with left and right foot pedals to control the rudder, and a single central stick between the pilot's legs that moved left and right to bank and forward and back to pitch the nose down and up.)

But as the Wright brothers should have remembered from the many hours it had taken them to master the art of piloting gliders, flying a plane was much harder than riding a bicycle. By 1913 twelve of the first forty-eight U.S. Army flying officers had been killed in training accidents. It was very much a sink-or-swim system.

The French training system was more cautious and methodical, but it also took much longer to get pilots off the ground at all. Students began, by themselves, in a low-powered plane with clipped wings that they would taxi around the field until they learned to keep it heading in a straight line by proper use of the rudder. These training planes were known as *rouleurs*—"rollers," since they could not generate enough lift to do anything else—or "penguins," for the waddling gait they usually assumed as their student pilots alternately overcorrected to the left and the right as they taxied. Once this task was mastered, students were permitted to cautiously raise the tail off the ground by moving the stick forward while taxiing. After graduating from the *rouleur,* the aviators-to-be would be placed in a slightly higher-powered trainer that they could taxi across a dip in the middle of the field to experience a brief airborne moment. This would be followed by longer airborne hops and eventually a low, straight flight of a hundred feet straight down the field. The whole process took six months and about fifty hours of flying time.

With the outbreak of war, the qualified pilots who served as instructors were naturally in immediate demand at the front. The French simply shut down their schools, assuming the war would be over before any new pilots would be needed. By the end of 1914 they had already reversed that decision. In Britain, the Central Flying School at Upavon, which trained pilots for both the RFC and the Royal Naval Air Service, had room for only forty students at a time, and it lost not just its instructors but most of its airplanes when the Expeditionary Force deployed to France. Civilian flying schools were pressed into service to help train military fliers, but as the urgent demand for new—and replacement—pilots rose, it was not uncommon in the early days of the war for pilots with a total of five hours' solo flying time to be deemed "qualified" for service in France.

The training that many new pilots received was all the more inadequate and erratic because the instructors the military flight schools were left with were typically pilots who had cracked under the pressure of combat and had been sent home for a rest. British officials privately conceded that the situation was serious; there was the one instructor who spoke with an uncontrollable stammer, and there were many others, resenting the drudgery of training, who habitually referred to their students as "odious Huns." There was also, especially in Britain and France, an absurd amount of time taken away from practice at the controls to make the recruits jump through the hoops that military tradition had prescribed from time immemorial. French pilot candidates were given a strong dose of close-order drill to remind them that they were soldiers. British cadets got two months of pure military studies before they were allowed near an airplane.

The losses of the Somme battle, in which the RFC suffered a 100 percent casualty rate in eighteen weeks, plunged this already bad situation into crisis. Accident rates soared as replacement pilots were rushed through their courses. In the last two weeks of September 1916, training accidents killed forty-three British pilots and injured forty-eight. By November 1916, squadrons were reporting that the new scout pilots who were arriving from England had received so little training that they could do little more than take off and land. Nonetheless, they were frequently thrown into combat at once. Trenchard insisted that "short training was a consequence of the number of casualties and not the casualties of the shortness of training." He noted that the RFC took only one casualty for every hundred airplanes that crossed the lines, versus one in three infantrymen. As the historian John Morrow pointed out, that may have been "a strong argument for joining the air corps" but it was hardly an answer to the vicious circle of attrition and inadequate training in which the RFC was now caught.

By the end of 1916 the near-suicidal madness of this policy was becoming evident even to the dullest minds, and it was clear that much more rigorous training would be needed for combat pilots. It was not enough just to be able to control an airplane; that was merely the first step of a long education. The French began adding specialized courses on top of an initial twenty-five hours of primary training; a scout pilot would be sent to two additional schools, one for aerobatics and one for gunnery, for a total of forty-five flying hours. The British did much the same, and specifically required that pilots have a minimum of five hours in the model they would fly at the front before going overseas.

Experience in combat had revealed that a common mistake inexperienced pilots made was to open fire from much too far away. To correct this problem before it might cost them their lives, student pilots were now given the chance to engage in mock combats in planes fitted with "camera guns" that shot pictures instead of bullets when the trigger was pulled. By the end of the war, fighter-pilot trainees were also receiving eight hours of practice in formation flying, four hours in offensive and defensive tactics, and two hours in high-altitude patrol work.

Throughout the war there would be pressures to skimp on needed training, especially when new models of aircraft were delivered to operational squadrons and there was no time to rotate pilots through the *écoles de perfectionnement* that were meant to provide the necessary transition to the new aircraft. When the American 1st Pursuit Group received its first Spad XIIIs in late July 1918, "in some cases the test ride in the new machine was also a flight in the battle zone," the historian James Hudson noted; on these occasions the Spads were usually accompanied by the older Nieuports in the squadron to give the pilots of the new machine some protection while they familiarized themselves with its controls and handling. But all in all, training became much more rigorous and realistic as the war progressed.

Yet organizing and managing an effective training system proved to be a huge logistic operation, far bigger than anyone had imagined. The British estimated it cost a staggering $25,000 to train each pilot, this at a time when a fighter plane such as the SE5a cost $4,000. By the end of the war, training in Britain and France would involve scores of specialized facilities, hundreds of training squadrons, thousands of aircraft, and tens of thousands of instructors and support staff. France trained 134 pilots in all of 1914; by 1918 it was turning out five times that many a month, the Germans about the same. By the end of the war the French had trained 18,000 pilots, the British 22,000.

America's entry into the war in April 1917 would provide a humbling lesson in just what it took to put together a training program from scratch. On the day the United States declared war on Germany, its air service—still the Aviation Section of the Signal Corps—had 26 trained pilots and possessed 55 trainer aircraft, of which General John J. Pershing observed "51 were obsolete and the other 4 were obsolescent." A month later, showing admirable American spirit if nothing else, the United States announced it would supply five thousand pilots for service on the Western Front by the spring of 1918.

George Squier, by this time a general and newly appointed to the post of Chief Signal Officer, was innovative enough to shelve military tradition and bring in "useful people" from civilian life now that war was declared, even if they were lacking in military training or military bearing. And so he did not hesitate to take up the telegraphed offer from a Yale history professor, Hiram Bingham, who asked if he might be useful as an instructor. Bingham was well qualified for the job. In addition to being a college teacher (and a famous South American explorer; he discovered the Incan city of Machu Picchu in 1911), Bingham had "caught the disease" of flying in 1916 and, at age forty-one, taken lessons from Glenn Curtiss. Squier immediately commissioned Bingham a major and placed him in charge of a new organization: the United States Schools of Military Aeronautics. Eight universities were recruited to provide ground-school instruction to cadets. By May 21, 1917, the first students were arriving.

Organizing flight schools was another matter, however, and here, as Bingham himself admitted, the hope of mass-producing thousands of qualified pilots became "a genuine expression of the 'American Idea,' defined by Strunsky in one of his charming essays as 'splendid courage accompanied by a high degree of disorder.' " By the Armistice the ground schools had admitted 22,689 candidates and graduated 17,540, yet only 767 of them were serving as pilots on November 11, 1918. The worst fiasco had occurred when the honor graduates of the first ground-school courses were specially selected to be sent to France for flight training at once while the American flight schools were being organized. The trouble was the French schools could not take anywhere near all of the 1,800 Americans, and most ended up being sent to Issoudun, where the Americans were just beginning to build an advanced flight school of their own. As the winter dragged on they found themselves confined for months in camps far from an airplane, forced to study the same subjects that "they had mastered with so much enthusiasm at American ground schools" and being "treated by despairing officers as though they were draft dodgers who needed military discipline and who deserved reprobation rather than sympathy," Bingham recalled. The thousand idled cadets came to be known as the "Million Dollar Guard," since at their $100-a-month pay and with the ten-month wait that loomed ahead of them, that was what they were costing Uncle Sam. Morale plummeted; at times there was near mutiny as the cadets were assigned construction duty and menial labor in the muddy camp.

"There was worse to come, however," Bingham continued, "for in the spring of 1918 there began to arrive in France as First Lieutenants, wearing wings, and speedily to be placed in positions of authority, the very classmates of these unfortunate cadets, who had not been quite keen enough to graduate with honors from the ground schools, and had accordingly been sent to American flying schools." One young lieutenant who was shot down later in the war was asked what the treatment was like as a prisoner of the Germans. "A damn sight better than I got in France as a cadet," he replied.

Finally the Secretary of War, Newton D. Baker, learned of the injustice and ordered that all of the cadets be commissioned at once. But the problems were more severe than this. American flight schools were receiving thousands of the JN-4 Curtiss "Jenny" dual-control trainers; in principle this American-made airplane was an excellent and sturdy machine, but it had almost nothing to do with the machines flying on the front lines. The instructors often were pilots who had themselves just learned to fly and were no more than one step ahead of their charges. "When an instructor thought that you could land without cracking up the airplane," recalled one student, "he turned you loose and from then on you were pretty much on your own and taught yourself to fly." The pilots trained on the Jennies arrived in France to find they almost had to learn to fly all over again. They also had to be given a completely new course in navigation; American pilots who had had no trouble finding their way cross-country at home repeatedly got lost over the French countryside's crazy quilt of fields, small villages, and identical-looking straight white roads.

The French and British by this time were already using more sophisticated primary trainers that better prepared students for what they would experience in action. One of the hardest things for American pilots trained on the Jenny was learning how to handle rotary-engined machines, especially the high-powered Sopwith Camel. "They were so small the torque of the big hundred-and-forty-horsepower rotary motor would turn them over if a pilot took his hands or mind off the controls," recalled the American pilot Elliott White Springs. Another American, John McGavock Grider, was at the RFC aerial gunnery school in Ayr, Scotland, with a group of American fliers when six more American pilots arrived from France to take the course and were placed in Camels. Three of them were killed in crashes in a single week. Then four more pilots were killed in quick succession going into right-hand spins in Camels. The mood at the camp was glum; the colonel "tried to put pep in the boys" by putting on a display of low-altitude aerobatics:

> Then he made all the instructors go up in Camels and do the same thing. It was a wonderful exhibition and then he made us a little speech and told us there was nothing to worry about, to go to it. Several of the boys were so encouraged that they took off in Camels and tried to do the same thing. Only one was killed.

While the military authorities wrestled with the considerable challenge of how to train men to be pilots, they also began to wrestle with the challenge of figuring out which type of men were trainable. Medical screenings and the deliberately high-pressure competition of ground school could identify the obvious misfits, but predicting who would make a good pilot proved surprisingly difficult. The problem was complicated, especially in America, by what Hiram Bingham recalled as the incessant pressures to offer a second chance to students who had washed out: "Over and over senators, representatives, distinguished citizens, and depressed parents came to beg special consideration for sons, nephews, cousins, friends, acquaintances." The typical American view, he said, seemed to be that any patriotic young man willing to risk his life flying had an inalienable right to do so.

It was easy to weed out cases such as the candidate who asked to be readmitted to the course on the grounds that "my fainting has always been due to a mental shock I receive when I allow my mind to ponder upon pain." But many candidates who seemed to be fine prospects and who passed all the preliminary hurdles never could become even minimally competent fliers, while some who were borderline on medical evaluations or educational background became the deadliest fighter pilots.

Toward the end of the war two British officers, a medical doctor and an experienced pilot, published in the distinguished medical journal *The Lancet* the first psychological study of aviators. There were some touches of traditional British class consciousness in their conclusion that enthusiastic "sportsmen" made the best pilots (it was especially helpful to have been an expert horse-rider in civilian life, they said, in order to have the natural "hands" needed to sense and control an airplane's motion: "We have never known of a man who has consistently been in the first flight in the hunting field making anything but a good pilot"). But the authors' overall conclusion was that by far the most important factor that distinguished successful from unsuccessful pilots was the possession of a certain temperament that cut across class lines: "Flying is not now confined to the public school boy, the cavalry officer, or the athlete. We take many of our pilots at present from the lower middle classes and some from the artisan class." Among the sixty successful pilots they surveyed were farmers, draftsmen, engineers, teachers, accountants, barristers, and one "window designer and dresser." The most consistent characteristics they shared were high spirits, quick judgment, a sense of humor, and "little or no imagination"—by which the authors apparently meant that the best pilots did not dwell much on what might happen to them. Indeed, the authors concluded that the *less* a fighter pilot "knows about his machine from a mechanical point of view the better," since such knowledge could inhibit risk-taking. When asked to list their "favorite amusement," most of the pilots put down sports, motoring, or women, though reading, theater, natural history, and music also appeared. (One wag responded: "Killing Huns and dancing.")

As a new and technical specialty, flying by its very nature tended to attract men whose common bond was less the ties of military tradition and social class that characterized the officers' corps of most of the world's armies at the time: it was an enthusiasm for flying that brought them together, and flying enthusiasts came in all types, including many who never would have found a place, nor cared to, in the regular army. Squadron 15 of the RFC was described as a mix of "public school boys, Canadians, Australians, a cross-country jockey, a man who had had half an ear shot off in some American brawl, and a New Zealander who read Homer in the original Greek." Eddie Rickenbacker, who would become America's top-scoring ace in the Great War, had dropped out of school at age twelve and eventually become a top race-car driver. Raoul Lufbery, the top ace of the Lafayette Escadrille, had spent his early years wandering the world, working as a waiter in Constantinople, a baker in New Orleans, a ticket-taker on the Bombay railroad, and finally talking himself into a job—with no experience or qualification—as a mechanic for a French exhibition flier he ran into in Saigon in 1912.

The letters and diaries of First World War fliers brim with a naïve enthusiasm that in part may simply reflect the times but mostly seems to be an expression of the kind of genuine fearlessness and "lack of imagination" that the authors of *The Lancet* study had identified. Even while risking their lives in combat, pilots, notably unlike their counterparts in the trenches, enjoyed the great compensation that they were getting to do something they loved to do, namely fly in airplanes at a time when few ordinary mortals had that privilege. "We got through the clouds at 10,000 feet," wrote one observer who flew in a two-seat DH4 on a long and exceedingly dangerous daylight bombing mission in October 1917, "and the sight was a glorious one, with the big white wool-packs surging up all around us. At 10,500 we flew up a beautiful cloud valley with the white peaks towering on either side, white as snow." There were no such compensations in the trenches, where death and danger were relieved only by the company of mud and lice and half-buried rotting corpses.

There is no doubting the genuineness of the boyish enthusiasm of Lieutenant Y. E. S. Kirkpatrick, who wrote to his family from training school at Gosport, England, in July of 1916: "I am having an absolutely ripping time here and I simply love flying. Flying gives you a ripping feeling like ski-ing; in fact it's jolly like ski-ing except that you get a nicer view." A year later, posted to France as a scout pilot, he was still sending home a steady stream of bantering letters. "I've been having quite an exciting time lately with several scraps," he reported in one letter. One of those "scraps" included a narrow escape with a half-running engine and a bullet in the fuel tank. Another involved being hit by enemy fire: "I expect by the time you get this letter you'll be colossally thrilled to hear I've been wounded."

The dangers were real enough. Sixteen percent of French pilots were killed during the war, and death rates for British and German fliers were about the same; among the most aggressive pilots it was closer to 25 percent, which equaled that of the infantry. The odds of escaping death, injury, or capture altogether were certainly no better than fifty-fifty for pilots in the air services of the major combatant nations.

Being killed outright was one thing, but death in the air could take some particularly gruesome forms. The exposed gas tanks and fuel lines that were a standard feature of all airplanes were a constant reminder of "what we all feared most," recalled Captain H. Brokensha of the RFC, "to be set on fire in the air." Parachutes were heavy and generally furnished only to balloonists, not aviators; pilots talked about whether they would rather jump to their death or burn to death if it came to it. And while foot soldiers had their share of special horrors— especially poison gas, and the horrific disfigurement that shrapnel wounds from high-explosive artillery shells could cause—only airmen had the experience of being on the receiving end of explosive and incendiary bullets. The American flier Waldo Heinrichs, who miraculously survived being jumped by seven Fokkers (he went on to teach history at Middlebury College in Vermont after the war) recalled his encounter with this form of ordnance:

> An explosive bullet hit me in the left cheek and knocked out sixteen
> teeth, breaking both jaws and then tearing through the windshield,

breaking it also. I remembered spitting out teeth and blood and turned again for our lines. . . . Two more explosive bullets hit me in the left arm, tearing through and breaking the left elbow. Two broke in the right hand and nearly took off the right small finger. Another hit me in the left thigh, one in the left ankle, and one in the right heel. Two more hit me in the leg. Six of the ten wounds were from explosive bullets. . . . The blood which I spat out blinded my goggles so I threw them over my helmet. Dove for the ground and pulled out just before I crashed into the woods and fortunately found an open field. . . . I loosed my safety belt and as I moved up on the edge of the cockpit saw a pool of blood swishing around in the bottom of the plane.

Even without the enemy to contend with, flying was physically demanding in ways that are simply unfamiliar to pilots today. The airstream in an open cockpit could be bitterly cold; pilots regularly suffered frostbite and took to smearing their faces with a thick layer of grease as protection. Oxygen masks did not come into use until the very end of the war; flying at high altitudes often caused dizziness and blackouts. The ubiquitous rotary engines were lubricated by injecting castor oil into the fuel-air mixture, and a steady spray of the oil flew into the pilots' faces, often inducing chronic nausea and diarrhea. Just holding a plane in steady and level flight could be grueling work, requiring a constant physical exertion against the stick; again, the rotary-engine planes were the most punishing, since the gyroscopic force of the engine had to be countered.

As formation flying increasingly became a tactical necessity for survival, another set of demanding skills came into play. Keeping one's station during turns took considerable practice and judgment; pilots on the inside of the formation had to turn sharply while throttling down, and it was all too easy for them to stall or fall into a spin; pilots on the outside of the formation had to speed up. All of this was especially challenging in rotary-equipped planes, as they had no carburetor or throttle, and power could be varied only slightly, by adjusting the fuel-air mixture. (Landing was tricky without a throttle, too; the rotary engines were equipped with a "blipper" button so the pilot could momentarily cut the ignition as a way to reduce power during a landing.)

Pilots developed a whole vocabulary of breezy, jokey terms that made light of the dangers. In both the French and German air services, having a wreck was known as "splitting wood"; in the RFC a mission over the lines was a "stunt." Second Lieutenant H. G. Downing of the RFC wrote home in July 1917 about making a forced landing after his engine quit and missing telegraph wires by a few inches: "Ah! Well we don't get much money, but we do see life." In another letter he added almost casually toward the end:

By the way, I unfortunately smashed a machine the other day. I landed on my nose by mistake in the middle of the aerodrome. I did not hurt myself though & my C.O. only laughed & suggested mildly that I should land on my wheels another time. It was very funny because two more did exactly the same thing five minutes later.

The jokes that circulated among the squadrons perfectly reflected this devil-may-care spirit, often with heavy-handed gallows humor. The base newspaper of an American squadron at Tours published an article entitled "How to Come out of a Flat Spin, Dead or Crippled for Life." The newspaper of the U.S. 1st Pursuit Group (jocular name of paper: *Out of Control*) ran this small item under the headline TRUTH IN FLYING:

> Machine Gun Instructor Sikking: "What is the first thing to do upon landing after a battle in the air?"
> Student-Lieut. Young (thoughtfully): "Thank God!"

And then there were the poems and songs that every American aviator in France knew by heart:

> *The Caudron's a terror, the Farman's a wreck,*
> *They just took a Sop engine off'n my neck,*
> *If this sort of thing doesn't stop,*
> *They'll be wiping me up with a mop.*

And:

> *The young aviator lay dying,*
> *And as 'neath the wreckage he lay (he lay)*
> *To the mechanics assembled around him*
> *These last parting words he did say:*
>
> *Two valve springs you'll find in my stomach*
> *Three spark plugs are safe in my lung (my lung),*
> *The prop is in splinters inside me*
> *To the joy stick my fingers have clung*
>
> *Oh, take the cylinders out of my kidneys,*
> *The connecting rods out of my brain,*
> *From the small of my back get the crank shaft*
> *And assemble the engine again.*

And:

> *Oh - - - - - -*
> *The bells of hell go ting-a-ling a-ling*
> > *For you but not for me*
> *The blessed angels sing-a-ling a-ling*
> > *Through all eternity*
> *Oh, death, where is thy sting-a-ling a-ling*
> > *And grave, thy victory*
> *No ting-a-ling a-ling*
> > *No sting-a-ling a-ling*
> *But sing-a-ling a-ling*
> > *For me.*

Elliott White Spring's thinly fictionalized *Nocturne Militaire* is full of tales of hell-raising and cynical observations about death from the pilots of his squadron. Funerals for American pilots killed in a string of training crashes have to be spread out because the unit has only one American flag. Two fliers stage a duel with flare guns at fifty yards. Wild parties are organized; one member of the squadron "had made, bought, or stolen an elaborate set of American bar utensils, which he carried around with him wherever he went. . . . He would be sent to a station where gin-and-bitters was the only mixed drink and by the time he left even the village loafers would be calling for a Remas fizz or a Fishhouse punch." For one famous binge they rent an entire hotel in Dieppe. "Eleven court-martials grew out of it and two majors were killed in a wreck on the way home. Carol drank the mayor of Dieppe under the table and then went off with his wife."

Some of this was forced gaiety, but mostly it was a reflection of the self-selected personality types who gravitated to flying. The ability not just to overcome fear but to be truly fearless—to live completely for the adventure of the moment—was all the more apparent in the relatively small number of pilots in each squadron who scored the most kills. And it *was* a small number; the available data suggest that probably no more than a few percent of fighter pilots shot down more than half of all enemy planes accounted for. A few of these best pilots, one senses, really were killers; they made a point of aiming directly at the cockpit of an enemy aircraft and could write with satisfaction of seeing the blood of their victims spray over their own windscreen. But to most fighter aces the killing seemed almost incidental, a necessary fact of war but one on which they did not especially dwell—the "lack of imagination" at work again—and which they took no particular pride in. Georges Guynemer, a leading French ace, once narrowly missed hitting a woman on the street with his car. "Assassin!" she shouted at him. "Madam," he replied, "you don't know how right you are."

The best fighter pilots had an intense need to do what they did, but most expressed it in terms of an almost pure exhilaration and intensity that transcended the sordid facts of killing and war. "God, it was great!" Elliott White Springs wrote home to his sister after his first aerial combat. "Sherman was all wrong. He was unlucky enough to be in the wrong branch of the service." A British pilot described the feeling of having "every atom of personality, mental and physical . . . conscripted into the task" in a dogfight. There was certainly none of the ennui of the officers and men in the trenches who felt they were mere pawns in a huge, mechanical slaughter; once in the air, a pilot could at least believe that he alone held his destiny in his hands. The sensation of being vitally alive in moments of aerial combat is a recurrent theme in the memoirs of First World War pilots. In Spring's postwar novel *Leave Me with a Smile,* the narrator contrasts his wartime and peacetime selves: "He had been happy then, looking death in the face and laughing. . . . But today, secure, safe and smug, he was miserable."

Attempts to impose military discipline on the fliers' high jinks and independence of spirit only added to their self-defined sense of being pioneers, members of a breed apart. Benjamin Foulois, the U.S. Army's first active pilot, appears to have started the unofficial but resilient tradition of what he called "indoctrinating" the more conservative branches of the Army, buzzing cavalry units on parade and the division headquarters' latrine, at dawn, at twenty-five feet. Carl Spaatz, later to become one of the senior Second World War commanders of the U.S. Army Air Forces, said that fliers had viewed themselves this way from the start: "We flew through the air," he said, "and other people walked on the ground. It was as simple as that."

Traditional Army commanders viewed the fliers as "more or less of a mob," as one American general put it after the war: "No people in this war needed discipline more than aviators and none had less." Pershing sent an order to the aviation schools reminding them that "slovenly, unmilitary, careless habits" would not be tolerated. The aviators retaliated by becoming even more slovenly—tales abounded of pilots shot down wearing their pajamas—and scoffing at the idiocy of the "tin majors" who ordered them to wear spurs with their boots, or stiff 1865-style collars while flying. Such anecdotes about nonflying commanders, true or not, became the stuff of legend: A white-moustached cavalry colonel, newly appointed as commanding officer of a flying school, demanded to know why so many nearly brand-new airplanes were badly damaged. "Rough landings," he was told by his adjutant. "I will remedy that," replied the colonel. "Take a memorandum: There will be no more rough landings at this field."

The sense of being a breed apart was reinforced, too, by the rather intoxicating fact that that was how an adoring public saw them. At the beginning of the war, the armies of the conflict stuck to the military tradition of anonymity in reporting developments at the front; only when a soldier had done something truly extraordinary was he entitled to be "mentioned in dispatches" by name. The Germans were the first to see the propaganda value of making heroes of their aviators; by the summer of 1916 the French followed suit; and even the

THE AIR SERVICE SOLDIER,-- IN ONE POSE AND
SEVERAL SNAPSHOTS. (UNPOSED).

A cartoon in a U.S. Air Service newspaper in France makes fun of the ubiquitous photographs of pilots suavely posing in front of their machines. *(Author's photo/National Archives)*

British, who sniffed that this was a breach of military decorum, were eventually shamed into it by a clamor in the press and Parliament, particularly after the Army had refused to make public the name of the RFC aviator who shot down the great German ace Max Immelmann.

To a public desperate for heroes and a glimmer of good news, the battles in the skies seemed somehow "cleaner" than the horrors of the trenches. Not that those at home learned until later about what trench warfare was truly like, but what they did know was that the conventional images of military heroism were sorely wanting in this new kind of collective annihilation. The trenches were a "troglodyte" world, as the French poet Blaise Cendrars called it; soldiers were no longer brightly clad champions on horseback but drab-colored moles literally burrowing into the ground and seeking to be as inconspicuous as possible. "Death was anonymous and the sacrifices were made collectively," writes Lee Kennett of the battles of the Western Front; "such battles were not made for individual feats of heroism that could fire the public's imagination." Pilots, especially fighter pilots, were the very antithesis of this collective anonymity; they fought as single champions; they were often personally flamboyant; they soared into the skies while other men hid in the mud. They were, in short, chivalrous in a world where chivalry was dead.

Freud wrote of the "narcissism of small differences." It was inevitable that even within the air services pilots would split into cliques with their own separate cultures and identities. The fighter pilots increasingly saw themselves as the stars; the bomber pilots grumbled about their colleagues almost as much as the regular Army did and complained that they took the same risks themselves, even greater risks, but never received the same credit. A ditty written by one anonymous bomber pilot asked: "Who never gets to be an ace? / Who is the man who stops the lead? / Who is the man who comes back dead? / Who is it never dies in bed? / THE BOMBER."

But newspapers knew a good story when they saw one, and they relentlessly focused on the exploits of the more flamboyant fighter pilots. Many of the accounts were fatuous, such as the *Sunday Herald*'s story of a British flier who, about to go on leave and wanting to "enjoy his little holiday with relish," shot down five Huns before luncheon. John McGavock Grider and his pilot comrades talked of forming "The Society for the Extermination of Amateur Aerial Authors," the purpose of which "will be to protect the public from a flood of bunk."

The chivalry of the "knights of the air" was not entirely an invention of the purple-penned press. Whether it was because airmen could get away with it where the men in the trenches could not, or whether it was because they themselves had begun to believe the myths that were written about them, the aviators did make the occasional chivalrous gesture to their foes. Not infrequently, both German and Allied aviators flew over one another's aerodromes and dropped notes to report on the fate of missing men, whether they had been killed or taken prisoner. The more flamboyant gestures, such as the time Georges Guynemer spared the life of the German ace Ernst Udet, breaking off their combat when he saw Udet's gun had jammed, were unusual to the point of being unique. Udet was so shaken by Guynemer's magnanimity that he could not fly for months afterward.

For the public, though, the "bunk" and the reality seamlessly blended, and the effect was overwhelming. Rickenbacker, Lufbery, Boelcke, von Richthofen,

Acts of chivalry by the "knights of the air" were more the stuff of journalistic invention than reality, but they did occur. A message listing the fate of British pilots recently shot down was dropped by a German pilot over an RFC airfield in France in February 1917. *(Public Record Office)*

Guynemer, and Fonck became household names. Aristocratic Frenchwomen donated their furs to be made into aviators' flight suits. After Lieutenant William Leefe Robinson shot down a Zeppelin in September 1916, he was mobbed wherever he went. Five artists vied to paint his portrait for the Royal Academy. "The police salute me, the waiters, hall porters, and pages of hotels and restaurants bow and scrape, visitors turn and stare," Robinson wrote his parents. An ace was now the unquestioned thing to be. For years after the war, the American writer William Faulkner walked with a limp and told hair-raising tales of dogfights and crashes and cruising over France in his Spad "with a crock of bourbon in the cockpit." Actually, he had never made it farther than flight school in Canada, but he was hardly alone in desperately wanting to be known as a member of this new band of heroes. Thirty-eight Americans served in the Lafayette Escadrille, the squadron of volunteer fliers formed in France before America's entry into the war; over the years some four thousand people claimed to have belonged to this famous unit. Whatever its impact in the war, the impact of aviation in the public imagination was becoming everything that H. G. Wells had foreseen it to be.

4

GRAND PLANS

Fifteen miles up the Norfolk coast from Yarmouth is a small town called Ingham, and it was there at 7:40 on the night of January 29, 1915, that an observer looking out to the sea spotted what he said looked "like two bright stars moving." The lights came closer; fifteen minutes later, as they crossed over land, they separated, one turning up the coast to the northeast, the other to the southwest. That this was not another practical joke by the village toughs floating balloons with Chinese lanterns was made unmistakably clear at 8:25, when the bombs started falling.

The first German Zeppelin raids on England were not a spectacular success. The airship L 3, which had turned southwest, dropped nine 110-pound bombs on Yarmouth, killing one man and one woman and wrecking several dwellings and a warehouse. Her companion, the L 4, meandered over the Norfolk countryside scattering her payload on the town of King's Lynn and a few smaller villages; these bombs hit a church and some cottages and killed one more man and one more woman. It wasn't until he read an English newspaper several days later that the commander of the L 4 realized he had been sixty miles from where he thought he was. Both ships iced up on their return across the North Sea and nearly did not make it back.

Germany had launched previous attacks on Allied civilian targets, even previous attacks on English soil. The lone German Taube aircraft that dropped a token bomb on Paris every afternoon for several weeks in the fall had provoked a warning from American President Woodrow Wilson that Germany was tarnishing her image by such acts, though among Frenchmen the "five o'clock Taube" provoked little more than a Gallic shrug. In December German warships had bombarded several English coastal cities, inflicting a much greater toll, with 137 killed and almost 600 injured.

Yet this strike from Germany's Zeppelins caused a sensation around the world. The story filled the entire front page of the *New York Times* the next day. It was widely assumed, and reported as fact, that the Germans' target had been the royal palace at Sandringham. In Germany, the raid was greeted with a joy that bordered on the delirious. "It has come to pass, that which the English have long feared and repeatedly have contemplated with terror," declared the *Kölnische Zeitung*. "The most modern air arm, a triumph of German inventiveness and the sole possession of the German Army, has shown itself capable of crossing the sea and carrying the war to the soil of Old England!"

It had actually been the German Navy, not the Army, that had carried out the raids, and this was no happenstance. The German Navy had been chafing for a way to challenge the vastly superior British fleet, and many of its officers saw in both the Zeppelin and the U-boat the secret weapons that could slip by the Royal Navy's impenetrable wall of ships. Grand Admiral Alfred von Tirpitz, commander of the German High Seas Fleet, thought that it was "odious" to scatter "single bombs from flying machines" and "hit and kill old women." But, he admitted, "if one could set fire to London in thirty places, then what in a small way is odious would retire before something fine and powerful." Tirpitz added: "The measure of the success will lie not only in the injury which will be caused the enemy but also in the significant effect it will have in diminishing the enemy's determination to prosecute the war."

But it was Rear Admiral Paul Behncke, deputy chief of the naval staff, who became the most passionate and outspoken advocate for what the bombing of London could do. Left in charge at the Navy's headquarters in Berlin when the war began, Behncke wasted no time drafting a memorandum urging the all-out strategic bombardment of Britain. Such attacks, he wrote on August 20, 1914, "may be expected, whether they involve London or the neighborhood of London, to cause panic in the population which may possibly render it doubtful that the war can be continued." Behncke fought battle after bureaucratic battle seeking approval for his plans. The naval high command feared that moving one of its Zeppelins to Belgium, in closer striking distance of the English coast, might be risky because a hostile Belgian civilian might take a potshot at it on takeoff, but even more because "the Army High Command will try to commandeer it for its own purposes." Others objected that the military value of the Zeppelin raids was nil, and that the ships were needed for their primary and intended mission of fleet reconnaissance. And then the Kaiser himself expressed horror at the idea that an attack on London might kill one of his royal cousins.

Behncke spent the fall drawing up more precise plans and target lists; he singled out the Admiralty building with its radio station, the War Office, the Mint, the Central Telegraph office, the Bank of England, and the London docks. By the end of the year official resistance was weakening, in part because of French bombing attacks on Freiburg that had caused civilian casualties, and the Kaiser at last relented so far as to allow attacks outside of London.

From that moment what would follow was inevitable. By February 1915 the Kaiser agreed that the London docks could be struck, although "no attack is to be made on the residential areas of London, or above all on royal palaces." By May he agreed that all of "London east of the longitude of the Tower" could be considered a legitimate target. By July he gave way completely to the naval staff; now all of London, save for palaces and monuments (which would be "spared as far as possible"), was approved for air attack.

London was hit for the first time on the night of May 31. This time it was the Army's turn, and there was little doubt that the object was to sow as much chaos and panic as possible. The airship LZ 38 carried a load of thirty small bombs, each no bigger than a grenade, plus ninety 25-pound incendiaries. Again the damage was slight but the reaction considerable. Seven people in East London and surrounding suburbs were killed, but reporters wrote grossly exaggerated accounts playing up the lurid details of a baby burned to death and a couple who were found dead kneeling by their bed "in an attitude of prayer." Society women arrived in chauffeured limousines to view the damaged houses. A coroner's inquest brought in a verdict of willful murder.

Desultory raids in other parts of the country meanwhile set off an epidemic of false Zeppelin sightings and rumors of enemy spies "signaling" to German airships with lights from the ground. The authorities quickly had their hands full investigating the reports. In nearly every case they turned out to be railway signal lights changing from red to green, headlights of cars ascending a steep hill, flickering fires from lime kilns, shepherds moving across their fields at night with lanterns at lambing time, or flare guns fired by British aircraft. (In 22 percent of the cases investigated, "suspicion appeared to be solely aroused by the nationality of the alleged signallers.") In cities such as Hull that had received visits from real Zeppelins, urgent meetings were convened by prominent businessmen, and deputations were sent to the Home Office Defence Headquarters demanding "protection against Air Raids."

There was no protection to offer. When the Expeditionary Force embarked for France in August 1914, the Director of Military Aeronautics, General Sir David Henderson, was appointed to command the air force in the field; he took with him as his deputy Frederick Sykes, the RFC's chief of staff; Sykes in turn stripped his headquarters at Aldershot of every pilot, airplane, and mechanic, even every typewriter. Sefton Brancker, then a junior officer, was hastily promoted to the temporary rank of lieutenant colonel and left in nominal charge of things at home as Assistant Director of Military Aeronautics. No sooner was he installed than he found himself summoned into the august presence of Lord Kitchener—field marshal, hero of Khartoum and the Boer War, now Secretary of State for War. Kitchener explained to Brancker that he had just informed the Cabinet that the Army would assume responsibility for patrolling "quite a large portion of the British coast" with its aircraft. "My jaw dropped," Brancker recalled:

> I pointed out that every man & aeroplane at my disposal at home was
> required for training & building up new units. . . . K. looked sur-

prised—put on his most terrifying frown—& told me not to talk non-sense but to go away & do what I was told!

I did nothing. I could think of nothing to do and the next day went to see him again and told him frankly that I had done nothing. . . . K. glared his worst. I expected an explosion but slowly his face relaxed and he said—"Well—perhaps you are right but you've got to do *something* in order to carry out this bargain although it may be use-less." I sent one pilot & one small & very feeble single-seater to fly in the neighborhood of the Forth.

A few weeks later Kitchener, facing up to the reality, offered to turn all responsibility for air defense over to the Admiralty, and the First Lord, Winston Churchill, agreed at once and threw himself into the task with his trademark enthusiasm and energy. Churchill, who had taken up flying himself before the war until his wife pleaded with him to give it up—he was said to be a competent pilot once in the air, but could never be trusted to take off or land by himself—had been well aware for several years that the RFC's meager home-defense plans left a large hole that the Navy might move into. In 1912 and 1913 the Admiralty had begun to form its own units of airplanes and seaplanes specifically for "the aerial protection of our naval harbors, oil tanks, and vulnerable points." At that time, the War Office had strenuously objected to this poaching on its preserve; traditionalists in the Royal Navy simultaneously objected that the Navy should stay out of aviation altogether. Now, in the fall of 1914, Churchill's past critics were clamoring for action. "They have pissed on Churchill's plant for three years," dryly observed Richard Davies, a Royal Navy lieutenant who would later become a famous naval aviator, and "now they expect blooms in a month."

Despite Churchill's previous support for air-defense efforts, there was not a lot he could offer at once. When the war began there were in Britain a total of only thirty-three antiaircraft guns, and twenty-eight of those were one-pounder "pom-poms" that, Churchill said, were of "very small value." This was an under-statement; they were of no value. Truth be told, even superior guns would prove less effective in shooting down enemy aircraft than in creating an impressive spectacle for the citizenry: subsequent analysis would show that it took some eight thousand shells fired for each hit scored.

Churchill recognized that, under the circumstances, a good offense was not merely the best defense but the only defense, and he pressed hard for the Royal Naval Air Service to strike directly at the German Zeppelins at their bases. The raids the RNAS carried out in the fall of 1914 were a triumph of surprise and daring in the face of inadequate technology. Dropping tiny bombs from low alti-tudes, British naval pilots managed to destroy a Zeppelin in its shed at Düssel-dorf on October 8. Then on November 21, three of the Royal Navy's Avro 504 biplanes took off from Belfort, France, near the Swiss border, where they had been brought on trucks under cover of darkness and great secrecy. Skimming over Lake Constance at a height of ten feet after their 125-mile flight across the Black Forest Mountains, the British pilots achieved complete surprise and managed to destroy a Zeppelin and the gasworks at the main Zeppelin base in Friedrichshafen.

Avro 504
bomber
engine: 80-hp rotary
top speed: 80 mph
armament: four 24-lb. bombs

But there was only so much that aircraft armed with four 24-pound bombs apiece could do, and dive-bombing high-value targets from altitudes of a few hundred feet was not a tactic one could expect to live to repeat more than once or twice. Immediately after the Friedrichshafen raid, the Germans greatly increased protection around Zeppelin bases, adding guns, searchlights, and bomb nets over the sheds.

Nor was it a tactic that could survive the flak the Royal Navy now found itself drawing from the Army. The War Office may have been willing to turn over air defenses at home to the Admiralty, but this sort of preemptive action was not what it had in mind; and the RFC, sensing both an institutional threat and competition for aircraft production, repeatedly demanded that the Navy get out of the strategic-bombing business. By May 1915 Churchill was forced to resign as First Lord following the Gallipoli disaster, and with him went the Admiralty's most tenacious political infighter and the strongest voice for naval aviation.

Although throughout the war the Navy would remain ahead of the RFC in developing the specialized aircraft, navigation instruments, and bombsights needed for long-range bombing, the Army always held the lead on the field of bureaucratic skirmishing. At the time of the Somme fighting, the RFC, pleading urgent necessity, succeeded in diverting sixty new Sopwith 1½ Strutters from the Navy, just as they were about to go into action as the nucleus of a new RNAS strategic-bombing force.

On the other hand, the British authorities were well aware that the Zeppelins could inflict little real harm, defense or no defense: an intelligence analysis in January 1916 concluded that "air raids have not hitherto succeeded in doing any damage of naval or military significance." Even the new million-cubic-foot airships that carried two tons of bombs, and even the two-million-cubic-foot "Super Zeppelins" that appeared in mid-1916 and could carry more than four tons, were little more than harassment devices. Many of the Zeppelin raids had to be aborted because of mechanical difficulties or unfavorable weather; navigation remained far more a matter of guesswork than it should have been; and the radio direction-finding system the Germans brought into service in April 1915 was both inaccurate and a security gaffe of monumental proportions. To ascertain its location, a Zeppelin would broadcast to shore stations, which would take bearings on the signal and radio back the fix. The British Admiralty's radio intelligence service had no difficulty listening in and decoding the messages. Bombs were frequently dropped on cloudy nights using radio fixes alone, and these would not uncommonly be thirty or forty miles off target.

But none of this shook the German faith in the great and terrible things the Zeppelins would do, nor did it reassure the British populace. Just as the attackers saw bombing in terms of its psychological effect, so the defenders did; it was becoming increasingly important to make a show to reassure civilians they were being protected. A book published in London in 1916, *Zeppelins and Super-Zeppelins,* stated: "It is particularly humiliating to allow an enemy to come over your capital city and hurl bombs upon it. His aim may be very bad, the casualties may be few, but the moral effect is wholly undesirable. When the Zeppelins came to London they could have scored a galling technical triumph over us if they had showered us with confetti."

Something—anything—would have to be done to deny the Germans their triumph. Blackouts of cities were ordered. H. G. Wells proposed that any British flier who succeeded in shooting down a Zeppelin be awarded a knighthood, "if he gets down alive." A network of sound detectors, searchlights, ground observers, and command posts was gradually expanded. Fighter pilots, armed with newly developed explosive and incendiary bullets, began flying dangerous night missions; following preset patrol lines, they would hope—a very long shot indeed—to spot a Zeppelin caught in the searchlights or—an even longer shot— to come within perhaps 200 yards of their target, at which point the naked eye on a starlit night might be able to make out the enemy craft as a dark shadow.

When it finally happened, the effect was every bit as stunning as the jolt the British psyche had endured from the German bombs. On the night of September 2, 1916, Lieutenant William Leefe Robinson, flying alone in one of the much-maligned British two-seater BE2s, caught up with the German Army airship SL 11 over the capital. Robinson had taken off at 11:30 p.m. from Sutton's Farm near London; twice he caught sight of airships only to lose them in the clouds. One technique was to cut the engine and glide for a few minutes to try to listen for the airship's motors. Robinson tried that: nothing.

He had been in the air for almost two and a half hours and was about to turn back when his attention was caught by a reddish glow to the northeast of the city where incendiary bombs had fallen from one of the German raiders. Rapidly closing in on his target, he dodged both friendly antiaircraft shells and machine-gun fire from the airship's guns, then passed directly under the enemy and emptied a full drum of explosive and incendiary bullets into his target, raking it from bow to stern. It had no effect. He made a second pass, emptying another drum of ammunition along the side of the airship: again no effect. Finally, he maneuvered to within five hundred feet of the airship's tail and concentrated all of his fire just ahead of the tail section. After a few seconds a red glow appeared, and a few seconds later the ship erupted in flames, lighting up the entire sky as it turned on end and slowly fell.

The spectacle was witnessed by millions in the city below. "Cheers thundered all around us from every direction," recalled one woman who had raced to her window at the sound of the explosion. "It was magnificent—the most thrilling scene imaginable." The next day crowds of Londoners in cars and carts, on bicycles and on foot, jammed the lanes of the tiny farming hamlet of Cuffley, a few miles north of the city, to see the wreckage of the airship where it had fallen in a beet field behind the village pub, the Plough Inn. Robinson had to settle for the Victoria Cross instead of the knighthood that H. G. Wells would have awarded him. The German Army decided to call it quits. The German Navy tried several more times, but at a terrible cost; by the end of 1916 a total of seven airships had been shot down, along with their irreplaceable commanders and crews.

Yet in this strange psychological war that had begun in the skies over England, even the British successes could be taken as a sign of encouragement by the Germans. "How disturbing these attacks are for the English is shown by the tremendous expense they go to prevent these attacks, and by the jubilation demonstrated when an airship is destroyed," insisted Peter Strasser, the commander of the German Naval Airship Division, in mid-1917. Strasser would die a year later leading the final Zeppelin raid of the war, still determined to bring the war home to the enemy populace.

The German Army, though it had given up on airships, had not given up on the idea that had animated the airship raids. By May 1917 the Army had a new weapon with which it believed it could shock the British public, and its government, into bowing out of the war: the huge twin-engined Gotha bomber, modeled on the Russian Il'ya Muromets. The plan was to strike London, particularly its government centers, in highly visible daylight raids designed to be seen by millions of Londoners. By September the Gothas were joined by the even bigger R-planes, four-engined behemoths that could carry a ton of ordnance and fly at fourteen thousand feet.

Gotha G.V
long-range bomber
engines: 260-hp six
cylinder (2)
top speed: 87 mph
armament: 1,100-lb.
bomb load; 2 to 3
machine guns

From such altitudes the chances of hitting the Bank of England or the Admiralty were nil. Eventually the Gotha and R-plane crews were told that their target was simply "the morale of the English people."

Faced with the fast-moving Gothas, the British system for reporting and tracking raids quickly broke down. On June 13, 1917, fourteen of the German planes made their first daylight raid over London, dropping a hundred bombs; 162 people were killed and 426 injured, the worst single raid of the war. The Liverpool Street railway station was badly damaged. A second raid on July 7 caused an even greater shock to the British capital; although the loss of life was considerably less, some four hundred houses and shops were wrecked and the Central Telegraph office was hit.

The public clamor swelled. A series of nighttime raids on London in September further shook the government. The Allies were already on edge about what a collapse of morale might mean. In May tens of thousands of French soldiers had mutinied, refusing to continue a hopeless offensive. One French infantry regiment took over a whole town and proclaimed an antiwar "government." Elsewhere in France there were outbreaks of civil unrest and strikes among factory workers. In London, the government was increasingly uneasy about the reaction to the Gotha and R-plane raids among the teeming masses of the East End, "a large proportion of whom were aliens," as one official report noted.

"Aliens" meant Jews, and the not uncommon view among the not uncommonly anti-Semitic British elite was that they represented an excitable and cowardly presence in the heart of the metropolis that might easily panic: they certainly lacked the courage and discipline of good British military men under fire. Although *The Lancet* opined that from a psychological view the populace was doing well, reacting "in bulk with bravery and prudence," the authorities were not so sure. They were particularly concerned that during the September raids 100,000 East Enders had sought refuge in the underground stations; police estimated that on one night at the end of September the Old Street station alone was packed with 10,000 people, 3,000 of them on the platform. On several nights crowds started flocking into the stations as early as 5:40 p.m. without even waiting for an air-raid warning; in fact no raids came on those nights. Tens of thousands more Londoners began making an exodus to the suburbs each night to escape the German bombers.

In Parliament and the press, demands for retaliation surged. *Flight* called for Berlin to be bombed: "Until we can bring home to the originators of bombing raids on towns not ordinarily in the military zone the horror and injustice of it, these raids will continue." A British industrialist offered a prize of £1,000 for the first British aviator to drop a bomb on the enemy capital. David Lloyd George, the Prime Minister, hastened to promise that vengeance was coming: "We will give it back to them and we will give it to them soon. We shall bomb Germany with compound interest."

The British reaction was in fact nothing short of extraordinary. During the whole of the war, as Lee Kennett observed, "the Germans dropped on England less than 300 tons of bombs, killing 1,400 people and injuring 4,800 more. These are the kinds of totals one would find reported for a single 'quiet' day on the western front. Total property damage was put at slightly over £2 million—less than half what the Great War cost the British each day." Yet following the first London raid the Cabinet took all of one day to order a doubling in the size of the RFC, from 108 to 200 squadrons. Four days after the second raid, Lloyd George appointed a Cabinet committee to examine the entire question of air defenses and air operations. Lloyd George was nominally the chairman, but it was General J. C. Smuts, the South African soldier and statesman and member of the War Cabinet, who wrote the report, and it was he who was responsible for its sweeping recommendations. It was now, Smuts said, beyond question that the air arm "can be used as an independent means of war operations." Nobody who witnessed the second Gotha raid on London "could have any doubt on that point."

> As far as can at present be foreseen, there is absolutely no limit to the scale of its future independent war use. And the day may not be far off when aerial operations with their devastation of enemy lands and destruction of industrial and populous centres on a vast scale may become the principal operations of war, to which the older forms of military and naval operations may become secondary and subordinate. . . . It requires some imagination to realise that next summer, while our Western front may still be moving forward at snail's pace in Belgium and France, the Air battle front will be far behind the Rhine, and that its continuous and intense pressure against the chief industrial centres of the enemy as well as on his lines of communication may form the determining factor in bringing about peace.

Smuts's two recommendations went hand in hand: An all-out strategic-bombing campaign against Germany should be waged; and a new, independent air force, reporting to its own ministry, should be the one to wage it. The birth of the Royal Air Force as a third service, equal to and fully independent from the Army and the Navy, owed its conception directly to the lust for vengeance that inflamed Britain in the summer of 1917.

To undertake such a sweeping military and governmental reorganization in the midst of a war was a radical, indeed almost mad, step and the Cabinet hesitated, but sentiment in Parliament was unstoppable. A bill establishing the RAF and an Air Ministry was drafted and passed with scant debate, becoming law on November 29, 1917.

The man chosen in January 1918 to head the new service as Chief of Air Staff was Major General Hugh Trenchard, brought back from command of RFC field forces in France to take up the post. It was a profoundly mixed blessing. If, as is often said, Trenchard was the "father of the RAF," he was—as Lord Beaverbrook observed many years later—"a father who tried to strangle the infant at birth though he got credit for the grown man." Trenchard was an efficient and well-respected leader of men and an able administrator, but for all of his enthusiasm for aviation he was very much an old-school general. The son of an infantry captain, he had been privately educated as a boy and failed the entrance examination to the Army on several tries; he finally received his commission after entering the regular army through the militia, an avenue that depended more on family and county connections than on ability. He served in various posts around the

British Empire with distinction and obvious energy, however, and when he was ordered in 1912 to learn to fly he found he had only ten days to pass the tests and get his aviator's certificate before he would be over the age limit. He did it in a week. Trenchard was a man of vigor but not an intellectual, and not broadly educated even in military matters, much less technical ones.

Much of the inconsistency and friction in RAF policy over the next several months flowed directly from Trenchard's personality. He in many ways embodied the stereotype of the narrow-minded British general of the era. He was loyal to his military comrades and regimental traditions, he believed in the bold attack, but his thinking often did not seem to go much deeper than that. Trenchard never developed a clear strategy or purpose for the strategic-bombing mission the RAF had been formed to carry out. He seemed to have little time for technical details. His own actions were full of contradictions: he had opposed the formation of the RAF, fearing it would weaken the support the air arm could provide to the Army on the ground, yet he agreed to become its head; he opposed the notion of bombing Germany simply as retaliation, yet drafted a plan that emphasized hitting a blow at German "morale" and repeatedly evaded orders to focus the strikes more purposefully on German industrial targets; he told the general who succeeded him as commander of flying units in France that one should never resign in wartime, and three months later proceeded to do just that himself; and after taking personal command of the RAF's independent bombing force a month after that, in May 1918, he complained of it, in his diary, that "a more gigantic waste of effort and personnel there has never been in any war."

Trenchard's inconsistent and at times petulant stance regarding strategic-bombing policy seemed to have less to do with any clearly defined philosophical difference with the civilian authorities than with his resentment of civilian authority in general. His resignation as Chief of Air Staff came after repeated clashes with Lord Rothermere, the first Secretary of State for Air, who Trenchard believed should have left the decisions to him. Rothermere found Trenchard's behavior to border on the insubordinate and described him as an intolerant know-it-all, a man "of dull, unimaginative mind."

Trenchard's resignation led to a cascade of other changes, culminating in the appointment of the much-disliked Frederick Sykes, now a major general, in his place as Chief of Air Staff; the resignation of Rothermere; and the wooing back of Trenchard by Rothermere's successor, William Weir. The civilian government had, after all, created the RAF to hit back at Germany, and Weir was determined to see that happen; he was willing to turn a blind eye to Trenchard's insubordination and complaining in exchange for the political and institutional value of having someone of his stature at the head of the bombing program. Trenchard was offered his choice of four posts: inspector general of the RAF overseas; inspector general of RAF areas at home; commander in chief of all air force units in the Middle East; or commander of long-range bombing forces in France. Trenchard agreed to accept the last, but the price he exacted was huge: as commander of the RAF's "Independent Force" of long-range bombers, Trenchard would be permitted to report directly to Weir, bypassing Sykes altogether. As far as Trenchard was concerned, this meant he did not report to anyone. He then created his own staff that duplicated the Air Staff and proceeded to ignore not just the Air Ministry but the entire War Cabinet, too.

One legitimate reason Trenchard was granted such extraordinary independence was that it would be necessary for him to deal directly with the French military authorities in order to secure bases to operate from. In response to the pressing demands from Whitehall to strike back against Germany in the fall of 1917, the RFC had formed its first separate bombing unit, the 41st Wing, but it did not have much to work with. None of its aircraft could reach Berlin; none could reach targets in Germany at all except from forward bases in France. The most advanced bombers it could call on were ten Handley Page O/100s, commandeered once again from the Navy, and these dated back to January 1915—they had been ordered during the Navy's original, Churchill-inspired era of enthusiasm for strategic bombing. (The Admiralty's superintendent of aircraft construction, the irascible Murray Sueter, had rejected the firm's first drawings for a bomber by saying, "Look, Mr. Page—what I want is a bloody paralyser, not a toy.") The Handley Page was Britain's first twin-engine bomber, and it could carry nearly a ton of explosives, but by now it was years behind the state of the art, certainly nothing like the Gotha in its performance or range. The Handley Page was so slow and such a poor climber that it could be safely used only for night missions. (It also didn't help that one of the first operational models had been promptly delivered to the Germans when its pilot lost his way in the fog and landed at a German airfield.)

Handley Page O/100
bomber
engines: 250-hp V-12 (2)
top speed: 85 mph
armament: 1,800-lb. bomb
load; 4 to 5 machine
guns

The only other planes the 41st Wing had were one squadron of faster but considerably smaller single-engine De Havilland DH4s, which could outrun most fighters but carried only a few hundred pounds of bombs; and one squadron of the now-obsolete FE2 pusher fighters, pressed into service as small night bombers. Weir, who in the fall of 1917 was serving as controller of aircraft supply in the Ministry of Munitions (under the byzantine British aviation bureaucracy that made him responsible for all aircraft production at the time), tried to press the Cabinet's policy forward by ordering Handley Page to begin work on a huge four-engine bomber that could "seriously worry Germany." But the first of these V/1500 bombers was still being readied for a raid on Berlin when the Armistice arrived.

The French for their part were less than enthusiastic about the British lust for revenge. Morale cut two ways, and with French cities much closer to the front than German cities, France would likely bear the brunt of escalating reprisals. A French Air Service note on aerial bombardment drafted in December 1917 warned that it was best "not to let public opinion get carried away with the idea that reprisals are worthwhile unless we are sure that the morale of our civilians is ten or twenty times stronger than that of German civilians, for if a battle of this sort starts, that is the proportion in which the explosives will come."

DH4
observation/light bomber
engine: 375-hp V-12
speed: 130 mph
armament: 320-lb. bomb
load; 2 to 4 machine guns

Trenchard decided to take the route of least resistance between Cabinet pressure from home, French reluctance in the field, and the technological inadequacy of his aircraft. The result was a total muddle. Trenchard repeatedly gave way to French requests to divert the Independent Force to tactical missions. To fend off the meddling civilians back home, he increasingly rationalized the desultory performance of the strategic raids his bombers did carry out by asserting that even when his bombers did not hit their targets, and even when the raids were sporadic and dispersed, they were really hitting enemy morale all the harder. "Actual experience goes to show that the moral effect of bombing industrial towns may be great, even though the material effect is, in fact, small," Trenchard had insisted in a memorandum in late November 1917. By May 1918 he went even further: "The anxiety as to whether an attack is likely to take place is probably just as demoralising to the industrial population as the attack itself." The best plan, he explained, was thus not to concentrate on wiping a target off the map, which he declared to be impossible anyway, but quite the opposite: to scatter bombs all over Germany so that workers all over the country would lie awake worrying if they were going to be next.

It was a bizarre argument, but it had the virtue of establishing the proposition that it didn't matter where his bombs fell; any bomb that hit the soil across the German frontier "counted" by this reasoning. The French Air Service had produced a policy paper concluding that a fundamental principle for bombing operations was that they must adhere to a plan and be aimed at achieving a precise and well-defined military result. *On ne bombarde pas pour bombarder,* it warned: One does not bomb just for the sake of bombing. Trenchard was not just bombing for the sake of bombing; he was making a grand fetish out of bombing for the sake of bombing. And "morale" was such an ill-defined measure of success that almost anything could be claimed as a success. Any word of grumbling in intercepted letters written by German soldiers or civilians, any scrap of gossip about discontent in reports from agents or travelers from neutral countries, was seized on in British intelligence reports as proof of "growing dread and excellent effect of our raids on Germany."

Fliegeralarm!

Langgezogene, auf- und abschwellende, zwei Minuten dauernde Sirenentöne
und Abschüsse von Signalbomben bedeuten:

Fliegeralarm!

Sämtliche Strassenbahnen halten.

Sofort Deckung suchen.

Ein gleichmäßiger, eine Minute dauernder Sirenenton bedeutet:

Fliegeralarm aufgehoben.

Jeden Abend 6 Uhr wird zur Prüfung der Anlage während ¼ Minute das Signal
„Fliegeralarm aufgehoben" gegeben.

Düren, den 18. August 1918.   Der Oberbürgermeister.

Orders posted by the mayor of Düren, near Cologne, instruct
trams to halt and people to take cover when an air-raid warning
signal is sounded. The orders were issued after an RAF bombing
attack on August 1, 1918; as it happened, the town was not
struck again. *(Public Record Office)*

In this game, even the fact that British workers had *not* exhibited wholesale
panic as a result of German bombing raids could be twisted into a proof of the
proposition that morale bombing was effective. One RAF staff memorandum in
late 1917 explained: "The Germans in raiding England expect to create a panic
amongst the working classes. This has not been the case and their supposition is
evidently based on the fact that they consider the psychology of the British
workingman as being the same as that of the German—hence it may be assumed
that the ordinary working German is liable to panic."

Reasoning about enemy morale was like looking into a hall of mirrors, and it
was not surprising that the British belief that Jews of the East End were the weak

link in British morale would cast a reflection in assumptions made about Germany; this same memorandum observed that because "a majority of German chemists" are recruited from the ranks of "the German Jew and the Polish Jew," who are "not usually brave," bombs dropped in the area of German chemical works would have a particularly strong effect on the morale of the workforce of those industries.

Trenchard, of course, was hardly alone in exaggerating the effect that bombing the enemy would have. One RAF officer said that throughout the war, staff officers who knew perfectly well how many artillery shells it took to destroy a target "appeared to think that some magic in the air would enable them to gain decisive results with one-hundredth the part of the necessary weights, provided it was in the form of aircraft bombs." It was hardly uncommon for the theories of peacetime to be proved wrong when they collided with the realities of war. Yet in contrast to the rapid adjustments that the major air forces made to the emerging realities of air-to-air fighting, the prewar theories about bombing were astonishingly resistant to reality. The public's almost mystical faith in the potency of this terrible new weapon spread an intoxicating cloud that fogged the minds of politicians and even military officers.

But the problem went deeper, for to measure the effect (or rather lack of effect) that aerial bombardment was having, and to pinpoint what it would take to make it more effective (or rather effective at all), required a new kind of scientific thinking that was frankly beyond the ken of a traditional military man of Trenchard's background and outlook. The same was true of the German Navy and Army staffs. It is human nature to equate the effort with the result: merely launching a long-range bombing mission was a logistic challenge of such stupefying proportions that it was hard to admit the possibility that the effort might be wasted, and official communiqués often made no distinction between dropping a bomb on a target and destroying a target.

One of the first attempts to scientifically analyze the results of bombing had been carried out in July 1915 by the British General Staff. The staff had examined the records of 483 bomb-dropping operations conducted by British and French forces from March 1 to June 20, and the results were nothing short of abysmal. Some 4,000 bombs had been dropped; almost none had hit anything. Worst of all were the many attempts to bomb railroad stations and junctions. "The results attained hitherto have been negligible": 991 bombs dropped on 141 occasions, and a total of 3 instances in which any military result whatever was achieved, according to agents' reports. Even when railroad tracks were damaged they had been "easily and quickly" repaired, usually within a couple of hours. The GHQ ordered that attacks would be restricted in the future to enemy headquarters, telephone exchanges, and factories producing ammunition and poison gas, but this hardly got to the root of the problem.

Neither did the French efforts to address what at least some of their officers had candidly admitted was a woefully disappointing performance. The French had been the first to form a specialized bomber unit, the Groupe de Bombardement No. 1 (GB 1), made up of three escadrilles of Voisin pushers each capable of lugging one hundred pounds of bombs. In late May 1915 GB 1 began operations, hitting the Badische Anilin und Soda Fabrik in Mannheim, believed to be the source of the chlorine gas introduced into trench warfare for the first time by the Germans in April. This was followed by some small tit-for-tat reprisal raids against German cities. But the small payload and the increasing losses from German fighters and antiaircraft guns quickly began to make the high command wonder if the sacrifice was worth the results.

Attempts to improve the performance of bombers were made, but they were uncoordinated and piecemeal, and often worked at cross-purposes. One obvious step was to build a bigger bomber, another was to shift to night bombing, and yet another was to see what could be done to improve the aim of bombardiers. What no one quite grasped for some time was that each of these measures affected the others. Night flying made the bombers less vulnerable but in turn created new problems of bombing accuracy and navigation. Bombsights were under development, but to use them generally required the pilot to fly a straight and level course directly upwind or downwind for a considerable distance in approaching the target, which could make the aircraft a sitting duck.

It was the British work on scientific development and testing of bombsights, however, that held the seeds of a more systematic approach to targeting: the all-encompassing system of breaking down cause and effect in a military operation that, in the Second World War, would come to be known as operational analysis. It enjoyed a brief, evanescent existence in the First World War. For the first months of the war the RFC had conducted almost no experimental research. But by December 1914 an "Experimental Flight" had been formed at the Central Flying School in Upavon, and by the summer of 1915 the unit had begun to conduct a series of groundbreaking experiments on aircraft performance and bomb ballistics.

The rather unlikely hero of this story was Henry Tizard, a lecturer in chemistry at Oxford when the war broke out. Tizard had a strong mathematical background and moreover possessed that quality rare even among scientists, a true independence of mind. As a student at Oxford, he had been given a strong push in this direction by his chemistry tutor, who had one day suggested that Tizard ought to settle down to some serious reading in the subject during the upcoming long vacation. "Observing perhaps a lack of enthusiasm on my face," Tizard recalled, his tutor casually pulled a ploy that would have a lifetime effect upon him:

> He turned to his shelves and picked out a new-looking book entitled *Chemical Statics and Dynamics.* . . . "This is an important subject," he said, "but the book is rather too mathematical for me. I wish you'd

take it away and see how many mistakes you can find in it." I believe this was the first time that any senior man had indicated that I might know more than he did about something that was worth knowing, and the first time that anyone had suggested to me that there might be mistakes in a printed book of science.

Tizard quickly found one mistake, which aroused "the detective spirit," and by the time he was done he had a long list of corrections, which he sent the author. He had also learned a valuable lesson in trusting to his own resources.

At the outbreak of war Tizard entered the Army as an artillery lieutenant, whereupon, among other things, he developed his own completely unorthodox but highly effective method of training recruits to use the Maxim gun. It was in early 1915 that an old Oxford friend now in the RFC, Robert Bourdillon, wrote to him asking if he would transfer to Upavon to assist him in developing an idea he had for a new bombsight.

The bombsight itself was not a new concept. As Riley Scott's prewar experiments had proved, the laws of physics made dropping a bomb from a moving airplane a very different problem from shooting a rifle at a bull's-eye target. When a bomb was released from an airplane, it carried with it the forward momentum of the airplane's motion; thus by the time the airplane had passed over its intended target it was already too late to drop a bomb. A pilot had to release the bomb at some distance ahead of the target, in other words, and just how far ahead depended on both the plane's altitude (the higher up, the longer time the bomb had to travel forward before reaching the ground) and its speed relative to the ground (the faster the plane was moving, the faster the bomb would travel when released).

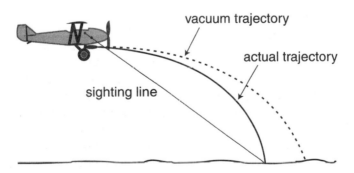

The simple bombsight: the pilot would release the bomb when the target lined up with two aiming bars. Most sights had sliding adjustments calibrated for speed and altitude to set the sighting angle. But experimentally determined correction factors were still needed to compensate for air resistance.

At first, improvised bombsights were widely used that consisted of nothing more than two nails driven into the side of the fuselage alongside the cockpit. The pilot would squint along the two nails until the target was lined up and then pull the bomb-release lever. To set the nails in position, an air speed and height would be chosen for the mission and the wind speed guessed at, and then it was a matter of algebra and trigonometry to line up the nails at the correct angle. A slightly improved version was the "lever sight," in which the two nails were on a rotating lever that could be adjusted to any angle.

There were several serious drawbacks with these crude sights. One was that it was necessary to guess at wind speed in order to calculate the actual speed of the airplane over the ground. Another was that real bombs traveling through real air don't perfectly obey the theoretical laws of ballistics. In a perfect vacuum, all free-falling objects describe the same parabolic arc regardless of their shape or size or weight. In the air, however, wind resistance comes into play, and the aerodynamic properties of the bomb matter a great deal in determining its trajectory as it falls.

Bourdillon's idea was to use the sight itself, along with a stopwatch, to measure the airplane's ground speed directly: the bombardier would click his stopwatch when a fixed reference point on the ground passed through the sight, then click it off when that same point passed directly beneath the airplane. The problem of wind resistance and the resulting "ground lag" between the vacuum trajectory and the real trajectory would meanwhile be solved by conducting bomb-dropping tests and calibrating an adjustable height and time scale on the bombsight to match the data.

Tizard arrived at Upavon in the summer of 1915 in time to play a crucial part in devising what proved to be a brilliant system for measuring bomb ballistics with an unprecedented degree of precision. He rigged a camera obscura, pointing straight up, that took a series of photographs at fixed time intervals as a test plane flew directly overhead. From the apparent size of the plane's wings in the photo and the focal length of the camera one could calculate the plane's altitude; from the movement of the aircraft from frame to frame one could calculate its ground speed; and from the position of the plane in the photo at the moment of bomb release one could calculate its exact downrange distance at that moment. The test pilot was instructed to release the bomb when he judged himself to be approximately over the camera, but the precise instant was recorded by a wireless signal transmitted automatically when the pilot pulled the release lever. On the ground, a pen recorder was hooked to the radio receiver to record this moment; the time the bomb struck the ground was also recorded when its explosion caused the pen to jiggle.

Tizard also was instrumental in developing a systematic—and, for the first time, consistent and reproducible—method of measuring performance of aircraft in flight. The CFS 4 bombsight that grew out of this work began to reach operational squadrons in early 1916. By the end of the year more than two thousand sights were in use, up from 325 in 1915.

The CFS 4 was better than nothing, but Tizard had reached the stark conclusion from his experiments that it actually was not much better: the errors in bombing accuracy caused by failing to properly adjust for ground lag or wind speed were actually of far less significance than the errors introduced by the routine gyrations of the aircraft itself in flight. Even under calm winds and ideal flying conditions the interaction of man and machine through the control stick, Tizard found, inevitably introduced a slow periodic oscillation in pitch and roll. Though scarcely perceptible to pilot or passenger, it was enough to play havoc with bombing accuracy. Even in an aircraft with inherent stability such as the BE2, oscillations in pitch and roll of about one degree every twenty seconds would occur. From an altitude of 6,000 feet, a one degree error would cause a bomb to miss its aim point by 100 feet. This was the first glimmer that the only way to get a handle on bombing accuracy was to look at the whole bomb-delivery system as a *system,* in which each component affected every other component.

Tizard had been rejected by the Navy, in which his father had served, because of poor eyesight, the result of a boyhood accident. But he nonetheless decided that if he was to get to the bottom of the problem of bombing accuracy he would have to learn to fly himself. He asked permission and was told that the Central Flying School's instructors could not waste their time training him since they had to train pilots for the war. Tizard finally proposed that he be taught only on days when the weather was too bad for the real students to go up, and this was agreed. He was also given special permission to wear his glasses underneath his flying goggles.

By May 1916 he had earned his wings, and a few months later he was sent to France to observe Allied bombing operations. The absurd reception he received was a taste of the difficulties he would be up against in dealing with the RFC command. Upon arriving at the front, he presented a letter of introduction to Colonel H. R. M. Brooke-Popham. But as soon as he began poking around and asking questions, he was detained—there were rumors of German spies in Allied uniform operating behind the lines—and brought in to see Trenchard himself. Trenchard asked what he was doing and demanded to see his credentials. Tizard said he had given his credentials to Brooke-Popham, but explained his mission and said that he had been sent by the War Office. Trenchard picked up his phone and called the War Office. A brief conversation ensued. "They say they have never heard of you," Trenchard announced.

Brooke-Popham then arrived, which would have seemed to solve the problem, except that Brooke-Popham, though he remembered Tizard, said he did not remember any letter. Finally Tizard spoke up: "If he doesn't mind my saying so, Sir, I noticed that Colonel Brooke-Popham's table was very untidy. Could you send someone to look through his papers, Sir?" The letter was found.

It did not take long for Tizard to conclude that the chance of a bomb actually hitting its intended target was "small." The problem went far beyond the technical issue of bombing accuracy. There were problems at every step of the way that it took to get a bomb delivered to its target: navigation, target identification by pilots, the difficulties of flying at night and through clouds, the use of photoreconnaissance to select targets and assess results, and the limitations imposed by the aircraft itself.

The RFC had scarcely given any attention to navigation at all. In training pilots to fly cross-country in England, the standard reference were the maps in Bradshaw's railway guide; pilots would sometimes find their way by diving down to read the names of the stations. A British memorandum in September 1917 noted with a certain ruefulness: "Experience has shown that it is quite easy for 5 squadrons to set out to bomb a particular target and for only one of those 5 ever to reach the objective, while the other 4, in the honest belief that they had done so, have bombed 4 different villages which bore little if any resemblance to the one they desired to attack."

Even when fliers managed to find the right target, the realities of combat flying frequently prevented the bombsights from being used as intended. Trying to fly over a hostile target in a steady pattern and at a fixed speed from a prescribed direction for up to two minutes while manipulating a stopwatch was too much for many air crews. "In a great many cases this has caused the bomber to ignore the fact that a sight was on the machine and simply drop the bombs when he thought best," noted an American report toward the end of the war.

The Royal Navy was developing a much improved "course-setting" bombsight that would factor in the wind vector to allow the bomber to attack from any direction, but the RFC showed little interest in it; it began to reach operational squadrons only at the end of the war. The RFC's experimental unit was also doing research on cloud flying and the instruments that would be required both to navigate and to permit a pilot to keep control of his airplane when he could not orient himself with a visual horizon. But here, too, the scientific work was well ahead of the knowledge or concerns of operational commanders, and few aircraft on the line were fitted with any but the most rudimentary instruments.

But there was a larger problem that was also going unanswered, despite the best efforts of Tizard and a few other scientists to focus attention on it. This was how to get the biggest bang for the buck, as a later generation would put it. What industries or other targets were most vulnerable to aerial attack? Which were most vital to the German war effort? What effect would bombs of various kinds have on them? Was the payoff greater, on average, in attacking a closer (or more easy to spot) target of lesser importance or a more distant (or less visible) target of greater importance?

It may seem astonishing that the military men were not troubling themselves with these questions, but Trenchard's emphasis on morale bombing had conveniently shoved this kind of critical thinking aside. The chief advocate of a scientific approach to targeting was a naval officer, Lord Tiverton, Tizard's counter-part in the Royal Naval Air Service. Tiverton had in 1915 and 1916 conducted experiments involving dropping unarmed bombs of various sizes down deep mine shafts to record their ballistic characteristics, and by 1917 he had begun to carry out statistical calculations on the numbers of bombs that would be required to achieve a desired level of damage on various key German targets. At the behest of the War Cabinet he also began a study of the specific German indus-tries that should be targeted. In sharp contrast to Trenchard, he concluded that it was necessary to "obliterate" factories, not merely scare the workers; even the rough estimates he produced showed that this would be an effort of monumental proportions, far greater than anyone had anticipated. To destroy the Badische chemical factory, for example, Tiverton calculated it would take 2,500 bombs; to obliterate the Bosch magneto factory would take 1,000. These figures translated into 1,500 sorties by DH4 bombers.

But Tiverton acknowledged that "I have a shrewd suspicion that very little more than a guess can at present be made on the quantity of air strength neces-sary to bomb Germany intensively and hence I am not too optimistic about the state of knowledge." In April 1918 he strongly urged that experiments be carried out on actual buildings to see what kind of bombs worked best and in what pat-terns; he suggested that civil-engineering experts who had experience using explosives to demolish buildings be consulted; he proposed that the effects of German raids on London be carefully analyzed. He suggested that full-scale out-lines of German target factories be laid out in bombing ranges on Salisbury Plain so pilots could practice making bombing runs and a realistic prediction could be made of the results that might be expected. Only through such a scientific approach could one properly weigh all of the competing factors.

He got almost nowhere. Throughout the spring and summer of 1918, Tiver-ton, who was now serving in the new Air Ministry's Directorate of Flying Oper-ations, wrote memo after memo complaining that Trenchard's Independent Force was frittering away its efforts. The Air Ministry had by now unambigu-ously ordered Trenchard to focus his attacks on German chemical factories and iron and steel works; Trenchard as usual chose to ignore what the civilians told him. In late May Tiverton dryly observed in a memorandum to Sykes that bomb-ing has two possible objectives: "(a) A serious attempt to end the war. (b) Merely to keep our own unenlightened populace quiet." It was clear that he feared that "b" was the policy the RAF was pursuing.

By August he had grown even more frustrated:

> A systematic choice of key industries was begun upon paper and the chemical targets definitely chosen for priority. In fact the I.F. have not systematically attacked them, or even attempted to do so. No sys-tematic use has been made of the right type of bomb so that both the Technical Department and the technical officers of the Independent

Force are in despair. Although models of all targets have been offered by the War Trades Intelligence, no intensive training in knowledge of such targets has been undertaken. In fact, therefore, nothing has been done to carry out the promised policy.

A few of the British scientists, in effect calling Trenchard's bluff, put forth some bloodcurdling proposals that eerily foreshadowed what was to come in the next great war. If indeed the policy is to strike at the enemy's morale, they seemed to be saying, then here is how we could *really* do that. Mervyn O'Gorman, the superintendent of the Royal Aircraft Factory, suggested starving out the Germans by setting fire to the country's corn and hay crops from the air. The War Cabinet went so far as to order a detailed report on spreading a crop blight over Germany. Tiverton at one point put forward the idea of dropping delayed-action bombs; he reported that experiments had been carried out and proposed that "the fuse can also be arranged that any attempt to unscrew it will ensure immediate detonation." (Tiverton did allow that one drawback was that it might give the Germans the same idea; one such bomb in Piccadilly, he admitted, would "produce a situation extremely difficult to grapple with.")

Nothing came of these proposals; but then nothing came of the more realistic proposals of the scientists, either, to concentrate on key industrial targets—even when they had the full backing of the Air Ministry and the War Cabinet. Trenchard's response to all suggestions and criticisms was to maintain an air of almost majestic indifference. During the first half of 1918 only 5 percent of the RAF's bombing raids were against factories; more than half were aimed at railroad stations and sidings, in spite of the mounting evidence of the ineffectiveness and extremely high cost of such raids. One RAF study found that even when a 112-pound bomb scored a direct hit on a railroad line, it would cut only one rail, and the damage could be completely repaired by a party of ten men in an hour and a half. In the month of May alone 58 percent of the RAF's day bombers and 39 percent of its night bombers were lost.

Trenchard did not care. By September and October the Independent Force was directing 85 percent of its bombing missions against railroads and airfields. "I particularly dislike," Trenchard declared in one memorandum, "the expression 'Scientific and Methodical Attack.' "

By the fourth year of war an argument was beginning to be heard that the air weapon, far from magically delivering a knockout blow to the enemy's will to resist, had only exacerbated the terrible stalemate of the trenches. Orville Wright wrote to a colleague: "Neither side has been able to win on account of the part the aeroplane has played. Both sides know exactly what the other is doing. The two sides are apparently nearly equal in their aerial equipment, and it seems to me that unless present conditions can be changed, the war will continue for years." The only hope to end the war, Orville proposed, was if the Allies were to achieve such overwhelming superiority in the air that the Germans' "eyes can be put out."

America's entry into the war in April 1917 seemed to offer just such a possibility—or at least many Americans so believed. General Squier, the U.S. Army's top aviation officer, declared that by "building an army in the air, regiments and brigades of winged cavalry on gas driven flying horses," America would "put the Yankee punch in the war" and sweep the German lines, "smothering their trenches with a storm of lead." The rhetoric that swelled in Washington that spring was a strange combination of Wellsian futurism and old-fashioned faith in American know-how. Secretary of War Newton Baker said that a huge American aviation program would be an expression of "America's traditions of doing things on a splendid scale." Or, as Baker would later put it with ironic hindsight: "We were dealing with a miracle. The airplane itself was too wonderful and too new, too positive a denial of previous experience, to brook the application of any prudential restraints."

If America was going to assemble a gigantic air armada, then one American air officer named William Mitchell knew just the man to lead it: William Mitchell. Thirty-seven years old, "ramrod straight," Major Mitchell was aggressive, self-promoting, intolerant of inefficiency, strong-willed to the point of insubordination, and extremely capable. Though he had grown up in privilege as the son of a wealthy Wisconsin family—his grandfather had made a fortune in banking and railroads, and his father was a United States Senator—"Billy" Mitchell had sought out difficult assignments as well as intellectual challenges since running off to enlist in the Army as a private during the Spanish-American War. (He was a private for only a few days, until his father caught up with him and quickly arranged a commission as an officer.) Mitchell worked in Alaska constructing the Signal Corps' telegraph lines, was assigned to help in the reconstruction of San Francisco following the 1906 earthquake and fire, and read everything he could find on engineering in his spare time. In 1913 he was assigned to the General Staff in Washington, becoming the youngest officer ever to serve in such a post. There he learned to fly on Sundays, traveling by boat down the Potomac to Newport News, Virginia, every Saturday night and returning in time to be back in the office Monday morning.

As early as 1906 he had become a convert to the cause of aviation, writing an article for the *Cavalry Journal* in which he declared with his trademark certainty, "Conflicts no doubt will be carried out in the future in the air." In the spring of 1917 Mitchell was sent to Europe as a military observer to learn about air operations. The American declaration of war found him in Spain; he at once jumped on a train to Paris where, acting on his own authority, he opened an office as the senior American air officer on the scene, got the French government to supply him two military aides, and began issuing orders and drawing up dramatic plans. When he could not get Washington to pay attention to repeated cables that laid out what Billy Mitchell thought should be the manpower and manufacturing goals for the entire United States aviation program, Mitchell—or so he would always claim—pulled strings with the French government, and in short order French Premier Alexandre Ribot himself was cabling President Wilson directly with a proposal for a huge American contribution to Allied air forces. Mitchell's role in preparing the Ribot cable seems to have been minor at best, but in any case the French request for the United States to deliver 4,500 airplanes, 5,000 pilots, and 50,000 mechanics by the spring of 1918 caught Washington's attention.

The American General Staff thought it was preposterous. The President, the press, and the Congress did not. "GREATEST OF AERIAL FLEETS TO CRUSH THE TEU-TONS," declared the *New York Herald* when the bill approving the plan came before Congress. In July the House passed it on a voice vote. It was the largest single appropriation in United States history: $640,000,000.

Never would so much be spent for so little. Even with the help of a quick decision to manufacture only existing European designs rather than attempt to design its own models, no amount of money could overcome the fact that the capacity and capability of America's aviation industry were woefully inadequate to the task, and years behind Europe. In the previous year a half-dozen small American airplane companies had managed to deliver a total of eighty-seven machines to the U.S. government. President Wilson brought in an automobile-industry executive, Howard E. Coffin, to mobilize aircraft production and apply the American automakers' assembly-line methods to airplanes. Coffin grandiosely told Hap Arnold, then a colonel assigned to Coffin's Aircraft Production Board, that he would deliver forty thousand airplanes within a year. A few months later Coffin wrote an article for *The Saturday Evening Post* that promised, in the words of its headline, "FIFTY THOUSAND OPEN ROADS TO BERLIN!"

The Board did achieve some prodigies of production and organization. To overcome the shortage of spruce lumber that was vital to airframe construction, 27,000 soldiers were formed into Spruce Production Regiments that built logging camps and railroads in the Pacific Northwest. And the spectacular success of the American Liberty engine showed what could be achieved by applying assembly-line methods to the aviation industry. Designed to answer the need for a more powerful engine, the Liberty was the work of two leading auto-engine designers, J. G. Vincent of the Packard Motor Car Company and E. J. Hall of the Hall-Scott Motor Car Company. They were summoned to Washington on May 29, 1917, and facetiously told they would be locked in a hotel room until they came up with a workable design. With the help of draftsmen loaned by the government's National Bureau of Standards, they had it done in five days. In their design they gave as much consideration to simplicity of manufacture as performance. The first complete engine arrived in Washington for testing on July 3; the first production models came off the assembly lines in December.

Building the airplanes themselves was another matter. The decision to standardize a few chosen European models—the DH4, the Spad, the Bristol Fighter, and the Italian Caproni bomber—seemed to make much practical sense; it would take years for American aeronautical engineering and design know-how to catch up with the state of the art, and the war would not wait. But the reality was that manufacturing and design were inseparable. Translating European plans to American shop methods and converting from metric units led to countless delays. The degree of craftsmanship and meticulous hand finishing of components that was the norm in European aircraft shops, and which was inherent in the European designs, proved remarkably resistant to American mass-production methods and their reliance on relatively unskilled labor.

The basic inability of the American makers to innovate meant that, even had they succeeded in producing the British, French, and Italian aircraft designs, they would have been condemned to swift obsolescence given the rapid pace of technological change at the front. As it was, they wasted tens of millions of dollars alone in trying to adapt the foreign airframes to the homegrown Liberty engine. An order for two thousand Liberty-powered Bristol Fighters from the Curtiss company was canceled when two of the test models crashed. Curtiss was ordered to manufacture the SE5 fighter as well; he managed to produce a total of one by the time of the Armistice. Attempts to build a Liberty-powered Caproni bomber ended in failure when it proved to be seriously underpowered.

The only American-built planes that reached France were 1,400 DH4 day bombers, powered by the 400-horsepower Liberty V-12. But once these were in service, pilots found they could not run the engine at full throttle without running the risk of shaking the frame to pieces, as it had been designed for a much smaller engine. In the end, 80 percent of the U.S. Air Service's combat planes would be supplied by French manufacturers.

Mitchell excoriated the shortsightedness of the bureaucrats responsible for this mess, but without a large industry that possessed both the design knowledge and the plant capacity and experience, there was little the government could do. "No amount of money will buy time," Charles Walcott, secretary of the Smithsonian and chairman of America's newly formed National Advisory Committee for Aeronautics, had warned in April 1917, on the eve of the American declaration of war.

American-made planes or no American-made planes, Mitchell was determined that he would show what his country's penchant for "doing things on a splendid scale" could achieve. And he was determined that he would be the man to do it. When Brigadier General Benjamin Foulois was named Chief of the Air Service and arrived in France in the late fall of 1917 to take charge with a ready-made staff of one hundred officers and three hundred men, Mitchell loudly griped that as "incompetent" a bunch of "carpetbaggers" had never set foot in a war zone. None of them, Mitchell would claim with his usual outrageous exaggeration, had even seen an airplane in his life; it was typical, he fumed, of the way the conservative, hidebound Army was always trying to stifle the air weapon.

Of course, it was also a threat to Mitchell's personal ambitions, but the distinction between the interests of air power and the interests of his own ambitions was not one Mitchell would ever take pains to observe. The following summer Pershing ordered Foulois to assume the top operational air command in the field, in the newly formed American First Army; Mitchell, who had temporarily held the post, was to step down and take command of the air units of an army corps. When Foulois arrived at the office, Mitchell was his surly and abusive worst. He told Foulois there was no way a newcomer could acquire the necessary knowledge in such a short time. He announced that the entire office staff, and even the office furnishings—desks and chairs, maps on the walls, telephones—belonged to him and would move with him to his new post. The next day the two men went on an inspection tour and Mitchell continued his performance, complaining to Foulois in front of junior officers about the incompetence of Foulois's officers and staff. The day after that Foulois wrote to Pershing demanding that Mitchell be relieved from duty at once for insubordination and sent back to the United States.

But Pershing had already learned that it was sometimes necessary to overlook the hotheadedness of his airmen. Through his chief of staff he sent a diplomatic reply to Foulois acknowledging the difficulties but asking that he "meet Colonel Mitchell even more than half way" in this case: "We must put aside absolutely all questions of personality and use in the proper place the men who can give the results we seek. If Colonel Mitchell can do this we cannot afford to put him aside." Foulois swallowed his pride, and in late July 1918, in a gesture of even greater magnanimity, stepped aside entirely to allow Mitchell to have his place as the top tactical commander of all United States air forces.

A few weeks later, the American First Army was at last given its own sector of the front, which Pershing had been pressing for from the start. It was the Saint-Mihiel salient, a twenty-four-mile-long bulge in the lines that the Germans had held since their 1914 drive toward Verdun. The Americans would be given the job of straightening out the bulge. It was Pershing's big chance. It was also Mitchell's.

Planning for the air operations that would support the American attack began at once. Mitchell would have under his command the largest air force ever assembled, 1,481 planes, half of them French squadrons. In addition to sweeping aside German fighters so the Allied artillery-spotting aircraft could operate freely, the concentrated Allied forces would conduct relentless attacks against enemy troops and artillery positions on the ground and the supply trains that supported them. Using fighters as "trench-strafers" had proved an increasingly pivotal role for the Allied air forces throughout 1917 and 1918. At the Battle of Cambrai in November 1917, which had seen the first major use of tanks, British DH5s and Sopwith Camels had been fitted with bomb racks to neutralize what was already seen as the major threat to armor on the battlefield: direct fire from enemy artillery. At Cambrai the British fighter-bombers spotted and swiftly neutralized two batteries but failed to notice a third—which proceeded to knock out sixty-five tanks on the first day of the battle. But by the summer of 1918 the Allied system of aerial counter-antitank support, in the form of machine-gun fire and 20- or 25-pound bombs dropped from massed forces of small, fast airplanes like the Camel, the SE5, and the Bristol Fighter, was having a devastating effect on German resistance.

Air commanders were also becoming aware that the psychological impact of air attacks on troops could be immense: ironically, far greater than the impact on civilians subjected to "morale" bombing. Strafing attacks on German troops at the Somme had at times caused "a condition akin to panic," according to a postwar German assessment. Panicky American soldiers were so given to opening fire on any airplane that passed overhead that a color leaflet was quickly printed depicting the Air Service's roundel insignia and advising the troops not to shoot at their own airplanes.

For the Battle of Saint-Mihiel, Mitchell wanted to be sure that the troops would have all the help they could get. Despite a soaking rain that had turned the

Incidents in which panicky soldiers opened fire on aircraft were common on both sides. American authorities issued this leaflet to inform troops how to recognize their own airplanes. *(Author's photo/U.S. Air Force Museum)*

airfields to a sea of mud, a strong southwest wind that made formation flying exceedingly hazardous, and low clouds that cut visibility to less than a mile,

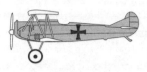

Fokker D.VII
single-seat fighter
engine: 160-hp six
    cylinder
top speed: 120 mph
armament: 2 synchronized
    7.92mm machine guns

Mitchell's force went into the air with the start of the offensive the morning of September 12. Heavy guns had been firing at the German trenches since 2 a.m. Nearly half of Mitchell's force was made up of fighters, and they faced a formidable German force equipped with the new Fokker D.VII. Losses on the first day were serious. But the Allied fighters were still able to deliver a steady pounding to retreating German columns of troops, guns, and transport caught on the road from Vigneulles to St. Benoit. The American ace Eddie Rickenbacker recalled his pass over the half-mile-long convoy:

> Dipping down at the head of the column I sprinkled a few bullets over the leading teams. Horses fell right and left. One driver leaped from his seat and started running for the ditch. Half-way across the road he threw up his arms and rolled over upon his face. He had stepped full in front of my stream of machine-gun bullets. All down the line we continued our fire. . . . The whole column was thrown into the wildest confusion. Horses plunged and broke away. Some were killed and fell in their tracks. Most of the drivers and gunners had taken to the trees before we reached them.

Mitchell repeated these tactics in the Meuse-Argonne offensive the following month. On October 9 he sent two hundred bombers and a hundred fighters to attack German troops on the ground in the largest single daytime raid.

Adding to the psychological pressure from the air was a barrage of leaflets dropped on the troops. One showed a map of the Saint-Mihiel advance along with a chart of the growing might of the American Army: "The salient which the Germans held for four years was taken by the Americans in 27 hours," it read. Another included a copy of Pershing's order for humane treatment of POWs and assured any Germans considering surrender that "the rations of Prisoners of War are *exactly the same* as those of American troops. White Bread, meat, bacon, lard, marmalade, tobacco, etc. are included in the daily rations of the American soldiers and, therefore, are also in the rations of the prisoners." As an American report noted: "Prisoners came over in increasing numbers with the American leaflets concealed on their person, in some cases held in their hands as passports." Hindenburg personally issued a proclamation denouncing the leaflets and ordering punishment for soldiers who read them or failed to turn them over to an officer. The Austrian high command threatened to hang aviators captured carrying propaganda.

A renewed American offensive on the Meuse on November 1 was again supported by masses of low-flying aircraft that machine-gunned German positions which had survived a pre-assault bombardment that included three-quarter-ton shells fired from fourteen-inch naval guns mounted on railway cars. The German lines actually broke and fled. The Allied advance was so rapid that in the confusion Douglas MacArthur, commanding an infantry brigade, was briefly taken prisoner by American sentries who thought he was a German officer. Behind the lines the German Navy and twenty thousand garrison troops mutinied. Troops in Berlin deserted en masse. On November 9 the Kaiser abdicated; the next day the Armistice was signed; and at 11 a.m. on November 11 the guns fell silent.

It has often been said that the effect of the First World War on military aviation was greater than the effect of military aviation on the war. Aviation played a vital but never a decisive role, though it came the closest in the last great battles, when the Allies gained overwhelming air superiority through concentration and force of sheer numbers. Had the war gone on another year, some of the controversies over doctrine, command, and organization of air forces that would embroil the world's militaries in the interwar years might have been settled in the clarifying purgatory of combat. The war had provided enough evidence to erase any doubt that mastery of the air was indispensable to the operations of armies. But when it came to drawing more precise lessons, things became much hazier; exactly how aircraft could best be employed to attain that essential mastery in future conflicts, and to what end that mastery could best be used, were questions that would attract bitterly competing theories. Partisans could point to evidence from the incomplete battlefield experiences of the Great War to support widely divergent conclusions.

However much airmen complained that they had been neglected, however many times they told the story of the blundering cavalry colonel who had insisted on inspecting his airfield on horseback and went into a rage when the planes had scared his horse, the war had been an unvarnished triumph for the air services in the internal bureaucratic battles they spent so much of their time and energy fighting. The Allied airmen may not have single-handedly vanquished the Hun, but they had delivered a walloping blow to the forces of stodgy conservatism on their own side. The three major belligerent powers, France, Britain, and Germany, had each begun the war with a hundred or fewer operational aircraft apiece on the Western Front, and four years later over that same front they were each fielding air forces of two to three thousand aircraft. No military officer of 1918 would be taken seriously if he suggested, like Marshal Foch in 1913, that the contribution of the airplane to combat was *zéro*.

Even more important, the war catapulted airmen into public and professional prominence, high rank, and authority. From his command post at Souilly, Billy Mitchell "practically held court," Hap Arnold recalled. Wearing a flamboyant uniform he had designed himself, issuing a steady stream of orders to aides, driving himself about the countryside at breakneck speed in his personal Mercedes (reputed to be the fastest car in France), Mitchell was a celebrity. "American, British, and French fliers—and generals—streamed in to see him," said Arnold. On the night of the Armistice, Mitchell's car was mobbed in Paris by an exuberant crowd of French pilots who pushed aside a cordon of gendarmes that had closed the boulevards to vehicles; Mitchell's admirers paraded his car up and down the packed streets screaming over and over, "Vive nôtre Général Américain!"

The aircraft industry was another front where the forces of military aviation had carried the field. Whatever corners had been cut to get new machines produced, the war had done for the aircraft and aero-engine industries what nothing else could have done in four years. The war created a critical mass of manufacturing capacity, expertise, and experience. France and Britain had each produced a staggering fifty thousand aircraft during the war. France alone had produced nearly a hundred thousand aero-engines. The general-purpose observation craft of 1914 had rapidly evolved into a series of specialized types that spanned the entire spectrum of air operations: offensive fighting, ground attack, reconnaissance, day bombing, night bombing, defensive interception, and escort.

And at least indirectly, the war had spurred a revolution on a third front that would prove to be of the greatest significance to the future of military aviation— even though it would take a decade for the aviation world to realize that a revolution had occurred. The Allies by 1918 held an insuperable lead in quantity and had mobilized an unstoppable edge in industrial capacity, but it was the Germans who had the lead in quality. The Germans had done something no one else had: they had made aircraft design a science. At the time it seemed a footnote to the conflict. In years to come it would loom as a major consequence.

The Germans' secret weapon that had made this possible was a "dark, heavy, medium-sized man with a goatee." When in 1909 British Secretary of State for War Richard Burdon Haldane had spoken admiringly of the example set by "my old university at Göttingen" in creating a chair in aeronautics, he spoke to a future that even he could not have anticipated. Ludwig Prandtl, the scientist chosen for the chair, would come to be known to later generations as the father of aerodynamics. His students Theodore von Kármán, Max Munk, and Adolf Busemann would become leaders in aerodynamic research in Germany and America in the 1920s, '30s, and '40s. The mathematical methods he developed for calculating the lift and drag of wings are still in use today.

Prandtl was apolitical, and innocent about worldly matters to the point of naïveté; the coming of the war and the ending of the war seem to have made little difference to him. He was a pure scientist, but a pure scientist with a practical mechanical sense that led him to focus on the connection between theory and engineering—the discipline that is today known as engineering science, the search for scientifically rigorous, first-principles solutions to matters of engineering importance.

Von Kármán recalled his teacher as a man whose absorption with his work could lead him to be at turns aloof and impulsive. He was "a tedious lecturer," since he could never make a statement without immediately qualifying it. Von Kármán's sister was once seated next to Prandtl at a formal dinner at an international scientific conference and asked him a question about the mechanics of flight; she watched in amazement, and then in embarrassment, as Prandtl demonstrated a point by folding his menu card into a paper airplane and launching it into the air. It landed on the shirtfront of the French Minister of Education. When Prandtl found himself nearing the age of forty, he decided he should get married but didn't know how to go about finding a wife, and finally hit upon the solution of writing to the wife of his former professor and asking for the hand of one of her daughters. He did not specify which one. The family decided that the older daughter, in her late twenties, would be the best match, and by all accounts it was a successful marriage.

As early as 1904 Prandtl had begun to lay the theoretical groundwork for twentieth-century aerodynamic science. By 1915 he had conducted the first wind-tunnel tests since the Wright brothers' that systematically examined the performance of different airfoil shapes. By 1917 he and his assistant Albert Betz had worked out the basic theory for calculating the lift and drag of wings of various planform shapes.

Up until this time the design of airplanes remained almost a black art. It is no exaggeration to say that Prandtl's work made it a science. His 1904 paper on boundary-layer theory provided a means of calculating the drag created by the friction of air moving over a surface of any given shape. His 1915 wind-tunnel tests revealed the counterintuitive fact that much thicker airfoils than were commonly used actually generated a much greater lift-to-drag ratio. And his 1917 work on the so-called lifting-line theory established how the lift-and-drag coefficients of a wing needed to be adjusted to take into account the wing's aspect ratio, its ratio of length to fore-and-aft width.

The impact of his work on the design of German fighter aircraft was immediate and stunning. The Fokker D.VII, which had harassed Allied fighter pilots from its introduction in combat in April 1918 to the end of the war, incorporated all of these advances in basic aerodynamic science. Lifting-line theory unmistakably proved that wings with a higher aspect ratio generate more lift with less drag, and it explained why by focusing on the aerodynamic role played by the tips of the wings. Whenever lift is generated, the pressure above the wings is lower than that below. But this means that the airflow near the tips of the wings tends to curl up around the tip, forming a tornado-like vortex that spirals back from each wing tip. These vortexes of turbulent air create a low-pressure wake that exerts a backward tug on each wing: drag. A long wing, however, minimizes the percentage of the total wing surface that is affected by the tip vortexes.

The Wright brothers' wind-tunnel tests in 1901 had shown exactly the same result: high-aspect-ratio wings have a better lift-to-drag ratio. But they had no theoretical explanation for why this should be so, and their crucial experimental finding seems to have almost vanished from scientific knowledge. The Wright Flyer had an aspect ratio of 6.4; most First World War aircraft had much shorter, stubbier wings—the Sopwith Camel, for example, had an aspect ratio of 4.1, while the Spad XIII's was even lower, about 3.7. Anthony Fokker designed his D.VII with considerably longer, narrower wings, having an aspect ratio of 4.7.

Prandtl's discovery that thick airfoils with a wide, blunt leading edge performed better than thin airfoils was even more of a departure from accepted practice. The Wrights' wind-tunnel tests had suggested that thin airfoils were best; this also seemed to make intuitive sense, as a thin, knife-like edge would seem better able to slice through the air without meeting resistance. But the small scale-model wings the Wrights had used, and the low speed of the air moving through their tunnel, had in fact disguised a devastating flaw in thin airfoils; when scaled up to actual flight conditions, a sharp leading edge can cause what in modern aerodynamic terminology is known as massive flow separation: instead of smoothly flowing in parallel to the contour of the airfoil, the airstream sharply diverges, leaving a large pocket of stagnant air behind the wing. Like the tip vortexes, such a low-pressure void exerts a huge drag force on the wing, resisting its forward motion through the air.

The Göttingen laboratory quickly designed a series of new airfoils that were a revolutionary departure from the existing state of the art. The Göttingen pattern

Wright 1908

Royal Aircraft Factory 1915

Göttingen 1917

Ludwig Prandtl at Göttingen made the counterintuitive discovery that thick airfoils generate less drag and more lift than thin ones. Thick wings coincidentally made it possible to substitute internal bracing for high-drag external spars and wires.

was first used in Fokker's famous Dr.I triplane, flown most notably by the famous ace Manfred von Richthofen; it also was the major secret of the D.VII's spectacular performance. The Göttingen airfoil not only generated more lift than the thin British and French airfoils but generated *much* more lift at a high angle of attack. That made the wing less susceptible to stalling, and thus made the Fokkers much more maneuverable in an acrobatic dogfight, as well as much safer for inexperienced pilots to fly. It also meant that planes using the Göttingen airfoils could greatly outclimb the Camel and Spad, even with a lower-horsepower engine.

The thick airfoil shape offered another considerable advantage: it allowed the wings to be completely, or almost completely, self-supporting, since the thicker section made room for a strong internal beam, doing away with external wire bracing and struts. The Fokker triplane had at first been built without external struts or braces at all, but a single strut was later added—purely for psychological reasons—to reassure pilots who could not believe the plane's cantilevered wings would not snap off in midflight. The D.VII had simple N-shaped struts but no wire bracing.

The Allies may have won the World War, but in the Armistice they paid a bitter compliment to the scientists and engineers who had bested them in the aeronautical-engineering war. Article 4 of the Armistice listed the war materiel that Germany was required to turn over at once to the Allies. On the list a single airplane was mentioned by name: The Fokker D.VII. Some were turned over. Anthony Fokker himself, however, fled to Holland with six trainloads of parts, including 120 D.VIIs that were under construction in his factory when the war ended, and more than 400 engines. The Versailles Treaty forbade Germany to retain an air force; the Allied Control Commission in 1920 went further and ordered German factories to cease aircraft production altogether. What the victors could not eradicate through the solemn instruments of international law was the scientific knowledge and technical expertise that Germany now possessed, an unbottleable genie if there ever was one.

PART TWO

VERSAILLES TO MADRID, 1919-1939

5

LESSONS LEARNED AND MISLEARNED

It took all of two weeks for the airmen to suspect that peace was going to be hell. Billy Mitchell expressed the obligatory sentiment of relief that the killing and suffering had ended, but at the same time could not conceal his disappointment: "I was sure that if the war lasted, air power would decide it," he said. Leaving a celebratory dinner with the 94th Aero Squadron at Longuyon in late November, Mitchell felt he was saying good-bye to America's air force: "It would soon fade into oblivion, and probably never be resuscitated until the time came for another war. Then we would start out again by making terrible mistakes and perhaps be defeated before we began," he wrote in his memoirs.

The great strides that military aviation had made during the war made the march back to peace all the more heartbreaking. In the United States Army the ground force had swelled to twenty times its peacetime size, but the Air Service had grown two hundred times. Shrinking back to "normal" peacetime levels, Mitchell asserted, would mean the end of the air arm altogether.

But it was the impending collapse of the aircraft industry that really threatened to drive a stake through the heart of postwar military aviation. The Allies had purchased more than a billion dollars' worth of aircraft during the four years of fighting. Three days after the Armistice, the United States government cancelled $100 million worth of airplane contracts. American aircraft production plummeted from 14,000 airplanes in 1918 to 328 in 1920. Similar cancellations were announced in France and Britain, and within a year hundreds of thousands of skilled aircraft workers in Europe and America were out of a job. The workforce of the French aircraft plants fell from 183,000 in 1918 to 5,200 in 1920. Renault let its entire aviation design staff go. The Hispano-Suiza engine factory gave up on the airplane business and started building luxury automobiles. The small civilian market that might have at least cushioned the blow was all but wiped out by a flood of war-surplus inventory. The U.S. Army drew up lists of planes and engines that it sold off at bargain-basement prices: twelve-cylinder Liberty engines that had cost $4,000 apiece went for $250.

In simplest terms, the problem came down to money. "Not a dollar is available for the purchase of new aircraft," the head of the U.S. Air Service reported in late 1919. Nor was there any reason to hope that would change anytime soon. Appealing to the public to hand over millions of more dollars to arms makers, whom many blamed for the war, was hardly a winning political strategy in 1919. The entire world was demobilizing as fast as it could, with the United States at the head of the pack. The country was eager to put the war behind it, and Europe back where it belonged in the national psyche, at the far side of a wide ocean. Fifteen thousand troops were being sent home a day; why should the Air Service be singled out for special dispensation? And in fact the aircraft companies had become particular villains in the American public eye. Congress, looking for someone to blame for the failure of the American manufacturers to deliver on the great promise of putting the "Yankee punch in the air," held hearings that vilified the "aircraft trust" and unearthed tales of bloated profits, shady dealings, and shoddy workmanship.

But if the aircraft firms were villains, Billy Mitchell and his airmen were heroes, and Mitchell had the instincts of a natural-born showman to exploit that heroic image to the hilt. He pulled strings to have himself ordered back to Washington and at once began to show his aptitude in the game of bureaucratic intrigue and publicity. His particular genius was to cast what was manifestly a demand for special treatment for the air arm as an innocent appeal for mere equality.

Mitchell insisted, at least at first, that he would conduct his campaign on "a high plane." In 1919 the leading hothead in the Air Service was the man Mitchell had sparred with in France, Benjamin Foulois. That fall, sounding more like Mitchell than Mitchell himself, Foulois got himself into immediate hot water by testifying before Congress that the Army General Staff had "utterly failed to appreciate the full value" of air power. "Based on practical experience in Army aviation since its birth in 1908," Foulois declared, "I can frankly state that the War Department has earned no right or title to claim further control over aviation." Foulois decided he had better get out of town fast after that, and volunteered to serve as military attaché in Berlin. He spent the next four years there, among his other duties becoming a drinking companion of the German aces Hermann Göring and Ernst Udet.

Mitchell's flirtation with tact and restraint lasted about as long as Foulois remained on the Washington scene. Soon Mitchell was back in form. He publicly railed against the decision to put an infantryman in charge of the Air Service—a post he, not incidentally, thought was due him. Mitchell was instead given the job of assistant chief, and was tactless enough to promptly send his boss a memorandum declaring that infantry, cavalry, and artillery officers were "entirely incapable of handling air units"; in fact, he said, "practically all our trouble in the past has been due to officers with little or no knowledge of air matters being detailed to positions of trust and responsibility in the Air Service, for which they were not fitted." He urged Congress to require by law that the Air Service be staffed entirely by officers with flying experience during the Great War; non-fliers should be "entirely excluded." He ignored the chain of command to lobby the Secretary of War directly for a multimillion-dollar budget for new aircraft.

But most of all Mitchell knew how to make headlines. He had a perfect ear for the newspaper clichés of the day. His battle was always against the "bureau-cracies," the "establishment," the hidebound forces of "conservatism" who, he explained, had *always* opposed aviation, had *never* grasped its importance.

And he knew how to stage the sort of stunts that newspapers just ate up. In fall 1919 two Army pilots flew a Martin bomber 9,823 miles around the perime-ter of the United States. Mitchell constantly pushed his pilots to set new speed and altitude records. Hap Arnold even accepted, on behalf of the Army, the chal-lenge of a San Francisco newspaper that had speculated about whether pigeons or airplanes were faster. Flying a DH4 biplane, Arnold arrived in Portland, Ore-gon, with a coop of carrier pigeons, picked up the governor of Oregon, and the two men then raced the pigeons back to San Francisco. The newspaper posted regular bulletins on the progress of the competitors. Bookies did a brisk business among the crowds who jammed Market Street waiting for the latest report. Arnold won.

Mitchell's blend of showmanship and tell-it-like-it-is brashness struck an immediate chord with the public. His self-cultivated image of the Young Turk fit perfectly with the heroic picture of the airman that the Great War had already consecrated. It also fit perfectly with another image that Americans had always loved: the plucky individual who defies the rules, tells off the boss, does what's right and the consequences be damned. If the generals thought Mitchell behaved at times like "a spoiled brat," as one put it, the public loved him all the more for his thumbing his nose at authority. And if his unceasing claim to speak for the forces of progress doing battle against outmoded tradition was hyper-bolic, it nonetheless encapsulated everything people still felt about the wonders of flight.

The greatest ally of Mitchell and his fellow air-power proponents, in fact, was the romantic image of Great War aviation that only intensified as it passed into legend in the 1920s and 1930s. While Mitchell bemoaned the fact that avia-tion had been denied the chance to prove its "decisive" potential, this would have come as news to many members of the public who were already convinced that aviation *had* been decisive in the war. At least that was the message of a string of Hollywood movies that did much to shape the myths and memories of the war. Hollywood did not invent the romantic legend of the Great War fliers, but it had a knack for distilling legends into a few emotionally powerful images and phrases that would stick in people's minds. That was nowhere more the case than in the First World War flying movies.

Movies were *the* medium of popular culture in America between the wars. By the 1930s, even in the midst of the Depression, the number of tickets sold by movie theaters each week equaled 60 percent of the entire population. It was the time "When the Movies Really Counted," as the historian Arthur M. Schlesinger, Jr., once put it in an essay of that title. The flying-ace movies were often high-budget spectaculars that drew huge audiences with their dramatic midair collisions, dogfights, Zeppelins coming down in flames, and airplanes crash-landing into trees. Howard Hughes's 1930 *Hell's Angels* was the most expensive movie ever made at the time; he used 137 pilots in filming the climactic dogfight scene. Hollywood operated a veritable air force of war-surplus and captured German planes. One German Pfalz D.XII fighter logged many more hours flying in Hollywood movies than it had in the war itself.

The Hollywood portrayal of the war in the air inevitably focused on its most naturally dramatic aspect, the aerial combat between fighter planes, and the plots were mostly the melodramatic standards found in all early movies: love triangles, rivalry between brothers, the coward who must overcome his fears. None were propaganda movies in any normal sense of the term; they were not deliberate efforts to promote the cause of air power or air-mindedness. It is a stretch to claim that the movies shaped air policy in any direct way. But implicitly, the Hollywood versions reflected and amplified several themes that formed an essential part of a widespread public support for aviation that had emerged from the war.

One theme was that aviation somehow did defy the grim rules of modern combat. In both *Hell's Angels* and the 1927 silent movie *Wings,* the hero single-handedly paves the way for a battlefield breakthrough by bombing, with unerring precision and devastating effect, a supply dump or machine-gun nest holding up the Allied advance. Not only does air power decisively end the stalemate of the trenches by clearing the way for the big push, but acts of chivalry abound among the airmen on both sides. In *Wings,* the hero's machine gun jams in the midst of combat. "But there was chivalry among the knights of the air," the title card declares, and the German breaks off the fight with a salute. In *Suzy,* a German pilot flies overhead during the funeral ceremony for his French rival and drops a memorial wreath.

Even two films with a strikingly antiwar message, *Ace of Aces* and *The Dawn Patrol,* portray aviation as a noble field of combat that escapes the taint of futile slaughter that colors the war as a whole. In *Ace of Aces,* the hero, Rocky Thorne, is a sensitive, well-to-do, soon-to-be-married sculptor (played by the silent-film star Richard Dix) who announces he will not be swayed by "flag-waving hysteria" to join the "lemmings" marching to certain and pointless death. But when his fiancée goads him into enlisting by calling him "yellow," Rocky joins up to be a fighter pilot. Quickly overcoming an initial reluctance to pull the trigger on a German plane he has lined up in his sights, Rocky soon becomes a mechanical, cold-blooded killer, methodically racking up his score while growing ever more callous and cynical about love and life.

Yet even as he mocks the idea of chivalry in the air ("Which knight of the Round Table are you?" he retorts to a fellow pilot who accuses him of acting in an "unsporting" manner toward the Germans), Rocky's very cynicism is used to drive home the message that being "sporting" *is* the norm—a norm that he alone is defying. Rocky's crisis, when he begins to see what the war has done to him, comes when he shoots down a German cadet piloting an unarmed plane, which he has flown over the American airfield on a typically chivalrous mission—to drop a note informing them that one of their comrades has been taken a prisoner of war.

The Dawn Patrol takes an even darker view of war and what it does to men. As the story begins, a high-strung RFC squadron commander (played brilliantly by Basil Rathbone) is cracking under the strain of having to send woefully undertrained replacement pilots to near-certain death day after day. "You know what this place is?" he shouts. "It's a slaughterhouse, and I'm the butcher." The squadron's top pilots (Erroll Flynn and David Niven) are feeling the strain, too, but even in their disillusionment ("it's a big, noisy, stupid game that doesn't make any sense at all," Flynn's character declares) they continue to play the game with wild abandon. When the famous German ace "von Richter" moves in across the lines and issues a challenge to the British airmen, the pair immediately disobey orders and take off to strafe the German air base (and to return the pair of trench boots that von Richter had insultingly dropped along with a note suggesting that the British flying officers would be safer on the ground). In the final scene, Flynn, now promoted to squadron commander and enduring the same torments of his predecessor, disobeys orders once again and takes it upon himself to carry out the dangerous solo bombing mission his squadron has been assigned. On the way back (after bombing his objective with the usual deadly accuracy), he is shot down by von Richter himself, to whom he gives a final salute as he spirals down to his death.

Throughout *The Dawn Patrol* the war is portrayed as senseless, futile, the doing of "some criminal idiots sitting around a table." Yet the pilots defy their fate through the heroism they unflinchingly show, through their high spirits and valor, and even by the swiftness of the "clean" death they meet in the air, so marked a contrast to the protracted agony of the trenches.

More generally, all of the Hollywood portrayals reinforced the theme of the airman as a new breed, a modern man unencumbered by the hoary traditions and rigid discipline of all those dull-witted, red-faced, by-the-book generals who sent waves upon waves of soldiers over the top to be cut down by machine-gun fire. The fliers, even as they live up to the traditional military values of heroism, are notably "unmilitary," a wild, hell-raising, undisciplined bunch, a slap in the face to the traditional army and the old ways of thinking. To a world still numb from the slaughter of the Great War, these were the kind of men one wanted to have in charge if the unthinkable were to happen again.

Not long after the Armistice, a small commission of British experts began showing up in cities, at factories, and at railroad stations across western Germany. They had come to dig up whatever firsthand evidence they could on the effects of the bombing campaign carried out by Independent Force, and its predecessor, the 8th Brigade, during the last year of the war. The air-intelligence staff had already pored over maps and aerial photographs, and in many cases had been able to plot the exact location where every single bomb had fallen during a British raid. The British commissioners now sought to fill out that picture by interviewing factory workers and directors, examining municipal records and photographs that had been taken by the local authorities at the time of the raids, and studying countermeasures that had been instituted by the Germans.

The results were not encouraging. In fact, they were abysmal. The heavily bombed Badische works had almost completely escaped damage; a large proportion of the bombs had dropped between buildings. Blast furnaces had suffered a few hits, but most of the damage had been superficial: masonry dislodged, roofs and windows broken. "Only in a very few cases were vital points hit," the British investigators reported. The aim of the attackers was "very erratic both day and night." Workers were puzzled as to what portion of the works the British had been trying to hit. "Generally speaking, the directors did not attach much importance to air raids. . . . Damage had invariably been repaired at once without any difficulty, and in very few cases had any stoppage of work resulted." Indeed, the director of the Badische works "expressed surprise" when asked "if the works had been stopped as a result of bombing"; he said he thought it would be "an impossible undertaking" to shut the factory down by aerial attack.

Despite the heavy concentration of attacks on railroad lines and stations by the Independent Force, those raids seemed to have had equally little impact. Just hitting the target was hard enough: a survey carried out by the French Army, when it reoccupied the Briey Basin, found that of 1,300 bombs dropped on Thionville station, only 100 had landed on the target at all, and a quarter of those had failed to explode. The British survey found, furthermore, that even when the bombs did fall in the right place and did manage to explode, the resulting interruption of rail traffic was minimal: "The enemy worked out large schemes for diverting the traffic in case of necessity" around damaged points; trains could be rerouted at "a moment's notice." On the rare occasions when a truly vital junction was hit, the movement of trains was delayed for at most forty-eight hours while the damage was repaired. Railroad officials, signalmen, and linesmen were interviewed, and "the consensus of opinion of our bombing by the German officials is summed up in the word 'annoying.' "

A number of simple German countermeasures to aerial attack had proved remarkably effective. Dummy blast furnaces consisting of lights and "a certain amount of masonry" had been constructed a thousand meters or so from the real plants at Bous and Völklingen and had successfully drawn off British bombers. The real blast furnaces were reinforced with earthen roofs that afforded considerable protection from the bombs that did hit.

This was exactly the sort of objective data that scientists like Tiverton and Tizard had been insisting on during the war, and it was exactly the sort of data that could have been used as the basis of scientifically sound operational concepts for the future. But the written report that the survey commission issued suffered from an almost schizophrenic disconnect between the detailed evidence it had gathered and its overall conclusions. "The accumulation of evidence from all quarters of Germany," the report declared on page one, "provide indisputable proof of the efficacy of air raids during the period under review."

It was perhaps just human nature to accentuate the positive; no one wanted to come right out and say that so much effort, and so much expenditure of British blood and treasure, had been nothing but a colossal waste. It was especially human nature to do so when that was what one's boss unmistakably wanted to hear. On the first of January 1919 Trenchard had already published an official statement on the activities of the Independent Force, and it was clear that his belief in the effectiveness of morale bombing was well on its way to becoming an unquestionable tenet of RAF policy. "At present the moral effect of bombing stands undoubtedly to the material effect in a proportion of 20 to 1," he asserted in a statement that combined dogmatic certainty, numerical precision, and utter meaninglessness. He gave no clue as to what this ratio meant—20 to 1 in terms of *what* exactly—nor did he give any clue as to how he had arrived at this mathematical principle. What he did make clear was that this was the policy he had pursued, and that it had worked.

The British bombing survey commission had gotten the message. And so while all of its hard evidence pointed to the ineffectiveness of the Independent Force's campaign, it dutifully reported every scrap of soft anecdote, speculation, or theorizing that established the "moral" effect of the campaign. "It is not possible to estimate to what extent the mental state of the workmen" affected German war production, they noted; "but there is no doubt that in some cases it is very considerable." The report speculated that workers were constantly worried that they might be bombed. At one point it even seemed to argue that *not* bombing factories was more effective than bombing them: "It would not be an exaggeration to say that more moral effect with consequent loss of production was caused by alarms than by the actual dropping of bombs on the works." As further proof of the effect on morale, the commissioners quoted letters citizens had written during the raids on cities stating that the bombing had been frightening or had gotten "badly on one's nerves."

An American investigation in occupied territory came to exactly the opposite conclusion about the effectiveness of targeting enemy morale. The British and French bombing of cities, including Cologne, Frankfurt, Bonn, and Wiesbaden, had been "not productive," the Americans found: "This investigation has decidedly shown that the enemy's morale was not sufficiently affected to handicap the enemy's fighting forces in the field. The policy as followed by the British and French in the present war of bombing a target once or twice and then skipping to another target is erroneous."

Trenchard's embrace of morale bombing during the war had begun reluctantly. But upon his return as Chief of Air Staff in 1919 he became increasingly insistent on not just its effectiveness in the war but its "correctness" as a matter of fundamental principle. The gospel was to be repeated over and over in Air Staff memoranda. "The ultimate objective of air attack," explained Trenchard's staff in a typical incantation of the formula, "is largely achieved by influencing the morale of the enemy population and the maximum effect will be achieved by aerial bombardment of legitimate objectives in his great centres of production."

There were several powerful forces at work turning what was at best a theory—and a theory that, on the face of it, had a lot going against it—into dogma. One was that morale, at least in the broadest sense, really had seemed to be the key to Germany's sudden collapse in the fall of 1918. Had her soldiers and sailors not mutinied, had the public not risen up and demanded a civilian voice in the Kaiser's military-dominated government, the Germans could well have gone on fighting. The war had also seen the mobilization of the entire industrial society of the warring nations. It seemed clearer than ever that the true source of a combatant's power was the stamina of its populace and the productive capacity of its economy.

Throughout the 1920s Trenchard and his staff penned memorandum after memorandum stressing that civilian society was not only the fundamental source of an enemy's power but also its Achilles' heel. "The Army policy was to defeat the enemy Army—ours to defeat the enemy nation," Trenchard opined at a staff meeting. "The Army only defeated the enemy Army because they could not get at the enemy nation." In another paper (whose unusual articulateness suggests it was ghostwritten by one of the younger and brighter members of his staff), Trenchard explained that attacking the enemy's armed forces, even its air force, is not only more difficult but less effective in wearing down its military strength than is striking directly at the source of these arms:

> The rifleman or the sailor is protected, armed and disciplined, and will stand under fire. The great centres of manufacture, transport and communications cannot be wholly protected. The personnel . . . who man them are not armed and cannot shoot back. They are not disciplined and it cannot be expected of them that they will stick stolidly to their lathes and benches under the recurring threat of air bombardment.

The General Strike of 1926 only reinforced the view of Britain's ruling elite that order and stability in modern industrial society were held in place by only the weakest of props, which might give way under the slightest pressure. "Who does not know that if another great war comes our civilisation will fall with as great a crash as that of Rome?" Prime Minister Stanley Baldwin asked that same year. Whether it was East End Jews, or socialist revolutionaries, or the working classes as a whole that were regarded as the weak reed, it seemed alarmingly clear to Britain's rulers that the urban social order could not be counted on to hold up in a crisis.

Striking directly at the enemy's centers was not only more effective than battling its armed forces on land and sea, according to the Air Staff, but it was cheaper in terms of lives and money. To those who challenged the morality of terror attacks directed against the enemy populace, the RAF commanders retorted (much in the vein of the air-power visionaries of two decades earlier) that anything that would shorten war would make war more humane; it was in any case infinitely preferable to "morons volunteering to get hung in the wire and shot in the stomach in the mud of Flanders," as one RAF commander later put it savagely and succinctly.

Another force behind the dogma was the consideration that if Trenchard's theory was correct, it was an unassailable case for the continued independence of the RAF—and for a hefty chunk of the defense budget. The debt of the Great War weighed heavily on British finances and politics in the postwar years, and cutting the size of the armed forces seemed the easiest way to trim the budget while sparing popularly supported domestic programs. As the newest and smallest of the armed services, the RAF knew it was the most vulnerable. In the years after the war, the British Army and Navy had repeatedly questioned the need for a separate air force, arguing that it led to wasteful duplication of staff and needless expenditure since the air force was merely a "supplementary" force that supported sea and ground operations. In January 1919 Winston Churchill was named Secretary of State for both War and Air with the clear expectation that the two services would shortly be united as well, the RAF losing its autonomy and reverting to its previous status as a part of the Army.

Trenchard bristled, pointing out that he had worked hard to build up an esprit de corps, and all would be lost if the RAF were now to be broken up. More to the point, the Air Staff argued that subordinating the air force to the Army would destroy its ability to conduct strategic operations, since the Army would inevitably demand aircraft to directly support ground operations—which was the last thing the RAF wanted to contemplate now, recalling the murderous attrition rates it had suffered in trench strafing during the war; throughout the Battle of Cambrai, for example, 30 percent of fighters sent off on close-support missions had been shot down each time they crossed the lines.

And finally, there was the argument that if Britain herself lay vulnerable to the aerial striking force of a foe, then she had no answer but the threat of reprisal on like terms. This argument was far less dogmatic or self-serving than the others. The Great War had apparently proven that little could be done to stop bombers from getting through. Britain's home defenses had shot down only 21 aircraft and 9 airships over England during the whole war, 18 of these by fighters (11 aircraft and 7 airships) and 12 by antiaircraft guns (10 aircraft and 2 airships). AA guns had proved particularly ineffective: of the 4,600 British airplanes shot down over the Western Front, antiaircraft guns had brought down about 5 percent; likewise only about 3 percent of the German planes accounted for on the Western Front had succumbed to AA fire.

At best, fighter interceptors and guns could make things harder for a foe, forcing his bombers to fly higher and take evasive action; as one Air Staff paper noted, if Britain had no defenses at all, then an enemy could simply hire a commercial airline company to bomb London. But in general the notion of defense was viewed as "insidious." Building up defenses would inevitably come at the expense of the retaliatory force, which was the only thing that could have a real effect on the enemy. And futile defense measures, rather than bucking up civilian morale by showing that the air force was doing something to protect them, might have the opposite effect by raising false hopes that would then be dashed. The Air Staff argued that military morale might in turn suffer if fighter pilots were blamed and vilified for their inevitable failure to stop enemy bombers.

In 1922 the theoretical "enemy" that most worried the British government was France (usually referred to elliptically as the "Continental Air Menace"). Amidst the ashes of war, France was indeed the only continental power that remained. The RAF had a total of three home squadrons. France's "striking force" had some 600 airplanes; and extrapolating from casualty statistics of the Great War—17 killed and 33 wounded per ton of bombs dropped by German raiders over England—the Air Staff calculated that a French attack, estimated at 100 tons of bombs a day, would produce 40,000 casualties in London during the first week of a war alone.

In response, the Air Staff proposed to expand what it called the "Home Defence" forces, but *defence* had a loaded meaning: well over half of the twenty new home squadrons Trenchard proposed would be the retaliatory bomber force. The only effective defense, the Air Staff insisted—and with some justice in the mid-1920s—was a *very* good offense. In July 1922 the Cabinet approved the expansion, to be completed by 1925.

The British government's theoretical worries about the continental balance of power suddenly grew much more tangible in 1923 when France, over Britain's strenuous objections, reoccupied the Ruhr to pressure Germany over reparation payments. Tensions across the Channel grew. Worse, Britain realized that its military weakness had emasculated its diplomatic influence in Europe's affairs: Britain had become a country that could be ignored. A subcommittee of the Committee of Imperial Defence was appointed under Lord Salisbury to examine the state of Britain's air forces, and it recommended a further expansion of the Home Defence force to fifty-two squadrons by 1928. The Air Staff now set a formula of two bomber squadrons for every fighter squadron in the Home Defence forces. This formula was arbitrary, but it reflected Trenchard's dogma. "Attacking France with forty-eight extra bombers would have a considerable moral effect," Trenchard argued during one staff meeting when the question of adding four extra fighter squadrons was raised, "whereas bringing down a few more of their bombers would have very little effect on the French nation, who would probably never hear of it." The two-to-one ratio had the appeal of mathematical simplicity even if it was based on no mathematical analysis.

Trenchard explicitly referred to the bomber force as a "deterrent," and in an eerie foreshadowing of the theories of mutually assured destruction that occupied Cold War force planners, there was almost an existential dimension to the thinking that went on in the Air Staff in the 1920s and 1930s. The whole idea of a deterrent was that you would never actually have to use it; its mere existence would suffice to keep an enemy in check. But that in turn cast a sense of make-believe and abstraction over the whole exercise. In the Air Staff papers that discuss the offensive role of the Home Defence forces, the target was rarely anything more specific than "France," as in the 1930s it would be "Germany." As the historian Malcolm Smith observed, the Air Staff appeared "so enamored with the apparent simplicity of its theory of strategic airpower that careful and detailed planning seemed unnecessary." Trenchard at one point stated his belief that, in the event of a bombing contest with France, "the French would squeal before we did." This was about as far as he got in analyzing the details of offensive theory. It was not until the eve of the Second World War that the Air Staff began even to identify specific targets in Germany and compile them into a list.

The lack of specifics in the theory led to a corresponding void in tactical and operational development. The RAF continued its well-established tradition of neglecting engineering, logistics, and even such basic skills as navigation: as of January 1933 only 38 out of 1,346 officers from the rank of flight lieutenant to group captain had passed the basic specialty course in navigation. (By May 1940, on the eve of the Battle of Britain, it was 109 out of 2,742.) In creating a deterrent, it is front-line strength that counts: the "shop window," again. And so, in building up as large a display of force as possible, the RAF all but abandoned any worries about developing a system for logistics, supply, and repair that could operate under realistic wartime conditions. Even if war came, the belief in the short, fast knockout blow seemed to argue for the policy tersely enunciated by one of Trenchard's successors: "In war there will be no repair."

The preoccupation with display was made all the worse by Trenchard's insistence upon re-creating in the RAF squadron the traditional pomp of the old-line British Army regiment. Pilots were expected to fly with walking sticks and razors in their kit so that if they had to make a forced landing on a cross-country flight they could turn out properly for inspection. Because it would upset the "regimental model" if certain positions in the organization were reserved for officers with specialized technical training, Trenchard insisted that there would be no technical specialties; any officer could perform any role.

But the doctrine had worse effects yet. Because the policy of deterrence depended for its effectiveness on the impression of awe it would create in the mind of the foe, any public questioning of the policy might undermine the deterrent. It thus became heresy in the RAF to doubt that a nation could be brought to its knees by morale bombing. Applicants to the RAF Staff College were given a qualifying exam in which, year after year, one question concerned the "correct" doctrine to be applied in a large-scale air war. As Scot Robertson, a defense analyst who reviewed the examiners' reports found, it was clear that they were seeking one answer and one answer only: "namely, that the only appropriate use for airpower lay in the offensive against enemy morale."

It was dogma; it was also magnificently vague dogma. The Air Staff's disinclination to delve into the question of exactly *what* to bomb was, in part, ordinary bureaucratic self-defense; vagueness is always a refuge against criticism. But mostly it reflected the broader political priorities of the RAF at the time, above all protecting its turf against encroachment from the other services. Trenchard grandly declared, "The air is one and indivisible," which sounded as though it meant something even if it didn't; what it really meant was that a bureaucratic tail was wagging a doctrinal dog. In a 1923 address, Trenchard asserted that "since it can attack several classes of objective, it is not surprising that the object of an air force is not easy to define in precise terms. Does this matter? Not if we have nimble minds on the Air Staff ready to apply a somewhat general definition to a particular case." The trouble was that everything from the type of armament to the type of airplane to the type of training for pilots depended critically on the "particular case." Creative ambiguity—or just plain intellectual mushiness—that helped to shield the RAF from political potshots also prevented it from developing a clearly defined strategy and a practical-minded doctrine.

People who did not have an air force to run, or a ministry or parliament to worry about, were much freer to expound their ideas about air power in vivid details and with comprehensive sweep. The 1920s and 1930s were a heyday for theorists who began to publicize the doctrine of morale bombing in books, popular magazines, and newspapers, and in specialized military journals as well. With their visions of victory won through a crushing attack from the air, they were in some ways taking up where H. G. Wells had left off; to this they added selective evidence drawn from the Great War that seemed to bear out these ideas.

The theorists did not so much shape policy as give it a respectable intellectual face for the public; in their hands the sometimes inconsistent, tentative, contradictory, and incoherent policies of real air forces were molded into an ideal, all-encompassing credo of air power as a decisive and transformative force in warfare. In fact, Trenchard and other senior RAF leaders in the 1920s and 1930s apparently never even read the work of the man who would become the most celebrated of the air-power "prophets": the Italian Giulio Douhet.

Douhet's name is now legendary in the history of air power; he produced the most sweeping and unwavering manifesto in support of the doctrine of victory through strategic bombardment. Though a military officer (he eventually held the rank of general in the Italian Army), Douhet had an odd career that seems to have involved little experience with or responsibility for aviation. There is more than a little evidence that he never learned to fly himself. Most of his writing was done during a period of bitter estrangement from the Italian Army, and from society altogether, for that matter. During the Great War, Douhet had been hauled up before a court-martial and sentenced to a year's imprisonment for having sent directly to the Italian war minister a long memorandum in which he not only proposed breaking the stalemate on the Austrian front by launching an attack against the enemy's cities with five hundred bombers, but also denounced the entire military hierarchy as incompetent.

In 1920 he was formally exonerated when a military tribunal set aside his court-martial conviction and declared that he had acted for "the good of the country" in violating military discipline, and the following year he published what would become his most well-known work, *The Command of the Air.* Many of the arguments Douhet advanced in *The Command of the Air* were familiar. But few writers had ever laid them out in such bold and categorical terms. Modern war had taken on "a character of national totality. . . . The entire population and all the resources of a nation are sucked into the maw of war." At the same time, the nature of the air as a new theater of military operations had erased the battlefront and the distinction between soldiers and civilians. Nor could airplanes be stopped, either in the air or from the ground: "Nothing man can do on the surface of the earth can interfere with a plane in flight, moving freely in the third dimension. All the influences which had conditioned and characterized warfare from the beginning are powerless to affect aerial action." A modern force of bombers armed with explosives, incendiaries, gas bombs, and perhaps even bacteriological weapons could destroy a city in an instant. "Normal life would be impossible in this constant nightmare of imminent death and destruction." And so, to "put an end to horror and suffering, the people themselves . . . would rise up and demand an end to the war—this before their army and navy had time to mobilize at all!"

All of these considerations led to one inexorable conclusion: Whoever launches the swiftest and most ruthless aerial assault on the enemy's society, at the commencement of hostilities, will prevail. The only answer to enemy attack is to resign oneself to the blow and be prepared "to inflict even heavier damage upon him."

In the 1927 edition of his book Douhet went even further in pressing home what he saw as the brutal truth. Treaties limiting the use of any weapons, "no matter how inhuman and atrocious," were pointless, mere "scraps of paper," "demagogic hypocrisies." In a fight to the death, a nation will use every means available. "There is no use burying one's head in the sand"; the purpose of war is to harm the enemy as much as possible, and "he is a fool if not a patricide" who hesitates to use every available weapon, however terrible. Indeed, Douhet asserted, if one side does not use poison gas first in an all-out attack on enemy cities, it will merely be ceding victory to its enemy.

Douhet's prophecies were translated into French, English, German, and Russian after his death in 1930, but even before then they were well known in some circles outside of Italy. Most of the world's air staffs, including many in the Italian Air Force, did not and could not completely subscribe to his apocalyptic vision of aerial destruction, but they still found his ringing endorsement of the revolutionary significance of air power congenial. In his 1927 edition Douhet declared, as few serving officers could, that the air weapon rendered ground and sea forces largely irrelevant, except for passive defense and perhaps as a sort of mopping-up police force. Diverting aviation units to support the fight on the ground and at sea was accordingly "worthless, superfluous, harmful." The war would be won in the air alone; only by putting every available resource into a force that could launch the greatest, most instantaneous, and most devastating attack on enemy cities would enemy resistance be crushed.

To air commanders seeking recognition, funding, and independence from the older services, it was hard not to cite such an articulate advocate. Among the air forces of the world, the United States Air Service was probably the least "Douhetist" in theory and practice; yet in 1933 the commander of what had, by then, become the Army Air Corps sent thirty mimeographed copies of a translation of *The Command of the Air* to members of the House Committee on Military Affairs with a cover letter stating that Douhet's book "presents an excellent exposition of certain principles of air warfare."

American air leaders also frequently quoted two prominent British military authorities who wrote in much the same sweeping terms. One was General P. R. C. Groves, who in a much-cited article in *The Atlantic Monthly* in 1924 wrote that the air weapon had become "the key weapon of war"; at the outbreak of the next European war, Groves asserted, "each side will at once strike at the heart and nerve centers of its opponent: at his dockyards, arsenals, munition factories, mobilization centers, and at those nerve ganglia of national morale—the great cities." Even more well known was Captain B. H. Liddell Hart, a prolific author and the military correspondent for the *Daily Telegraph*. In his 1925 book *Paris: Or the Future of War,* Liddell Hart argued that the air weapon was so powerful, and civilian society so vulnerable, that just the threat of aerial attack might suffice. The real target in modern war was not an enemy's territory, or its army, or its industries; the real target was the enemy's will to resist and its policies, and when faced with the threat of annihilation, rational men would choose surrender over a "fight to the death." Both Groves and Liddell Hart thus saw the air weapon as a powerful tool of diplomatic coercion that would be of incalculable influence even in peacetime.

But such publicity cut two ways. On the one hand, it greatly enhanced the prestige of the air arm as the possessor of a threat of unimaginable proportions. On the other hand, it became a threat that nations saw in a mirror as well. The "shadow of the bomber" would hang over Europe, and Britain in particular, throughout the 1930s. Harold Macmillan recalled years later that "we thought of air warfare in 1938 rather as people think of nuclear warfare today." Major General H. L. Ismay of the Committee of Imperial Defence warned in a Cabinet paper in September 1936 that it was possible and "indeed probable" that "the dominating feature of the German war plan is the delivery of a knock out blow to England from the air; that this blow will be immediate, continuous, and of an intensity that has not perhaps been visualized; and that it may be delivered earlier than we, at present, have reason to anticipate." Such fears, whatever their basis in reality, cast a shadow not only upon the man on the street but also in the halls of diplomacy throughout the 1930s. (Following the Munich crisis in 1938, President Roosevelt's ambassador to France would report dryly: "If you have enough airplanes you don't have to go to Berchtesgaden"—referring to Hitler's summer retreat, where British Prime Minister Neville Chamberlain had hastened in order to reassure the Führer that, to avoid war, Britain would not oppose the Nazi government's annexation of German-speaking parts of Czechoslovakia.)

Passing around mimeographed copies of Douhet's views cut two ways in another sense, too. Air-power proponents could underline the passages extolling the revolutionary import of the air weapon and the need for an air force independent from the army and navy, but members of those older services could quote Douhet as well—to paint this new weapon as barbarous, unspeakable, an affront to all civilized norms. Douhet certainly did not mince words, and they were not words that put the air forces of the world in a chivalrous light. In the 1930s a Royal Navy spokesman observed that in seeking to build bombers the Air Ministry obviously intended to use them against cities, "a method of warfare which is revolting and un-English." A U.S. Navy captain declared that the bomber enthusiasts in the U.S. Army's air service wanted to turn the United States into a nation of "baby-killers." And the U.S. Army's *Infantry Journal* in one of its issues in 1937 laid out a two-page spread of photographs showing piles of dead bodies and ruined buildings from the Japanese bombing of China. The headline was simple and pointed:

WAR À LA DOUHET

Among the many financial burdens that were looming large in postwar Britain was that of maintaining the Empire. From Africa to India, Britain's imperial burdens had grown only larger with the end of the war. Along with new responsibilities, including mandates over Palestine, Iraq, and other territories ceded by the defeated Turks, there was old unfinished business that had been put on hold during the Great War. Notably, there was the "Mad Mullah" of British Somaliland.

Mohammed bin Abdullah Hassan had been a thorn in the side of British rule for decades. Inspired by tales of the Mahdi's dervishes in Sudan who had fought a holy war against the British, and imbued with the doctrines of a tiny, puritanical Islamic sect he had joined during a pilgrimage to Mecca, Hassan began preaching his own fanatical brand of Islam among the Somali clans upon his return to his country in 1895. With a following that at times grew to more than ten thousand, he declared a jihad to drive the infidel from the shores and to purge the faith of its corruptions.

But what began as a nationalistic and religious movement soon degenerated into sheer carnage. The Mullah's men raided livestock from and slaughtered the members of rival tribes that had accepted British rule. The Mullah became delusional in his power, claiming for himself magical powers that sounded like something out of *A Thousand and One Nights*: he could push whole towns into the sea with his feet and fell his foes by uttering incantations. Desertion from his force was punished by the slaughter of the deserter's entire extended family. The Mullah once put three hundred women to death because he had dreamed they had refused to pray.

Four British expeditions had been mounted against him, the last, in 1904, fielding five thousand regular troops and ten thousand irregulars under the command of a major general. Each time the Mullah lost thousands of men in battle; each time he survived his defeat and regrouped to resume his raiding. In 1909 the British administration of the territory basically gave up and withdrew from the interior of the country. An orgy of destruction followed, in which a third of the population was slaughtered.

In late 1918 the War Office dispatched a general to Berbera to assess what it would take to deal with the Mullah once and for all. He reported back that it would be a matter of a difficult three-month campaign against the Mullah's well-equipped troops. The War Office immediately balked at the cost of such an undertaking and refused to pay. Whereupon the Colonial Office, fearful that it would get stuck with the bill, privately approached Hugh Trenchard to ask if perhaps airplanes might be able to do the job a bit more cheaply.

Churchill was immediately enthusiastic. He had always liked airplanes, and now he also saw an opportunity to make a splash as a forward-looking, efficient administrator. Demonstrating that modern technology could not only do a job better than the old services but do it more cheaply to boot would suit the politics of the moment. Churchill began incessantly pressing the Cabinet to let the RAF have a crack at Somaliland. At first the War Office strenuously objected, but it at last gave way in October of 1919 after securing a promise that "under no conceivable circumstances" would regular Army reinforcements be demanded to bail out the RAF if it got into a jam.

And so on January 21, 1920, the Mullah was at his compound at Medishe when he and his men caught sight of five machines in the sky heading their way. One of the Mullah's men suggested that perhaps they were messengers from

Istanbul come to inform the Mullah of Turkish victory in the Great War. (News traveled slowly.) A more imaginative aide suggested they were chariots sent from Allah to carry the Mullah to heaven. Whether he was being sycophantic or wry is unclear; in any case those who suspected the truth of what the airplanes' appearance portended knew better than to say so directly, for the Mullah, true to the practice of despots from time immemorial, regularly put to death those who bore ill tidings.

Failing to spot the Mullah's compound, the flight passed by and struck a nearby fort. But then one plane was seen returning. The Mullah went outside to watch and was leaning on the arm of his vizier when the first bomb fell. The vizier was killed, the clothes of the great man himself singed.

Over the next three days, the RAF's Z Unit—36 officers, 189 enlisted men, and one flight of six DH9 bombers plus six spares, all transported on three warships—bombed Medishe and the Mullah's nearby fort twice daily. When word then came that two of his other forts had fallen to British-led native troops, the Mullah fled with about seven hundred riflemen. The British forces pursued, and a month later it was all over and the Z Unit on its way home. The Mullah escaped to Abyssinia, where he and a small number of remaining followers died a month later of disease and malnourishment. Total British casualties were two native soldiers killed.

It was the money that spoke loudest, though: Churchill told the House of Commons that the last conventional land expedition against the Mullah had cost six million pounds. The RAF had done it for seventy thousand. *The Aeroplane* crowed that the operation was proof that henceforth the infantryman's role would be degraded from that of "the first line of attack to the position of mere 'mopper up.' "

The War Office and the Army, needless to say, were unconvinced. H. L. Ismay, then a colonel and in command of the Somaliland Field Force, grumbled that "Somaliland had been somewhat of a hoax on the part of the Air Ministry." Some military men complained that by driving the Mullah's forces to flee from their forts and disperse across the countryside the RAF had only made the Army's job harder. Others pointed out that the Mullah's forces had already been weakened by six years of skirmishes with the British-commanded Somali Camel Corps, by factional tribal fighting, and by a Royal Navy blockade that cut off supplies of modern arms.

But the RAF could—and at nearly every opportunity did—point to the dispatch of the Governor of Somaliland, who praised the RAF as "the main instrument of attack and the decisive factor." The air raids, he said, "exercised an immediate and tremendous moral effect over the Dervishes, who in the ordinary course are good fighting men, demoralising them in the first few days."

A declaration of jihad by the Emir of Afghanistan against the British forces in the northwest frontier of India in 1919 had given the RAF the opportunity to score what seemed another remarkable victory for its new policy of "air policing" the Empire. Three of the huge four-engine Handley Page V/1500 bombers that had been developed to bomb Berlin in the Great War had just been completed when the war ended. In January 1919 the RAF shipped one to Karachi. After being placed on exhibit to the public for several months to help build up the proper "moral" effect, the plane took off on May 24 for a three-hour flight over the mountains to Kabul. There it dropped four 112-pound and sixteen 20-pound bombs.

The Air Staff subsequently claimed that the raid "caused a panic in the capital and the evacuation of about half the inhabitants" and was the decisive factor in the Emir's suing for peace. The RAF's supporters enthusiastically agreed. Liddell Hart wrote, "Napoleon's presence was said to be worth an army corps, but this aeroplane seems to have achieved more than 60,000 men did."

The Army once again dismissed the RAF's claims as nonsense. The only panic, they said, had been in the Emir's harem.

What was increasingly clear, though, was that positions were hardening along service lines. Both the RAF and the Army realized that the success or failure of the policy of air policing was likely to decide the continued independence of the air force. And now the Air Ministry was ready to up the stakes. In February 1920 Churchill, in a speech to Parliament, raised the possibility of the air force's taking full responsibility for the military control of Iraq. The immediate trouble in Iraq was that the British administration was trying to extend its control over—and collect taxes from—the vast rural areas that the Turks had more prudently left to their own devices. A garrison of 60,000 Anglo-Indian troops was stationed in the country, but it was not clear that even this sizeable force could do the job; moreover, it was costing the treasury some 18 million pounds a year. Churchill once again pressed the argument of economy and efficiency; in a note to Trenchard, he suggested that it ought to be possible for the RAF to do the job for a mere 5 to 7 million pounds a year.

While the Cabinet wrangled over the plan, an open rebellion broke out in the summer of 1920 among the Kurds and Shi'ite Muslims who formed a majority of the population, which gave the Air Ministry a chance to say "I told you so." When it was over, the government had spent 40 million pounds in a ground war that had cost more than a thousand dead among the British forces. Trenchard, per form, insisted that had air power been used from the start, a sufficient "moral effect" would have been achieved at once and quenched the rebellion. Lloyd George mulled over the idea of saving even more blood and treasure simply by installing an amenable potentate and paying him off with a million pounds a year to cast a friendly eye upon British interests, but in the end the air force proposal won out. Field Marshal Sir Henry Wilson, Chief of the Imperial General Staff, consoled himself in his diary with the thought that perhaps it was best to give the RAF the rope with which to hang itself: "The sooner the Air Force crashes the better."

The Cabinet's approval of the "Air Control" scheme for Iraq, to go into force on October 1, 1922, was in itself a remarkable boost to the prestige and influence of the RAF. An air officer would assume top command of all British military forces in Iraq, with a commensurate hike in the RAF budget to cover the expenses of the operation. The RAF would command eight squadrons of aircraft, nine battalions of Anglo-Indian infantry, a contingent of local troops, and four companies of RAF armored cars.

Air control basically meant dropping bombs on recalcitrant villages that rebelled against governmental authority or refused to pay taxes. As one RAF officer explained it, the aim was "to establish a tradition" of action that the natives would learn to respect.

> One objective must be selected—preferably the most inaccessible village of the most prominent tribe which it is desired to punish. . . . The attack with bombs and machine guns must be relentless and unremitting and carried on continuously by day and night, on houses, inhabitants, crops and cattle. No news travels like bad news. The news of the punishment will spread like wildfire. . . . This sounds brutal, I know, but it must be made brutal to start with. The threat alone in the future will prove efficacious if the lesson is once properly learnt.

The actual success of air control in the Middle East was decidedly mixed. Field Marshal Wilson sarcastically summed it up as the RAF's "appearing from God knows where, dropping their bombs on God knows what, and going off again God knows where." But the RAF was assiduous in promoting its own flattering version of events in professional journals and parliamentary reports, and soon, as far as the public and the politicians were concerned, it *was* a success. Tales of marauding cross-border raiders who fled on the mere rumor that aircraft would pay a visit, or of tribes that rebelled in the morning but paid their fines in the afternoon after a few well-placed bombs fell on their villages, became staples.

And there was no denying that, for whatever reason, the cost of policing did drop significantly from the time the RAF took charge. The Air Ministry claimed that it was doing the job in Iraq for 85 percent less than had been budgeted for the Army; in Palestine and Transjordan costs had been cut 90 percent. In Palestine that certainly had more to do with luck of timing than skill; a few years later the Arab revolt in Palestine would erupt and troops would have to be rushed back in.

Yet the apparent successes had secured the future of the RAF. "I cannot emphasize too much the value your successful command in Iraq has been to us," Trenchard wrote John Salmond, the Air Officer Commanding in Iraq, after a special subcommittee of the Committee of Imperial Defence recommended preserving the independence of the air service in 1923.

Salmond was just one of many high-ranking RAF officers holding key commands during or just before the Second World War who served in air-control missions in the 1920s and 1930s: Arthur Harris and Charles Portal, who would each head Bomber Command during the war, and Edward Ellington, Chief of Air Staff just before the war, were among the others. At a minimum, the experience reinforced their confidence in the soundness of the doctrine of morale bombing; at worst, the growing role that air control was playing in keeping the RAF alive through a difficult period further stifled any inclinations to dissent from the party line.

Although some air officers thought air control might be a special case, that the "natives" were especially susceptible to the threat of aircraft because of their primitive and superstitious natures, the RAF tended to officially downplay such arguments. (It was, interestingly, usually the foes of air control who invoked the most racist arguments; for every RAF officer who explained that "the only thing the Arab understands is the heavy hand," there was an Army officer who countered that natives were immune to the effects of sustained bombing because of their dull, insensitive, and primitive natures.) Indeed John Salmond made a point of insisting that the lessons of Iraq and Afghanistan were fully applicable to Europe: "Humanity was the same the world over," he said; what worked in Kabul would work in Paris.

Salmond and other top officers certainly never hesitated to extract broad implications for the theory of morale bombing from their experiences in the Middle East. In a lecture on "Air Strategy" at the Royal United Service Institution in 1924, Wing Commander C. H. K. Edmonds explained: "Just as in the small war our continuous bombing made the tribe's life intolerable and brought him to heel, so in the case of the big war our object is to destroy the enemy's *morale*—we must make him feel that life has become so impossible that he prefers to accept peace on our terms." Patently ad hoc theories about why bombing sometimes persuaded the natives and sometimes did not swiftly became enshrined in RAF dogma. Salmond concluded from his experience in Iraq that people under bombardment go through three phases: first panic, then resignation, then finally weariness with the continued disruption and inconvenience to everyday life—and a desire for peace. Soon this "three-stage" theory became another bit of unquestioned wisdom dutifully repeated by staff officers. But the most perfect bit of ad hockery was the frequent suggestion that the interplay between bombardment and its effect on morale was so subtle, and had to be so finely judged, that only an experienced air officer could be entrusted to call the shots, deciding what to bomb, when to bomb it, and with how much force. This was an art, or maybe even a religion, and no one after all expected an artist or a high priest to explain himself with mathematical equations.

From the start the RAF had also enthusiastically viewed the air-control operations in Iraq and elsewhere as a way for its crews to obtain realistic combat experience; pilots waggishly referred to Sheikh Mahmud of Kurdistan as "the

Director of R.A.F. training." Reports of air operations in Iraq were printed up and circulated at the Staff College and among home squadrons. But it was an odd kind of training for the demands of modern aerial warfare. For one thing, it was absurdly one-sided. By 1932 the RAF had lost only fourteen pilots in all of its air-policing operations throughout the Empire, from Aden to India. Against virtually undefended targets there was little need for formation flying or night flying, and little of either was done. Weather in the Middle East was rarely a problem. Navigation was taken care of by having a convoy of vehicles drive across the desert, leaving a track visible from the air for the pilots to follow. Bombing from hundreds or even tens of feet off the ground eliminated the need for accurate bombsights. Indeed, the Great War–era Bristol Fighter, widely used in air-policing operations, had no bombsight at all.

Aircraft that were slow, reliable, rugged, and able to operate from rough landing fields were favored over all others. A series of oddly named and often ungainly biplanes began to emerge from British factories to fill the requirements of duty in the Empire: the Vildebeest, the Wapiti, the Hart, the Hound, the Hinaidi, the Hyderabad. Financial and technological constraints would have limited the development and purchase of advanced types of aircraft in the 1920s in any case, but the torpor of British aircraft engineering in this period was deepened by the undemanding nature of air policing, which offered little incentive to the manufacturers to try anything new. In 1927, when the RAF finally decided to replace the World War–era DH9 day bomber, it specified that the new plane had to use as many of the parts from the DH9 as possible. The maker of the plane that was chosen, the Westland Wapiti, proposed to recycle the entire wings and tail unit of its predecessor. Likewise the 1929 Handley Page Hinaidi, which replaced the 1925 Hyderabad night bomber, basically just took the plan of the older plane and replaced some wooden components with metal.

Unsurprisingly, by the time many of these unremarkable aircraft reached production they were obsolete even for their intended roles. And bombing accuracy, as always, was greatly overstated. The RAF claimed it could single out the house of a particular sheikh for attack when bombing a village, but there were many documented cases in which pilots failed to get even half of their bombs to land within the targeted village at all. Yet Trenchard continued to be oblivious to the technical limitations of his aircraft. At one point he insisted that the obsolescent Bristol Fighter would have "great power" in air-policing operations if only it were "used energetically." Rather than providing realistic lessons, air control was merely reaffirming old prejudices in doctrine, operations, tactics, and technology.

Just as the published theories of Douhet and his fellow prophets of air power had provoked objections, both sincere and calculated, that morale bombing was immoral and a violation of traditional principles of warfare, so its employment in practice in imperial policing set off occasional waves of unease. Churchill was aghast upon receiving a report of one action in Iraq in which pilots had strafed women and children who had fled from a village into a nearby lake. (Churchill had also averted a public-relations disaster when, in 1920, Trenchard had blithely drafted a paper suggesting that aircraft might be used to put down "industrial disturbances" in "settled countries" such as India, Egypt, Ireland, and even England; Churchill told him to remove the references to Ireland and England and never put that suggestion in writing again.) The Labour government that came to power in 1924 and the growing tide of public opinion favoring international disarmament cast a slight pall over the RAF's enthusiastic descriptions of teaching the natives a lesson. Some critics even began to describe its methods as "Hunnish."

In response, the Air Staff became increasingly vague in reporting casualties and increasingly emphatic that it really took very little destruction to make a tribe come to heel: that was the essence, after all, of targeting "morale," Salmond explained. An RAF officer speaking at the Royal United Service Institution in 1928 insisted that bombing people from the air generated less "animosity" than sending troops in on the ground, and he sarcastically said of those critics who claimed that women and children were always hit, "I really cannot see why an air bomb should be thought to have inherently some sort of fatal attraction toward the fair sex."

But for the most part imperial policing served the RAF well as both a bureaucratic strategy and a public-relations tool; if one was going to sell a policy of bombing people, it was hard to pick better people to bomb than the Mad Mullah or cutthroat desert raiders. Beginning in 1920, the RAF each year staged a huge aerial display at its airfield in Hendon near London. The spectacle drew hundreds of thousands of spectators, and the culmination of the Hendon Pageant was always a mock battle demonstrating some aspect of air power in action. For the first two years, the grand finale had drawn on themes from the Great War. But the 1922 pageant featured an "Eastern Drama": A tribe of "Wottnots" (consisting of airmen in blackface) occupied a fort with hundred-foot-high towers. RAF bombers swooped down and, after a sufficient amount of suspenseful buildup, set the fort ablaze with incendiary bombs. For the 1927 show, the setting was a village in "Irquestine" where European women and children were being held hostage; this time the RAF came to the rescue with airlifted troops, supplies dropped by parachute, and bombing runs on the enemy strongholds.

In later years the RAF tried to soft-pedal the message with what it thought might be more socially acceptable displays of the peaceful uses of aircraft, but the air enthusiasts were converts by this time and would have none of it.

"Personally," opined *Flight* magazine, "we were much more excited when our aircraft swooped down upon hordes of many-coloured 'Wott Knotts,' scattering them in all directions, and then blowing *everything* up with terrific bangs."

If the RAF would police the Empire, the United States Air Service would defend the homeland. At least that was the cover story, and it proved to be a felicitous invention on several counts. For one thing, the defense of the United States was a mission that clearly asserted the air arm's independence from the Army, yet in the nicest possible way. While such a mission could scarcely be subordinated to official Army doctrine that placed the infantry firmly on top, with aviation in a supporting role only, it seemed to be the Navy's ox that was being gored the most by the airmen's insistence that they could take over the defense of the nation's shores. Officially, most senior Army leaders opposed anything that might divert the Air Service's attention away from infantry support, and a seaborne invasion of the United States in truth seemed a remote possibility; unofficially, the Army took an undeniable pleasure in watching the Navy squirm.

In the isolationist spirit that closed back over America after the war, it was also better politics for American airmen to talk about protecting the homeland than about blasting foreign cities to ruins à la Douhet. But the truth was that the coastal-defense mission demanded almost exactly the same skills, armaments, instruments, airplanes, training, and tactics that would be required for the long-range bombing missions that all air-power enthusiasts were convinced—whether it was prudent to say so or not—held the key to the successful employment of air power. Intercepting an enemy fleet steaming toward American shores would require planes that could carry large weapon loads at considerable distances; it would require powerful bombs that could be delivered with accuracy against targets heavily defended by antiaircraft guns and fighter planes; it would require pilots who knew how to navigate in good weather and bad without the aid of landmarks. It was not a perfect match, but it promised to provide a far better justification for building a modern long-range bomber force, and far better impetus for learning how to use it under realistic conditions, than the RAF's air-control operations did.

Recasting the fight for air force independence as an epic battle of air versus sea also had the virtue of reducing a complex bureaucratic story into a simple melodrama of the kind newspapers and their readers loved. And that was right up Billy Mitchell's alley.

Starting in 1920, Mitchell began a relentless campaign aimed at goading the Navy into accepting a test of the ability of aircraft to attack large ships, a test that Mitchell was confident he could portray as a decisive showdown of air power against sea power, the new against the old, innovation against tradition. In testimony before Congress, in speeches, in articles, Mitchell sank the Navy thousands of times. He declared that the day of the battleship was over; this at a time when the Navy was pressing for a capital fleet "second to none."

The Navy's position was never so uncompromising or backward as Mitchell managed to portray it, but then the Navy was simply no match for Mitchell when it came to public-relations skills. Few could match him. Many naval officers were keenly aware of the significance of air power, both as a means for projecting force at sea in the form of the aircraft carrier (which had made its tentative debut during the Great War) and as a possibly revolutionary factor in naval strat-

egy that might require a complete rethinking of the concept of surface operations. To evaluate the effect that aerial explosives would have on ships, the Navy
itself conducted a test in October 1920 in which charges of various size were
placed inside the hulk of an old battleship, *Indiana,* and set off to see what kind
of damage they could do.

The Navy was interested in gathering objective data. Mitchell was invited
along with other Air Service officers to observe the tests, on the condition that
they would not discuss the results or release any information without the Navy's
permission. Two months later, pictures of the wrecked battleship appeared in the
newspapers. It was not hard to imagine how they had got there. The *New York
Times* accompanied a spread of seven photographs of the *Indiana* with a variety
of sarcastic barbs aimed at the Navy Department, which, "for reasons but known
to itself," the paper declared, had forbidden "a free or thorough discussion as to
the effects of the new weapons upon naval warfare."

If planes versus carrier pigeons was a story, then planes versus battleships
definitely was one. The press went to town, and Mitchell was there to meet them.
In January 1921, while testifying before Congress on the Army appropriation
bill, Mitchell challenged the Navy to turn over captured German warships for a
bombing trial against Army planes: "Give us warships to attack and come and
watch it," he taunted. The Navy was cornered and knew it had to accept the challenge. Secretary of the Navy Josephus Daniels shot back that he would stand
bareheaded on the bridge of any ship Army aviators tried to bomb. It was just as
well that he was out of office before he had to make good on that promise, but in
the meanwhile it only fed the story and played into Mitchell's efforts to portray
the trials as a decisive test of air versus sea. Meanwhile, too, Mitchell quietly
ordered that a new bombsight and a new one-ton bomb be prepared specially for
the upcoming trials.

The drama played out to its inevitable stage-managed conclusion. In July
1921 Mitchell's First Provisional Air Brigade, based at Langley Field, Virginia,
began to take on the targets the Navy provided. The culmination was to be an
attack on the German battleship *Ostfriesland,* a 24,000-ton dreadnought. The
Navy was in command of the trials and had drawn up a schedule specifying a
carefully programmed sequence of strikes, each to be followed by a pause during
which an inspection party would board the ship and assess the damage. On the
first day of bombing, July 20, things went more or less as agreed. Most of the
nineteen 550- and 600-pound bombs that were dropped missed; all but two of
the six hits or near-misses were duds. When the inspection party boarded, it
found several compartments filling with water but concluded that had the ship
been manned and under power, damage-control parties and pumps could easily
have coped with the situation. That ended the first day's work.

The next day the rules called for the Army bombers to drop 1,000- and 2,000-
pound bombs singly, halting after any hit to allow the inspection party to come
aboard again. After the first day's rather disappointing performance, and with the

movie cameras rolling, Mitchell decided that staging a show was far more important than obeying orders. Just after noon he sent his planes in one after another without letup. Mitchell clearly wanted to send the *Ostfriesland* to the bottom as quickly as possible. The Navy command vessel tried to wave the planes off using all the agreed-upon signals—making heavy smoke, taking down an all-clear flag—but the Army fliers ignored them and kept on coming. The admiral in charge sent an outraged radio message to the Army planes to hold off; to the admiral's astonishment, an Army aviator coolly replied that he would let the admiral know when they were done. The first bomb fell at 12:15 p.m.; at 12:33 p.m. the *Ostfriesland* began to list heavily to port; three minutes later her stern disappeared; and four minutes after that she slipped beneath the waves.

The Navy was furious, and quite correctly pointed out that the test was heavily loaded, if not rigged outright: a battleship in wartime would have been able to maneuver and fire back at her attackers. The official report of the trials, issued by the Joint Army and Navy Board a month later and bearing the signature of General Pershing, concluded that while aircraft, like mines, destroyers, and submarines, had added to the dangers that capital ships faced at sea, they had hardly made the battleship obsolete: "The battleship is still the backbone of the fleet."

In the court of public opinion none of these judicious conclusions held water. Senator William Borah declared that the battleship *had* been proved obsolete. Mitchell wrote his own report about the trials and, although his superiors ordered him to keep it confidential, it was mysteriously leaked to the press just as the *Indiana* pictures had been. Mitchell's report declared with characteristic certainty that "the problem of the destruction of seacraft by Air Forces has been solved, and is finished," and that the time had come for the creation of an independent air force to take control over coastal defenses within two hundred miles of land.

In subsequent books and speeches, Mitchell and his supporters added masterful bits of embroidery to the story of the *Ostfriesland* test. She was not just a battleship; she was an "unsinkable" super-battleship that had survived a brutal pounding from naval guns during the Battle of Jutland, only to succumb to the few bombs dropped from Mitchell's bombers. In a biography of her brother, Ruth Mitchell recounted how tough admirals and "old sea dogs" "wept aloud" as the great German ship went down. Neither claim was remotely true: the *Ostfriesland* was virtually untouched at Jutland and was an ordinary warship of nearly obsolescent design; and somehow no one who repeated the story of the weeping admirals and sea dogs was able to identify any of them by name. But the sinking of the *Ostfriesland* was well on its way to becoming a legend.

In the middle of the *Ostfriesland* tests Mitchell's wealthy wife, Caroline, left him and charged that he had been behaving in an "erratic" manner. His enemies spread the word he was unstable, and the Army went so far as to order him to report to Walter Reed Hospital for a mental evaluation. He passed with flying colors and came back roaring. In articles, speeches, and testimony to Congress,

Mitchell hammered away at his theme in colorful and vivid terms. "No seacraft whatever has a chance against the combined attack of aircraft, in any way, shape, or form." For the cost of one battleship, fifteen hundred aircraft could be purchased. If the Navy persisted in building battleships, it ought to provide the crew with parachutes so they can "come down slowly when blown up in the air." The Navy's backwardness had left it at the mercy of a "third-class foe."

By 1925 Mitchell's frustration with what he perceived as the impossibility of budging the status quo had built up to an explosive level. Appearing before a special House committee on aviation in January, he let loose with a blast that was volcanic even by his standards. "The Army and Navy know nothing about flying and won't learn," he thundered. The War Department was simply "incapable." "Vested interests" were strangling aviation progress. As Hugh Trenchard observed in one of his more perceptive comments, Mitchell "tried to convert his enemies by killing them first." Secretary of War John Weeks decided that this was not a way to win friends in Washington; indeed, as he subsequently wrote in a letter to the President, Mitchell's "whole course has been so lawless, so contrary to the building up of an efficient organization, so lacking in teamwork, so indicative of a personal desire for publicity at the expense of everyone with whom he associated that his actions render him unfit for a high administrative post such as he now occupies." In March, Weeks ordered that Mitchell's appointment as Assistant Chief of the Air Service not be renewed. Mitchell reverted in rank to colonel and was shipped off to Fort Sam Houston, Texas.

Mitchell's fellow fliers, the press, the nation responded with an outpouring of support. From coast to coast, editorial opinion was solidly behind the gadfly who dared to defy the establishment. Newspaper cartoonists depicted Mitchell spilling the beans; Mitchell letting the cat out of the bag; Mitchell standing shoulder to shoulder with the ghosts of Columbus, Fulton, and Morse before the desk of a functionary with a long beard who declares, "It Can't be Done." Army and Navy "bureaucrats" were depicted as Rip van Winkles sleeping through the dawn of air power—or as antiaircraft gunners congratulating one another as they shoot General Mitchell's plane out of the sky. Telegrams and letters of encouragement flooded in from civic groups, businessmen, active and retired airmen. The World War ace Eddie Rickenbacker sent a cable signed by two thousand fellow veterans of the Air Service praising Mitchell as "the patriot who has told the truth."

Buoyed by this tidal wave of support, and chafing all the more in his exile in Texas, Mitchell decided to stage one last, dramatic showdown. In September 1925, after a Navy dirigible crashed during an exhibition at an Ohio fairground, he issued a statement to the press that the administration of the Army and Navy air services was "almost treasonable." Airmen were so "bluffed and bulldozed" that they dared not speak out while their "bureaucratic superiors" distort the facts and "openly tell falsehoods about aviation to the people and to the Congress."

The Wrights made a crucial advance in airfoil design with the help of wind tunnel tests that led to a greatly improved lift-to-drag ratio. The results are apparent in photographs of the 1901 and 1902 gliders being flown as kites. The 1901 glider (*left*) struggles at a high angle of attack as drag keeps the tether lines close to horizontal; the 1902 glider (*below*) soars easily and almost directly overhead.
(Library of Congress)

"One of the most famous photographs ever taken": Orville at the controls as Wilbur, running alongside to balance the machine, releases his hold at the moment of takeoff.
(Library of Congress)

Air exhibitions drew huge crowds and helped to generate wide public enthusiasm for flight. This image of the National Air Meet, held in Indianapolis, June 1910, is a slight fake: the airplanes were added from other negatives. *(Library of Congress)*

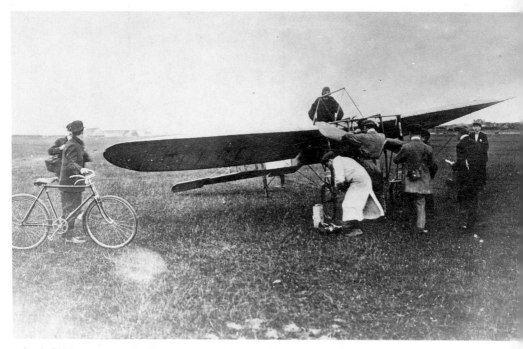

Louis Blériot prepares for his 1909 cross-channel flight. His feat won him a £1,000 prize and caused a sensation in Britain; a politician denounced government neglect of aviation while "Frenchmen are landing like migratory birds on our shores!"
(Library of Congress)

The slow, highly stable BE2s remained a mainstay of British air forces in the Great War even after they began to be shot out of the sky by far more maneuverable opponents. *(Public Record Office)*

Igor Sikorsky's extraordinary four-engine Il'ya Muromets lands near St. Petersburg in February 1914.

Observation remained the fundamental role of aviation throughout the First World War. The British alone took more than 400,000 aerial photographs, such as this view of German trenches.
(above: National Archives; below: Public Record Office)

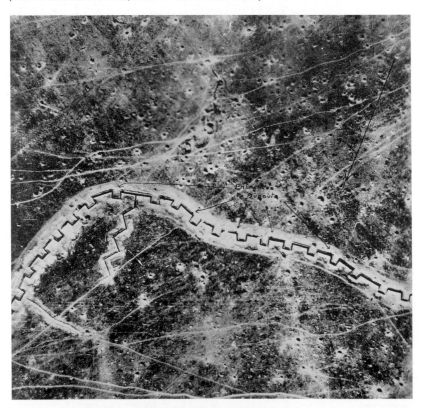

A German aviator drops a bomb over the Western Front.
(National Archives)

Blériot Works, England, 1914. Although experiments in mass production were tried during the war, aircraft continued to be made almost entirely in small shops, by hand.
(Royul Air Force Museum)

Major General Hugo Sperrle (standing at center) with Spanish and German officers, observing from a hillside as the Condor Legion obliterates a Basque position, spring 1937.
(Courtesy James S. Corum)

Condor Legion heavy 88mm flak gun in action in Spain.
(Courtesy James S. Corum)

Three weeks later Mitchell was charged with insubordination and conduct prejudicial to good order and discipline.

His court-martial opened on October 28 in Washington. Mitchell's attorney, Congressman Frank Reid, won an early ruling that the defense would be permitted to introduce testimony to establish that Mitchell's statements were true—an extraordinary concession, for there was no legal basis to the theory that insubordinate statements were any less insubordinate merely by virtue of being true. Mitchell tried to use that opening to turn the tables on the Army and Navy, but it was his undoing; week after week the court was treated to often hairsplitting testimony about what precisely Mitchell had said about the dirigible accident and other matters and whether his words were technically accurate. Refusing to acknowledge his obvious guilt of the charges against him, Mitchell came off as more the evasive schoolboy than the courageous martyr. Editorial support began to dwindle.

Grandstanding by Mitchell and Reid then made even his supporters cringe. Reid threatened to subpoena the President himself to testify. Mitchell declared he should be tried by "representative Americans instead of members of the Army and Navy bureaucracy." And of course his exploitation of the Navy's fatal dirigible crash had made many naval officers' blood boil. Rear Admiral William A. Moffett, head of the Navy's Bureau of Aeronautics, testified that Mitchell was of "unsound mind" and suffering from "delusions of grandeur." Pointing a finger at Mitchell from the witness stand, he melodramatically declared: "Instead of an eagle soaring aloft with eyes for the country's defense, we have, instead, one who really played the part of a vulture swooping on its prey once it is down."

After seven weeks of testimony, it took the court all of thirty minutes to convict Mitchell on all counts. An enterprising reporter fished through the trash cans and concluded from a scrap of paper that Mitchell's boyhood friend, General Douglas MacArthur, had cast the lone dissenting vote. On February 1, 1926, Mitchell resigned from the Army, denouncing the "blind opposition" to aviation of the "bow and arrow men" and "self-perpetuating oligarchies" that ran the Army and Navy.

In retirement he found that his words no longer made headlines the way they had when he was a brigadier general. Yet within the Air Service there were many ready to take up the fight where Mitchell had left off. Hap Arnold had been a firm friend and acolyte of Mitchell's ever since Mitchell had pushed him back into flying in 1916, three years after a crash had convinced Arnold to give up flying and return to the infantry. Arnold had organized a farewell luncheon for Mitchell when he left Washington for his banishment to Texas and had led a delegation of well-wishers to greet him at Union Station upon his return to face his court-martial. Now Arnold narrowly avoided a court-martial himself when an investigation found he had been surreptitiously carrying on Mitchell's campaign, lobbying members of Congress and Air Service reservists to generate support for greater air force independence. Given a choice between resigning his commission or facing a military trial, Arnold coolly said he would stand trial; the Air Service, eager to avoid another embarrassment so soon after Mitchell's court-martial, decided instead to post him to the Army's cavalry headquarters at Fort Riley, Kansas, with a fitness report reprimanding him for displaying "below average" judgment and common sense and expressing doubt that he could be entrusted in the future with "any important mission."

Within the halls of power and the court of public opinion, Mitchell's legacy was also still very much alive. With the passage of the Air Corps Act of 1926, Congress granted an extraordinary degree of autonomy to the service. It was renamed the Air Corps; a new post of Assistant Secretary of the Army for Aviation was created; and in a striking departure from time-honored principles that no member of the General Staff represented a specific branch of the service, a separate air section was created within each of the five divisions of the staff. It was not the full independence that Mitchell had fallen on his sword in the hopes of attaining, but it offered some consolation that he had not sacrificed himself entirely in vain.

6

THE QUEST FOR PRECISION

Hap Arnold believed that the United States would have a real air force only when the right combination of technological advances and international events made it both possible and necessary, and that combination did not even remotely exist in 1926. People were "so used to saying that Billy Mitchell was years ahead of his time that they sometimes forget it is true," he wryly observed. For all of the grand visions of air power propounded by Mitchell and his fellow "prophets," the state of the art of aircraft manufacture in the 1920s was nowhere close to delivering a machine that could fulfill even the basic requirements of payload, range, and speed implied by their theories.

The tiny budgets available to the world's air forces for the purchase of new military aircraft certainly didn't help the moribund state of technological progress. But it would take more than money for the aircraft builders to escape the technological dead end at which they found themselves during the years immediately following the Great War. It would also take imagination, and a willingness to throw out nearly every tried-and-true design principle and every manufacturing technique in the aircraft builder's book.

The limits of the wood-and-fabric biplane had already begun to be apparent during the last year of the war. As speeds increase, stress forces acting on the wings and fuselage of an aircraft increase in proportion to speed squared: doubling the speed increases stresses by a factor of four. The incremental improvements in traditional biplane design made during the war had led to great advances in performance, but the merciless mathematics of the square law ensured that it was only a matter of time before a point of rapidly diminishing returns would be reached. By 1918 it was becoming a challenge to assemble a wood-and-fabric structure that was strong enough to carry the mass of the ever more powerful engines that were being mated to the fuselages of fighter airplanes. No breakthrough in speed, power, or carrying capacity would be possible without a radically new approach to aircraft structures.

Drag forces also increase in proportion to speed squared; this meant that the wartime expediency of covering a multitude of design sins with higher-horsepower engines was swiftly approaching the point of diminishing returns as well. Airplane structures needed to be not only stronger but also much more aerodynamically efficient, shaped to generate far less drag.

The configuration of the biplane tied these two problems together. The biplane was inherently strong but inherently high-drag. A single wing offered a way to leap to an entirely new aerodynamic regime: with the monoplane's smaller wing area, drag would be reduced, allowing higher speed; higher speed in turn meant that a smaller wing-surface area could still generate adequate lift to keep the plane flying. (Lift increases with speed squared just as does drag.)

But a fundamental structural barrier blocked the way to the single wing. A smaller wing generating more lift has to support more weight per square foot of wing, a factor known as "wing loading." It is difficult to make a single wing as strong as a biplane to begin with; it is much harder as wing loading increases. The revolution would have to wait for a way to make the entire airframe better able to resist the forces that tend to cause a fuselage and its attached monoplane wing to flex, twist, or even crack under the stresses of flight.

At the end of the war Anthony Fokker had prefigured the critical structural breakthrough that would turn the monoplane from risky novelty to commonplace standard. A fabric-covered monoplane wing, even with the thick internal girder that Prandtl's Göttingen airfoil made possible, is only as strong as its internal framework of braces and spars. For his D.VIII monoplane fighter that would appear over the skies of France just before the Armistice, however, Fokker chose to cover the wing surfaces with plywood sheets. This so-called stressed-skin construction turned the wing covering itself into a structural component that provided additional stiffness and helped the structure resist bending forces. Fokker had a picture taken that shows him posing in front of a D.VIII while twenty-four men are sitting and standing on its wing to demonstrate the strength of the design.

A closely related structural advance came in an earlier German fighter, the 1915 Albatross. The traditional fuselage in fabric-covered planes derived its strength from a series of wide bulkheads tied together by girders and braced by wires. The Albatross fighter was among the first to employ what was loosely termed monocoque construction; as in a stressed-skin wing, the plywood covering of the monocoque fuselage bore most of the stress. This not only allowed for lighter internal framing—essentially no more than a skeleton framework to provide attachment points for the shell—but also meant that the much thinner internal frame members ate up much less of the interior space needed for cockpit and equipment, allowing the width and height of the external fuselage to be shrunk significantly. That paid off in reduced drag as well.

Yet these new construction techniques were slow to spread through the aircraft industry. Innovation is always risky, especially when it means replacing well-developed skills, crafts, and shop procedures with new techniques. Most aircraft firms were content to stick with the customary fabric-covered frames.

Even slower to spread to the manufacturers was the other great German wartime advance: the scientific and engineering methods developed at Göttingen, which offered an even more fundamental solution to the drag problem by showing how to reshape all of an airplane's components according to the basic laws of airflow. As early as 1920, however, aerodynamicists in Britain and the United States were becoming aware of just how far ahead the Germans were in basic aerodynamic science. Detailed technical descriptions of Prandtl's work began appearing in English-language aeronautics journals, and in America especially the scientists who could grasp its significance were at once appalled at how completely their own country, the nation that had invented the airplane, had fallen behind in the science of flight.

Joseph Ames, a professor of physics at Johns Hopkins University and chairman of the still-embryonic National Advisory Committee for Aeronautics (NACA) that had been established by Congress in 1915, now began a personal crusade to close the research gap. "Aeronautics is in no sense a function of an engineer or constructor or aviator," he asserted in one of the many speeches he gave to scientific audiences around the country in the immediate postwar years. "It is a branch of pure science. Those countries have developed the best airships and airplanes which have devoted the most thought, time and money to the underlying scientific studies."

NACA had begun with a budget of $5,000 a year, which was not enough to support more than what the committee's name implied: offering advice. But it had the authority, at least on paper, to carry out its own research "in the event of a laboratory or laboratories" being placed under its control. When America's involvement in the war seemed imminent, Congress had voted a $50,000 appropriation for the first NACA laboratory, and in 1920 the Langley Memorial Aeronautical Laboratory was dedicated at Langley Field, Virginia. That same year brought an event that would prove of much more far-reaching import for the coming second American aeronautics revolution. As a result of lobbying by Ames and other American scientists, Max Munk, one of Prandtl's most gifted students, was hired by NACA to become its chief theorist. Munk brought with him not just the Göttingen group's body of theoretical aerodynamics but also his own idea for a new and revolutionary wind tunnel.

The basic aerodynamic equations that Prandtl and others had derived had revealed an extremely important fact about wind-tunnel testing. Airflows acting on a small-scale model in a wind tunnel would be representative of what happens to a real airplane in flight only when an aerodynamic parameter known as the Reynolds number was the same in both situations. Unfortunately, the Reynolds number depended directly on the size of the surface. Typical scale models were one-twentieth the size of the real thing, which meant the Reynolds number was twenty times smaller as well. This problem was at the heart of the misleading results the Wright brothers had obtained for their thin airfoils.

Munk noted, however, that the Reynolds number was also proportional to air pressure. Thus if a wind tunnel could be built in which the air pressure was pumped up to twenty times normal atmospheric pressure, it would be possible to obtain realistic data using scale models. Munk had tried without success to interest the Zeppelin Company in his idea while working there immediately after the war. At Langley he immediately began working to sell his new employers on the idea, and a year and a half later the Variable Density Tunnel (VDT) was in operation. Its pressure vessel, made of two-inch-thick steel plate, was constructed by a nearby shipbuilding company using techniques normally employed in ships' steam boilers.

Munk was by all accounts impossible to work with. Along with the scientific ideas he brought from Germany, he brought a very German attitude about proper academic social hierarchy, and he treated the scientists of his division as underlings who were to do his bidding without question. He drove the engineers building the VDT to distraction and near rebellion by constantly meddling in construction details. "Dr. Munk does not seem to have any clear idea as to what he wishes in the engineering design, excepting that he is sure that he does not want anything that we suggest," complained the head of NACA's aerodynamics section. When Munk was named the laboratory's chief of aeronautics in 1926, the section heads under him resigned en masse. Offended, Munk himself then resigned, whereupon the section heads all returned.

His personal arrogance aside, Munk's scientific approach was fundamentally at odds with the very engineering-heavy culture of American aeronautics; the highly mathematical nature of his papers was simply beyond most American aircraft designers at the time. Until American universities began to produce aeronautical engineers with the necessary background in mathematics and physics, this was unlikely to change.

But the practical results of his methods began to pay off at once. With Munk's new wind tunnel, the United States in a stroke leapfrogged over the competition; the VDT alone "made the United States the undisputed leader in applied aerodynamics for the next fifteen years," wrote the aviation historian John Anderson. Munk put the VDT to immediate use, developing a systematic series of airfoil shapes that builders could incorporate into their wing designs, confident that they would perform exactly as calculated. The idea was extended in the early 1930s into NACA's famous "four-digit series" of airfoils, which became a veritable "designer's bible," used by aircraft manufacturers throughout the United States, Europe, and Japan.

NACA built other radically new wind tunnels in the late 1920s and 1930s that kept America in the lead and provided the crucial technical underpinning to complete the transition from biplane to monoplane in the 1930s. In 1929 B. Melvill Jones, the first professor to hold a newly endowed chair in aeronautical engineering at Cambridge University, gave a lecture before the Royal Aeronautical Society that would become a classic in the history of aeronautics. Jones began with a simple and direct condemnation of aircraft designers' failure to deal with the drag problem:

> Ever since I first began to study Aeronautics I have been annoyed by the vast gap which has existed between the power actually expended on mechanical flight and the power ultimately necessary for flight in a correctly shaped aeroplane. . . . There is a natural tendency to decide on one day that the gain—say 20 per cent on the total drag, or 7 per cent on the speed—to be had by spending endless trouble on improving the undercarriage design is not worth the trouble; on the next day to come to a similar conclusion about the drag of the engine cooling apparatus; on the next day about the wire, struts and minor excrescences . . . omitting to notice that if all the improvements were made at once the total gain . . . might reduce power consumption to a small fraction of its original value and so extend the range and usefulness of the aeroplane into realms that would otherwise be unattainable.

Jones reviewed the sources of unnecessary drag and concluded that an astonishing two-thirds of the power expended in a typical airplane of the day was being thrown away; a properly "streamlined" airplane whose parts were shaped to conform to airflows could fly sixty miles per hour faster without any increase in engine power.

It was a call to arms. NACA had already scored one well-publicized coup using one of its advanced wind tunnels to achieve drag reduction in aircraft. The Propeller Research Tunnel allowed testing of full-size propellers and engine housings, and a major research effort had been made on reducing the drag associated with air-cooled engines. Air-cooled radial engines were rapidly becoming the standard in postwar America; with fewer moving parts, they were more reliable than water-cooled engines and easier to maintain; the short crankshaft of the radial configuration was robust and able to take the strain of rough landings, a consideration that loomed large for naval carrier operations and for the fledgling commercial airline fleets that had to cope with a variety of semiprimitive conditions across a vast country. The fixed-cylinder radial configuration overcame centrifugal-force problems and horsepower limitations of the spinning-cylinder rotary while retaining the advantages of lighter weight, and weight was a major consideration in commercial transport, where every pound of engine was a pound less of paying freight. Takeoff weight also was the critical factor for carrier-based planes, which faced the immutable constraint of having to get airborne in a very short distance.

But by definition air-cooled engines required cooling air to flow over cylinders protruding into the airstream, and that was a significant source of drag. No one knew just how significant that drag was, but at the behest of commercial manufacturers NACA was asked to study the problem in 1927. By the next year the NACA researchers, working largely through trial and error using the Propeller Research Tunnel, were able to announce an astonishing result: a properly shaped cowling that covered the cylinders could reduce the drag of the entire fuselage by 60 percent, while actually improving engine cooling by efficiently directing airflow over the cylinders. In a nonstop flight from Los Angeles to New York, a Lockheed Vega fitted with the NACA cowl increased its top speed from 157 to 177 miles per hour. The invention was widely publicized and widely adopted, and helped to put the issue of drag reduction on the map. It also significantly narrowed the advantage that liquid-cooled engines had always offered by virtue of their slimmer cross section and correspondingly smaller head resistance.

In 1931 NACA posted another major milestone in the art and science of drag reduction when it opened a huge wind tunnel with a thirty-by-sixty-foot cross section that could accommodate a full-size aircraft. The Full Scale Tunnel allowed systematic "drag clean-up" trials in which individual components, everything from canopies to radiator inlets to antenna poles, could be added or removed or modified and the effect on drag measured each step of the way.

By the 1930s American aircraft had begun to take on a distinctly "modern" appearance: sweepingly curved bodies, streamlined canopies, tapered fillets smoothing the abrupt angles, fully enclosed engines, retractable landing gear. The biplane with its boxy protuberances suddenly began to look every bit as quaint as it would in the jet age a half century later. A new age in flight had truly begun.

Little of the groundbreaking aeronautical research of this period was aimed specifically at military application. Contracts from the military services in America and Britain did play an essential part in the development of engines with ever higher horsepower by firms such as Wright, Curtiss, Pratt & Whitney, and Rolls-Royce throughout the interwar period, and research in the U.S. Army's own labs helped achieve the initial breakthrough in cylinder design in the 1920s that made high-powered radial air-cooled engines feasible in the first place. U.S. Army scientists also pioneered the use of ethylene glycol as a coolant in liquid-cooled engines, a crucial advance incorporated by Rolls-Royce in its famous Merlin engine of the 1930s. The U.S. Army was also the prime mover in the development of higher-octane fuels, needed to prevent knocking in the high-temperature, high-boost military engines. In 1930 the Army adopted 87-octane fuel as its standard; in 1935 this was increased to 100 octane. During the 1930s the R-1820 Wright Cyclone engine went from 500 to 1,200 horsepower, with a significant portion of this increase directly attributable to the use of higher-octane fuel. The civilian aviation industry was a secondary beneficiary of much of this military-directed engine and fuel research.

Yet the very opposite was true of almost every other major development that lay behind the monoplane revolution of the 1930s. It was the growing needs of civilian air transport that would time and again prove the motive force that carried these critical technologies from the wind tunnels of Langley Field or the blackboards of Caltech to the aircraft factories of New York, Seattle, and southern California. If the civilian market benefited from the military's development of powerful and reliable engines, military aeronautics benefited many times over from the developments spurred by the sudden burst of commercial air travel that swept America in the late 1920s and early 1930s.

In the 1920s, civilian air travel was still a novelty rather than a practical means of transportation or a reasonable way of making money. Transport planes rarely had a top speed of more than 120 miles per hour. Costs were extremely high; the scattered attempts at launching commercial airline services throughout the world nearly all relied on heavy government subsidies.

As early as 1922, the French builder Louis Breguet called attention to the connection between improved aerodynamic performance and commercial viability; he calculated that doubling the lift-to-drag ratio of existing transport aircraft would cut per-pound, per-mile costs by a factor of five.

In that year American aircraft companies had built three hundred airplanes, a postwar nadir. But in 1926 came a new law requiring the Post Office to contract with private enterprise for the carriage of airmail. The airmail contracts were the catalyst the airline industry needed, and they marked the start of a boom such as few industries have ever seen. That year American factories built 1,186 planes. The incorporation of the new structural and aerodynamic technologies began to offer hope for real cuts in operating costs. And then came Charles Lindbergh.

In 1919 a New York City hotel owner had offered a $25,000 prize for the first nonstop flight between New York and Paris. By May 1927 it remained unclaimed—though six men had died trying. Of those left in the race that spring, the "dark horse," the press reported, was an unknown twenty-five-year-old "St. Louis mail pilot named C. A. Lindbergh." The other competitors planning to attempt the hazardous flight were flying planes built by well-known makers: Fokker, Wright, Sikorsky. Lindbergh's was built by Ryan Airlines of San Diego, a small firm about as unknown as the pilot. To cut weight to the bare minimum, Lindbergh was proposing to do something that most aviation experts considered borderline suicidal: he would make the 3,600-mile flight by himself. "I do not think that a man can stay awake thirty-six hours by himself with nothing but the sea, sky, and air as an environment and a motor roaring away monotonously," warned Colonel Clifford B. Harmon, the president of the International League of Aviators.

One by one the competitors dropped out. Two crashed. One was grounded by a court injunction filed against the plane's owner by one of his pilots, who had been fired and replaced on the eve of the planned takeoff. On May 12 the press began to take the dark horse more seriously when he landed in New York at the end of a solo, one-stop, coast-to-coast flight from San Diego—setting a new transcontinental record in the bargain. At 7:52 a.m. on May 20, Lindbergh's *Spirit of St. Louis* took off from Roosevelt Field on Long Island. By nightfall he was reported to have circled over Saint John's, Newfoundland, then headed out to sea.

The afternoon of the next day William L. Shirer, a cub reporter covering the French tennis championships for the Paris *Tribune,* noticed the American ambassador leaving his box in a hurry. He went to find out what was going on: the ambassador said he had just got word from the embassy that Lindbergh had been spotted over Ireland. Shirer jumped in a taxi and soon found what seemed like every car in Paris on the road heading for Le Bourget field. Two miles from the field he gave up, paid the driver, and made the rest of the way on foot. By ten o'clock the crowd had grown to half a million. Soldiers with fixed bayonets had to push the mob off the airstrip. And then the sound of a plane was heard; a small searchlight picked it up, and it was Lindbergh beyond doubt. When he clambered out of the cockpit the crowd almost tore him apart in their rapture; only a quick-witted French aviator saved the day, fighting his way to Lindbergh's side and creating a diversion by tossing his helmet into the crowd, gaining enough time to whisk him into a car and have his plane surrounded by military trucks.

The sensation that Lindbergh's feat aroused defied rational explanation. He was on the front page of every newspaper in the Western world. France awarded him the Legion of Honor. In England he was invited to Buckingham Palace and presented to both Houses of Parliament. President Coolidge sent an American warship to bring him home. Arriving in Washington he was greeted by the largest crowd ever to gather in the nation's capital. In Coolidge's welcoming speech "the usually laconic Yankee President let loose with torrents of words," recalled Shirer; Lindbergh was a "wholesome, earnest, fearless, courageous product of America," a "genial, modest American youth, with the naturalness, the simplicity, and the poise of true greatness." On and on Coolidge went, in one of the longest speeches of his career. In New York City four million people turned out for a ticker-tape parade to honor the new hero; the *New York Times* the next day devoted its first sixteen pages to the event. By July Lindbergh had received three and a half million letters and a hundred thousand telegrams from well-wishers. "He had better get the money now," sardonically commented Gertrude Ederle, who had received her own headlines and ticker-tape parades a year before as the first woman to swim the English Channel, but who now was almost forgotten.

But the public did not forget Lindbergh. Nor did they ever really forget the excitement he aroused for aviation. The *Spirit of St. Louis* was no great technical innovation, but Lindbergh's flight set off a huge surge of interest in flying. The "Lindbergh boom," they called it. The air carriers plying the mail routes began to find they had paying passengers in numbers that had been inconceivable just months earlier. From fewer than 9,000 passengers carried in all of 1927, the figure leapt to 48,000 in 1928; 162,000 in 1929; 385,000 in 1930.

By 1929 output from American aircraft factories had grown to 6,193—90 percent of these civilian aircraft. And by 1933, a fateful year in the history of both aviation and world events, American plants had begun to produce airplanes that at last reflected the full state of the art of structural and aerodynamic science. In that year Boeing delivered its first Model 247 transport, a ten-passenger, all-metal, twin-engine monoplane with monocoque construction, retractable gear, engines with NACA cowlings, and engine wing-mountings designed by NACA to reduce drag as well. Boeing received orders for fifty-nine before the first one even flew. The first 247 to fly, in fact, was a production model—so confident were the engineers, they did not bother to build a test model.

The 247 had a cruising speed of 190 miles per hour, 70 miles per hour faster than most of the existing competition. Boeing had its own airline company, Boeing Air Transport (it would later become United Airlines), and it naturally was at the top of the list of airlines that were waiting to receive the 247. TWA, seeing that it might have to wait years to get one of the sleek new planes that had proved such an instant success with the flying public, accordingly wrote to the Douglas Aircraft Company asking if Douglas might be able to build something comparable. The result, three years later, would be one of the most successful airplanes ever built: the DC-3.

That the Nazi rise to power on January 30, 1933, coincided almost precisely with such a fateful era in American aviation history was not entirely coincidental. The same forces of violent anti-Semitic nationalism that would help bring Adolf Hitler to power in Germany would hasten the departure to America of scientists who would revolutionize American aerodynamic science and its practical applications.

Like Max Munk, the man who would play a crucial part in the DC-3's successful design was trained at Göttingen and came from an upper-middle-class Jewish family. Theodore von Kármán's family included a long line of scholars; among them, von Kármán always proudly asserted, was the sixteenth-century rabbi and mathematician who was said to have invented the first mechanical robot, the "Golem of Prague." Von Kármán's father had acquired the "von" from Emperor Franz Josef as a result of his work reorganizing Hungary's antiquated educational system, which in turn had led him to be called upon to supervise the education of the Emperor's young cousin, Archduke Albrecht.

A child prodigy in mathematics, von Kármán had somewhat alarmed his father by being able at age six to multiply six-digit numbers in his head; worried that his child would become "some kind of a freak," his father henceforth forbade him to think about mathematics and made him read books of history, poetry, and geography. But von Kármán returned to mathematics with a passion in his teenage years, and went on to win a national prize in mathematics and science while attending the Minta, or "Model" Gymnasium, the secondary school his father had founded as a revolutionary experiment in education.

The Minta encouraged reasoning over rote learning and emphasized drawing examples from everyday life. Von Kármán would surely have been a prodigy no matter where he grew up, but the burst of enlightenment that turn-of-the-century Budapest enjoyed, and that the Minta embodied, provided an atmosphere of support and intellectual nourishment that was in many ways unparalleled in the world at the time. In an agriculturally rich but backward nation, social reforms had swung open the doors of opportunity and prosperity, and the Jewish middle classes eagerly rushed in. By 1910, 50 percent of Hungary's lawyers and 60 percent of its doctors were Jewish. Even more extraordinary was the concentration of scientific genius that emerged from this Hungarian Jewish enlightenment. Besides von Kármán, six of the century's greatest physicists, including some who would play a crucial role in America's invention of the atomic bomb, were Hungarian Jews who would flee rising anti-Semitism: George de Hevesy, Michael Polanyi, Leo Szilard, Eugene Wigner, John von Neumann, and Edward Teller. Other physicists joked that the Hungarians must be men from Mars who had adopted the cover story of being Hungarian to explain away their accented English; there seemed to be no other explanation for this galaxy of brilliance.

Von Kármán, born in 1881, was the first of these men from Mars to land on the planet. After earning a degree in mechanical engineering from the Polytechnic University in Budapest, he won a fellowship to Göttingen; there he developed an explanation for form drag—the drag caused by flow separation—that would become a classic in aerodynamic science. Von Kármán was an "internationalist" by instinct and upbringing; he had grown up speaking German and French in addition to his native Hungarian. There were already signs of the disasters to come for Jews in Germany, where he had become a professor at Aachen, when a cable reached him while he was vacationing in the summer of 1926. "WHAT IS THE FIRST BOAT YOU CAN TAKE TO COME HERE? MILLIKAN," it read.

Millikan was Dr. Robert A. Millikan, Nobel Prize winner in physics, president of the California Institute of Technology. This much von Kármán knew, but the rest of the message seemed obscure. Ten days later a letter arrived: Caltech was inviting him to come to Pasadena to help the university set up an aeronautical laboratory. That fall he took the first of a series of leaves of absence from Aachen that would have an enormous impact on American aviation.

Millikan was convinced that southern California was the place where the aircraft industry in America would take off, and he was convinced that forging ties between industry and academia was the key to the success of both. For his part, von Kármán said, "I had always believed that the goal of my life was to eliminate the gap between scientific theory and application." Von Kármán spent longer and longer stretches at Caltech, helping to plan a wind tunnel and hire staff, but remained reluctant to accept the offer of a permanent position that Caltech repeatedly pressed on him. The coming of the Nazis settled the matter. In January 1934 he received a stiff letter from a bureaucrat in the German Ministry of Education informing him that his leave of absence would not be extended and that if he did not resume his full-time post at Aachen he would be required to resign and forfeit his pension. Von Kármán at once wrote back with his resignation, scornfully informing the bureaucrat, "I hope in future years you will be able to do as much for German science as you have done for foreign science this year." Writing to his old mentor Prandtl—who had courageously protested the dismissal of Jewish scientists from Göttingen, but who would become increasingly co-opted by the Nazi regime as war approached—von Kármán was a bit more coy but no less sarcastic: "The German academic life has some advantages, for instance a definitely better beer than here, but I think you will agree with me that this is not sufficient reason for me to neglect the disadvantages." Later that year von Kármán, to his surprise, was approached by the Nazi government with a proposal that he return to Germany to become a consultant to the rapidly expanding Aviation Ministry. Von Kármán knew that the ministry's head, Hermann Göring, was reputed to have said, "Who is or is not a Jew is up to me to decide." But, von Kármán recalled, "I had a good laugh and remained in Pasadena."

The Douglas Aircraft Company, which had set up operations in Santa Monica, California, not far from Caltech, had already approached the university for help on several small projects. The design of the prototype DC-1 was of a different order of magnitude, however, and it also represented an ideal opportunity for von Kármán and Millikan to demonstrate how the gap between science and practical application could be bridged. One crucial problem had cropped up almost immediately, and was as swiftly solved by von Kármán. Wind-tunnel tests showed that as speeds increased, the DC-1's low-wing, monoplane configuration gave rise to potentially fatal eddies that broke from the leading edge of the wing and then struck smack on the tail, causing the entire tail unit to buffet violently. Von Kármán concluded that the problem was due to the sharp angle where the wing met the fuselage, and that it could be cured by adding a small fillet to smooth the corner. Von Kármán recalled his hands-on approach:

> I enjoyed climbing into the ten-foot wind tunnel with a wad of putty, and imagining myself being the airplane as I tried to feel where I might be pressed by an element of air. To show up the flow of air on the model, we used wool tufts and fine threads attached with tape to different parts of the model. We tried to observe where the tufts and threads ran smooth, and where they flapped violently, indicating that air was separating from the surface and wasting its energy in turbulent motion. In this way we found, as we suspected, that the air had rough going when it went past the corner between the wing and fuselage. As soon as the corner was puttied in, the air smoothed out. It was really wonderful to see how well the fairing worked.

Von Kármán's discovery that the effective load-carrying capacity of thin metal sheets could be greatly increased by running stiffeners along them was another crucial insight that helped Douglas pioneer the aircraft industry's transition from wood to all-metal construction.

The DC-3 that followed swiftly from the DC-1 prototype carried twice as many passengers as the Boeing 247 and pushed cruising speeds to 205 miles per hour. Its range when fully loaded was nearly a thousand miles. With a 95-foot wingspan, the DC-3's lift-to-drag ratio was 14.7, nearly double that of the typical World War biplane. Its wing loading of 25 pounds per square foot was more than three times greater than the typical biplane's. In time, Douglas would build more than ten thousand DC-3s.

The American civil aviation boom set off aftershocks that jolted loose some other technological developments that, although less obvious, would prove no less vital to the realization of the air force vision of a long-range strategic bombing force. Some were purely serendipitous. Pilots flying Boeing's 1930 forerunner to the 247, the all-metal single-engine Monomail, had found they were barely able to get airborne in the thin air of Cheyenne, Wyoming, high in the Rocky Mountains. The problem turned out to be a fundamental one. High wing loadings exacerbated a tradeoff that was inherent in all propeller designs. When a plane is cruising at high speed, the optimum angle for the propeller blade—its "pitch"—is fairly coarse. But during takeoff, the slow forward motion of the airplane results in a net flow of air over the blade at such a high angle of attack that the air will stall over parts of the blade, causing a drastic loss of thrust. The only recourse the pilot had was to cut engine rpm during takeoff, making the propeller turn more slowly, thereby reducing the effective angle of attack and preventing a propeller stall—but also sacrificing power and greatly extending the length of the takeoff run.

As cruising speeds increased in the new fast monoplanes, so did wing loadings, and that in turn made for still longer takeoff runs; optimizing the propeller pitch to the new planes' higher cruising speeds exacerbated the takeoff problem still more. Boeing's solution was an idea that had generally been considered far too complex to put into practice: an adjustable-pitch propeller. The idea was to fasten the blades to the spinning propeller hub via a mechanism that could feather them to different angles. For cruising speed, the optimal coarse pitch could be maintained, but during takeoff the propeller would be reset to a fine pitch to match the airplane's slower forward speed.

The first adjustable propeller Boeing tried had just two settings. But, stimulated by the new and potentially vast commercial market for the device, a wave of innovative thinking and tinkering was set off, and within a year a greatly improved, continuously variable-pitch prop was available. It automatically adjusted to maximize energy efficiency at any selected engine speed. The continuously adjustable prop was retrofitted into most of the 247s to create the 247D model. The variable-pitch prop was one of those small and seemingly arcane details of technical innovation that changed everything. Few people looking at an airplane fitted with a variable-pitch propeller would have noticed the differ-

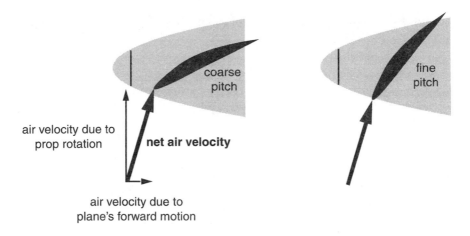

air velocity due to
prop rotation   net air velocity

air velocity due to
plane's forward motion

coarse
pitch

fine
pitch

The angle at which air strikes a propeller blade is the combined result of forward motion and prop rotation speed. Slow takeoff speed produces a very high angle of attack, causing air to stall over the blades. The variable-pitch prop solved the problem by rotating the blades to fine pitch for takeoff to reduce angle of attack, returning to coarse pitch for cruising speed.

ence, but this one piece of technology was literally the making of the fast mono-plane—whether commercial transport, bomber, or fighter.

The higher speeds and greater wing loadings of the new monoplanes created another problem that was similarly solved through the commercialization of inventions that had been awaiting realization. Higher wing loading—which is another way of saying smaller wings—made for greater efficiency during high-speed cruising but caused problems during the slower speeds of takeoff and landing, when the wings would cease to generate enough lift to keep the plane flying. The solution was a variety of adjustable high-lift devices ("flaps" and "slats") that could be extended from the leading or trailing edges of the wing to effectively expand the wing area and generate more lift when needed on takeoff and landing, but retracted during flight to reduce drag.

For a plane to go fast, it also had to be able to cope with a wider range of speeds than was ever the case before. And so one set of technological advances called forth further advances. Variable-pitch propellers and flaps were the way an airplane could still perform at low speeds for takeoff and landing without sac-rificing the aerodynamic gains that had made the breakthrough to high-speed flight possible.

In the American factories that now were quickly establishing themselves as the world's leaders, the air-power revolution was riding the back of commerce.

And had it not been for a fortuitous brush with the world of commerce in 1934, the United States Army Air Corps might well have gone the way of the interwar RAF in blithely neglecting the essentials of navigation and bad-weather flying.

The United States owed its technological edge in navigation equipment fundamentally to the growth of American civilian aviation. But indirectly, the credit was due to one of those slightly eccentric and very American philanthropic foundations that sprang up in such abundant variety in the early decades of the twentieth century. In 1936 the sociologist Eduard Lindeman published a study of one hundred American foundations, representing roughly a third of those then in existence, and found that all but six had been founded since 1900. Most had been endowed by wealthy industrialists; 61 percent of their activities fell in the broad category of "education," mainly support to institutions of higher learning.

Daniel Guggenheim was typical of the industrialists who sought to invest— or as their detractors would put it, launder—their excess wealth in this manner. Guggenheim's father had made a fortune in mining ventures. In 1880 he paid a friend a few thousand dollars for a one-third interest in several speculative mines. By 1890 the mines were valued at $14.6 million. The Guggenheim family business invested in new mines and smelters, and by 1923 Daniel Guggenheim was able to sell his Chile Copper Company to Anaconda for $70 million. He then settled down to giving his money away. In 1924 he and his wife established the Daniel and Florence Guggenheim Foundation with the stated aim of advancing "the well-being of mankind throughout the world."

It was a rather more narrow cause that soon attracted his interest. His son Harry had been a naval aviator during the war and had continued to fly as a private pilot, and he was well aware of America's dearth of first-rate aeronautical scientists and research centers. In 1926 Daniel Guggenheim announced the establishment of a new fund, the Guggenheim Fund for the Promotion of Aeronautics, endowed with $2.5 million of the family's money. Over the next several years, the fund would establish six schools of aeronautical engineering at universities around the country. It played a crucial role in underwriting Caltech's aeronautics program and, in particular, agreed to foot the bill for Millikan's offer to von Kármán of a munificent $10,000 a year to direct the program; it set up a "Model Air Line" to provide air service between Los Angeles and San Francisco; and it launched a highly focused research effort aimed at solving the problem of "fog flying," as it was then known.

There were three separate problems in flying under poor visibility. One was simply navigating from point to point across country without getting lost. The second was coming down at the right slope and direction to end up on a runway when the ground was fogged in. And third, in many ways the most perplexing and daunting, was controlling the plane while flying blind, without the visual feedback of seeing the horizon.

From the earliest days of flight, pilots had been taught to rely on the feel of the wind and the seat of the pants. Basic instruments such as the altimeter, air-

speed indicator, compass, and inertial turn-and-bank indicator were mostly standard by the end of the World War, but as Charles Lindbergh recalled, the
accepted wisdom was that "a good pilot doesn't depend on his instruments." The
only catch was that "good" pilots would often fly into clouds and immediately
become disoriented when they lost sight of the horizon. It was not unusual for
pilots who survived such an ordeal to believe they were flying straight and level
the whole while, only to emerge from the cloud to see the ground spinning rapidly toward them. It was also not unusual for pilots not to survive the ordeal.

A fundamental fact of physics, as counterintuitive as it seemed to pilots, is
that a steady turn feels no different from level flight. In 1926 an Air Corps pilot,
Major David Myers, did a simple experiment to prove it. He placed a pilot in a
revolving chair and spun him around at a steady rate. If the man in the chair had
his eyes shut, he was unable to accurately state even which way the chair was
spinning. If the rotation was slowed, the man would sense the deceleration but
usually insist that the chair had now stopped spinning altogether; if the chair was
stopped, the man would say the chair was now spinning in the other direction.
When pilots were placed in the spinning chair with a hood over their heads but
supplied with a compass and turn-and-bank indicator that they could consult, the
pilots flatly refused to believe what the instruments were reporting.

In the late 1920s many mail pilots were discovering the hard way what Myers
had found sitting in an office swivel chair. Fog and low clouds in the famous
"Hell Stretch" over the Allegheny Mountains between New York and Ohio took
a constant toll of pilots. By late 1927 the National Bureau of Standards had
worked out a scheme using radio beacons to guide pilots flying on the New York
City–to–Cleveland route when ground landmarks were invisible. It proved a brilliant success, and soon beacons were guiding pilots on major airways all over the
country. That solved problem number one.

The other two problems were much thornier. In August 1928 Harry Guggenheim announced that the matter had grown so pressing that the fund would take
the unusual step of setting up its own "Full Flight Laboratory" to study fog flying under realistic conditions. Guggenheim believed that what was at stake was
public confidence in flight itself; the commercial airline industry would never
succeed unless people believed that flying was safe. But he knew that the experiments would need an experienced test pilot, and so it was to the chief of the Air
Corps that he turned for help: Would the Air Corps be willing to assign a pilot to
the project for a year? The Air Corps was willing. The man chosen for the job
was Lieutenant James H. "Jimmy" Doolittle.

Doolittle was already famous. He was the first man to fly an outside loop, the
first man to fly across country in a single day, and the winner of the 1925 Schneider Trophy for setting that year's world speed record. That same year he had also
received a doctorate in aeronautical engineering from MIT.

Initial tests at Mitchel Field on Long Island convinced Doolittle that even if
pilots could be trained to trust their instruments, the instruments of the day were

scarcely trustworthy. Flying from the hooded rear cockpit of a Consolidated NY-2 using instruments alone to guide him—a "check pilot" manned the controls in front as a safety backup—Doolittle had little difficulty identifying the inadequacies of the instruments. The inertial bank indicator was not much help in telling a pilot what his plane was doing. Consisting of a little metal ball in a U-shaped tube filled with a viscous fluid, it was actually little more than a mechanical analogue of the seat of the pants; it indicated if the plane was correctly banked to match the centrifugal force of a turn, but as for where the little ball ended up, a correctly banked turn looked no different from straight-and-level flight: in either case, the inertial force pushed the ball straight down toward the floor of the cockpit. A change in compass reading could in theory tell a pilot if he was in a turn; the catch was that magnetic compasses were slow to react, and would get stuck altogether while the plane was banking. Altimeters, which indicated height by measuring barometric pressure, lagged too, often hundreds of feet behind the true reading as their mechanical components caught up with changing pressure while climbing or descending.

Doolittle took upon himself the job of seeing whether instrument makers could improve this bleak situation. He went to see Elmer A. Sperry of the Sperry Gyroscope Company in New York, and Sperry proposed outfitting the cockpit with two entirely new instruments. One would be a gyroscopic compass. Unlike a magnetic compass, a gyroscope set on the ground to point true north would continue to faithfully point that direction no matter how the plane banked or climbed or dived; it would also react instantly, without any appreciable time lag. The other instrument Sperry proposed was a gyroscopically maintained artificial horizon. Also calibrated before takeoff, this instrument would give the pilot a true picture of the plane's attitude at any instant, showing whether its wings were banked, and its nose pointing up or down, by providing a reference line that always remained parallel to the true horizon.

On September 24, 1929, flying out of Mitchel Field on a fogbound morning in the hooded cockpit, Doolittle took off and made a landing using nothing but instruments. A radio beacon provided a line for the final approach, and a marker beacon across the approach line sent a second signal indicating when he was in range to cut his engine for the final glide down the runway. The next day's *New York Times* ran a story declaring that "aviation had perhaps taken its greatest single step in safety."

Despite Doolittle's starring role in these tests, the Air Corps showed remarkably little interest in the results until an unexpected circumstance shook its complacency to the core. In February 1934, President Franklin D. Roosevelt, believing that the Post Office airmail contracts had been awarded fraudulently by the Hoover administration, sent a quiet inquiry to the Air Corps: Would it be able to take over the job? Benjamin Foulois, whose years of lobbying for the top job in the Air Corps finally had paid off two years earlier, interpreted this presidential inquiry as an order and not a request. He met quickly with his staff and replied yes, sir; asked when his pilots would be ready, he shot from the hip again and said a week to ten days. Douglas MacArthur, the Army Chief of Staff, was taken aback by Foulois's precipitous reply, but felt he had to put on a bold front, too. He at once declared that of course the Army would do a "magnificent" job of delivering the mails. With that assurance, Roosevelt canceled the contracts and directed the Army to take over the job until new competitive contracts could be awarded.

It was a disaster. Just as operations began on February 19, the country was hit with some of the worst late-winter weather ever recorded: snow, ice, dense fog, gales. The Army pilots almost to a man lacked experience flying at night and in bad weather. Gyro compasses, artificial horizons, and radios were hastily installed in the Army planes designated for mail duty, but pilots were unfamiliar with their use and mechanics with their maintenance. In the first week six Army pilots were killed in crashes. By the time commercial airmail service was restored on June 1, the toll had reached twelve dead in sixty-six crashes.

The President was furious. But this costly and embarrassing episode taught a lesson the Air Corps never forgot. In mid-March the Army instituted a course in instrument flying and began an immediate research program on automated and assisted landing systems. The fiasco had also driven home the importance of radio communications, and the Army began constructing a nationwide network of control towers, navigation beacons, and radio stations that would, in the Second World War, become a worldwide system to control and guide its aircraft.

Delivering the mail was not war, but it was a lot closer to the real thing than the routine the Air Corps had complacently settled into. In the words of the official history of the United States Air Force, the Air Corps "had become the prisoner of ease and habit." It took a civilian job to break the habit and disturb the ease—and to awaken the Air Corps to the fact that civilian technologies had passed them by.

Whatever an American bombardment force was going to do, Americans were convinced it would need to be done with precision and deftness. Certainly the emerging raison d'être for heavy-bombardment aviation that the Air Corps was touting in the 1920s—repelling an invading sea force—demanded high accuracy; even the more narrowly defined Air Corps mission as laid down by official postwar Army doctrine—supporting the infantry with direct attacks on enemy positions—demanded high accuracy.

Yet, beyond these rationales, something more elemental fostered the faith in the power of precision that began to loom large in Air Corps thinking from the 1920s on. Even before anything approaching a formal doctrine or operational concept for the employment of bombers began to gel, American air leaders had turned their attention to finding a technical solution to the problem of bombing accuracy. If morale bombing was to be the peculiarly British obsession between the wars, then precision bombing was to be America's.

It may have been naïve or self-righteous, but it was an obsession honestly come by. In some ways the American belief in the virtues of precision warfare was a cultural shibboleth that transcended any immediate developments in air-power doctrine. Americans, as the post–Second World War official history of the Army Air Forces noted, had a "traditional reverence for marksmanship" dating back to the days of the frontier, back to Lexington and Concord. Myth and legend or not, and the slaughter at Gettysburg and the "pacification" of the Philippines argued the former, Americans liked to believe that in a fight they got results with a minimum of well-directed force. It was the stuff of American legend: Minutemen taking deadly aim behind stone walls, the Kentucky rifleman who could knock a squirrel out of a tree a half mile away, the western sheriff who could shoot a gun out of a desperado's hand before anybody got hurt.

Most Americans of the nineteenth and early twentieth centuries were also firmly convinced that the United States held an unquestioned moral superiority to the decadent Old World when it came to both the purpose and means of warfare. European nations fought for territorial gains and other such ignoble ends; America fought for right. It was this attitude on the part of Woodrow Wilson that so galled French President Georges Clemenceau during postwar negotiations: As the French sought to carve up Europe to their satisfaction, Wilson pressed his Fourteen Points. (God, Clemenceau observed aloud, had only required ten. But then talking to Wilson, he added, was rather "like talking to Jesus Christ.")

In much the same vein, Trenchard's advocacy of morale bombing at the end of the war had met instant opposition from many American military and political leaders, on both practical and ethical grounds. Colonel Edgar S. Gorrell, the head of the Air Service's technical section, prepared an American proposal for strategic bombing in late November 1917 that parted ways with Trenchard's thinking at several key points. Gorrell saw specific German war factories as the "shank of the drill" that supported the enemy army at its point; if the shank were weakened, the drill would break. In contrast to Trenchard's notion of scattering bombs around Germany to keep as many workers awake and rattled as possible, Gorrell advocated concentrating the entire weight of the American bombardment force on one precise objective at a time. "There are a few certain indispensable targets without which Germany cannot carry on the war," he wrote. For instance, German gas shells were "being fired at Allied troops and positions over a large area of the Front; but the manufacture of these shells is dependent upon the output of a few specific, well-known factories." Knock out the chemical plants and the production of gas shells will cease. The same was true of aircraft, which were dependent on the Bosch magneto factory and Mercedes engine plants in Stuttgart.

The war ended before the United States could assemble a bomber force large enough to test Gorrell's theories. But Secretary of War Newton Baker spoke for the consensus of American feeling when, in his 1919 Annual Report, he rejected as a matter of principle any strategy that would make civilian areas a deliberate target of air attack. Although he made direct mention only of the German attacks upon London and Paris—and not the RAF and French attacks upon Cologne and Frankfurt—Baker argued that the results showed that this strategy should have no place in America's planning for future war. Such attacks, he wrote, "constituted an abandonment of the time-honored practice among civilized peoples of restricting bombardment to fortified places or to places from which the civilian population had an opportunity to be removed." They were also ineffective. Rather than weakening British resolve to continue the fight, the German raids had only underscored to the British public the necessity of defeating "so brutal a foe," who had shown his willingness "casually to slaughter women and children and to destroy property of no military value." Baker concluded that, far from sowing defeatism, "history will record these manifestations of inhumanity as the most powerful aids to recruitment in the nations against which they were made." On both moral and practical grounds, morale bombing was not an admissible use of air power. If America's air commanders wanted to develop a bombardment force, the burden would be on them to prove that it could be something other than a weapon of indiscriminate terror.

The RAF had basically dealt with the dismal inaccuracy of bombing during the war by declaring that accuracy didn't matter under the doctrine of morale bombing. The American air force had no such easy out: it needed to show that it really could blow up what it was aiming at.

Tests during the war had reported that bombs dropped from 8,000 feet fell an average of 800 feet from the target even under ideal training conditions; under the stress of actual combat, things went rapidly downhill from there. An Army board reviewing the lessons of the war in 1919 concluded that "the most important problem" that would have to be solved to improve the accuracy of aerial bombing was the development of a markedly better bombsight. The best bombsight the Army had in 1919 was the Mark I, a British model developed by the Royal Navy in 1917 as an incremental improvement on the standard CFS 4 model. Instead of requiring the bombardier to use a stopwatch to measure ground speed, the Royal Navy's Mark I "drift" sight automatically adjusted for the effect of a headwind or tailwind through a simple geometric calculation: To calibrate the sight for wind speed, the pilot would first fly directly across the wind while the bombardier lined up an angled "drift bar" with the apparent track followed by an object on the ground. The greater the crosswind, the greater the ground track would angle away from the direction the nose of the plane was pointing. The drift bar was mechanically linked to the bombsight's speed-setting scale to correct it for the wind velocity thus determined. The pilot would then turn ninety degrees and fly directly upwind or downwind for the actual bombing run.

The limitations of the World War sights were no mystery. While the drift concept as a way of directly ascertaining true ground speed was sound in theory, the

Mark I was a crude realization of the idea. Having to fly directly across the wind for a calibration run and then directly parallel to the wind for the bomb run imposed unrealistic tactical constraints for a bomber flying in a war zone. The lack of a stabilization system that would keep the sight mechanism parallel to the ground even as the plane banked from left to right or the nose bucked up and down was, as Henry Tizard had found back in 1915, a major source of bombing error. And the Mark I and its ilk made no provision whatever for addressing the basic matter of steering the plane on the correct course during the bombing run. The standard method during the war had been for the bombardier to tap the pilot on the left or right shoulder to give steering instructions—or, in airplanes that put the pilot in the rear seat, to tug on strings tied to the pilot's arms.

Billy Mitchell was the first to light a fire under the Army's technical men. In March 1921 he held a meeting in Washington with Lawrence B. Sperry of the Sperry Gyroscope Company and asked him if he could start work at once on a gyroscopically stabilized bombsight to be used in the naval bombing tests that summer.

It would end up taking three years rather than three months for Sperry to deliver. The complex gyroscopic mechanism turned out to be the least of the difficulties of the Sperry bombsight. An electromechanical computer that was supposed to automate most of the bomb-aiming calculation proved expensive, temperamental, and cumbersome. Sperry informed the Army that if they ordered fifty of his C-1 sights, the price would be $12,500 apiece—more than the per-unit cost of many airplanes. The calculator mechanism suffered constant breakdowns. It used so much electricity that two turbine-driven electric generators had to be mounted under the wings to provide the power for it, and it then proved so sensitive to fluctuations in the current flow that the plane had to fly at a precise and unvarying speed to keep the generators spinning at the right rate.

The Army tried developing its own bombsights in-house with somewhat better success; at least these D models were cheaper and much more reliable, and while they lacked all of the automated features of Sperry's models they still seemed to promise an improvement over the Mark I. An official Air Corps publication released December 12, 1927, boasted that the D-1 sight had cut the average bombing error at 8,000 feet from 795 feet to 160 feet.

The timing of such a confident statement could not have been worse. One week later the Air Corps began a full-scale live bombing exercise that made it abundantly and embarrassingly clear that almost nothing had changed since the war. The Pee Dee River Bridge near Albemarle, North Carolina, was only two years old but was soon to be flooded by a new dam, and the Army was given the chance to use it for target practice first. In a week of clear skies, flying at low altitudes, a detachment from the 2nd Bombardment Group lobbed miss after miss at the bridge until finally the air commander ordered his aircraft to fly in formation and all drop their loads simultaneously on cue from the lead aircraft. This was substituting a company of men with shotguns for a marksman with a rifle.

The Chief of the Air Corps, Major General James E. Fechet, was disgusted with the performance and the day after receiving the results sent an order to the Army's Materiel Division to do something about it at once. The present bomb-sight, he wrote, "falls far short of being the instrument of precision which is nec-essary in order to insure accurate bombing under service conditions from normal bombing altitudes. . . . I cannot too strongly emphasize the importance of a bomb sight of precision, since the ability of bombardment aviation to perform the mission of destruction is almost entirely dependent upon an accurate and practical bomb sight." An Army Bombardment Board echoed the chief's con-cerns. "The situation is exceeding acute," it reported. A bomber was no better than its bombsight, and so urgent was the need that in developing a workable sight "the question of cost" should not even be considered. What was needed was results.

The Army spent the money. It did not get the results. Then in late 1931, and almost by accident, the Army learned the astonishing news that the Navy had been developing an advanced bombsight of its own. And the Norden Mark XV bombsight was an advance that left the Army's efforts in the dust. This was hardly the first time the Navy had acted as though the real enemy it had to keep its secret weapons hidden from was not a potential foreign foe but the United States Army. The Navy always thought that the Army couldn't keep a secret. But in this case there was obviously more to it than that. With the two services jock-eying for control of the coastal-defense mission, the Navy was hardly eager to give the Army a leg up on such a crucial technological breakthrough.

The Sperry Gyroscope Company was a large and established firm, the recog-nized leader in gyroscopic instruments. Carl L. Norden, Incorporated, the maker of the Navy's Mark XV, was a tiny and virtually unknown outfit operating from a one-floor factory in New York City. The Norden company was a creature of the Navy. It had been founded only in 1927, and then solely at the Navy's behest, and it had agreed to do no commercial business, no business with foreign coun-tries, no business at all, in short, other than making bombsights for the Navy. Adding to the intrigue was the fact that Carl L. Norden, the engineer who con-tributed both his name and the technical expertise to the business, was a former Sperry employee who had been waging a relentless personal and professional feud with his former boss. Norden had quit in 1913 after receiving what he con-sidered an insulting $25-a-week raise for solving an oscillation problem with Sperry's gyroscopes. Elmer Sperry was convinced that Norden had swiped all of his ideas from him. Norden scornfully retorted that Sperry would claim he had invented gravity, and file a patent for it, if he thought he could get away with it.

Norden was the classic solitary inventor. He always worked by himself, often taking off to Switzerland for months at a time seeking creative contemplation. He was a perfectionist who saw the world in black and white, and to say he lacked social skills would be an understatement. The Navy called him "Old Man Dynamite." An employee once asked him why he had designed a part a certain way; he snapped, "I had ten thousand reasons, none of which are any of your damn business."

In 1918, while working as a consultant after his rift with Sperry, he had caught the attention of the Navy's Bureau of Ordnance with the logical and precise comments he had filed about problems with an aerial torpedo under development. Soon he was working full-time as a consulting engineer on the bombsight problem for the Navy.

The Mark XV was the culmination of a series of highly innovative designs that Norden had developed. Gyroscopically stabilized, the Norden sight automatically determined ground speed and crosswind using what was called the synchronous method. Looking through a telescopic lens, the bombardier would turn a knob that rotated the sight until the target followed the track marked by a vertical crosshair. He then turned a second knob, which controlled the speed at which a motor-driven mirror tracked along the airplane's ground path, adjusting it so that the target appeared to stand still in the sight. From those settings, together with altitude, airspeed, and bomb type, the bombsight calculated ground speed and crosswinds and determined the correct bomb release point. The mechanism could even be coupled directly to a gyroscopically controlled autopilot, so that the bombardier's act of lining up and synchronizing the target would send steering instructions to the airplane itself; on the final approach, the bombsight would "fly" the plane to the proper release point and send a signal at the right instant to drop the bombs automatically.

Although the Norden bombsight was, by its very nature, a highly sensitive instrument—basically, it used gears and wheels and currents to represent the mathematical equations of the trigonometric bomb-dropping problem, and any wear, slip, friction, or dirt affected the accuracy of the mathematical result that emerged—it was an order of magnitude ahead of the Army's models. In theory it could cut average bombing error to something like 100 feet. The entire package weighed a mere fifty pounds, an extraordinary engineering achievement. Bombardiers began to joke about the bombsight getting commissioned instead of them, but the feeling of awe the device evoked was more than a joke. "The more I found out about the bomb sight, the more ingenious and inhuman it seemed," one bombardier said. "It was something bigger, I kept thinking, than any one man was intended to comprehend. I ended up with a conviction . . . that a bombardier can't help feeling inferior to his bomb sight."

No one else in the world had anything like it, nor would they. Lieutenant Colonel Herbert Dargue of the Air Corps made eight trips to Europe in the 1920s and 1930s to check on rumors of advanced British, French, and German bombsights and each time found only slightly improved First World War models. Dargue reported in March 1936 that the British, lacking anything approaching a precision bombsight, were "resigned to the fact that horizontal bombing will be effective only against area targets." The Germans, meanwhile, were putting most of their faith in dive bombing.

In October 1931, a Navy plane flying at 5,000 feet achieved a 50 percent hit rate in a bombing trial on the anchored cruiser USS *Pittsburgh* using a Norden sight, a stunning improvement over previous accuracy rates. This time the test results were shared with the Army, and in April of the following year the Army joined the Navy in placing an initial production order for fifty-five Mark XVs, at $4,275 apiece.

The irony of it was that by 1932 the Navy had all but abandoned the idea of high-level precision bombing after tests had shown the promise of dive bombing as a way to evade antiaircraft fire while precisely attacking an enemy ship. Still, the Navy was hardly going to give up such a prize to its rival service, and it zealously blocked all subsequent attempts by the Army to establish its own separate business relationship with Norden. All contracts had to go through the Navy's Bureau of Ordnance. Even long after the Navy dropped its requirements for a precision high-altitude bombsight, it still insisted on a 50-50 split of the production. At one point well into the Second World War the Navy had thousands of spare Norden sights sitting on the shelf while Army bombers were flying combat missions with inferior models.

The Navy also exercised its jurisdiction over the program to impose a "Secret" classification on manuals, paperwork, and the bombsights themselves that proved to be an enormous burden on training and operations. The Army complained but was not entirely guiltless on this score, for it found that a certain amount of mystification could be inspiring to the troops, as well as insulation against public skepticism about the true precision of precision bombing. Presenting the Norden bombsight as a wonder weapon was a brilliant bit of public relations and morale building. At air bases the Nordens were kept under lock and key in secure vaults, escorted to their planes by armed guards, and shrouded in a canvas cover until after takeoff. Eventually the secret got out: in 1941 American authorities learned that a spy working as a draftsman at the Norden factory had delivered drawings of the Norden sight to Germany in 1938, and in 1942 a complete bombsight fell into German hands. But even after the Army eased security rules, it still kept up the PR smokescreen. A PR man at the Army's bombardier school in Midland, Texas, invented a "Bombardier's Oath" that, he solemnly assured reporters, all recruits had to swear to: "In full knowledge that I am guardian of one of my country's most priceless military assets . . . [I] swear to protect the secrecy of the American Bombsight, if need be, with my life itself."

It was another PR genius—quite possibly Norden's business partner Theodore H. Barth, who had a way with people and a gift of persuasion that Norden himself notably lacked—who coined the phrase that would be heard over and over about the Norden bombsight: It could put a bomb in a pickle barrel from twenty thousand feet.

The Norden bombsight was surely an extraordinary technological achievement, and there was no denying that it represented a vast practical leap over existing bombsights. There was also no denying the importance that the Army attached to it: the United States would spend $1.5 billion developing and producing the Norden sight, more than half the amount that would be spent to develop the atomic bomb. But perhaps inevitably, its combat performance, as events would only much later prove, was never as great as its myth. Once again, the magic aura of the air weapon was having a hypnotic effect.

Right up to the eve of the Second World War, American air commanders would insist with straight faces that they had nothing in their sights but a hostile fleet steaming toward America's shores. It is "utterly absurd," said Major General Frank M. Andrews in December 1936, that anyone would consider the American long-range bomber force as anything but purely defensive. At the time Andrews uttered this categorical opinion, however, the Air Corps' best and brightest rising young officers had for a solid decade been receiving indoctrination in the gospel of offensive bombardment as the alpha and omega of air power.

The Air Corps Tactical School had been established in 1920 to train the next generation of air leaders; it stood at the top rung of a dozen or so specialized training schools within the Air Corps, and only officers who had ten or more years of commissioned service were generally eligible to compete for one of the coveted spots at the school. Many of the lieutenants, captains, majors, and lieutenant colonels who graduated from the Tactical School would go on to become general officers: of the 320 generals serving in the Army Air Forces at the end of the Second World War, more than 80 percent would be Tactical School graduates.

For its first six years, the school's curriculum dutifully toed the official Army line that airplanes supported infantry, period. But by 1926 the signs of heterodoxy were already boldly in evidence. In that year's "Employment of Combined Air Force" course, students were told that air attack against the enemy's interior was the key to success: "It is a means of imposing will with the least possible loss by heavily striking vital points rather by gradually wearing down an enemy to exhaustion." By 1930 long-range bombardment's star had ascended even higher, to the zenith of the air power firmament. That year the "Air Force" course text left no doubt that bombardment was the "basic arm" of military aviation, and that except in the most unusual circumstances the weight of the air force would be deployed against "strategical objectives" in the enemy's interior, not on the battlefield or its immediate vicinity. By the mid-1930s air-ground cooperation was relegated to a single one-day course that counted for one-fortieth of the students' final grades, and while the title of the course was "Aviation in Support of Ground Forces," its main message was clearly "don't bother": students were told that "the most valuable contribution the air force can make to the ground campaign is the successful destruction of targets deep inside enemy territory."

As great an influence as the Tactical School was as a training ground for future air leaders, its influence went far beyond training; its faculty was the only group of air officers who had the time and the reason to give much attention to questions of basic doctrine and strategy, and it quickly became the de facto think tank and doctrine center for the entire Air Corps. With its move from Langley Field, Virginia, to Maxwell Field in Alabama in 1931, the Tactical School had greater freedom to develop its "forbidden doctrines" far from the eyes of the Army General Staff in Washington. And so it was there, two years later, that what would become the distinctly American theory of strategic bombing began to take shape.

In their fundamental assertion that air power held the key to a decisive or even independent victory through the bombardment of the enemy's interior and the consequent collapse of the enemy's will to resist, the Tactical School theorists were no different from advocates of morale bombing such as Trenchard and Douhet. The Tactical School's great innovation, however, was to substitute precision attack on key industries for indiscriminate terror bombing as the mechanism for inducing enemy collapse. The overwhelming American moral opposition to indiscriminate bombing; the breathtaking promise of the Norden bombsight (the Tactical School's newly arrived assistant commander in 1934 was Herbert Dargue, late commander of the Air Corps' premier bombardment group and close friend of Norden company president Theodore Barth); and the innovative thinking of several of the Tactical School's faculty members all helped ferment this new bombing doctrine. In retrospect, the same type of self-fulfilling logic that drew in the morale-bombing advocates was working its seductive force on the inventors of precision bombing. They saw a technological capability, then posited a theory to explain how that capability could translate into decisive effect. The American doctrine of precision bombing was all the more seductive for the intricacy and cleverness of the mechanism by which, according to its developers, it would produce victory.

The essential idea, which seems to have come to several of the Tactical School's instructors more or less simultaneously during the 1933–34 school year, was that modern industrial economies were not only highly interdependent but highly vulnerable to any disruption at key "bottlenecks" or "choke points." Major Donald Wilson, a Tactical School student who had joined the faculty with its move to Maxwell in 1931, had worked on civilian railroads and recalled observing that the lack of a key lubricating ingredient could bring an entire railroad to a halt. And so the hunt was on among instructors and students to find the weak threads in what Wilson termed the "industrial fabric" or "industrial web" of modern societies.

It was obviously both easier and more politic to use the United States as the case study; information about American industry was readily available, and in the 1930s it was still taboo to consider potential targets in a foreign country. The Tactical School staff combed press reports of manufacturing interruptions; they wrote to the Army Industrial College to ask what effect a loss of electric power in the Northeast would have on munitions plants. Soon the "classic example" of an industrial bottleneck "literally fell into our laps," recalled Haywood S. Hansell, Jr., a Tactical School instructor who would become a high-ranking air force commander in the Second World War. In March 1936 a flood in Pittsburgh hit a small factory that made a simple but unusually shaped spring needed for variable-pitch propellers. It was the only such plant in the country, and soon the entire American aircraft industry was suffering delays in production. An entire industry had "been nullified," said Hansell, just as effectively as if every aircraft factory in the country had been bombed.

Choke points could be physical or conceptual. Rail junctions were an obvious example, but detailed economic analyses revealed more subtle connections that held the industrial fabric together: basic commodities such as oil, ball bearings, brass, aluminum; networks such as electric power transmission lines; even services such as banking and communications. The Great Depression only seemed to confirm the notion of a domino effect that could spread through the entire economy starting from a single flick at the right spot. The stock-market crash and the ensuing chaos and dislocation that engulfed the world's mightiest industrial economy were ample proof of the delicate state of equilibrium and interdependence in which modern economic systems were precariously balanced.

Precision was a perfect match for the industrial-web theory. It would not take wanton destruction to achieve decisive results; it would not even be necessary to destroy all of the factories in a key industry. It was "for want of a nail" writ large: find the crucial industries or services that the entire economy depended upon; strike precisely at the single crucial factory or office or supply line in each of those key industries; and economic disruption would spread like a shock wave through the enemy society, far wider and faster than physical destruction itself could be inflicted.

It was never completely clear what was going to happen next, however. Some of the Tactical School lecturers implied that the ultimate purpose of unraveling the enemy industrial web was to bring the enemy's "war-making capacity" to a standstill; the enemy's will to resist would collapse because its military *means* to resist would be neutralized once armaments production was halted. But by the late 1930s another payoff of precision attack upon the enemy's industrial web was receiving at least equal attention from the Tactical School theorists: it was the most efficient and humane way "to destroy the will of the people at home" and thus pressure the civilian populace to surrender. Killing people was both immoral and inefficient. No one knew how many civilian deaths it would take to break morale, and such attacks might even backfire, stiffening the resolve of the populace. But a few well-placed bombs to sever the crucial strands of the industrial fabric could cause such disruption of day-to-day life that the resulting social disintegration and public clamor would force an enemy government to sue for peace.

In a lecture for the Tactical School's "Air Force" course in the late 1930s, Major Muir S. Fairchild made the point that the enemy's military-industrial mobilization was not so much the ultimate target of the precision-bombing campaign as it was an extra strain on the enemy nation's entire industrial web that made an overall economic collapse through bombing of vital nodes all the easier to achieve. Wartime production stretched the web taut, ever nearer its breaking point; when it snapped, the *whole* economic structure of modern urban life would go with it. "If we picture section after section of our great industrial system ceasing to produce all those numberless articles which are essential to life as we know it," Fairchild told his students, "we can form an idea of the pressure that would be exerted" upon the civilian population. As an illustration, he asked the students to imagine what would happen to New York City if vital points in the aqueducts, railways, and electric power lines feeding the city were cut. Refrigerated food would quickly spoil and the delivery of new food would be interrupted; sanitation would be undermined; the threat of fire would loom large; and in short order, "the area would become untenable and the population could have to be evacuated."

The Tactical School precision-bombing theory tied up all the loose ends in one neat bow. Bombs would kill things, not people; advanced technology and sophisticated economic knowledge would substitute for brute force; the precise application of military power to the specific vulnerabilities of modern society would replace the caveman tactic of raw intimidation. It was a scientific and humane version of Trenchard's and Douhet's theories—humanized through the wonders of American know-how.

Many of the Tactical School instructors did not hesitate to join Trenchard and Douhet in declaring theirs to be an independent war-winning strategy: by striking at the "economic and political heart" of a nation, the airplane could bring about defeat directly in a way soldiers and sailors never could. And so, they began to insist, the diversion of air power to fighting battles on the ground could only dissipate its force and weaken its most effective punch.

The Tactical School did few studies, however, that extrapolated actual data on bombing accuracy all the way through to their operational implications in order to see how many bombs and bombers it would actually take to destroy a given target. When one Tactical School instructor, Laurence S. Kuter, finally did look at the hard numbers in the late 1930s, he gulped. There was no shortage of test data on bombing accuracy: From 1930 to 1938 the Air Corps had dropped more than 200,000 bombs during training exercises. The results showed that from low altitudes, 5,000 to 8,000 feet, the bombers were sometimes able to hit within a hundred feet of their target using the Norden bombsight. But generally two, or three, or four hundred feet was more the norm, especially when the bombs were dropped from 10,000 feet or above. And this was under ideal conditions of clear skies, well-marked targets, and no one shooting back.

When Kuter began to analyze the data and factor in the known destructive radius of explosives, he saw that the implications were staggering. For example, the Tactical School had earlier assumed that a mere nine 300-pound bombs would be sufficient to put out of commission one model chokepoint they had identified, the Sault Ste. Marie locks on the Great Lakes. A single medium bomber could carry out the entire mission. But Kuter found that a miss of two hundred feet, even a hundred feet, would leave a hard target such as a canal lock unscathed. Even under ideal conditions—sandy soil, delayed fuzing so the bomb penetrated before exploding—a 2,000-pound bomb would blast a crater only fifty feet wide. (Destructive force dissipates very rapidly with distance. A 2,000-pound bomb might kill a man standing a hundred feet away but leave a man two hundred feet away with nothing worse than his hair mussed. The Army's Picatinny Arsenal in New Jersey had blown up in a cataclysmic accident in 1926—a *billion* pounds of TNT had gone up—but a watchman 1,500 feet away was only slightly hurt.)

Taking everything into account and using realistic data, Kuter now calculated that getting those required nine hits on the canal locks would require 120 bombers dropping more than a thousand bombs. The entire precision-bombing theory should have been cast into doubt by this unsettling discovery.

But by this point the seductive power of an idea had gained the added momentum of technological and institutional enthusiasm, and theory was once again in the vanguard, charging far ahead of technological reality. The year 1935 had been pivotal. In that year the Air Corps had won a major step toward full independence with the establishment of a "GHQ Air Force" that would report directly to the Chief of Staff in peacetime and to the commander of Army field forces in wartime. It was not the clean break that many airmen wanted, and the Army still aimed to keep a check on air force autonomy through an almost schizophrenic division of authority: although GHQ Air Force would command all operational air units, the Office of the Chief of the Air Corps, with its own entirely separate chain of command, would continue to control personnel, funds, and procurement. Nonetheless, the key principle of unified tactical command of all aviation units under a single air officer, long a central demand of airmen, had at last been recognized.

In 1935, too, the air force had seen the airplane that seemed the answer to its fondest air-power dreams: the B-17.

The aeronautical advances stimulated by the American civil-aviation boom quickly made their way into the military side of the business at Boeing and Martin and other American aircraft makers, setting off a parallel revolution in the design of bombers. Following closely on the heels of its civilian Monomail, Boeing produced the first American all-metal monoplane bomber, the B-9, which made its maiden flight in April 1931. With two engines, a streamlined fuselage, and retractable landing gear, the B-9 could carry 2,260 pounds of bombs in an internal bay at a maximum speed of 186 miles per hour, some 50 miles per hour faster than any of its predecessors. In stretching out the fuselage, however, Boeing's engineers had underestimated the longitudinal stresses that would be placed on the airframe. The B-9's metal skin became so wrinkled in flight that crews began calling it the "tissue-paper bomber." The following year Martin rolled out its prototype B-10, which proved more successful; the production version featured the first fully enclosed crew positions and gun turrets and a top speed of 207 miles per hour.

Like the rest of the government, the Air Corps was struggling with budget cuts imposed as federal revenues declined in the wake of the 1929 stock-market crash. In 1926, Congress had authorized a five-year expansion plan eventually leading to a force of 1,800 aircraft, but the completion date kept slipping as appropriations lagged; by 1933 the Air Corps had only about 1,200 aircraft that were usable for duty. But suddenly help came from an unexpected quarter. The day President Roosevelt was sworn in to office, March 4, 1933, he called Congress to an emergency session to begin just five days later. "This Nation asks for action, and action now," he declared. By June, Congress had passed the sweeping National Industrial Recovery Act; among its many other provisions to provide employment and stimulate the economy, it authorized $3.3 billion in public-works spending. While most of that money would go for projects such as schools, hospitals, municipal buildings, dams, and bridges, the Public Works Administration agreed to help the nation's struggling aircraft firms by transferring $7.5 million to the Air Corps' fiscal year 1934 budget for the purchase of new airplanes. With that windfall, the Air Corps was able to order ninety-two aircraft, a significant boost for both the service and the industry.

With its immediate budget crunch eased, the Air Corps decided to press ahead with a more ambitious program its engineers had been studying: in August 1934 the Air Corps mailed a circular to manufacturers requesting design proposals for an improved multiengine bomber that would have a range of 2,200 miles and a top speed of 250 miles per hour. A flying prototype would have to be submitted to the Air Corps in just twelve months.

All but one of the manufacturers that responded assumed "multi" meant two. Douglas's entry, which would become the B-18, was based on its DC-3; Martin proposed an improved version of its already successful B-10. Boeing alone decided to try a radical leap into new territory. The company had already been doing preliminary design work on a four-engine passenger plane, the model 300. Boeing President Claire Egtvedt flew directly to Dayton to ask the authorities at Wright Field if a four-engine design would be eligible: It would. The Boeing board voted to risk $275,000 of the company's money on the project.

Boeing's prototype, the model 299, which emerged just one year later, incorporated all of the best features of previous bombers as well as the streamlining advances of the 247 transport. But it was a huge airplane, with a wingspan of 104 feet and a weight of 35,000 pounds that was four times that of the B-10. It could carry 5,000 pounds of bombs for a range of 1,700 miles or 2,500 pounds for

2,300 miles. Newspaper reporters on hand at sunup on July 28, 1935, to witness the model's first flight saw the machine guns thrusting from its five streamlined turrets and decided it looked like a fort with wings; the nickname "Flying Fortress" stuck. Three weeks later, the 299 flew 2,100 miles nonstop from the Boeing factory in Seattle to Wright Field, making an average speed of 232 miles per hour.

This was the airplane the Air Corps had been waiting for, and nothing was going to stop them from getting it. In a strange echo of the events that had marked the Army's initiation into aviation twenty-seven years earlier with the crash of the Wright Flyer at Fort Myer, disaster struck—but as before, even a disaster was insufficient to deflect the momentum that had now built up. On October 30, with an Army test pilot at the controls and the Air Corps brass lined up to watch, the 299 lifted off from the runway at Wright Field, then immediately plunged downward, striking the ground and bursting into flames. Both pilots were killed and four crewmen injured. The plane was a total loss, and by the rules of the competition Boeing was out of the running; the production contract would have to go to Douglas's B-18. An investigation subsequently revealed that the pilot had taken off with the elevator controls locked.

But the Air Corps at once found a loophole to get the plane it wanted anyway. The Army could purchase a small number of planes for "service test" purposes without going through competitive bidding. In January 1936 thirteen YB-17s (the "Y" indicated a test model) were ordered from Boeing. Hap Arnold, who had worked off his disgrace after being exiled to Fort Riley following the Mitchell affair and had since returned to Washington as Assistant Chief of the Air Corps, was exultant over the B-17. "For the first time in history," he declared, here was "Air Power that you could put your hand on."

The B-17 was far and away the most advanced airplane in the world at that moment, and you didn't need to know a thing about airplanes to know it. Its smooth expanses of streamlined metal, its bristling armament, its four huge, 930-horsepower, nine-cylinder radial Wright Cyclone engines set it apart from anything else in the air. At every opportunity the Air Corps showed off its fleet of four-engined giants, sending them on "goodwill missions" to South America, flying them around the country, intercepting ocean liners in the mid Atlantic (with a photographer and an announcer from the NBC radio network along for the ride)—until the Navy complained loudly enough to make the Army Chief of Staff order the fliers to stay clear of the Navy's territory. Only no one could ever find the order in writing, and the Chief of Staff almost immediately authorized the commander of GHQ Air Force to approve "exceptions" to his "rule" limiting the planes to within a hundred miles of the coast.

The fourteenth YB-17, designated the B-17A, signifying the first "production" model, entered service in 1938; equipped with turbo-superchargers that boosted engine output to 1,000 horsepower and also allowed the engines to work at high altitudes by compressing the thin intake air, the B-17A had a top speed of more than 300 miles per hour and a service ceiling of 38,000 feet. Orders for thirty-nine further-improved B-17B models, at an unprecedented $100,000 apiece, followed that year. The Air Corps, and Boeing, had their eyes on an even bigger bomber, too, a behemoth with a 5,000-mile range that was already taking shape as the experimental XB-15.

And nowhere in the Air Corps did the B-17 have a profounder impact than at the Air Corps Tactical School. Its instructors could now point to a real-life airplane as the embodiment and proof of all of their theories. Here was a bomber that could cross vast distances, carry tons of bombs, outrun the fastest pursuit planes of the day; here indeed was air power "you could put your hand on."

7

THE FIGHT FOR THE FIGHTER

The winter of 1935 was a lonely time to be a fighter pilot, especially a fighter pilot in the United States Army, and especially one assigned to the Air Corps Tactical School. Claire L. Chennault was a West Texas cowboy turned fighter pilot, a pugnacious loner by nature, and his chronic bronchitis, exacerbated by eighty cigarettes a day and the mounting fury he felt in his isolation among the "bomber boys" at Maxwell Field, didn't improve his disposition.

The bomber's ascent had brought about the fighter's nadir. Douhet had asserted that bombers flying in formation, supporting one another with interlocking fields of fire, were invincible. The B-17's speed and firepower seemed to fully justify Douhet's prophecy. The fastest fighter in the United States force, the all-metal monoplane Boeing P-26, with a top speed of 235 miles per hour, was scarcely faster than the B-10's 207 miles per hour; it was roundly outpaced by the B-17. Hap Arnold concluded that even if a fighter could catch a bomber, the increasing speeds would be pushing the limits of human reaction time; a fighter pilot would have seconds, or maybe fractions of a second, to get off a shot.

Exercises that pitted fighters against bombers officially confirmed that the fighter plane was as good as dead. Chennault heard such talk all the time. "The fighter is obsolete," people were saying. The umpire of a 1931 exercise at Wright Field concluded: "Due to increased speeds and limitless space it is impossible for fighters to intercept bombers and therefore it is inconsistent with the employment of air force to develop fighters." At the Tactical School, Chennault grumbled, the bomber boys were telling their students that bombers were so invulnerable they would not even require fighter escorts. Captain Harold L. George, one of the Tactical School's leading bombardment enthusiasts, asserted, "The spectacle of huge air forces meeting in the air is the figment of the imagination of the uninitiated." Antiaircraft artillery was a "mild annoyance," nothing more. Lieutenant Kenneth N. Walker, another instructor, lectured that the only way to break up an enemy attack was to bomb the enemy air force while it was still on the ground; once in the air, a bomber force was virtually "impossible to stop," he said, with either fighters or flak.

Chennault was furious over all of this; he thought it was madness; a theory, an unproven theory, was being used to negate all of the experience of the World War. The exercises, he said, were ridiculous, artificial, rigged even. "All sorts of fantastic and arbitrary restrictions were placed on fighters" in the maneuvers to keep them from performing successfully. The P-26 was in any case obsolescent, the first monoplane fighter in the U.S. force; it had fixed landing gear and a relatively small 570-horsepower engine; and once new types were developed that incorporated all of the aerodynamic advantages of the B-17, fighters would once again have the edge over them. Nothing had arisen to shake the basic lesson of the war, Chennault said: Air supremacy gained by concentration of fighter aircraft was the sine qua non for the application of all air power.

In the class he taught on pursuit aviation, Chennault hammered away at this theme. The key to employing pursuit aviation was good intelligence about the enemy, and flexibility. In a report he wrote on the role of pursuit in the "Next Great War," he argued that the erroneous conclusion of "some of our military leaders" that "pursuit can never be effective for the denial of hostile bombardment" was largely due to a failure to analyze *why* fighters had failed to intercept bombers in exercises. In every case, he said, the "outstanding factor" was "the lack of definite, continuing information of the hostile force." Chennault doggedly worked out a plan for a warning network that could direct fighters to the right spot, en masse. He organized a counterdemonstration, an exercise at Fort Knox in 1933 in which fighters guided by a network of ground observers equipped with nothing more than binoculars and telephones succeeded in intercepting bombers. He lectured on tactics and formations. He identified speed, rate of climb, ceiling, firepower, and maneuverability as the key factors in fighter performance, placing speed above all; he firmly dismissed proposals for multiseat fighters as a step in the wrong direction, since their added weight would take a toll on speed and rate of climb, essentials for fighters that needed to get airborne quickly and intercept incoming bombers once a warning was received.

Chennault won few converts. The pursuit course counted for a small portion of students' grades, and many students were put off by his vehemence. The bombardment proponents could offer an all-encompassing theory whose internal logic was smooth, compelling, and attractive; Chennault frankly admitted that war was messy and that pursuit aviation did not fall into a neat box: "Pursuit has no 'normal' role in any phase of war," he said. "Pursuit is a weapon of opportunity and should be employed strictly in accordance with the situation at a given time."

Being deliberately undoctrinaire in an age of doctrinal certitude came at an inevitable price. During the 1935–36 school year, Chennault set down his views and critique of the establishment position in a blistering series of articles in the *Coast Artillery Journal,* a professional publication sympathetic to the idea that airplanes could be shot down. His colleagues at the Tactical School were taken aback by the bitterness of his words, and in the spring of 1936 at the annual faculty meeting a motion to drop the pursuit course from the school's curriculum was adopted. Soon Chennault was a virtual outcast. He moved his family off base and ate lunch alone or with small groups of enlisted men. The next year his ill health forced his retirement from the Air Corps, and he went into as complete a self-imposed exile as a retired American Army officer could arrange for himself in the 1930s: he moved to China to become an adviser to Chiang Kai-shek.

Belief in the bomber's invincibility and the impotence of the fighter only heightened the paradoxes of an air policy built around the doctrine of strategic obliteration of the enemy. If bombers were invincible, then surely this was ringing vindication of the decision to make the bomber the centerpiece of the air arm. This was how it looked to the airmen. To many others, however, the invincibility of the bomber led to an utterly different conclusion: If the bomber could not be stopped, the only hope for civilization lay in ridding the world of this scourge altogether. The war of mutual devastation—annihilation, even—that seemed the unavoidable consequence of a conflict between two nations armed with strategic bombers was hardly war at all, not a clash of arms but a suicide compact. If the bomber could not be neutralized through military means, some other means would have to be found to free the world of its threat.

It was in Britain that such thinking found the greatest voice in the 1930s. The feeling of vulnerability to air attack had taken root there in the earliest days of aviation, and since the war a belief in the awful power of the bomber had become a set part of the doctrine of British airmen. It was in Britain, too, where the shock of suffering and loss in the war was translated most passionately into agitation for disarmament to prevent such a catastrophe from ever recurring. Few doubted that the real cause of the World War had been a surfeit of arms, and throughout the 1920s and early 1930s the Liberal and Labour Parties staked their political fortunes on opposition to any and all proposals for rearmament. Successive governments had repeatedly deferred the expansion of the RAF to fifty-two squadrons that had been announced in 1923, but even the slowed expansion was too much for many. Clement Attlee, who in 1935 would become the leader of the Labour Party, denounced the planned addition of three squadrons called for in one year's budget, telling the House of Commons, "The Secretary of State for Air is very carefully laying the foundation for future wars."

By 1932 the pressures of the worldwide economic depression only added to the force of the disarmament arguments. The convening of an international conference on disarmament in Geneva that year was embraced across the British political spectrum. Winston Churchill, by then a backbench member of Parliament who had split with his own Tory party over its policy on India, recalled that "the virtues of disarmament were extolled in the House of Commons by all parties." When Germany demanded that it be freed from the restrictions of the Versailles Treaty as part of any new international arms-limitation agreement, the bastion of the British establishment, the *Times,* warmly endorsed the idea; it was nothing more than "the timely redress of inequality."

This was before Hitler came to power. But even the coming of the Nazi regime did little to make the supporters of disarmament question the rightness of their cause. Many in Britain's ruling elite distrusted France and felt that Hitler's insistence on equality of armaments was merely a matter of Germany's regaining her national self-respect. Lord Lothian said it was "unpatriotic" for Englishmen to "refuse to believe in the sincerity of Germany"; he for one welcomed "the offers made to the world by Herr Hitler." Most people would have called them demands, not offers, but then the *Times* reassured its readers that the "shouting and exaggeration" of the new German regime, even the daily public clubbings and murders on the streets of Berlin by storm troopers that went on throughout the spring of 1933, were "sheer revolutionary exuberance" and should not be "misconstrued" by those who truly understood the justness of Germany's case. Churchill's lonely, acerbic attack upon Prime Minister Ramsay MacDonald's proposal that France virtually abolish its large standing Army as a first step toward achieving parity with Germany was almost shouted down in Parliament. "Thank God for the French Army," Churchill had said, and he saw unmistakable looks of "pain and aversion" on the faces around him as he said it. As soon as he was done speaking, members of all three parties jumped to their feet to denounce Churchill as "a disappointed office seeker" who was trying to "poison" the spirit of amity and conciliation that MacDonald had worked so hard to forge at Geneva.

Even those who correctly perceived Hitler's intentions to rearm argued that this was all the more reason to put faith in international disarmament agreements. The Conservative Party's Stanley Baldwin—who held the antiquated and essentially meaningless title of Lord President of the Privy Council in the coalition government that had taken office in 1932, but who in truth was the de facto Prime Minister behind the weak MacDonald—warned the Cabinet in early 1933, shortly after Hitler became Chancellor, that the air arm would be the first branch of the armed forces the Nazis would build up. Baldwin said this clearly meant "we must have a convention prohibiting bombing." And it was Baldwin who, the previous fall, had ended a debate on disarmament in the House of Commons with the famous warning: "No power on earth can protect the man in the street from being bombed. Whatever people may tell him, the bomber will always get through."

The idea of prohibiting, or at least restricting, aerial bombardment through international law had never completely vanished even after the feckless efforts of the 1899 Hague conference. In 1923 a commission of international jurists at The Hague had drawn up draft rules on aerial bombardment; the rules forbade bombing "for the purpose of terrorizing the civilian population" and required that bombardment be limited to "the immediate neighborhood of the operations of land forces" and target only clearly defined military objectives. But the problems with such rules were apparent to everyone. James Spaight, a civilian official of the British Air Ministry who served as a technical adviser to the British delegation at The Hague, pointed out that it would be easy to obey the letter of the law and still devastate an enemy's cities: "The doctrine of the 'military objective' . . . will be no adequate protection. Any belligerent who chooses will be able to keep within the rules . . . and yet use his air arms for a purpose quite distinct from the destruction of objects of military importance, namely, for the creation of a moral, political, or psychological effect within the enemy country." Cities all contained targets that could be justified as militarily important and that the attacker could claim were his intended objective, whether he managed to hit them or not.

In February 1933 the British delegation at Geneva proposed a way out of these contradictions that reflected both the powerful momentum the aerial disarmament movement had gained and the lengths to which the world would perforce be driven if it truly wanted to abolish the threat of strategic bombardment. The Air Committee of the conference was "to examine the possibility of the entire abolition of military and naval [air] machines." If it was utopian, it was also the only logically consistent position.

The committee grappled with the idea but could see no way to close yet another loophole that immediately gaped open: how to ensure that civilian airliners would not be used for bombing. A month later, the British delegation tried another tack and introduced a draft treaty that would ban bombing rather than the bomber—subject only to one peculiarly British proviso that quickly became the butt of public ridicule: Bombing would be prohibited "except for police purposes in certain outlying regions." The British wanted to be sure they could still bomb the colonial natives who didn't pay their taxes.

Hitler played along with all of it. Germany was "perfectly ready to disband her entire military establishment." All she asked in return was simple equality. The *Times* once again called Germany's claim for equal treatment "irrefutable." Even Germany's abandonment of the Geneva conference and withdrawal from the League of Nations in October 1933 was not enough to extinguish the hopes of many in Britain that disarmament was the path to peace. During a by-election a few weeks later, the leader of the Labour Party, George Lansbury, declared that all nations "must disarm to the level of Germany as a preliminary to total disarmament." Labour won resoundingly, riding a tide of pacifist sentiment.

But quietly, within government, second thoughts were already beginning to percolate. The British Chiefs of Staff the following month warned that rising threats in Europe and Asia demanded an immediate commitment of £100 million over the next five years for new ships, completion of the naval base at Singapore, and expansion of the air force. Neville Chamberlain, the Chancellor of the Exchequer and future Prime Minister, had the year before insisted that the financial dangers facing the country outweighed the military dangers; now he reversed his view. The policy that emerged was a mass of contradictions and inconsistent half measures, but that was politics, which often demands hedging and trying to satisfy competing constituencies. Without abandoning its support for disarmament, the Cabinet agreed to propose some modest new expenditures for "the vital necessity of putting our defences in order."

And so in July 1934 the government presented to Parliament a proposal to complete the long-delayed expansion of the Home Defence Air Forces to fifty-two squadrons, with an additional twenty-three to follow. As in the original plan that Trenchard had devised, more than half of those new forces would be bombers. Parliament approved, but not without vehement debate: what was reassuring deterrence to some was a plunge into a mad arms race to others. Labour and the Liberals proposed censure motions. The only thing that united those who favored an expansion of the air force to counter the German threat and those who opposed it as a one-way ticket to Armageddon was the value both saw in portraying the air weapon as a thing of infinite terror.

During the debate, Winston Churchill took to the floor to declare that Britain's weakness in the air threatened the very stability of Europe. He revealed what he had learned through the network of confidential sources he had assiduously recruited in the Air Ministry and Foreign Office: Germany's aerial rearmament had been going on in secret for years, in violation of the Versailles Treaty; pilots had been trained, and hundreds of German-made bombers and fighters were already in service. If the current trends continued, Germany's still officially nonexistent air force would surpass the RAF in little more than a year.

Churchill realized one other crucial political and psychological fact that had escaped the air-power theorists of the RAF who had placed so much stock in the bomber offensive as the soul and essence of air power. As the reaction of the public had well shown, there was a fine line between deterrence and futility, strength and despair, when one was dealing with a weapon that promised swift and mutual destruction. Worse, even with equality of offensive power, a strategy based predominantly on the threat of attack could leave a democratic nation at the mercy of a more brutal foe.

Since the World War, Churchill had increasingly come to rely for scientific advice on Frederick Lindemann, a professor of physics at Oxford whom he had met in 1921 when Mrs. Churchill and Lindemann had been assigned to be partners in a charity tennis tournament. In many ways the men were opposites—Lindemann was, among other things, a nonsmoking, teetotaling vegetarian—but they shared physical courage, sharp minds, and a sense of acute alarm over the state of Britain's defenses.

It was Lindemann who now launched a well-timed gambit to change the terms of debate over air power. In a letter to the *Times* published August 8, 1934, under the headline "Science and Air Bombing," Lindemann denounced the "defeatist attitude" toward the aerial threat that had come to be conventional wisdom. "In the debate in the House of Commons on Monday on the proposed expansion of our Air Forces," Lindemann wrote, "it seemed to be taken for granted on all sides that there is, and can be, no defence against bombing aeroplanes and that we must rely entirely upon counter-attack and reprisals." That may be true at present, he acknowledged, but the idea that *no* method can ever be devised to protect cities against air attack "appears to me to be profoundly improbable." Lindemann called on the government to bring "all of the resources of science and invention" to bear on the problem. And he warned that a policy of defense through reprisal alone would leave "the heads of gangster Governments" with the power to blackmail the civilized nations of the world.

Seven months later Hermann Göring, head of the aviation ministry of the most prominent gangster government in the world, announced officially what everyone already knew: Germany now had an air force, the Luftwaffe. A few days later, on March 16, 1935, the Führer issued a decree establishing universal military conscription and the creation of a standing army of half a million men, thirty-six divisions.

The following day, a Sunday, William Shirer, now a foreign correspondent in Berlin, went to the ceremony at the State Opera House marking the national Heroes' Memorial Day. It was a scene, Shirer said, that Germany had not witnessed since 1914. "The entire lower floor was a sea of military uniforms, the faded gray uniforms and spiked helmets of the old Imperial Army mingling with the attire of the new Army, including the sky-blue uniforms of the Luftwaffe, which few had seen before." On the stage, Hitler stood beside Field Marshal von Mackensen, the last surviving field marshal of the Kaiser's army. "Ostensibly this was a ceremony to honor Germany's war dead," Shirer wrote. "It turned out to be a jubilant celebration of the death of Versailles."

The problem of defending against bombers was never so bleak as either the enthusiasts or the alarmists had it. Yearly war games that pitted the Air Defence of Great Britain (ADGB) Command against a mock force of marauding bombers found that a great many bombers would *not* get through. In the script for the 1932 exercise, war begins when "Southland" (which has built up a powerful striking force as a result of Northland's "pacifist policy") decides to "break Northland's national resistance in the early stages of the war by intensive aerial bombardment of the vital points in her economic and industrial system." But in the ensuing four-day exercise, Northland's fighters succeeded in intercepting 50 percent of the incoming raids by day and 25 percent by night. An analysis concluded that the day bombers would have lost a fifth to a third of their aircraft on each raid. The fighters were far from a perfect defense, but the games were far from a vindication of the view that bombers were "impossible to stop."

The system that the ADGB Command had developed to achieve these results had roots that went back to 1918. As Chennault had noted in his lectures and articles, the successful employment of fighters rested above all in getting them to the right place at the right time. This in turn required developing timely intelligence and having an integrated system that could put that intelligence to immediate use. The London Air Defence Area set up to deal with the Zeppelin and Gotha raids had not achieved a high success rate, but it had nonetheless identified and put in place all of the fundamental components needed to make an air-defense warning and control system work by the time the war ended. By 1918 its operations room in the Horse Guards was connected to a network of observers to the southeast and east of London; within half a minute of the time that an enemy aircraft was spotted, the information had been relayed over telephone lines to headquarters and an operator had placed a colored counter to mark the location of the sighting on a large table map. The commander of the London Air Defence Area sat in a raised gallery overlooking the table and could instantaneously follow the course of the raids as the counters crept across the map; another set of dedicated telephone lines connected the headquarters directly to squadron commanders at airfields around the city. Within two and a half to five minutes from the time an order was dispatched, aircraft that had been standing at the ready would be in the air.

The recognition that time was of the essence, and that the linchpin of the whole system was a central command post through which all information must be continuously funneled—so that limited resources could be directed where they were needed most—was a lesson that was never forgotten, even through the ensuing years of official neglect of the fighter arm. The operations-room system was carried over intact and steadily improved.

Such a defense manifestly depended on speed, so speed was what squadrons practiced. This training injected a note of realism notably lacking in the training of bomber crews. By the early 1930s, pilots and ground crews in fighter squadrons were drilled to get from hangar to field within four minutes of receiving a "stand-by" warning; they would have the engines warmed up in three minutes and be ready to take off in another two minutes. Some squadrons were achieving a breakneck eighty seconds from warning to takeoff. The staffs of the operations rooms that oversaw this sprawling network drilled with equal intensity to process incoming information and make quick and correct decisions. By the late 1920s, fighters were being equipped with radiotelephone gear in significant numbers; by 1934 the use of radiotelephones to send updates to fighters in flight and redirect them to newly spotted raids was becoming part of the routine.

The strength of the system had always suffered from the RAF dictum that (as the Air Staff had put it back in 1924) "the bombing squadrons should be as numerous as possible and the fighters as few as popular opinion and the necessity for defending vital objectives will permit." Only twelve regular fighter squadrons out of the nineteen originally called for were in existence by 1934. Yet all of the essential components of the system were in place. In principle there would be no difficulty expanding it, and indeed the 1934 expansion plan called for more than doubling the fighter force, to twenty-eight squadrons, by 1939.

There was one catch: Britain's geography. London, always assumed to be the prime target, lay perilously close to the coast; for that matter, nothing in Britain was more than seventy miles from the sea. An incoming bomber flying at 150 miles per hour could be over the capital in twelve minutes from the time it was spotted crossing the coast. The Hawker Fury biplane, the RAF's fastest fighter in regular service in 1934, took seven and a half minutes to reach twenty thousand feet. That didn't leave much room for error. Nor did it bode well for the future. In reviewing the results of the 1932 and 1933 exercises, the ADGB Command noted that the defenders had been driven farther and farther back toward London as warning times shrank. Faster bombers would cut warning times even more. Even if new fighters could keep pace with increasing bomber speeds, nothing could be done about the location of the English coastline. The Channel was a yawning gap in the warning system.

Lindemann's letter to the *Times* had not mentioned the particular problem of early warning, but he had urged in general terms that the full resources of science be brought to bear on the air-defense problem. The Air Ministry's Director of Scientific Research, H. E. Wimperis (he was best known for having invented the Royal Navy's famous Mark I "drift" bombsight during the war), had been thinking along the same lines. He asked his assistant, A. P. Rowe, to go through the files and see where things stood. As Rowe recalled:

> It was clear that the Air Staff had given conscientious thought and effort to the design of fighter aircraft, to methods of using them without early warning, and to balloon defences. It was also clear however that little or no effort had been made to call on science to find a way out. I therefore wrote a memorandum summarizing the unhappy position and . . . said that unless science evolved some new method of aiding air defence, we were likely to lose the next war if it started within ten years.

Wimperis at once set the bureaucratic ball in motion. In November 1934 he sent a long note to the Secretary of State for Air; Air Chief Marshal Edward Ellington, the Chief of the Air Staff; Air Marshal Hugh Dowding, the Air Council member who oversaw research and development; and several other key officials. "No avenue, however fantastic, must be left unexplored," Wimperis urged. Even the often proposed, and equally often debunked, notion of an engine-stopping ray, even a "death ray," ought not to be dismissed out of hand, given the advances that had taken place in the transmission of electrical radiation energy.

Wimperis's immediate suggestion was that a Committee for the Scientific Survey of Air Defence be empaneled, and he had just the man to chair it: Henry

Tizard, who after all was not only a distinguished scientist, now rector of Imperial College in London, but also a pilot himself who had worked on scientific problems of military aviation during the war.

Appointing a committee was the classic bureaucratic way to do nothing. But as Wimperis saw it, the choice of Tizard was a guarantee that the committee would mean business. Tizard had a remarkably deep network of contacts in the top ranks of the RAF, in the academic world, and in government research laboratories. Over years of serving on government panels, he had learned how to get things done pragmatically while never losing his genial disdain for conventional lines of authority. That the Air Defence Committee—the Tizard Committee, as it almost immediately came to be known—had no budget to speak of, no staff, and no administrative authority did not seem to give Tizard or Wimperis a moment's hesitation: they acted as though they had all the authority in the world and set to work at once.

Even before the first meeting of the committee on January 28, 1935, Wimperis phoned Robert Watson-Watt, a physicist who ran the National Physical Laboratory's Radio Research Station at Slough, west of London, and broached the "fantastic" question: Was it possible, he asked, to build a "death ray" that could disable an airplane or its crew from a distance?

The death-ray idea had a tedious history. For much of the previous decade the Admiralty and the Air Ministry had been beset by a never-ending stream of crank inventors who had claimed to have developed such a device. Most were outright frauds. None had come close to demonstrating anything workable. Watson-Watt was duly skeptical but asked his assistant, A. F. Wilkins, to work out some numbers: without telling him what it was all about, he asked Wilkins to calculate how much power a radio transmitter would need to emit in order to heat a certain amount of water to a certain temperature over a certain distance. Wilkins caught on immediately, noticing that the amount of water was "just about the amount of blood in a man's body." His calculation just as quickly settled the matter: the amount of energy needed was so vast as to be impossible.

Watson-Watt, Wilkins recalled, was "not at all surprised." But Watson-Watt at once proposed another idea: How much power, he asked, would it take to generate a detectable signal from a radio wave that struck a metal airplane and bounced back?

As is so often the case with scientific discovery, the idea of using radio waves to detect objects at a distance had occurred to others before, only to be dropped and forgotten. Marconi, the inventor of radio, had given a paper in 1922 before the Institute of Radio Engineers in New York in which he observed: "Electric waves can be completely reflected by conducting bodies. In some of my tests I have noticed the effects of reflection and deflection of these waves by metallic objects miles away." Marconi suggested that ships in fog or heavy weather could avoid collisions by using directional transmitters to detect the presence and bearing of other ships. More recently, scientists at the Bell Telephone Laboratories in

New Jersey and at Britain's General Post Office research station had noticed that short-wavelength signals fluttered whenever an airplane passed by. None of these findings had been pursued by the military authorities. The War Office and Admiralty had in fact turned down at least one direct approach from British scientists to conduct further research on the phenomenon.

Watson-Watt may not have been the first to invent what would come to be known in Britain as RDF ("radio direction finding," a deliberately misleading cover name) or in America—and by 1943, everywhere—as radar ("radio direction and ranging"). But he was the first to have a pipeline to people in the military establishment who would listen. He at once passed along the encouraging results he and Wilkins had worked out on paper, and Tizard and Wimperis reacted with a speed that was close to unprecedented in the history of either science or bureaucracy. The idea was discussed at the second meeting of the Tizard Committee on February 21; five days later, Watson-Watt, Wilkins, and Rowe were huddled around an oscilloscope watching the green trace rise and fall as an RAF Heyford bomber flew through the radio signal emanating from a nearby BBC shortwave station at Daventry.

A week later Wimperis sent Dowding a memorandum that began: "We now have in embryo a new and potent means of detecting the approach of hostile aircraft, one which will be independent of mist, cloud, fog, or nightfall." He proposed that the Air Ministry provide £10,000 at once to continue the research. Dowding approved it on the spot.

It was all an astonishingly ad hoc way to run a government research program, and there were numerous times when the lack of clear authority took its toll. Churchill and Lindemann felt they had a proprietary claim to the topic of air-defense research and bombarded Tizard with all manner of wild ideas and ridiculous schemes—an aerial "mine curtain" dropped on parachutes, a device for creating "artificial upcurrents" that would flip enemy bombers on their backs. The situation became even more confused, and personally acrimonious, when Churchill in early 1935 tried to wrest control away from the Air Ministry and persuaded the Prime Minister to establish a higher-level committee on air defense, with himself as a member. For a while the Tizard Committee existed in a state of administrative entanglement that only the British government could devise; it was simultaneously an autonomous advisory committee within the Air Ministry and a "technical subcommittee" of Churchill's higher-level committee, which itself was a subcommittee of the Committee of Imperial Defence, and no one was sure who answered to whom.

But Tizard's ability to cut through red tape with personal and informal connections continued to work wonders; the very lack of formal structure and authority let Tizard and his committee make decisions and get on with the job. At times the informality was almost comical. Tizard had to hire his own confidential secretary and paid him out of his own pocket. When he needed reports typed up, he relied on accommodating RAF commanders to have their assistants do the job. Many discussions took place in Tizard's flat in St. James's Court. Lindemann, who made a brief and grating appearance as a member of the Tizard Committee, made a great show of lambasting the Air Ministry for dragging its feet, as did Churchill at every opportunity ("I have never seen anything like the slow-motion picture which the work of this committee has presented," Churchill barked in a letter to the Secretary of State for Air), but actually the radar experiments were proceeding at breakneck pace. By mid-May 1935, an experimental radar transmitter was in operation at Orfordness; by July the range at which the radar was routinely picking up airplanes had grown to forty miles; by August experiments had begun on using radar to measure the altitude of aircraft; and by September the Treasury was being asked to approve the construction of the first of five operational stations covering the approaches to London up the Thames estuary: nine months from an idea in Watson-Watt's head to breaking ground for a working air-defense radar network.

The "Chain Home" cordon of radar transmitters and receivers that began to arise along Britain's coast that winter was a triumph of speed; it was no triumph of engineering elegance. Using relatively long wavelengths meant that standard radio components could be used; it also guaranteed that the installations would be huge—long, fixed antenna arrays supported by hulking 350-foot-high steel towers. The transmitting antennas simply sent out a blast of radio energy in a wide arc, painting the sky like a floodlight. An oscilloscope attached to a radio receiver plotted how long it took for an echo to come back. Unlike the two-dimensional maps of surrounding airspace that modern radar screens display, the oscilloscope traces that the operators of the Chain Home stations watched were nothing but an undulating green line with a spike marking when the echo had been received; the farther to the right the spike appeared on the screen, the longer the time it had taken to arrive, and thus the greater the distance away the target was. To determine *where* the target was, as opposed to just how far it was, the operator had to turn a dial that electrically steered the receiver antenna to whatever direction gave the strongest echo on the screen.

In terms of basic radar technology, the Germans were already far ahead, developing mobile, very-short-wavelength systems with rotating antennas that could sweep the whole sky in sequence and automatically display the precise range and bearing of multiple targets at once. No one in Britain knew of the German work. But Tizard was keenly aware from the start that any British advantage that would be gained by putting radar to use depended as much on properly incorporating the radar data into the fighter-control system as it did on the technology per se. In the summer of 1936 Tizard began to press the air commanders he knew to carry out a series of very realistic trials to test how well the operations rooms could handle what promised to be a flood of information from the radar stations, and translate it into actual interceptions.

The Biggin Hill tests, as they came to be known, were supposed to last two months. Committing pilots and planes to such a live experiment was a huge drain on resources, fuel, and personnel at a time when there was little of any to spare. But Tizard doggedly insisted that the only way to get valid data was to carry out a thorough and systematic series of trials, and he got his way; the experiment ended up stretching into early 1937, and in retrospect it constituted a breakthrough as great as Watson-Watt's original notion to work out the strength of a reflected radio wave.

The Biggin Hill experiment quickly showed that many business-as-usual procedures of the operations rooms were, as Tizard reported in October, "quite hopeless." Radar data, combined with radiotelephone communications to fighters in the air, now made it possible not just to give an early warning to get the fighters into the right general place but moreover to guide them directly and continuously to the precise spot where they could intercept an incoming bomber force, even as it changed course. The old procedure had been for observers' reports to go first to the headquarters operations room and then be passed by telephone to the sector rooms, each of which controlled one or two squadrons assigned to a fifteen-mile-wide block. Tizard saw that the extra delay in having headquarters calculate a track and ground speed and pass this information to the sectors was throwing everything off, and he suggested the radar data go simultaneously to headquarters and sector rooms. He also devised a remarkably simple way for the ops-room commander to calculate a new interception course for the fighters whenever the bombers changed direction. Tizard realized that what was in principle a complex trigonometric calculation could be done at a glance with a method that gave a rough but adequate result every time: Draw a line between the bombers and fighters; this formed the base of an isosceles triangle. Taking the bombers' new track as one of the two equal sides of the triangle, the other equal side automatically showed the new course the fighters should follow. Since the fighters were faster than the bombers, they would get to the new interception point first even if the calculation was not entirely precise. Years later the method was still known as the "Tizzy Angle."

Soon the system's elements began to click, and the results were nothing short of miraculous. Even when the bombers made frequent changes of course, the fighters could be directed to within three miles of them more than 90 percent of the time. Meanwhile the useful range of the radar sets continued to improve; they were now able to detect aircraft at a range of seventy-five miles or more. Along with course and altitude, operators found they could even get a rough idea of how many aircraft were flying together by the size of the spike the reflected signal traced out on the glowing green oscilloscope tube.

There was a hole in the system that remained, however: the fighters themselves. Here the problem was less one of not having the right basic technologies as of not knowing the optimal way to employ them for the maximum military effect.

This was the price of neglecting the fighter wing for so long. By the mid-

1930s aircraft makers had no difficulty figuring out how to apply to fighters the same aerodynamic advances that had ushered in the age of the fast, all-metal monoplane bomber: sheer technological momentum succeeded where air staffs failed. The year 1936 had seen the first truly modern fighters emerge from American factories, the Seversky P-35 and the Curtiss P-36, low-wing monoplanes equipped with retractable landing gear and the most powerful engines available, giving them a top speed of about 300 miles per hour, an enormous leap over the old P-26.

Even as the RAF clung to biplane fighters, technological advances forced their way to the notice of the Air Staff in Britain, too. In 1931 the RAF had issued a specification for an advanced, all-metal fighter having the fastest possible speed at 15,000 feet and the highest rate of climb. The winner was yet another biplane, the Gloster Gladiator. But one of the losers could not help catching the attention of Air Marshal Dowding, the director of the RAF's research and development. The plane's designer was R. J. Mitchell, famous for the series of monoplane racing seaplanes he had created. In 1931, his Supermarine S.6B had claimed for Britain the Schneider Trophy, awarded to the winner of an annual air race that attracted the world's fastest airplanes. (The wealthy, eccentric, and patriotic Dame Lucy Houston contributed £100,000 at the last minute to cover the cost of the RAF's participation in the race when the government had announced it could not afford to compete that year.) Two weeks after the race, an S.6B powered by an astonishing 2,530-horsepower Rolls-Royce engine broke the world speed record, and soon broke it again, hitting 408 miles per hour.

Mitchell's prototype fighter design shared the same sleek lines and high-performance potential of the S.6B, and while it had failed to meet one of the issued specifications—a low landing speed of seventy miles per hour—Dowding decided it was worth supporting as a line of experimental research. In January 1935 the Air Ministry issued a contract to Supermarine for an "Experimental High Speed Single Seat Fighter." The company would build one airplane. A similar one-airplane contract was issued to Hawker Aircraft for a second fast-fighter prototype.

The airframe and engine technology were now there. In particular, a crucial discovery by researchers at the Royal Aircraft Establishment at Farnborough in 1934 had given Rolls-Royce the solution it needed to put its liquid-cooled engines out in front once again for practical military application in fighter aircraft. The NACA cowl, which had dramatically cut the drag associated with cooling air flowing over the radial engine, had largely threatened to undercut the advantages of the reduced head resistance offered by the narrower V-12 liquid-cooled engine. A major source of drag that remained in liquid-cooled engines was from cooling air striking the radiator. Experiments using the Farnborough wind tunnels, however, showed that by enclosing the radiator inside a duct with aerodynamically designed inlets and outlets, the heat of the radiator could be harnessed to generate a sort of mini-ramjet as heated air rushed out of a nozzle to the rear. This thrust partially offset the cooling drag. The discovery came just in time to be incorporated into the new British monoplane fighter prototypes. (The Supermarine racing planes had circulated engine coolant through the wings, eliminating the radiator altogether. But this was not a practical solution for a military plane: a single bullet hole in the wing would cause a catastrophic coolant leak. Ethylene glycol also proved to have the annoying tendency to leak out through seams and rivet holes in the metal skin.)

But when it came to deciding on several crucial tradeoffs in the specifications for a fighter, the Air Staff confessed that it simply was unsure what a fighter needed to do best. For years, British air-defense commanders had been pleading for the establishment of a "fighter school" to study and develop realistic air-fighting tactics. In 1928, Air Marshal Brooke-Popham sent memo after memo to the Air Council urging that at least a small program to investigate fighter tactics be set up; he pointed out that fighter training manuals still were based almost entirely on the experience of the World War. And they dealt mainly with fighter-on-fighter tactics; there was almost nothing on how fighters should attack bomber formations, which was what the Air Defence of Great Britain Command's job was presumed to be. The Air Council at last deigned to respond: a school was not possible or necessary. Perhaps squadrons could each work out suitable tactics on their own, they suggested.

In 1932, Air Marshal Salmond tried again; this time he was told that logistical problems and "existing financial conditions must be regarded as precluding a project of this nature." Finally, two years later, approval was granted for two squadrons to be stationed at Northolt just outside of London to conduct the experiments the commanders had been begging for. Plenty of strings were attached. Above all, the squadrons would have no relief from regular operational duties; they would have to fit their tactical experiments in as time permitted. And this was to be strictly temporary. There simply was neither the time nor the money to establish a permanent fighter school. There would be no rotation from other squadrons of pilots in for instruction, either.

Air-defense commanders drew up a list of questions for the Northolt squadrons to address. To a fighter commander six years later, the questions would have seemed astonishing: they revealed how little was known about what would work and what would not. Should fighters be single-seaters or two-seaters? Should guns be movable or fixed? Perhaps even in single-seat fighters the pilot should have a control to steer the guns? Was speed or firepower most important? Or perhaps neither: perhaps pilot visibility was really the crucial factor? There wasn't much time to find out, and some of the conclusions that were drawn in the scant time available were disastrously wrong. Great hopes were placed in a powerfully armed two-seater with a gunner who sat in a powered, rotating turret. The resulting Defiant fighter would be massacred in the early days of the war to come; a flexible gun turned out to be no compensation for the severe penalty on maneuverability exacted by the added weight of a turret and second crewman. Highly scripted attack patterns were worked out on the theory that fighters needed to work as units to mass their fire against bombers, and that they would not have to worry about fancy aerobatics since they would not be engaging in air-to-air duels with enemy fighters. Pilots went through mechanical drills (with names such as "Fighter Attack No. 1"), learning to maneuver together as sections (three aircraft), flights (six aircraft), or entire squadrons (twelve aircraft) and to segue in and out of orderly queues to allow each plane or unit to fire on the target in turn, then fade back and let the next one have a go. It was almost exactly the wrong lesson for modern aerial combat. More than one RAF fighter pilot of the period would recall being actually forbidden by his squadron leader to practice aerobatics; it was deemed unnecessary—and it messed up the planes for inspection when oil leaked out of the engine during inverted flight.

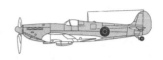

Spitfire I
single-seat fighter
engine: 1,030-hp V-12
top speed: 355 mph
armament: 8 wing-
     mounted .303-in.
     machine guns

But several of the experiments at Northolt and elsewhere paid off just in the nick of time. Most notable was a series of tests to determine how much damage a bullet could inflict on a modern bomber. While the speed and climb performance of even the biplane fighters of the early 1930s had posted huge gains over their World War counterparts, their armament had virtually stagnated. The biplane Fury, which in 1935 was the hottest fighter in the RAF, was equipped with two Vickers guns firing through the propeller, and its pilots were still equipped with a mallet to give the guns a good whack whenever they jammed. Surer firing mechanisms now made it possible to place the guns in the wings, which offered the added advantage that their rate of fire, no longer limited by the interrupter mechanism, could be increased to a thousand rounds per minute. But still the firepower designed for the Hawker and Supermarine fighters was modest:

Hurricane I
single-seat fighter
engine: 1,030-hp V-12
top speed: 324 mph
armament: 8 wing-
     mounted .303-in.
     machine guns

four .303-caliber wing guns for the Supermarine fighter, two wing guns plus two fuselage guns firing through the propeller for the Hawker.

While the planes were still in mock-up, however, the Air Staff learned what the firing studies had found. A minimum of 250 machine-gun bullets would have to strike a bomber to inflict fatal damage. And at the speeds that encounters could now be expected to occur between fighters and their prey, a fighter pilot would probably have no more than two seconds to fire before he would have to break off. Even when firing 1,000 rounds per minute, eight guns would thus be necessary to reach the lethal threshold during those scant two seconds. In April and May 1935 the Air Staff hurriedly issued modified specifications: Hawker and Supermarine would each have to find a way to get eight guns into their planes.

The prototype Supermarine Spitfire that made its test flight a year later, on March 5, 1936, had its eight guns. It was also one of the most memorably beautiful airplanes ever seen. Trim and aggressive as a barracuda with its distinctive elliptical wings and all-metal fuselage, powered by a 1,030-horsepower Rolls-Royce Merlin engine, the Spitfire flew at a blistering top speed of 355 miles per hour, enough to have won the Schneider Trophy just a few years earlier. The Hawker Hurricane prototype had flown a few months earlier; though it was built with fabric-covered wings and rear fuselage, which made it easier to manufacture using conventional factory procedures, its top speed of 324 miles per hour and other performance characteristics were not far behind the Spitfire's. In June 1936 the Air Ministry placed production orders for 310 Spitfires and 600 Hurricanes. It would not be a week too soon.

July 31, 1936, was a Friday, the end of a summer week in Berlin, and the stations were filled with the usual crowds setting off for summer holidays. It was about 11 a.m. when a caravan of buses pulled up at the small Lehrter rail station and ninety young men, all in civilian suits and carrying identical cheap suitcases, filed out and boarded a train bearing the banner *Reisegesellschaft Union,* the "Union Travel Society."

That evening, after arriving in Hamburg, the group waited until nightfall, then made its way to the harbor where the steamship *Usaramo* lay waiting. At about midnight the *Usaramo* weighed anchor and slipped out of the harbor, destined for Cadiz, Spain. Along with her complement of travel enthusiasts she carried 773 crates of bombs, antiaircraft ammunition, and the disassembled components of ten Junkers Ju 52 trimotor transport planes and six Heinkel He 51 biplane fighters.

Just five days before, Hitler, at Bayreuth for the annual Wagnerfest, had received two German businessmen from Spanish North Africa who had come to relay an urgent message from General Francisco Franco: The Spanish Army rebellion that had been launched against Spain's left-wing government was doomed to failure unless the thirty thousand tough and loyal Moorish and Foreign Legion troops Franco commanded in Morocco could be moved to the Spanish mainland. The Spanish Navy had thrown its lot with the government and was in control of the Straits of Gibraltar. What Franco needed was a German air bridge.

Germany's interests in getting sucked into the chaos of Spain were far from manifest. The Army rebellion had been triggered by the assassination of a prominent right-wing leader by government police, but it had not come out of the blue. In its five years of republican government, Spain had been torn by constant political upheaval, factional violence, separatist uprisings in the Basque region, Catalonia, and the mining district of Asturias, and an earlier failed coup attempt by the Army. Eleven prime ministers had come and gone; in the general election of February 1936 representatives of no fewer than thirty-two different parties, from anarchists to fascists, had been elected to parliament.

The Army rebels in the latest uprising had the backing of an amalgam of monarchists, right-wing parties, and the Catholic Church; Communists, labor unions, and many of Spain's poor workers and tenant farmers lined up on the side of the government. The hatred on both sides had been festering for years, and when it burst into open fighting the result was less a war than an orgy of almost tribal bloodlust. Death squads on both sides executed tens of thousands in the early fighting: priests, nuns, and landowners on the one side; labor union members, Communist leaders, and many ordinary people who had the misfortune to be living in a town captured by Franco's forces on the other.

Germany's strategic interests in Spain were slight, and Hitler was not particularly inspired by any feelings of ideological solidarity with Franco. But he at once saw a chance to discomfit France with a pro-German government on her border if the rebellion succeeded. Spain was, to be sure, also an important source of raw materials for Germany's arms industries. Hitler heard out Franco's envoys at their late-night meeting on July 25 and told Göring to go ahead immediately. The next morning Göring phoned Berlin, and within hours Lieutenant General Helmuth Wilberg, the Luftwaffe's master planner, strategist, and organizer, had set things in motion. Two days later, Special Staff W—W for Wilberg—was in operation, issuing orders and arranging for the movement of supplies. It was an astonishing feat of staff work. Even as the Union Travel Society's gear was being loaded in Hamburg harbor and its ninety pilots, mechanics, and ground crew were being issued their civilian clothing and cheap suitcases, ten additional Ju 52s were being rushed from the Junkers factory and prepared to be flown directly to Morocco to join another cover organization, the "Hispano-Moroccan Transport Company, Tetuán–Sevilla." Within three days of Hitler's go-ahead, the first military airlift in the history of warfare was under way. By the time it ended in mid-October, twenty thousand troops had been shuttled across the straits in Ju 52s flown by German and Spanish pilots. It would prove no less than the salvation of the rebellion.

Ju 52
transport/bomber
engines: 725-hp radial (3)
top speed: 170 mph
armament: 1,100-lb. bomb
  load

The original orders to the German "volunteers" strictly forbade combat action. But it was unavoidable that distinctions would blur, and within weeks German pilots had jury-rigged two of their Ju 52s to drop bombs; on August 13 one of these converted bombers scored a direct hit on a government battleship, *Jaime I,* which had been firing on the transport flights as they crossed the Bay of Malaga.

The arrival in Cartagena harbor on October 13 of the steamer *Stari Bolshevik* carrying a contingent of Soviet pilots and eighteen I-15 fighters to join the fight on the side of the government forces upped the stakes. Within weeks a hundred Soviet aircraft were in operation over the front around the capital of Madrid, including thirty-one I-16s, probably the most advanced monoplane fighters in the world at that moment, and thirty-one twin-engine "Katiuska" Tupolev SB-2 bombers. The Germans were now facing a highly trained and thoroughly modern air force. Avoiding combat would not be a realistic choice for much longer. (The SB-2 had acquired its nickname from the loyalist Republicans soon after its arrival in Spain; Katiuska was the name of a Russian character in a popular Spanish operetta. Another nickname came from Franco's Nationalists, who had originally believed the plane to be an American B-10, or a Russian knockoff of it, and accordingly dubbed it the "Martin Bomber." Franco's propagandists then took to calling it the "Martin Bomberg," which was supposed to be a clever allusion to the "Judaic-Bolshevik" conspiracy that, the Nationalists asserted, they were fighting to rid Spain of.)

After Hitler's initial decision to aid Franco, the Führer had showed remarkably little interest in the Spanish venture, and so it fell largely to the German high command to decide upon the next move. The generals were well aware of the risks in getting too deeply involved in Spain: French support for the government side carried the small but real danger that the conflict might erupt into a major European war, which Germany was hardly ready to fight in 1936, and the drain of military materiel to Spain threatened to delay the German rearmament program. But there were irresistible benefits. Above all, it would be a chance to test the ideas that the Luftwaffe and its covert predecessors had been developing over nearly two decades of intensive and secret planning.

I-16 Mosca
single-seat fighter
engine: 730-hp radial
top speed: 280 mph
armament: two 7.62mm
machine guns

By the end of October the high command had decided to press ahead. The Condor Legion, as the German expeditionary force in Spain would soon be known, would be immediately increased to 5,000 men and 100 first-line aircraft, including a special "experimental flight" that would test, under combat conditions, advanced models still in the prototype stage.

To provide official deniability—since Germany, along with the rest of Europe, had pledged to adhere to an embargo against arms shipments to either side—the German planes bore the markings of the Spanish Nationalist army, and the men of the Condor Legion were issued plain khaki uniforms; they traveled to Spain aboard ships of Panamanian registry and were permitted to write home to their families only via a cover address: "Max Winklet, Berlin, Post-office box." To maximize the number of Luftwaffe pilots who would gain the benefit of combat experience, crews were rotated out after six months. Volunteers were encouraged with the offer of an automatic promotion by one rank and a bonus of 1,200 marks per month in combat pay. By the spring of 1937 the Condor Legion was receiving, direct from the factories, the first production models of Germany's most advanced aircraft: the Heinkel He 111, Dornier Do 17, and Junkers Ju 86 twin-engine monoplane bombers, and the Messerschmitt Bf 109 fighter, an all-metal monoplane with a top speed of 292 miles per hour and an operational ceiling of 30,000 feet. And the Luftwaffe was already verifying and adjusting its tactics and doctrine based on lessons from the intense combat taking place in the skies over Spain.

He 111B
medium bomber
engines: 950-hp V-12 (2)
top speed: 230 mph
armament: 3,300-lb.
   bomb load; three
   7.9mm machine guns

As Churchill had pointed out during the 1936 parliamentary debates over air strength, Germany had been secretly reconstructing its air force even before Hitler came to power. In fact, well before German aircraft factories began to give material substance to Germany's aerial rearmament, a shadow Luftwaffe had been in existence in the minds and plans of a small but brilliant group of staff officers in Berlin. The German high command had put to good use its wilderness years following the Versailles Treaty; indeed, in some ways the punitive terms the Allies imposed worked to Germany's advantage. Operating in secrecy, freed from the conservatism and confusing lines of authority of the old Imperial army system, able to select only the best and most intelligent officers for its limited number of positions, the shadow general staff that began to operate—even before the ink on the Versailles Treaty clauses abolishing the German General Staff was dry—had soon developed the most far-reaching and flexible air doctrine of any army in the world.

Bf 109C
single-seat fighter
engine: 730-hp V-12
top speed: 292 mph
armament: four 7.92mm
   machine guns in nose
   and wings

Taking charge as senior air officer was Helmuth Wilberg. One of *die Alten Adler,* "the old eagles," Wilberg had learned to fly in 1910 and commanded the Fourth Army Aviation force during the great air battles over Flanders in 1917. Now, a scant year after the Armistice, Wilberg launched an exhaustive review of air lessons learned from the war. At the same time, Hans von Seeckt, commander in chief of the Army, instituted a series of telling changes in the regulations governing officer selection. Educational standards were sharply increased. To stay in the Army, all officers, of all ranks, were now required to pass the entrance examination for the school that had replaced the old General Staff Academy. All officers also had to demonstrate proficiency in a foreign language. Mastery of technical subjects was given a new respect. An engineering degree from a civilian technical university was now accepted in lieu of graduation from the Staff Academy for candidates for the General Staff. Professional competence increasingly replaced aristocratic connections as the sine qua non for promotion to high rank: even Wilberg's Jewish ancestry (his mother was Jewish, his father a distinctly nonaristocratic Prussian Protestant) had not hindered his ascent to the postwar General Staff.

By early 1920 some 130 experienced air officers and engineers were already at work on fifty committees, studying everything from industrial supply to tactics for combating enemy fighters. Throughout the 1920s the shadow air staff under Wilberg undertook thorough examinations of foreign air services and kept abreast of new technical developments. Under a secret agreement with the Soviets negotiated in 1924, the German Army built a modern air base at Lipetsk, 220 miles from Moscow, which became a clandestine school and laboratory for advanced fighter tactics. Operating under the cover of the "Fourth Squadron of the Red Air Force," the German presence grew to a permanent staff of sixty instructors plus fifty pilot trainees and as many as a hundred other technicians who traveled to the Soviet Union each summer as "tourists." Over the next decade the German staff at Lipetsk developed a comprehensive series of fighter manuals covering air-to-air tactics, ground attack, and cooperation with ground forces. Live-fire exercises with guns, bombs, and even poison gas were regularly carried out.

As a result, by 1934 the Luftwaffe, in notable contrast to the RAF and the U.S. Army Air Corps, was possessed of a thoroughly up-to-date fighter doctrine that reflected rigorous intellectual analysis bolstered by considerable practical experimentation. Although there was heated debate within the Luftwaffe over the role strategic bombing should play in overall air strategy, even the Luftwaffe's bomber enthusiasts never had any Douhetist illusions about the bomber's invincibility. All experience and analysis pointed to the conclusion that bombers could get shot down under various circumstances; whether the bomber got through depended entirely on the specifics of the tactical situation—above all, on air superiority vis-à-vis the enemy's fighter and antiaircraft defenses. A strong fighter force would always be needed as part of an effective air force to escort bombers, carry out air-superiority sweeps, and provide home defense.

And to attack enemy ground troops: while French, British, and American military theorists largely accepted as inevitable that the next war would see a repeat of the hopeless stalemate of the trenches, von Seeckt and other leading German strategists in the 1920s drew exactly the opposite conclusion. The World War had been an anomaly. The next war would be decided by speed, maneuver, technology, and mobility. Air power would be a vital component in this modern vision of battlefield victory, adding a third dimension from which firepower could be brought to bear at the decisive point in a fast-paced battle.

What all of this meant was that there were no pat formulas for command of the air. Lieutenant General Walter Wever, the Luftwaffe's first Chief of Staff, had distilled this thinking in Luftwaffe Regulation 16, "The Conduct of the Aerial War," issued in 1935. The document reflected Wever's principles, although it was actually written by the brilliant Wilberg. It stated that the air forces would be called upon to perform many tasks, from supporting the army and navy to striking strategic targets. Gaining air superiority was a prerequisite for all of these missions, but there were many places where the battle for air superiority would be decided: in attacks on enemy airfields and aircraft factories, in air-to-air combat, in the fire from flak units defending German troops and targets. Fortune would favor the side that could exploit technology, tactics, logistics, and surprise to the greatest advantage on each of these fronts. There was no reason to believe that the attacker or the defender, the bomber or the fighter, strategic strikes or battlefield attack, bombardment aviation or flak was a priori decisive; everything depended on developing detailed, realistic, and tested doctrines for employing each to its greatest effect under the many different conditions likely to be encountered in warfare.

The fighting in Spain offered swift confirmation of the soundness of most of the Luftwaffe's basic doctrinal concepts. An impetuous decision by a Nationalist commander to carry out a bombing raid with a group of unescorted Ju 52s in February 1937 left little doubt as to the wisdom of Luftwaffe doctrine emphasizing the need for strong fighter escort for bombers. ("Tomorrow we bomb, and whoever falls, falls!" the captain had berated his men. The captain was among those who fell.) In striking confirmation of the German doctrine of battlefield support, and at striking variance from American and British theories of strategic bombardment, the greatest effect of air power in Spain would again and again be in the ground battle. During the two-week Battle of Guadalajara in March 1937, the Republican air forces launched a series of pulverizing raids that deployed more than a hundred planes at a time against the fifty thousand Spanish and Italian troops driving toward the city. Sweeping the Italian and Nationalist fighters from the sky, the Russian-supplied and mostly Russian-flown I-15s and I-16s strafed and bombed the Italian mechanized columns in hours-long attacks. Flaming trucks barred the roads, soldiers fled in disorder into the fields, and what was to be a triumph for the Italian legions quickly became a rout. When it was all over, five hundred Italian troops were dead, two thousand wounded, a thousand vehicles and twenty-five artillery pieces destroyed. It was, as the historian James Corum concluded, "one of the most dramatic examples of the era of what air power could accomplish on the battlefield." (The Nationalists for their part were content to blame the arrogance of their allies for the defeat, and the lyrics of an instantly popular Spanish song mercilessly mocked the Italians: "Guadalajara is not Abyssinia.")

The Condor Legion, for their part, needed little convincing of the lessons of Guadalajara. With the arrival of Bf 109 fighters in increasing numbers as the spring wore on, the German air forces began to seize air superiority and turn the tables. The He 51s were reassigned to ground attack, and in the offensive that began against the Basque country on March 31, the Condor Legion carried out raid after devastating raid upon Republican forces. At Ochandiano, Condor Legion bombers dropped sixty tons of explosives on Basque positions in two minutes, killing two hundred men and leaving four hundred more too dazed to retreat as their position was quickly overrun by ground troops. On June 19 the major Basque city of Bilbao surrendered and resistance in the north collapsed.

He 51
ground attack
engine: 750-hp V-12
top speed: 205 mph
armament: two 7.92mm
machine guns

Many foreign observers drew the obvious conclusion about the effectiveness of ground attack and the primacy that tactical aviation had assumed in securing air superiority. Far from a "figment of the imagination of the uninitiated" or an outmoded relic of the Great War, huge aerial battles that pitted fighters against fighters were taking place day after day. The higher speeds and greater wing loadings of the new fighters meant that the radius of turns was greatly increased, and battles spread out across vast expanses of the sky. Frank Tinker, an American pilot who flew Russian I-15s and I-16s for the Republican side—a graduate of the U.S. Naval Academy, Tinker had been kicked out of the Navy for getting into one brawl too many and had then sold his services to the Spanish government for $1,500 a month plus $1,000 for every enemy plane shot down—earned a $2,000 bonus during the Battle of Brunete in July 1937 when he shot down two Bf 109s. In his memoirs he vividly recounted dogfights that lasted for an hour or more, involving scores of planes on each side; in one huge melee he encountered what seemed to be "practically the entire enemy air force."

The American military attaché in Valencia reported that Spanish aviators and commanders were unanimous in the view that pursuit aviation took precedence over the bomber force; without air superiority secured by fighters, neither battlefield attack nor deeper strikes were possible. "The peacetime theory of the complete invulnerability of the modern type of bombardment airplane no longer holds," the American attaché wrote in February 1937. "The increased speeds and modern armament of both the bombardment and pursuit plane have worked in favor of the pursuit. . . . The flying fortress died in Spain."

In the attack on Bilbao, the Condor Legion had found that even with overall air superiority, their bombers needed protection by escorts of their Bf 109 fighters to ward off even the ragged air defenses the Basques could muster. A French aviation captain on the scene, Didier Poulain, reported that "it has not been rare, and I have seen it myself, to observe the presence of thirty fighters to protect a flight of five bombers." He concluded: "Even the best bombers have no defence against fighters."

Poulain also had the chance to inspect the "Iron Ring" of fortifications surrounding Bilbao after the fighting was over and made another telling observation about the effect of air power on the ground battle. The reinforced-concrete bunkers and strong points, he found, were largely intact; like the Italians at Guadalajara, the Basque garrison had simply cracked under the demoralizing strain of continual air assault and the feeling of helplessness it had created. "Paradoxically," he added, "the civil population stands up very well to bombardment, and stubbornly refuses to evacuate towns bombed from the air."

To those who bothered to look closely, the failure of strategic bombardment was the other vital lesson of the air war in Spain. While air strikes designed to cut off reinforcements or supplies to troops on the battlefield had had devastating effects on both sides, the single attempt to apply Douhet's theories of decisive, morale-destroying devastation from the air was a fiasco. In the spring of 1938 Mussolini observed to his son-in-law that it was time for the Italians to be "horrifying the world with their aggressiveness, not charming it with their guitar." And

A Republican poster during the Spanish Civil War declares, "The art of Spain is a target of the fascist air forces." In fact, the Luftwaffe concluded that strategic bombing in Spain was largely counterproductive. *(Library of Congress)*

so, without Franco's approval, he ordered the Regia Aeronautica to carry out a massive bombing attack on the Republican-held city of Barcelona. For three days, from March 16 to 18, Italian bombers plastered the city with explosives. When they were done 1,300 people were dead and another 2,000 wounded.

All it did was stiffen Republican resistance and bring down an international outcry upon Italy and the Nationalists. A British observer on the scene recorded that all of the standard dogma about how the cowardly civilians would react to such an assault upon their morale was simply wrong: "There has been no stampede . . . no wild and unreasoned panic. No terrible and uncontrolled hysteria and certainly no thought of beseeching the government to seek an immediate and unconditional peace." Dr. E. B. Strauss, a British physician and psychologist whose specialty was the reaction of people to physical danger, obtained firsthand accounts of people who had undergone air attacks in Spain, and he agreed: "The morale of the civilian population in these cities is still high." The reason for this was simple, Strauss found:

> The bombed become automatically united in a common hatred and terror of the invisible and intangible enemy from the skies. Observers state that one of the most remarkable effects of the bombing of open towns in Government Spain had been the welding together into a formidable fighting force of groups of political factions who were previously at each other's throats. . . . The one thing which could be calculated to unite . . . countries spiritually and to give them the conviction that they were fighting a Holy War would be air-raid reprisals on the tit-for-tat principle.

The Condor Legion commanders largely agreed, viewing the Italian bombing as senseless and counterproductive. A Luftwaffe study at the end of the war found that civilian air-raid precautions, particularly the widespread construction of bomb shelters in Republican cities that came under aerial attack, had been quite effective both in limiting casualties and in maintaining civilian morale.

But none of these reasoned conclusions were the ones that most of the world, or even many of the world's air staffs, were prepared to draw from the fighting in Spain. They were all overshadowed by the events in one Basque town that would become indelibly etched on the world's consciousness. Even generations later, people who knew nothing else about the Spanish Civil War would know the name Guernica.

Lieutenant Colonel Wolfram von Richthofen represented both the old and the new breeds. A *Freiherr,* or baron, he was a distant cousin to the famous "Red Baron," Manfred von Richthofen of Great War fame. He had flown as a fighter pilot in his cousin's Jagdstaffel 1 during the war, and was credited with seven kills. Von Richthofen also was a technocrat of the new school. He had received a doctorate in engineering from Berlin University in 1929 and gone on to head the development and testing branch in the Air Ministry's Technical Office. Then in January 1937 he was appointed chief of staff of the Condor Legion, just in time to assume direction and planning of the northern offensive against the Basques.

Von Richthofen combined the unemotional self-discipline of a Prussian aristocrat with the unemotional logic of an engineer. He began each day with an unvarying regimen of exercises, running in place, push-ups. He prided himself on his self-control; once, he recalled, he became "physically ill" after being kissed on the cheeks by a Spanish officer, but nonetheless was able to conceal his feelings.

His approach to the use of air power in support of the Nationalist offensive was coldly ruthless and straightforward. Von Richthofen was a firm believer in the power of aerial bombardment on the battlefield when used in close coordination with infantry and artillery. His bombers and attack aircraft would accordingly pulverize the Basque positions wherever they were to be found, in town or countryside. At the start of the offensive on March 31, he posted an order in the operations room of his headquarters at the Fronton Hotel in Vitoria: it reminded "all concerned" that although they were to attack only military targets, they were to do so "without regard for the civilian population." He underscored this with a memorandum establishing his "golden rule" of bombing: If for any reason the assigned target could not be attacked, bombs were to be dropped anywhere over enemy territory, again "without regard for the civilian population."

Concentrated, continuously repeated air strikes were the key to crushing the morale of enemy troops, von Richthofen believed, and the Condor Legion's ground-attack squadrons were organized to carry out unbroken "shuttle" attacks in which one wave would be rearming while another struck the enemy's battlefield positions, each plane flying multiple sorties per day. Against Republican depots, troop concentrations, and headquarters in towns behind the front lines, von Richthofen sent his bombers. The converted Ju 52 transports could carry one and a half tons of bombs, but lacked sophisticated bombsighting equipment; jury-rigged to the underbelly was a retractable "dustbin" that would be cranked down in flight and occupied by a bombardier who, squatting on the thin metal floor as he peered through a simple bombsight, communicated to the pilot with push buttons that illuminated colored lights in the cockpit: red for left, green for right, white for straight. The Condor Legion commanders had no illusions about the accuracy of the Ju 52s. "It's a wonder we hit anything at all," said Lieutenant Hans Henning von Beust, who led one Ju 52 squadron. Von Richthofen made sure they hit plenty, though: carpet bombing, with a good load of incendiaries in the bomb mix, could make up for a multitude of imprecision when going after enemy positions in towns.

By the end of April the Republican forces had been blown out of town after town, and now reconnaissance photographs showed the enemy forces in retreat on the roads leading to the small town of Guernica just behind the front lines.

Two battalions had taken up position on the edge of Guernica. In his war diary entry of April 26, von Richthofen noted that at the entrance to the town the main road crossed a single key bridge; if that bridge were destroyed, the enemy would be "in the bag," unable to withdraw their heavy equipment.

At noon on the twenty-sixth von Richthofen issued the orders. Four of the new He 111 bombers from the "experimental bomber squadron," VB/88, would lead the attack, followed by all three squadrons of Ju 52s. That would make twenty-three bombers in all. Bf 109s would provide escort and follow up with strafing runs while He 51s bombed and machine-gunned at low level. Von Richthofen ordered the bombers to carry the "usual mix" of one-third incendiaries, two-thirds 250-kilogram high-explosive bombs, fifty tons in all.

To von Richthofen, it was clearly just another day's work. Four days later Nationalist, Moorish, and Italian troops entered the city and von Richthofen arrived to inspect his squadrons' handiwork. A "technical success," he noted. Unfortunately, the Ju 52s, bombing through the clouds of smoke left by the He 111s' earlier bombing runs, had not been able to see the target, and the bridge still stood. Still, all road traffic had been held up for twenty-four hours, and if only the Spanish ground troops had properly followed up the air attack, it would have been a complete success.

While touring the town that day, von Richthofen noted in his diary, he was shown the "Holy Oak" under which kings of Spain had traditionally come to swear an oath to respect the rights of the Basques. Interesting. On to the next town.

To the world, von Richthofen's "technical success" was something else: a deliberate massacre, a terror bombing of innocents. Within two days, reports of the attack appeared in newspapers around the world. The *Times* of London called the raid "unparalleled in military history" and stated that "the object of the bombardment was seemingly in the demoralization of the civil population and the destruction of the cradle of the Basque race." The foreign correspondents on the scene and their editors in New York and London and Paris all agreed that Guernica had been singled out for its symbolic importance to the Basques as the home of the Holy Oak—ruthlessly selected to be made an example of, to show what happens to those who dared to defy German might.

To those who witnessed the raid, it was certainly hard to believe civilians had not been deliberately targeted. The front of the Julián Hotel facing the railroad station was sliced off by a 250-kilogram bomb. Clusters of incendiary bombs landed among fifty girls tending vats of sugar in a candy factory, setting off a terrible inferno. The town's half-timbered houses soon were ablaze, and 70 percent of the city burned to the ground. Von Richthofen had told his men, as one would later recall, that "anything that moves on these roads or that bridge can be assumed to be unfriendly and should be attacked." Father Alberto de Onaindiá, a Basque priest, saw the He 51s strafing crowds in the marketplace and was convinced that the aim was "to murder poor innocent people."

Clumsy attempts by the Germans and Nationalists to quickly disassociate themselves from the attack only ensured that the story would grow to its subsequent legendary dimensions. Father Onaindiá traveled to Paris to tell what he had seen; the pro-Franco Vatican denounced him as a liar. The Nationalists' propagandist in chief, the "radio general" Gonzalo Queipo de Llano, told his listeners that the bombing was a myth; the fires had been started by "Asturian dynamiters, employed by the Marxists, in order afterward to attribute the crime to us." Queipo said that Onaindiá had been excommunicated and that a British newspaperman who had filed reports from the scene was a drunk. The Republicans responded by claiming that half of the "Holy City's" five thousand inhabitants had been killed or injured. In Paris, Pablo Picasso, who had agreed to do a painting for the Spanish Republic's pavilion at the international exhibition scheduled to open in Paris in June, but had so far lacked a theme, saw the photographs of Guernica in the newspapers on April 30 and the next day set to work with feverish intensity on a raging, gigantic, allegorical mural, titled simply *Guernica*, that proclaimed the martyrdom of the Spanish town.

Research decades later established that three hundred civilians were killed in the German bombing. But to the world it was a dread confirmed; Guernica was but a prelude to the terror bombing sure to come in the next great war.

The response of a public tutored by decades of foreboding alarms about the terrors of bombing was not surprising, but the uncomprehending reaction to events in Spain by air commanders, particularly in Britain and America, was harder to account for. Hap Arnold, who would soon become chief of the Air Corps, offered his views about the Spanish Civil War in May 1938 in an article in the semiofficial journal *U.S. Air Services:* "Anybody who draws any lessons" from the fighting in Spain, he wrote, "believing that they are equally applicable to two leading world powers in conflict makes a great mistake." It was particularly fallacious, Arnold insisted, to conclude that the Spanish experience had anything to say about the efficacy of strategic bombing, since strategic bombing had not really been tried there. In fact, Spain had not seen any proper test of air power at all, he insisted. Real air power *was* strategic bombing, and what had gone on in Spain was merely "support" of ground arms, not the use of air power that could "break the national will to fight, thus forcing governmental heads to suc for peace."

Major George Kenney of the Air Corps was even more dismissive. "Neither side has shown any real appreciation of Air Power," he flatly asserted. Bombardment, observation, and pursuit aircraft had all been repeatedly "diverted" from their proper roles in order to carry out attacks on dispersed frontline troops. The "alibi" given for such a frittering away of air power was a shortage of artillery on the battlefield, but the "more likely reason" was "a lack of any clear conception of the proper employment of air forces," in which such a diversion of resources is "a serious error."

In fact, Kenney went on, neither side had anything that could rightly be called an air force; each was just a "miscellaneous collection" of odd aircraft flown "mainly by foreign volunteers of many nationalities." And while "it has become quite fashionable among military and pseudo-military observers and critics" to claim that the Spanish Civil War had disproved Douhet's theories, Kenney concluded that, on the contrary, "due to the lack of any real air forces and the small numbers of aircraft employed, Air Power has not been tested in Spain to date." Q.E.D.

The RAF's reaction was even more striking: it simply ignored the war in Spain in its official publications and studies. One of the few times the subject seems to have intruded on the consciousness of the RAF leadership was at a meeting of the RAF's Bombing Committee in June 1937. The committee, chaired by the Deputy Chief of Air Staff, was charged with coordinating the development of new tactics and methods for bombers, and the minutes noted that there was brief discussion at this meeting of the question of fighter escorts, "interest in which has been revived owing to reports of the adoption of these tactics by both sides during the Spanish Civil War." But the committee quickly concluded there was no need to dig any deeper into the matter; it would be "unwise to base conclusions on these reports as the conditions were dissimilar in many ways to those expected in a major air war."

In fact, the British Air Staff and the RAF Staff College, in contrast to the intellectually rigorous examination of foreign developments that marked the Luftwaffe's general staff, paid scant attention to foreign aviation ideas or recent air operations in other countries at all. In fealty to the "regimental model" Trenchard had instituted, the RAF in the 1930s was "certainly the best air force in the world at polo and big game hunting," as James Corum concluded after examining the service journals from the period, but it was shot through with an intellectual laziness that made it seem more than anything "a sort of gentlemen-pilots club."

To the extent the British did pay any attention to the Luftwaffe, they assumed it was a mirror of the RAF. The Air Staff argued—"logically"—that Germany could not possibly be contemplating a land war against France, which would drain the nation economically; therefore, its air force must be organizing for a strategic knockout blow rather than the support of ground operations. British Air Intelligence was so shorthanded that in September 1938 it had to borrow German specialists from the War Office, but even so the Air Staff never seemed very interested in what they had to say. Air Marshal Victor Goddard, who headed the European Section of Air Intelligence from 1936 to 1939, said later that had anyone ever asked him, he would have said without hesitation that all indications pointed to the Luftwaffe's being organized to support land operations. But no one ever did ask him. The RAF's Operations and Planning Staff had set up its own intelligence "section," consisting of two officers, and declined to consult Goddard. The assistant military attaché in Berlin, an Army officer named Kenneth Strong, who would later serve as General Eisenhower's Chief of Intelligence, filed reports in the mid-1930s pointing to the same conclusion. The Air Ministry rejected his findings and told him he should butt out of air matters, which were not his business.

The Luftwaffe, unsurprisingly, did a far better job of analyzing and profiting from the lessons of air operations in Spain. Twice as many Condor Legion planes had been lost in accidents than in combat—160 versus 72—and most of the accidents had occurred at night or in bad weather; that led directly, and quickly, to a complete overhaul of pilot training throughout the Luftwaffe. Pilots were required to attain complete mastery of instrument flying and to practice taking off at night and forming into large groups of twenty-five or thirty aircraft in the darkness. Work shortly began on developing radio navigation aids to help direct bombers at night.

Many other practical matters were worked out during the fighting in Spain, notably the myriad technical details required to coordinate air attacks on the battlefield. Liaison officers ("Special Duty Air Commanders," or *Flivos* in the German abbreviation) were placed in frontline ground units with their own communications teams. Von Richthofen, who would go on to command a special Luftwaffe Close Battle Division that would spearhead the assault on Poland, fine-tuned the system, perfecting tactics, developing recognition signals to avoid friendly-fire accidents, and adjusting munitions loads for optimal effect. Procedures were worked out for coordinating aerial rendezvous between bombers and their fighter escorts. Armament, engines, radio gear, even the airframes and wings of many of the frontline bombers and fighters were adjusted to reflect combat realities. Devastating tactics for neutralizing enemy flak were devised: the instant that fire from an enemy battery was spotted, swarms of He 51s would swoop down to strike with bombs and guns, and a constant search was kept up for enemy flak command posts, which would be singled out for targeting.

Air-to-air tactics were also overhauled, in what would prove to be one of the most far-reaching results of the Luftwaffe's readiness to learn from its experiences in Spain. Like all air forces in the world in 1937, the Luftwaffe used the standard tight V formation for its fighters. One of the problems with the V formation, however, was that as speeds increased, staying in formation became ever more challenging, and an ever-increasing amount of the pilot's concentration was focused simply on trying not to crash into the other planes. Another problem was that during turns it became necessary for the number 2 and 3 planes to cross over and change places with each other; otherwise the inside plane would have to slow down and the outside plane speed up, which greatly limited how sharp a turn could be made. The moment of crossover, however, as many German fighter pilots found, was also a moment of maximum vulnerability that the Russian I-16 pilots would choose to swoop down on them.

But a quiet young Condor Legion lieutenant, Werner Mölders, devised a new tactic that instantly led to a huge increase in the flexibility—and lethality—of German fighter forces. Mölder's tactic was to have fighters fly in pairs (*Rotte* in German) and much farther apart. Separated by 600 feet, the two members of a *Rotte* could watch each other's blind spots and quickly turn inward to engage an attacker that appeared on the other's tail. Two *Rotte* could operate together, spaced out in height and position, to form a larger, "finger four" formation that offered similar advantages of flexibility, maneuverability, and mutual protection. In July 1938, using the new tactics and equipped with the latest model of the Bf 109 (the D model of what was now designated the Me 109), German squadrons shot down twenty-two Republican fighters without a single loss to themselves during one five-day period. Upon his return to Germany after his tour in Spain, Mölders was immediately set to work revising the Luftwaffe's fighter-tactics manual.

The Republicans' loss of air superiority against increasingly intense German air assaults was the beginning of the end for the government forces. A last-ditch Republican ground offensive across the Ebro River at the end of July 1938 was met with overwhelming air attacks; Republican positions during one five-day period in August were struck with more than ten thousand bombs a day. The Battle of the Ebro gave the Luftwaffe the opportunity to put another new weapon through its paces: the Ju 87 "Stuka," or dive-bomber, which was able to score a number of precise hits on bridges and other point targets where conventional bombers had failed.

By November the Republicans had been driven back across the river. The Soviets saw the writing on the wall and withdrew their support and forces. The final collapse came on March 31, 1939, when the last Republican resistance in Madrid ended.

By then the world's attention was elsewhere. Two weeks before, Hitler, who had declared the previous September at Munich that the German-speaking regions of Czechoslovakia were the "last territorial claim I have to make in Europe," sent his troops to occupy what was left of the country that had been sold out by Britain and France for "peace for our time." Neville Chamberlain, Prime Minister since 1937, architect of the policy of appeasement, was genuinely stunned by the Führer's perfidy.

"In spite of the hardness and ruthlessness I thought I saw in his face," Chamberlain said, "I got the impression that here was a man who could be relied upon when he had given his word."

Until Hitler's takeover of Czechoslovakia, the British government had continued its inconsistent policy of belated rearmament combined with hopes, even expectations, of continued peace. In his previous role as Chancellor of the Exchequer, Chamberlain had agreed that rearmament ought not to be permitted to interfere with the "course of normal trade," in order to avoid disruption to the economy, loss of exports, and the creation of manpower shortages. "No one is more convinced than I am of the necessity of rearmament," he had said, but his attitude was that the government needed to watch the military services like a hawk: "Seeing how good the going is now," they were inclined to ask for "100 percent or 200 percent" extra of everything. The sardonic joke among military men was that the treasury wanted to keep a tight rein on military spending so it would have enough money to pay the indemnity when the country lost the next war.

The German Anschluss of Austria in 1938 had put an end to the "course of normal trade" rule; the occupation of Prague broke down the last barrier, and in the spring of 1939 the British government agreed to double the size of its army, institute general conscription, and drop the "limited liability" doctrine that had envisioned restricting British involvement in a European war to naval and air support only.

As far as the RAF was concerned, though, if it did not have what it needed by this time it would have been hard to blame it entirely on the shortsightedness of government policy. Since the beginnings of rearmament in 1934 the RAF had been the favored service. Aircraft factories were working flat out. Yet they still had not caught up with the 1934 expansion plan, let alone the additions approved since. By the time of the Munich crisis only five of the twenty-nine existing fighter squadrons were equipped with Hurricanes; the rest still were flying biplanes. Of the pilot reserve of 2,500 men, only 200 were fit to go into service units immediately. Inefficiencies, breakdowns, and patched-together fixes continued to plague the entire fighter defense system. The Chain Home radar network, supplemented by a second network called Chain Home Low to provide low-altitude coverage, still had huge gaps. Complete coverage was available only from Dungeness, sixty miles southwest of the capital, to the Wash, ninety miles to the north. The system could not operate over land; once enemy bombers had crossed the coast, it would be up to the Observer Corps to spot and report their movements, and Tizard was concerned that the observers needed better equipment to do the job. Dowding agreed in principle but said it was impractical given the caliber of the men, all civilians, who manned the posts. "It is essential that the duties of the observers at the posts should be simple," Dowding insisted. "The basic idea is that a yokel in a field reports by telephone, 'I can see "X" aeroplanes,' or 'I can hear aeroplanes.' " Giving them height finders or sound locators was asking for trouble. "I have already gone beyond the limits . . . by giving them a squared plane-table with a telescope and range scale," he told Tizard, "but between you and me this is very largely to give them something to keep them amused and interested during the long periods when nothing is in sight or hearing."

During a home-defense exercise in May 1939, all of the telephone lines from the radar station at Stanmore failed. All air-defense circuits were in fact jeopardized by the incompletion of the new Main Ring system designed to reduce telephone congestion in London. To escape the expected German knockout blow, Fighter Command (as the ADGB Command was now called) had moved its headquarters to Bentley Priory, an eighteenth-century country mansion in Stanmore, but none of the group and sector operations rooms were protected against direct air attack, nor were the radar stations themselves.

And in the fall of 1938 only two sectors had in place a crucial component for making the entire system work with the speed, accuracy, and automatic efficiency required: a radio-location device that would allow the operations rooms to plot the track of the friendly fighters by radio bearings rather than by dead reckoning. Dead reckoning depended entirely upon making guesses about wind speed at various altitudes and then juggling complex geometric tasks to integrate the fighters' known course and speed with the effect of the wind. The radio method, by contrast, used a transmitter known as the Pipsqueak that automatically broadcast from one fighter plane in each three-plane section a periodic tone that could be triangulated by ground stations.

As the summer of 1939 edged toward its fate, the elements of Britain's home-defense shield finally, in hindsight almost miraculously, fell into place. By summer's end twenty-nine of the thirty-five fighter squadrons that would be available to defend the country had been equipped with Hurricanes and Spitfires. The radar chain was closed at last, covering an unbroken arc from Southampton to above the Forth in Scotland. The fighter radio-location system was operating.

In the last weeks of the peace, a Mr. F. Sidney Cotton, an Australian businessman, ex-pilot of the Great War, who ran a photographic company that had business dealings in Berlin, flew to the German capital aboard a Lockheed Model 12 airliner owned by something called the Aeronautical Research and Sales Corporation. He made two trips to talk business, in fact, on August 17 and again on August 22. Had the German government been aware of how Mr. Cotton was putting his photographic expertise to use, he would not have been permitted to return. For the Aeronautical Research and Sales Corporation was actually a front for the British Secret Intelligence Service, and the Lockheed was one of four such aircraft that were equipped for detailed photoreconnaissance. The August 1939 flights to Berlin that nosed along the northwest coast of Germany were a belated acknowledgment on the part of the Air Staff that while "bombing Germany" might be a fine strategy for a shop-window air force, in a real war it might be useful to know a bit more precisely what and where one was supposed to bomb.

With a curious devotion to justifying his actions to the very last, Hitler put his propaganda machine into overdrive in the final weeks of August. As the Führer sought to blame Poland for its coming annihilation, William Shirer jotted down in his diary the screaming headlines of the German press: COMPLETE CHAOS IN POLAND—GERMAN FAMILIES FLEE—POLISH SOLDIERS PUSH TO EDGE OF GERMAN BORDER—WHOLE OF POLAND IN WAR FEVER!

At dawn on September 1, 1939, Polish soldiers at the frontier looked up to see waves upon waves of German aircraft passing overhead. The Luftwaffe had committed 810 bombers, 340 Stukas, and 240 fighters to the assault on Poland. As bombs began to fall on railroad stations and airfields across the country, the vanguard of two army groups, a total of a million and a half men supported by two Panzer corps, rumbled across the frontier.

Chamberlain, vacillating to the last, waited two days. Then, on a sweltering Sunday morning, September 3, the people of Britain turned on their radios to hear their elected leader's weary, reedy voice tell them that for the second time in a generation, a state of war existed between Britain and Germany.

PART THREE

WARSAW TO NAGASAKI, 1939-1945

8

FINEST HOUR

"Glorious in revolt and ruin; squalid and shameful in triumph; the bravest of the brave, too often led by the vilest of the vile!" So Churchill described Poland on the eve of war. For most of the 1930s the junta that had ruled the nation since Marshal Józef Pilsudski's 1926 military coup believed it had much in common with Germany's new Nazi rulers—not least a shared antipathy to democracy, Bolshevism, and Jews. Just a year before the outbreak of the Second World War, the Polish colonels in Warsaw joined Hitler in the dismemberment of Czecho-slovakia's frontiers, grabbing its own scrap, hyena-like, as Germany occupied the Sudetenland.

An air of self-delusion gripped the Polish leaders to the very end. In the spring of 1939, as Germany's intentions became unmistakable and Britain and France moved to guarantee Poland's borders, the Poles haughtily rejected any discussion of a four-way pact that would align them with the hated Russians. Fearful of provoking the Germans, they delayed mobilizing their army of nearly two million. Surrounded by Germany on three sides, they insisted on making a stand at the frontiers, defending lines a thousand miles long that were militarily indefensible.

The one conceivable strategic justification for such a suicidal forward stand was that if its army could force the invaders to fight for every inch of Polish soil, the German war machine might bog down in the east while the French attacked from the west. If that was their calculation, the Poles reckoned neither with the hesitation of their allies nor with the Luftwaffe.

The German military had begun planning for the invasion of Poland—code-named *Fall Weiss,* "Case White"—just months earlier. Even Göring had learned of Hitler's decision to proceed with the invasion only on April 18, upon return-ing from a holiday in Italy. He protested; Hitler called him an "old woman"; but in fact the Luftwaffe remained far from prepared to go to war in 1939. Many squadrons were still equipped with obsolete aircraft, there were severe shortages of bombs, and the training of many combat pilots left much to be desired.

So did intelligence about the air forces of Germany's adversaries. The chief of the Luftwaffe's Intelligence Branch, Colonel Josef "Beppo" Schmid, was often occupied with other duties in Göring's ministerial office, spoke no foreign languages, and had a habit of telling Göring what he wanted to hear. A reluctance to deliver bad news was a valuable skill for surviving in the Nazi bureaucracy. It was not a sound operating principle for an intelligence agency.

Yet in war everything is relative, and the Luftwaffe's deficiencies paled against those of the Polish air force. Not only were the Poles vastly outnumbered and outclassed, with only about 460 first-line aircraft, many of them well behind the state of the art; they also lacked a coherent operational doctrine spelling out what those aircraft were to do. Schmid's fecklessness notwithstanding, from mid-April on the Luftwaffe succeeded in building up a reasonably accurate picture of the Polish air force through agents' reports, radio listening posts, and reconnaissance flights that routinely penetrated Polish airspace as deeply as one hundred miles. Since the early 1930s, the Luftwaffe had been developing general war plans that envisioned striking at the Polish rail system to disrupt mobilization and reinforcement. Above all, the Luftwaffe had an operational doctrine refined by experience in Spain that would prove devastating against the kind of First World War dispositions that Poland now put its entire faith in. In tempo, firepower, and sheer demoralizing effect, the blow that now fell upon Poland from the air was nothing short of overwhelming.

The Germans would spread the story that the Polish air force was destroyed on the ground in the opening hours of the war; so successful were they in planting this legend of German omnipotence that a half century later it is still repeated as gospel in countless histories of the Second World War. In fact, the Poles, anticipating Luftwaffe strikes against their airfields, had begun to construct dozens of secret satellite bases and had dispersed their aircraft by the time the bombs began falling. On Day One the Luftwaffe struck eighteen Polish air bases; on Cracow's airfields alone some two hundred tons of bombs were dropped in 150 sorties. Some first-line aircraft undergoing repair and maintenance were destroyed, but most of the casualties, probably only a few dozen planes in all, were obsolete and training craft.

Still, the attacks, carried out in faithful adherence to Luftwaffe doctrine that set the achievement of air superiority as the first task of any air campaign, were far from wasted. Communications facilities, fuel supplies, and spare parts were not as easy to relocate as airplanes, and the airfield strikes demolished these in droves. Many of the Polish satellite bases were not scheduled to be completed until 1940 or 1941, and landing accidents on the soft ground of unfinished airstrips took their own considerable toll on Polish aircraft as they dispersed; one squadron of bombers lost 50 percent of its strength in the first two days from such accidents.

As the German ground forces advanced, the Poles abandoned airfield after airfield, and with each move left more aircraft and supplies behind. By the end of the second day, little remained of the Polish air force's support structure. A week into the battle, the fuel shortage was so acute that the Poles began sending fighters out singly in what must still rank as one of the oddest reconnaissance missions ever assigned to an air force: the Polish fighter pilots were told to try to locate gasoline tank cars that had been left stranded by Luftwaffe attacks on the rail network; fuel trucks would then be dispatched to offload the precious supply and deliver it to the airfields. The Poles should at least have held a logistical advantage from operating on their home ground, but the Germans' logistics and transport system made a mockery of the Poles' disorganization. As it took over forward air fields, the Luftwaffe used its fleet of Ju 52 transports to quickly land men, bombs, fuel, and spare parts. It was a maneuver the Luftwaffe had practiced many times, and now it paid many times over.

To support the ground attack, half of the Stuka force—160 aircraft—along with 40 ground-attack Henschel Hs 123 biplanes and about twice that number of fighter escorts were assigned to a Close Battle Division under the command of Wolfram von Richthofen, now a major general. Its job was to help spearhead the thrust of the heavily armored and motorized Tenth Army at the *Schwerpunkt,* the decisive point of the German attack, opposite Lodz.

Using the techniques perfected in Spain, von Richthofen sent his dive-bombers screaming against Polish troop concentrations and fortifications that stood in the way of the advancing Panzers. His *Flivos,* air liaison officers, operated from mobile command posts, usually eight-wheeled armored cars that could maneuver with the vanguard of the mechanized ground force. As in Spain, von Richthofen dispatched his aircraft in wave upon wave that struck at enemy positions in an unremitting chain attack that was both physically and psychologically devastating. The Henschel pilots found they could produce a deafening roar by over-revving their engines as they came in at near-ground level, which often induced panic in the horse-equipped Polish units. They also could deliver panic of a far more substantial kind: in one five-minute assault the Henschels plastered a ten-square-mile area held by Polish forces with four thousand incendiaries.

Meanwhile, air strikes against railroads played havoc with Polish attempts to move up reinforcements. Several Luftwaffe groups assigned three to six aircraft to respond to radio reports of targets that had just been spotted, and troop trains were repeatedly hit by this means. By September 7 the Tenth Army's Panzer corps had broken into the open and was within thirty-five miles of Warsaw.

Despite its emphasis on supporting ground operations, the Luftwaffe had never ruled out strategic bombing as part of its overall mission. With more than a thousand bombers manned by crews that were well trained in navigation, formation flying, and cooperation with fighter escorts, the Luftwaffe of September 1939 was actually far more capable of striking strategic targets than the RAF and the Army Air Corps, air forces that had made strategic bombing the holy writ of air power. In his 1935 "Conduct of the Aerial War" regulations, Lieutenant General Wever had stated that while terror attacks against enemy civilians were to be avoided both on principle and on practical grounds—unless precisely calibrated to the enemy mood, he warned, such attacks might actually increase the will to resist—nonetheless a blow against the enemy capital at the decisive moment might disrupt the functioning of government, to the point of inducing collapse.

Eager to leave no aspect of air power untried in the Polish campaign, the Luftwaffe had included in its plans for the initial air assault a strategic strike against military and government targets in Warsaw, but poor weather over the capital had intervened. By the end of the first day's fighting, the battlefield and railroad attacks were going so well that commanders were loath to divert bombers to strategic targets. Now, with the Polish Army largely defeated but Warsaw refusing to surrender, Hitler demanded the city be bombed.

The first attempt was hastily and poorly organized, a "shambles" by one account, although the German bombers nonetheless managed to set the Jewish Ghetto ablaze with seven thousand incendiaries. It was a token, deliberately aimed or not, of Nazi intentions to come. On September 22 von Richthofen proposed that the effort against the capital be repeated, only this time done right. "Urgently request exploitation of last opportunity for large-scale experiment as devastation and terror raid," he signaled the high command. If given the job, he promised, "every effort will be made to completely eradicate Warsaw." The city, after all, "would be only a customs station in the future."

In a pattern that would come to characterize air strategy throughout the war, technological opportunity overrode theory and doctrine at the critical moment. Von Richthofen basically wanted to bomb Warsaw because he could bomb Warsaw. The Wehrmacht high command replied with more restrained orders: only those facilities essential for maintaining life in the city were to be destroyed. Von Richthofen, as usual, was not too concerned about precision targeting. On September 25 he sent in 400 bombers on 1,150 sorties; the attacking force included the old reliable Ju 52 transports, each supplied with two soldiers manning coal scuttles to shovel incendiaries out the open doors on either side of the airplane. The bombardment killed thousands of civilians, destroyed 10 percent of the city's buildings, and damaged 40 percent more. Luftwaffe losses in the attack were three aircraft, two of them Ju 52s. Warsaw held out for two days, then surrendered. By October 6, all organized resistance in Poland had ceased.

Yet the Polish fighting had shown that even a lopsided and swift aerial victory does not come cheaply. The Luftwaffe ruled the skies from day one; the remnant of the Polish air force fled to Romania on September 17; yet in those two and a half weeks the Luftwaffe lost some 260 aircraft, about 7 percent of its force, to a manifestly inferior opponent. Another 280 German aircraft were so badly damaged as to be written off as total losses. Among close-air-support aircraft, the loss rate was almost twice as high, 13 percent. Accidents and friendly-fire incidents when panicky German ground troops opened up with flak on Luftwaffe aircraft also took their toll.

Still: Hitler had gambled everything on a fast campaign, and he had won.

General Walther von Reichenau, the commander of the Tenth Army, had no doubt that the assault against Polish positions by von Richthofen's Close Battle Division had "led to the decision on the battlefield." The RAF and the French Armée de l'air would have nine months to absorb the lesson of Poland; there is little sign they did. Yet Neville Chamberlain, the British Prime Minister, had no trouble in locating the crux of the matter. On September 16 he wrote to Winston Churchill, who had just joined the government as First Lord of the Admiralty:

> To my mind the lesson of the Polish campaign is the power of the Air Force when it has obtained complete mastery of the air to paralyse the operations of land forces. The effects in this direction seem to me to have gone much beyond anything that we were led to expect by our Military Advisers.

On a Friday evening, April 5, 1940, the German ambassador to Norway invited a group of local dignitaries to the legation for the showing of a film. *Baptism of Fire* ran ninety minutes, and it was a masterpiece of Nazi propaganda. Soaring music accompanied scenes of the German war machine as it crushed the Polish Army and cities. The climax was footage of the terror bombing of Warsaw. As the film ended, a caption explained, "For this they could thank their English and French friends." The message was not meant to be subtle. Three days later, German forces invaded Denmark and Norway.

Hitler seems to have shared much of the Luftwaffe leaders' skepticism about the military effectiveness of strategic bombing, but he was, as the historian Richard Overy noted, "attracted to the political aspects of bombing." Whatever the reality of the Luftwaffe's strategy and actions, Hitler never missed a chance to play to the world's belief that the Nazi air machine was a terror weapon par excellence. Such posturing had worked before. Indeed, the threat was arguably far more potent than the act. In March 1939 when Hitler demanded the rest of Czechoslovakia, he thundered that if the Czech government refused to give in, "half of Prague would be in ruins from bombing within two hours." Hitler had been equally convinced that the threat to British and French cities would keep the western powers from making good on their pledge to come to the aid of Poland by declaring war.

In the strange stalemate of the fall of 1939 and spring of 1940—the "Phony War"—each side saw the threat of strategic bombing, and its own preparations to counter it, reflected in endless ghostly images. Civil defense cast the strangest reflections of all. On the one hand, both Germany and Britain had become convinced that protecting their populations against air attack was necessary to maintaining morale and diminishing the political leverage an enemy could exert through the threat of bombardment; on the other hand, taking mass air-raid precautions could have precisely the opposite effect, inducing panic in one's own populace and acknowledging fear to the enemy. During the Munich crisis in September 1938, the British government had hastily distributed millions of gas masks, set men to work digging slit trenches in central London, and published plans for the evacuation of two million people from the capital; none of this was lost on Hitler.

As the Phony War dragged on, Hitler's calculation appeared justified. The French Army advanced fourteen miles into Germany, looked around, and retreated behind the Maginot Line. The RAF dropped leaflets. The British magazine *The Aeroplane* consoled itself with a familiar argument: "The showers of leaflets on strongly defended German territory must have had a moral effect on the German people. They will realize that the leaflets could have been replaced by bombs."

The start of war had triggered the civil-defense plans that British authorities had devised to protect the populace against the knockout blow they were sure would accompany the commencement of hostilities, a blow that many fully expected would include gas attacks à la Douhet. A million and a quarter Londoners were evacuated to the countryside in the first three days of war, three-quarters of them unaccompanied schoolchildren enduring what would be a mostly heart-wrenching ordeal. Government ministries completed secret transfers of key personnel to outlying "war stations." A strict blackout was put into effect in cities. "Anderson" shelters, consisting of 900 pounds of corrugated steel sheeting that could be dug into a back garden to shelter four people, were distributed free of cost to families having an income of less than £250 a year. A half million volunteer air-raid wardens and 150,00 auxiliary firemen reported to their posts. Seventy percent of Londoners dutifully obeyed instructions and carried their government-issued gas masks with them at all times.

Then—nothing. As the months dragged on, a Gallup Poll reported, one in five Londoners had been injured in an accident in the blackout, stumbling on the sidewalks in the dark or being knocked down by a car; none had been injured by enemy bombs, for the simple reason that there hadn't been any. By the spring only 1 percent of Londoners could still be seen toting gas masks. The prewar theorizers of strategic air power had not contemplated the stalemate of mutual deterrence.

The RAF *did* carry out a few raids with bombs instead of leaflets. These were directed against naval targets, scrupulously avoiding the German mainland so as not to provoke retaliation against British cities. The bombers were simply slaughtered in the air. On September 4, fourteen Wellington and fifteen Blenheim bombers attacked German warships in a daylight raid on Wilhelmshaven; seven were shot down. On September 29, five of eleven attacking RAF bombers were lost in a raid on the Helgoland Bight. On December 18, twenty-four Wellingtons again tried to strike at Wilhelmshaven and half were shot down, three more crashing as they attempted to land upon their return. That was the end of RAF Bomber Command's attempts at daylight bombing. Forty years of theorizing were no match for a few hard encounters with German fighters and flak.

Especially flak: the effectiveness of the German antiaircraft guns against a few isolated bombing raids was a sharp taste of what was to come. Since the early 1930s the Luftwaffe had placed great emphasis on antiaircraft defenses and deploying AA guns as an integral part of the force structure, both for defending the homeland and for protecting forward-deployed air units and ground forces. In both quality and quantity, the German flak forces were now vastly superior to those deployed in the First World War. In Spain the Condor Legion flak battalion had certainly proved its worth, shooting down fifty-nine Republican aircraft. The star of the show was the 88mm flak gun, which would become something of a legendary weapon of the Second World War. It could shoot fifteen 20-pound high-explosive rounds a minute to an altitude of 35,000 feet; it could also ("to the horror of the experts in Berlin," von Richthofen had noted in his Spanish war diary) do double duty against ground targets when not immediately required for air defense: with its flat trajectory and extremely high muzzle velocity, the 88mm gun could send armor-piercing shells with an incredible wallop into tanks or fortifications.

Now, in preparation for the strike against the West, Göring assembled his flak force into two huge corps. On May 10, 1940, as two million men, 2,500 tanks, and 3,200 aircraft moved into their final positions for the invasion of the Low Countries and France, they were accompanied by a veritable mobile arsenal of flak artillery: 700 88mm guns, 180 37mm guns, and 800 20mm guns.

For the attack on France, von Richthofen's Close Battle Division had been expanded; it was now called the VIII Air Corps, and it now included a special quick-reaction force of eighty Stukas and fighters that were to be kept on alert at forward airfields. When a Panzer group requested air support, von Richthofen's Stukas could be there in forty-five to seventy-five minutes.

France was a rerun of Poland; actually it was worse than a rerun. Having learned nothing from the war in the East, the Allies played the part of hapless victim to perfection. At dawn on May 10, the Luftwaffe struck forty-seven French, fifteen Belgian, and ten Dutch airfields. Unlike the Poles, the Belgians and Dutch obligingly allowed their air forces to be caught on the ground and almost finished off in a single blow. An entire squadron of eighteen British Blenheim bombers was put out of action when the Germans hit the RAF base at Condé Vraux in France. All told, 210 Allied aircraft were destroyed on the ground that first day. Over the next five weeks, a full 30 percent of the aircraft the French would lose fell prey to Luftwaffe airfield attacks.

There was no single weak link that was the cause of the horrifyingly swift collapse of the French defenses that ensued. Every link—command, doctrine, organization, tactics, equipment, training, intelligence, morale—was weak, and each broke in turn as the strain fell upon it. Expecting a repeat of the last war, the French Army had become the master of the defensive. At once it was thrown into disarray by the speed of the German armored advance. By May 13, just four days into the battle, General Heinz Guderian's three Panzer divisions had already reached the banks of the Meuse River at Sedan, and von Richthofen's Stukas, heavily covered by Me 109 and Me 110 fighters, began pounding the French fortifications on the opposite shore. Air-driven sirens had been mounted to the dive-bombers to further intimidate troops on the ground, who, a French general later recalled, froze in place as wave after wave of Stukas plummeted at them:

> The gunners stopped firing and went to ground, the infantry cowered in their trenches, dazed by the crash of bombs and the shriek of the dive-bombers; they had not developed the instinctive reaction of running to their antiaircraft guns and firing back. Their only concern was to keep their heads well down. Five hours of this nightmare was enough to shatter their nerves.

By midnight of the thirteenth, German engineers had spanned the river and the first tanks were crossing. The next morning, in a desperate attempt to halt the German onslaught, the British and French sent in the only things they had, slow light and medium bombers, on what were for all intents and purposes suicide missions against the German bridgeheads.

Ju 87B Stuka
dive-bomber
engine: 1,100-hp V-12
top speed: 238 mph
armament: 1,540-lb. bomb
  load; two 7.92mm
  machine guns in wings,
  1 in rear cockpit

To evade German fighters, Allied bomber pilots had at first been told to fly low. That advice had lasted for the first three days of the battle, during which time nearly half of the British bombers in France, 63 out of 135, were ripped apart by flak and brought down. For the Sedan mission, pilots were now instructed to fly at medium height and head into their target in a shallow dive. The result was one of the war's most tragic displays of blind courage. The British planes were single-engine Fairey Battles and twin-engine Bristol Blenheims. The problem was not so much that they were slow and obsolete as that they were flying missions that no aircraft of their type were capable of performing. But the British had nothing else. Seventy-one British planes attacked at Sedan on the fourteenth; forty were shot down. Individual unit losses ranged from 66 to 100 percent. The French fared no better, especially when their antiquated Amiot 143 bombers were sent on a noon raid against the German bridgehead.

While the RAF had neglected battlefield-attack aviation as a matter of policy, the French had neglected it through muddling, political bickering, and bureaucratic inertia. With its single-minded emphasis on strategic bombardment, the RAF had simply shut its eyes to the battlefield support mission. (The experience of Poland, to be sure, had given someone in the RAF enough pause to make a token gesture; since September, the RAF had provided seven of its pilots the opportunity to practice dive-bombing. Each dropped an average of eight bombs during his training. This was Britain's feeble answer to von Richthofen's VIII Air Corps.)

France's was a more complex tale of woe and subterfuge. The air force had negotiated a separation from the Army in 1933 in a deal that has been rightly termed a "masterpiece of contrived ambiguity." The agreement required the Armée de l'air to continue to provide ground-cooperation aviation for the Army as a prime mission; the air force did so through nothing short of a masquerade, building a series of ground-cooperation planes that were merely strategic bombers in disguise. These "BCR" aircraft were supposed to do everything: *bombardement, combat, renseignement*—bombing, fighting, reconnaissance. It was a ridiculous notion, and the planes that emerged from these specifications were famous for their sluggish ungainliness, not to mention their sheer ugliness. The Amiot 143, Farman 221, and other BCR planes of the mid-1930s were slab-sided hulks with strut-braced high wings and an almost medieval appearance; with their angular fuselages and distinctly unstreamlined noses, they resembled a knight's tin-can helmet or a castle buttress more than anything that ought to be taking to the air in the twentieth century. The BCR planes did everything badly. But as ten-ton, 190-mile-per-hour, unspecialized craft, the thing they did least badly was strategic bombing. The Army correctly read the air force's indifference to battlefield support and responded with indifference in kind to the idea that aviation could play a significant role in the ground battle.

Air Minister Pierre Cot, who returned to office with the left-wing Popular Front government of Léon Blum in 1936, tried to rectify the huge discrepancies between air force doctrine and technology. A forthright advocate of a strategic-bombing force as a necessary underpinning of French foreign policy, Cot scrapped the BCR requirement and set new requirements for modern bombers. The French aircraft industry responded with a series of modern, sleek, fast, and highly capable aircraft with top speeds of 300 miles per hour. Cot's successor, Guy La Chambre, did the same for fighter aircraft, placing an order for a hundred Curtiss P-36s from the United States and launching a modern series of domestic designs. The Dewoitine 520 and Bloch 152 fighters, with top speeds in excess of 320 miles per hour and armed with two 20mm cannons in addition to 7.5mm machine guns, compared favorably with the British Hurricane and could hold their own against the Luftwaffe's Me 109. Only small numbers of these aircraft were ready in time, however, and many squadrons were caught in the midst of transitioning to their new planes when the fighting began.

Yet the BCR planes had been a symptom of a festering, deeper ill that even new aircraft could not cure: the French Army and air force were barely speaking to each other. Cot, in his enthusiasm for reorganizing the Armée de l'air as an "offensive" force, had issued regulations emphasizing the need to strike battlefield targets and enemy lines of communication as well as strategic targets;

a few top French air generals had tried to call attention to the lessons of the Spanish Civil War and had spoken enthusiastically of the need to emulate the Luftwaffe's use of air power on the battlefield. But the institutional barriers to army-air cooperation had grown too great, heightened by the French Army's resolutely defensive orientation. La Chambre sought to heal the rift by breaking up the strategic-bombing force and parceling out its modern bombers to units assigned to ground-support duties, but it was too much, too late; the reorganization managed to undermine the only mission the planes were capable of while doing nothing to build the kind of force that could make itself felt on the battlefield.

Even with the wrong planes, the French air force still might have been able to do something, especially if bombing raids against the German onslaught could have been concentrated with mass and adequate fighter support. But the chaos in both doctrine and command was insurmountable. The air staff, caught up in its never-ending fight with the Army for control over aviation units, had neglected detail after crucial detail of organization, command structure, and training needed to carry out ground-attack operations. Now, under the strain of battle, everything fell apart. Bombers and fighters missed their rendezvous; the piecemeal distribution of the air force among army groups prevented concentration; air zones and the ground zones they reported to became hopelessly confused as the French front collapsed; and many Army commanders had no idea about how to use aviation or even how to communicate to the aviation units such basic details as the location of enemy targets or the timing and direction of their planned ground maneuvers. D'Astier de la Vignerie, the air commander in the critical Northern Army Zone, said that "almost every evening" he would call the army commanders to ask if they had a job for his pilots; day after day they would answer, "Thanks very much but we haven't any work for them."

For their part, neither the Armée de l'air nor the RAF had ever practiced responding to urgent air-support requests. A postmortem on the action in France by Air Marshal Brooke-Popham found that from the time a "fleeting target" was spotted and reported to headquarters to the time orders were issued to a squadron to attack it, four to six hours had usually elapsed; on one occasion, it took twelve hours. (Brooke-Popham noted that the German dive-bombers were able to respond with "great rapidity" in similar circumstances; when the urgency of the battlefield situation dictated, the Germans would not even bother about radio security and simply send orders in plain language.)

Within this chaos, defeatism did its work. Sortie rates were appallingly low throughout the French air forces: 0.9 per plane per day for fighters, 0.25 for bombers, 0.04 for observation squadrons—this while many of the German planes were flying three or four missions a day.

The largest attack that the Armée de l'air would carry out in 1940 would not even take place during France's struggle for national existence against Nazi domination. It would come on September 25, when eighty-three bombers under the command of the puppet Vichy government struck the British base at Gibraltar in retaliation for a British attack upon the French naval base at Dakar.

At seven-thirty on the morning of May 15, 1940, Churchill, Prime Minister for five days, was awakened by an aide who told him the French Premier was on the telephone, urgently requesting to speak to him. Churchill lifted the receiver by his bedside and heard a distraught voice declare in English, "We have been defeated." Churchill hastened to offer reassurances. Things could not be that bad, so soon; a counterattack could still be organized against the German breakthrough at Sedan. He offered to fly to Paris at once to discuss the situation. There the following day, Churchill received what he called "one of the greatest surprises I have had in my life." At a conference with the top military commanders, he asked General Maurice Gamelin, the French commander in chief, where the French mobile strategic reserve was. "Où est la masse de manoeuvre?" Churchill inquired in his execrable French accent. Gamelin shrugged. "Aucune," he replied: There is none.

The RAF had sent to France six squadrons of Hurricane fighters, seventy-two aircraft along with spares; now the French were demanding more. Churchill, over Dowding's growing objections, agreed to send ten more squadrons. The rate at which Britain's last line of defense was melting away was appalling. During three weeks of fighting, close to 400 Hurricanes had been lost; British factories were producing only about 40 a week. Only thirty-six fighter squadrons were left at home, out of the fifty-two that the Air Staff had set as the minimum needed to defend the country. The only consolation was that the priceless Spitfires remained at home.

British and French fighters were able to engage German aircraft when they found them. Even French squadrons equipped with the older Curtiss and Morane fighters reported shooting down some Me 109s and 110s. But the French air-defense command system was almost nonexistent. Dowding had been ceremoniously shown the French system the previous fall on a trip to Lille. After a long and alcoholic luncheon, which his French counterparts enjoyed far more than the abstemious Dowding, the head of Fighter Command was ushered into a basement where a solitary French airman sat receiving messages via an ordinary telephone and drawing chalk arrows on a blackboard. Dowding was dismayed and furious; he dashed off a memorandum to his superior describing the French system as one of "pathetic inefficiency." Successfully challenging the German control of the skies required finding and engaging the fighters en masse, which was close to impossible with this system.

By the end of May the rout was nearly complete. Göring was delighted when on May 25 Hitler approved his request that the Luftwaffe be permitted to finish off the British forces bottled up at Dunkirk. Guderian's Panzers were ordered to halt while the Luftwaffe's bombers went to work.

But in that desperate moment for the Allied forces, when Nazi triumph seemed imminent, the Luftwaffe was handed its first serious setback. Spitfires taking off every fifty minutes from their bases in Britain joined the Hurricanes in patrolling the beaches where the last-ditch evacuation of 300,000 British and French troops was taking place, and the British fighters succeeded in gaining at least temporary and local air superiority. On May 26, RAF fighters shot down 37 Luftwaffe aircraft, with a loss of 6 of their own. RAF losses for the nine days from May 26 to June 3 were 177 versus 240 for the Luftwaffe. When the British could bring their mass of fighter power to bear, they were still a potent challenge to the Luftwaffe's hegemony of the air.

By the time of the French surrender in mid-June—which Hitler marked at Compiègne with a grotesquely reversed reenactment of the 1918 armistice ceremony that had ended the First World War—the Luftwaffe had lost a total of 1,400 aircraft, including 19 percent of its Me 109s and 30 percent of its Me 110s, Stukas, and bombers. France had lost 750 aircraft, the RAF 950. It was a much closer contest in the air than the swift and stunning Nazi victory implied.

Myth, legend, and several hundred books about the Battle of Britain have enshrined it in history as a discrete event with a clear beginning and end, conventionally given as July 10 to October 31 of 1940. In fact the fighting in France marked the start of a year of almost continuous aerial clashes between the RAF and the Luftwaffe. Conventional wisdom has it right, though, that by October the outcome had been decided. The Battle of Britain would be a supreme vindication of Dowding's system of fighter defense; it would mark the beginning of a long, slow, but certain death for the Luftwaffe in an inescapable war of attrition. And it would establish beyond all doubt—to Hitler, to America, and for that matter to the sometimes wavering populace of Britain itself—that Britain meant to fight.

There were doubts, even in high places. Lord Halifax, the Foreign Secretary, told the Cabinet in late May that Britain needed to face facts and that seeking peace terms from Hitler might become necessary. Privately, Halifax was contemptuous of what he saw as Churchill's bellicosity and schoolboyish heroics and fight-to-the-death rhetoric; he complained that Churchill "talked the most dreadful rot" while he, Halifax, advocated a foreign policy based on "common sense and not bravado." Halifax did not grasp the genuine inspiration that many ordinary people, on both sides of the Atlantic, took from Churchill's soaring words of courage and conviction at a moment of darkness and uncertainty. "We shall fight on the beaches, we shall fight on the landing grounds, we shall fight in the fields and in the streets, we shall fight in the hills; we shall never surrender," he declared in the House of Commons on June 4. At a time when public opinion was evenly divided over Britain's chances of victory if she fought on alone, such a display of resolution from the top went far in steeling national resolve.

Hitler for his part thought it was all bluff. Britain could not possibly go on; all Germany asked was a free hand on the continent, he declared; he had no quarrel with England. Flush with victory, the Nazi leader was further encouraged by intelligence reports that predicted the imminent collapse of the Churchill government and by offers of mediation by neutral countries. Pope Pius XII had confidentially offered his good offices to conclude a "just and honorable peace" between the warring parties. "I can see no reason why this war must go on," Hitler smoothly announced in a long "appeal to reason and common sense" that he delivered on July 19. For the occasion, the Führer even shed his usual frenzied shouting for the tone of the magnanimous conqueror; it was entirely up to Britain, he offered, to choose between peace and war.

The Nazi leadership and the German public were genuinely astonished when the British government without hesitation chose war. (The response came in a swift reply delivered by a spokesman; Churchill disdained to answer personally, he explained, because he was not "on speaking terms" with the German dictator.) Still, Hitler refused to believe it. In the months to come, the operational direction of the Luftwaffe's war upon Britain would continuously be overshadowed by Hitler's political calculation that Britain was on the verge of surrender. If the Battle of Britain was the first clash between two major powers to be fought and decided solely in the air, it was also the first such clash in which air power was applied chiefly to produce a political rather than a military effect. That the Luftwaffe failed to produce the desired effect was the result of a complex blend of strategic muddle, operational overconfidence, and a few small, surprising, and very telling tactical lapses.

The strategic muddle came straight from the top. On May 21 the German Navy had presented Hitler a staff study on invading Britain; Hitler showed little interest. A month later the Navy was again told that the Führer had made no decision. At the end of June, Göring issued orders designating the whole of the RAF, not just Fighter Command, as the Luftwaffe's prime target; bases, support units, and the British aircraft industry would be attacked. Göring made no mention of preparing for a seaborne invasion.

In any case, Hitler was busy exulting in France over his victory and apparently thought it would be a simple matter to keep the pressure on Britain until she came to her senses and came to terms. The Luftwaffe, still recovering and refitting after its losses in France, showed a corresponding absence of urgency in commencing its assault on Britain.

Operations began in July in an almost leisurely fashion with armed reconnaissance flights and probing attacks against Channel shipping, the closest British target that presented itself to the German bombers and fighters based in northern France. These raids were not inconsequential: once again the Ju 87 Stukas proved their accuracy and lethality, and 30,000 tons of British shipping went down in four weeks. The bombers were always accompanied by heavy fighter protection; it was not unusual for the RAF fighters to encounter twenty bombers accompanied by an equal number of Me 109s and another twenty Me 110s.

Yet the gradual start to the clash gave RAF Fighter Command a much-needed opportunity to work bugs out of the radar system and learn tactical lessons that would be crucial to its later victory in the far more intense clashes of August and September, when the battle shifted to the decisive ground of London and to the heart of Fighter Command itself. Immediately the radar system proved its worth. At the beginning of July, Dowding had only twelve Hurricane and six Spitfire squadrons in No. 11 Group, assigned to the frontline area in the southeast of England that was destined to take the brunt of the Luftwaffe onslaught. Fewer than 300 fighters faced the 1,000 German bombers and 400 fighters of Luftflotte 2 sta-

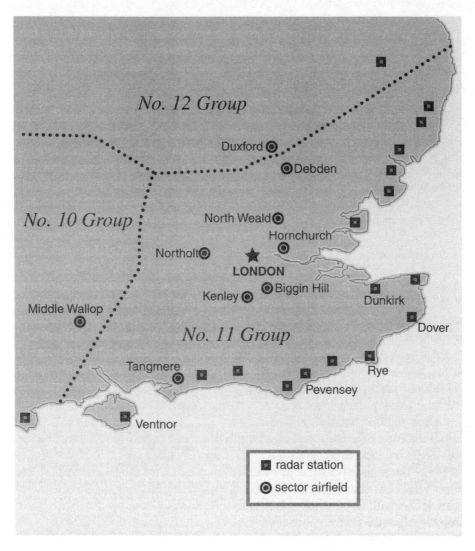

The British fighter defense system in summer 1940. Although the radar chain and key airfields were extremely vulnerable, German intelligence consistently failed to grasp their significance as part of an integrated network.

tioned immediately across the Channel. All told, the Germans had 800 Me 109s, 280 Me 110s, and 1,600 bombers and dive-bombers in France, Norway, and Denmark within striking distance of some part of Britain. Fighter Command had a total of about 750 Spitfires and Hurricanes to oppose them. Yet throughout July and early August, the RAF fighters—with the advantage of radar warning and control that allowed them to husband their resources, mass precisely when and where needed, and avoid equipment wear and pilot fatigue—were able to stay ahead of the attrition game. For the period of July 10 to August 12, losses stood at 148 for the RAF versus 286 for the Luftwaffe.

The tactical lessons that the Fighter Command pilots learned in those four weeks would pay off again and again. The Stukas proved to be extremely vulnerable to fighters that could penetrate their escort screen. They were slow, with a top speed of only about 250 miles per hour, and their defensive armament (two 7.9mm machine guns in the wings and one in the rear cockpit) was no match for the eight-gun fighters that could readily maneuver into position on their tails, especially once the Stukas began their slow and predictable bombing dives. By mid-August the Stukas would be withdrawn from the battle altogether after suffering a merciless pounding. The twin-engine Me 110 escort fighter also turned out to be no match for the British fighters. In theory it was an extremely innovative aircraft, a heavily armed "destroyer" with the range to accompany bombers on missions of 600 miles or more. But to achieve that range it had to be heavy, about three times the loaded weight of the Me 109; and weight meant high wing loading, which meant a large turning radius. The Hurricanes and Spitfires could easily outcircle them, get on their tails, and tear them to pieces.

Me 110C
two-seat escort fighter
  ("destroyer")
engines: 1,100-hp V-12
  (2)
top speed: 349 mph
armament: two 20mm
  canon, four 7.92mm
  machine guns in nose
  and wings, 1 gun in
  rear cockpit

The RAF pilots were learning some less orthodox lessons, too. Nearly every fighter squadron abandoned the choreographed, numbered attacks that Dowding had so painstakingly worked out before the war. A few squadrons began to imitate the German pair and "finger-four" formations, quietly ignoring Dowding's insistence that they follow the official and outmoded V-shaped flight-of-three pattern. Some also defied Dowding's rule that the eight guns on each fighter be set so their fire converged on a point 400 yards ahead; in practice, pilots found it was necessary to get much closer in order to aim with any certainty, and many had their armorers reset the convergence point to 250 yards. (As an aid to judging distance, British fighters were equipped with a gun sight that reflected a circular ring and bars onto a clear glass plate in the pilot's line of sight; the gap between the bars could be adjusted with calibrated knobs that specified range and target wingspan, so that when the wingtips of the target just touched the bars, the pilot knew he was in range to open fire. Although a vast improvement over the old fixed gun sights, the reflector sight still did nothing to solve the complex problem of "deflection" firing: unless a pilot were flying perfectly straight or in a perfect coordinated turn at the moment he fired, the nose of his plane would not be pointing the same direction that the plane itself was traveling through space, and the track of the bullets would be deflected accordingly. Only in 1944 did the RAF widely adopt a gyroscopic sight that compensated for deflection forces.)

If Dowding was in some ways an aloof, by-the-book commander, he was one who defied the stereotypes that people usually hold about military men who answer to that description. On the one hand he was touchy, a loner, with a streak of something very close to paranoia. This side of his personality was no doubt exacerbated by a humiliating series of blunders by the Air Staff, which repeatedly announced and then deferred his retirement; he was first told his retirement date would be June 1939, then March 1940, then July 15, 1940, then October 31, 1940, then finally some unspecified date in the future.

Yet Dowding had a mastery of technical detail that was extremely rare for top airmen at the time; he deserved most of the credit for getting the RAF to build the Spitfire and Hurricane, not to mention the vital radar chain. He also had a compassionate streak that belied his rigidity. As a squadron leader during the Somme fighting in the First World War, he had asked that his pilots be relieved periodically; Trenchard, furious, called him a "dismal Jimmy" who was "obsessed by fear of further casualties" and ordered him home. Dowding's pedantic conception of fighter tactics in the summer of 1940 may have been responsible for many needless casualties, but he also pressed for small but crucial technical improvements that saved many lives: notably the retrofitting of self-sealing fuel tanks for the Hurricanes and the installation of armor plates to protect the heads of pilots in both fighter models.

There was almost no stereotype of Nazi leaders that Hermann Göring did not fulfill. Fat, vain, pompous, corrupt, the Luftwaffe commander was blissfully ignorant of technical advances that had taken place in aviation since his own days as First World War ace. He once summarily dismissed radio navigation equipment as "boxes with coils—and I don't like boxes with coils." Göring owed his position entirely to his personal ties to Hitler and the political weight he pulled within the Nazi Party. Hitler cherished Göring's loyalty as an "old fighter" in the Party and extolled him as a "second Wagner" for his breadth of culture. It was a curious quality to look for in an air commander, even if it had been an apt characterization of the man, which it distinctly was not: Göring's "culture" consisted mainly in looting artworks from occupied lands. As a commander, he was given to issuing grandiloquently vague orders and crudely abusing his generals and pilots.

Göring's style of leadership magnified the strategic confusion and overconfidence that began to undo the Luftwaffe as the battle entered its crucial phase. On August 1, 1940, Hitler issued Führer Order No. 17: "In order to establish the necessary conditions for the final conquest of England, [the Luftwaffe] will overpower the English air force with all the forces at its command in the shortest possible time." The threat of invasion was to be made real by securing the air superiority necessary to cover a Channel crossing; preparations for a seaborne assault were to be completed by September 15. RAF airfields, aircraft factories, and factories producing antiaircraft munitions were to be targeted.

Göring followed with his own vague and pompous order:

FROM REICHSMARSCHALL GÖRING TO ALL UNITS OF AIR FLEETS 2, 3, 5. OPERATION EAGLE. WITHIN A SHORT PERIOD YOU WILL WIPE THE BRITISH AIR FORCE FROM THE SKY. HEIL HITLER.

Aldertag—"Eagle Day"—was set for August 13, and Göring confidently predicted it would take four days to eliminate the RAF now that the battle was to be joined in earnest. It was a grotesque overstatement, reflecting not just the Luftwaffe leader's bluster but also the failure of German intelligence to understand the RAF's organization—and the propensity of the Luftwaffe's intelligence chief to tell Göring and Hitler what they wanted to hear. A report by Colonel Schmid on July 16 was wildly inaccurate in its assessment of the relative strengths of German and British fighters, claiming that the Me 110 was superior to the Hurricane, and even to the Spitfire, except when the latter was "skillfully handled." Schmid underestimated by half the production rate of Spitfires and Hurricanes by British factories. He also ranked the Spitfire as inferior to the Me 109, when in fact the two were closely matched. (The Messerschmitt did have an advantage at high altitude, thanks to a more effective supercharger; and because the German engines were fuel-injected, Me 109 pilots could sometimes shake off a pursuing Spitfire by heading into a steep negative-gravity dive that would cause the carburetor-equipped British engines to cut out. But the Spitfire was significantly faster at 18,000 feet—354 versus 334 miles per hour—and could out-turn the latest model of the Messerschmitt, the Me 109E.)

Nowhere in his July 16 report did Schmid mention the existence of British radar. On August 8 he issued another report, this one acknowledging Britain's radar system—and accounting its every strength as a weakness. German intelligence had detected British radar signals, observed the Chain Home towers (which were visible from across the Channel), and intercepted radiotelephone communications between RAF pilots and their ground controllers; yet Schmid drew precisely the wrong conclusions. Failing to grasp that Fighter Command's radar system was a network that funneled information to a central command point, which could then vector fighters from any sector to concentrate wherever they were needed, he stated:

> As the British fighters are controlled from the ground by radio-telephone, their forces are tied to their respective ground stations and are thereby restricted in mobility. . . . Consequently the assembly of strong fighter forces at determined points and at short notice is not to be expected. A massed German attack on a target area can therefore count on the same conditions of light fighter opposition as in attacks on widely scattered targets.

Schmid's errors were not entirely the product of sycophancy. It was only in the spring of 1939 that the Third Reich political leadership had first mentioned Britain as a possible target to the Luftwaffe command. A little over a year was scarcely enough time to amass a competent intelligence picture of a complex enemy.

The outright mistakes that the Luftwaffe proceeded to make were nonetheless astonishing, given how close it came to paralyzing Fighter Command. The new phase of the battle began with massive Luftwaffe strikes against RAF airfields and coastal radar installations. The Chain Home radar system was a sitting duck. The transmitting towers, jutting straight up from the coast, were unmistakable. The equipment and control rooms were housed aboveground, often in unreinforced huts. On August 12, in preparation for Aldertag, 16 Me 110s, now acting as fighter-bombers and armed with two half-ton bombs apiece, struck at the radar stations at Dover, Pevensey, and Rye on the Sussex coast, and at Dunkirk in northern Kent. Meanwhile, a huge air armada of 100 Ju 88 bombers, escorted by 120 Me 110s and 25 Me 109s, headed for the radar station at Ventnor and the Spitfire factory at Woolston.

In the ensuing days, the fighter sector stations and airfields were hammered again and again. The fighter stations of 11 Group were hit thirty times in all. At midday on August 18, most of the airmen and women at Kenley had just sat down to their Sunday dinner when an announcement blared over the speakers: "Air attack imminent: all personnel not on defense duties to the shelters." Flying below the range of even the Chain Home Low radars, nine Dornier Do 17 bombers crossed the coast and roared in at treetop level over the base. Three minutes later came Stukas and high-level He 111 bombers. When the smoke cleared, three hangars lay in ruins, the base hospital was destroyed, and the cratered field was littered with twenty-four unexploded bombs. Kenley was a sector headquarters, and its vital operations room was also put out of action for two days by some of the hundred bombs the Luftwaffe had delivered. The sector station at Biggin Hill was struck later that afternoon. Tangmere, on the coast, the most forward and exposed of the sector stations in 11 Group, had been hit hard two days before, with a loss of seven Hurricanes on the ground; in the tangle of incoming Stukas and outgoing fighters, the antiaircraft gunners defending the field had been unable to get off a clear shot and had to hold fire. All three of these sector stations would come under repeated attack as August wore on.

As the fighting in the air grew even more desperate, British fighter pilots improvised new tactics. Though officially discouraged from doing so, some Hurricane pilots began attacking bomber formations head-on in order to engage the enemy as quickly as possible while avoiding their rear guns. There was an element of raw psychological intimidation in the calculation that bomber pilots would break ranks when they saw an enemy fighter coming straight at them at a combined speed of 600 miles per hour, about 300 yards per second.

The British pilots were always struggling to gain enough altitude in time to engage their targets, especially when the Me 109 escorts flew a high top cover: the Messerschmitts would come in well above the bombers, at 20,000 or even 30,000 feet, and wait to dive on the British planes once they were occupied with the German bombers. The usual division of labor sent the Hurricanes against bombers while the more capable Spitfires took on the fighters. But it took even a Spitfire thirteen minutes to climb to 20,000 feet; some Spitfire pilots took the initiative of flying inland, 180 degrees off the course assigned by their controller, to gain altitude before turning around and facing the enemy.

Aerial combat at such speeds was often so confusing that, for the majority of pilots, kills seemed more a matter of chance and opportunity than the result of tactical foresight. Fighter Command procedure was to vector fighters by radio instruction to the point of visual contact—in venerable British practice, the section leader would call "Tally ho" when he had spotted the enemy—and then they were on their own. Many pilots recalled turning and diving in a melee and suddenly finding themselves in empty skies miles from the battle. One novice Spitfire pilot, Sergeant C. S. Bamberger, said he was credited with shooting down an Me 109 in the Battle of Britain yet could not recall anything about it: "If you fired at something . . . what someone with my experience did was immediately break away, because someone else may be shooting at you."

In every unit there was a small number of pilots who possessed the magical combination of instincts, timing, eyesight, and three-dimensional imagination to score a consistently large share of the kills. For the rest, it came down to probabilities. The technological apparatus that would get them to the right place, a plane that could turn fast and deliver a lethal dose of ammunition if circumstances favored, was the statistical edge in a contest where so much was up to chance.

As the Luftwaffe's airfield attacks intensified, it was the closest to defeat Britain came. In the month of August, Fighter Command lost a quarter of its pilots. Although the inventory of spare aircraft was sufficient for the moment to make good the loss of machines, Spitfires and Hurricanes were being shot out of the sky or blown up on the ground twice as fast as factories could build new ones. Replacement pilots were yanked from bomber crews, and the advanced training course for fighter pilots was cut to half its length. Bamberger was a "surplus aerophotographer" in a bomber unit when he was thrown into a pilot training program; as of August 7 he had flown nothing but a biplane; on August 20 he

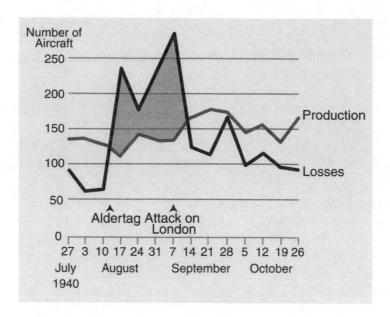

Weekly production and losses of Spitfires and Hurricanes during the
critical phases of the German assault. *(Basic data from Hough and
Richards,* Battle of Britain*)*

was taking on the Luftwaffe in a Spitfire, having had twenty-five hours' experience with the aircraft. "On one training flight only they told me I could fire my
guns," he recalled. "The next time I fired was at an enemy aircraft."

But the Luftwaffe did not "wipe the British air force from the sky." The
August 12 raids had knocked several of the radar stations off the air. British engineers worked around the clock to repair the damage and hustled a mobile transmitter to Ventnor, which had sustained the worst damage; the receivers there
were still out of commission, but getting a signal back on the air immediately,
they calculated, might fool the Germans into thinking their raids had been a
waste of time. An impatient Göring fell for their ploy. At a conference on August 15, he ordered that raids against the radar stations be discontinued "in view
of the fact that not one of those attacked has so far been put out of action." The
Germans shifted their attention elsewhere. A few more days of concerted attacks
on the radar system would have left Fighter Command in dire straits; indeed, as
the operations officer of the Luftwaffe's Luftflotte 2 later concluded, radar had
"doubled the efficiency" of Fighter Command. Abandoning the radar station
attacks was perhaps the Germans' worst strategic blunder of the war to date.

More strategic errors and intelligence failures followed. Just as the airfield
attacks of late August had Fighter Command on the ropes, Hitler and Göring
again abruptly shifted strategy. Encouraged by overoptimistic intelligence
reports, one of them placing Fighter Command's strength of serviceable fighters

at a mere 100, the Luftwaffe decided that the time had come for the decisive showdown. Hitler had always reserved for himself the decision on when to strike London. Now a blow to Nazi pride gave him just the occasion for ordering the attack that the Luftwaffe had in fact long been preparing for.

Two relatively small night raids on Berlin had been carried out in late August by Bomber Command; these were in retaliation for a series of Luftwaffe night attacks on British cities, including several on outlying areas of London. Nonetheless Berliners were, as war correspondent William Shirer recalled, genuinely indignant, even stunned that their city had been touched. Göring had boasted before the war that not a single British bomber would make it across the Ruhr; if one did, he declared, "My name is not Hermann Göring. You can call me Meier!"

Hitler now played up the "outrage" of the RAF raid on the German capital. The Berlin newspapers nearly all carried the same headline, Shirer reported: COWARDLY BRITISH ATTACK. On September 4, Hitler gave a speech dripping with contempt and vows of revenge. "When the British Air Force drops two or three or four thousand kilograms of bombs, then we will in one night drop 150-, 230-, 300- or 400,000 kilograms!" he thundered, and his audience, mostly a group of women nurses and social workers, interrupted him with tumultuous applause. "When they declare that they will increase their attacks on our cities, then we will *raze* their cities to the ground! The hour will come when one of us will break, and it will not be National Socialist Germany!" The next day Hitler ordered that the "revenge" attacks on London begin.

But in a lecture on the Battle of Britain delivered in Berlin in 1944, a Luftwaffe officer, Captain Otto Bechtle, explained what in truth lay behind the revision in strategy:

> The German High Command decided in September to switch the main weight of the air offensive to London, the heart of the enemy power. Incomparably greater success than hitherto [attained] could be anticipated from this policy. For while the main objective of wearing down the British fighter forces was not abandoned, economic war from the air could be embarked upon with full fury, and the morale of the civilian population subjected at the same time to a heavy strain.

By early September, Fighter Command was hard-pressed but not in the death throes that Göring believed. During the month of August each side had lost almost exactly the same number of fighters, about 440. The Luftwaffe had lost in addition an almost equal number of bombers. And Fighter Command still had its masterful fighter-control system intact.

On September 7, Göring announced over the radio, "I have taken personal command of the Luftwaffe in its war against England." He sent a thousand planes against London in the largest daylight raid ever. Another huge raid came a week later, on September 15. It was a huge strategic blunder to take the pressure off Fighter Command—"Thank God!" was the reaction of 11 Group's commander—and it was also a total hash of strategic thinking. Were the attacks on London intended to win air superiority? To bring political pressure by attacking economic targets? To break popular morale? To destroy commercial docks and thus tighten the U-boat blockade on British imports? To whittle away at British naval strength and so pave the way for a cross-Channel invasion? Or to make an invasion unnecessary? All of the above, apparently.

Operationally, too, the shift to London was a blunder. For an air force that had made attention to detail the hallmark of its prowess, there was one unaccountable lapse. Flying over London put the Me 109s at the very limit of their range. The fighters could carry only enough fuel to remain over the British capital about ten minutes before they had to head for home. In Spain the Luftwaffe had experimented with auxiliary fuel tanks for the Me 109; for some reason they were never produced for use in 1940. From September 7 to 15, the Luftwaffe lost 298 aircraft, 99 of them fighters, to Fighter Command's loss of 120. That marked the end of the Luftwaffe's daylight attacks on Britain. As the long summer of 1940 came to its close, the battleground shifted to the night skies and a different kind of air war.

The heroic lore of the outnumbered RAF persevering against the Luftwaffe obscured the fact that if the numerical odds were stacked against Fighter Command, this was in large part a deliberate choice by Dowding to husband his resources; he never put more fighters into the fight than he absolutely needed to, and the brilliance of the radar defense system was that it gave him the luxury to do so.

But there was more than a germ of truth to the myth of British coolness under fire that came to define Fighter Command, from its pilots to its commanders. The diaries and letters of Battle of Britain fighter pilots are full of boyish exuberance, flip macabre jokes, and an utter refusal to take themselves or the direness of their situation seriously—an attitude that seems almost incredible to a modern reader, but whose sincerity cannot be doubted.

Plunged into a war that would indiscriminately chew up millions of lives in its all-consuming totality, the British public was faced with the almost medieval spectacle of a battle fought by champions on its behalf. The British fighter pilots became celebrities, as had their counterparts in the First World War with far less reason, and their courage seemed infectious: a survey in September by the British Home Intelligence unit found the public facing Luftwaffe air raids with "confidence and calmness." The ready acceptance of what would turn out to be wildly exaggerated tallies of Luftwaffe losses versus RAF losses reported at the time was less a cause of the British public's confidence than one of the many enthusiastic symptoms of it. When the *Daily Telegraph* reported on September 16 the incredible—and in this case, completely true—story that one of Fighter Command's leading pilots was Douglas Bader, who had lost both legs in

"All right then—loser pays for a Spitfire."

"Spitfire Funds" proved immensely popular with the British public; for
£5,000 a group could "buy" a Spitfire and the right to name it.
From *Punch,* July 26, 1940. *(Reproduced with permission of Punch Ltd.)*

a flying accident nine years earlier and now led his squadron using a pair of
metal legs, it seemed par for the course.

The Spitfire had become something of a celebrity, too. Contributions poured
in to the "Spitfire Funds" that the Minister for Aircraft Production and press
magnate Lord Beaverbrook had cannily launched; a group could "buy" a Spitfire
and the right to name it for £5,000. By the following spring £13 million had
been raised. Subscribers could also buy Hurricanes, but they proved far less
popular.

The business-as-usual aplomb and confidence of the public was encapsulated
in one absurd moment during the height of the battle. Pilot Officer Kenneth Lee
was shot down near Whitstable on the coast and managed to bail out over a golf
course. Bloodied, in shirtsleeves, Lee was taken to the clubhouse. On the way, he
heard some players muttering at the last hole that the air battle overhead was dis-
tracting them from their putting. Once inside, waiting for an ambulance to arrive,

Lee's attention was caught by a more insistent voice: "Who's that scruffy look-ing chap at the bar?" it demanded. "I don't think he's a member."

After the war, the victorious American and British air forces, crowing over the apparent triumph of their theories of strategic air warfare, would argue that the Battle of Britain had been not so much won by the British fighters as lost by the German bombers: the Luftwaffe had failed because it wasn't a real strategic air force; it lacked a mighty four-engined bomber like the B-17; it was distracted by an obsession with ground support. This was rubbish. The Luftwaffe's twin-engine bombers were perfectly capable machines, and its considerable superior-ity in navigation, radio aids, night operations, and escort doctrine and tactics put German aircrews far ahead of the British or the Americans in being able to place bombs on strategic targets. The Blitz, the night bombing of British cities that began in earnest in September 1940 and continued through the spring of 1941, left no doubt as to the German bombers' capacity to sow destruction and inflict pain in the purest Douhetian fashion. Eighteen thousand tons of high-explosive bombs fell on London alone; nearly two thousand tons each on Liverpool and Birmingham; a thousand tons each on Plymouth, Bristol, Glasgow, and Coven-try; a half a thousand tons each on Southampton, Portsmouth, Hull, Manchester, Belfast, Sheffield. What the Luftwaffe's peformance should have raised doubts about was the theory of strategic bombing itself—the idea that a country could be defeated by striking at its morale or economic underpinnings.

Whatever muddle of Nazi grand strategy sent tons of high explosives and incendiaries raining down upon London in September 1940, the RAF and the Luftwaffe soon found themselves locked in a duel in which revenge and the abandonment of restraints swiftly transcended strict military calculation. In June the British War Cabinet had directed Bomber Command that targets in Germany could be struck only if they had been clearly identified by the aircrews; if cloud obscured the target area the planes were to return with their bomb load. Now bomber crews were told exactly the opposite: in light of the "cruel, wanton, and indiscriminate" German attacks, RAF crews were "not to return home with their bombs if they failed to locate the targets they were detailed to attack."

Both sides would claim that they aimed to bomb only military targets. The Luftwaffe night bombers did successfully strike docks, shipyards, aircraft facto-ries, gasworks, warehouses, railroad stations. But raids in which a hundred thou-sand incendiaries might be dropped in a single night could hardly limit their damage to military targets. And even with their radio navigation aids and special pathfinder units to lead the attacks and mark the aim point with incendiaries or flares, the Luftwaffe bombers measured the accuracy of their night raids in miles, not feet. By the time the Blitz came to an end, more than forty thousand British civilians had been killed and fifty thousand seriously injured; hundreds of thousands of houses lay in ruins and millions more were damaged. Whatever their initial intentions, each side would attribute the worst motives to the other as civilian casualties mounted. And both the logic of retaliation and the technological realities ensured that, increasingly, they would be right: inflicting civilian casualties *would* become the de facto purpose of it all. That was one thing, at any rate, the bombers could do well.

The road to hell had been well surveyed. Looking ahead to how Britain could possibly defeat Hitler with few remaining ground troops and no way to engage the Nazi war machine on the Continent, Churchill in early July 1940 had written in a memorandum that he saw only one way to victory: "an absolutely devastating, exterminating attack by very heavy bombers from this country upon the Nazi homeland." The Luftwaffe attack on London that inaugurated the night Blitz now set both sides irrevocably on this road.

When the all-clear sounded at 6:10 p.m. following the thousand-plane German daytime raid of September 7, fires were raging across the East End of the city. After nightfall the Luftwaffe returned. The Thames docks now became infernos as the bombers poured hundreds of more tons of high explosives and incendiaries into their now brilliantly illuminated target. The bombing went on throughout the night. Lumber, sugar, paint, rubber, and wheat that crammed hundreds of acres of warehouses fueled the blazes; barrels of rum exploded; paint blistered off fireboats moored across the river.

The next afternoon, Churchill drove to the center of the devastated areas. Everywhere crowds of East Enders greeted him with the same cry: "We can take it, but give it 'em back!"

Overall, the British public did take it extraordinarily well. A professor of preventive medicine at the University of Glasgow surveyed the mental-health effects of the bombing and reported that "psychiatrists were surprised to find how bravely most people stood up to the test."

But there were cracks in the facade of composure that at times caused Churchill and other officials grave concern. Britain's Home Intelligence unit reported in late September that thousands of working-class people had fled Southampton and Plymouth following the raids on those cities and were living in rough camps in the countryside. Many Londoners had also fled; every room within seventy miles of the capital was reported taken. There were many complaints about "unsanitary messes" and "improper behavior" in the crowded public air-raid shelters, and about the poorly organized rest shelters that struggled to provide temporary housing for thousands left homeless. And there were small but distinctly alarming signs of social fissures. Following a well-worn path, anti-Semitic rumors about Jews who monopolized air-raid shelters circulated widely in the East End. A story reported by American newspapermen, and subsequently played up with great éclat in Germany, told of a hundred East Enders led by a few Communists who forced their way into the air-raid shelter at the posh Savoy Hotel.

Largely unreported at the time were stories of grislier doings. A bomb hit the

Café de Paris on the night of March 8, 1941, killing thirty-four of London's well-to-do. In an autobiography published decades later, the novelist Nicholas Monsarrat described what happened next:

> The first thing which the rescue squads and firemen saw, as their torches poked through the gloom and the smoke and the bloody pit which had lately been the most chic cellar in London, was a frieze of other shadowy men, night-creatures who had scuttled within as soon as the echoes ceased, crouching over any dead or wounded women, any soigné corpse they could find, and ripping off its necklace, or earrings, or brooch. . . . It was not the first air-raid looting I had heard of, but for some reason it seemed the worst.

The England experts in the German Foreign Office were convinced that battering the East End would further drive the working classes across the "social fault line" of London and terrify the ruling elite with the specter of revolution, forcing a rapid conclusion to the war. With perhaps equal naïveté, Nazi officials also apparently had high hopes that British anti-Semitic rumblings could be exploited. Clearly in their element, German propagandists produced aerial leaflets that named six allegedly Jewish members of the British Cabinet and concluded, with a typical bit of Nazi vileness, "Sing the Jews' favorite hymn! Onward Christian Soldiers!"

Yet there was more than just sanitized official optimism in the observation of many in Britain that the bombing had forged a feeling of common purpose and necessity, of being in it together, that had the opposite effect from what the Nazis hoped. "Hitler is doing what centuries of English history have not accomplished," the *New York Herald Tribune*'s correspondent reported. "He is breaking down the class structure of England."

In a contest that pitted civilian morale against civilian morale, to "give it 'em back" was more than a merely emotional act of revenge; it was a strategic calculation aimed at giving the folks at home something to buck them up. And since it was easier to come up with clever ways to make raids more brutal than it was to make them more precise, each new twist of the knife inevitably brought a still more violent lunge in return. Churchill was particularly incensed at the German use of one-ton naval mines fitted with impact fuses and dropped by parachute; such a weapon wafting down from five thousand feet, he noted, could not possibly be aimed, and Churchill badgered Bomber Command to retaliate in kind. Another German innovation was a delayed-action fuse that incorporated a booby trap solely intended to kill any bomb-disposal team that tried to defuse it. This was a long way from the genteel leaflet raids and mutual restraint of one year before.

The pressure to give it back was all the greater because there was precious little that the British could do to stop the night raiders in the fall of 1940. If the shift to night attacks meant that the Luftwaffe's bombers could no longer pose a threat to Fighter Command, neither could Fighter Command pose much of a threat to the bombers.

As early as 1936, at the second meeting of Tizard's air-defense committee, the idea of building a radar unit small enough to fit into an airplane came up. With considerable prescience, Tizard argued that if the radar defense system proved successful, the Germans would almost certainly respond by shifting to night attacks. Ground-based radar could steer the fighters to within three to four miles of their target, which was fine for daylight operations, but at night a pilot would need to come within 500 to 1,000 feet of his quarry in order to make visual contact. Something would be needed to fill the gap between four miles and 500 feet. An airborne radar set was one obvious solution, at least in theory.

But cramming an entire radar system into a night fighter was an enormous technical challenge. It was not until June 1939 that the first airborne test of the system, designated "RDF 2," was conducted. The weight of the equipment and the need for a second crewman to operate it made a twin-engine plane a minimum necessity, and the only available plane that fit the bill as of summer 1940 was the slow Blenheim light bomber. The new, faster Beaufighter was supposed to replace it but was beset by technical problems.

So were the radar sets themselves. Where Dowding had been quick to see the necessity of cutting through red tape to get Watson-Watt's original experiments with radar started in 1936, he had been slow to see that the ad hoc experimental organization that had been cobbled together for that job was incapable of the much more complex task of integrating electronic components into an airframe and arranging for their mass production. Watson-Watt kept trying to run everything himself, even having his scientists manufacture the operational sets. The research scientists tapped for the war effort were known as "boffins," and the way the term was spoken by military men always seemed to combine awe for their genius with at least a hint of exasperation at their everyday impracticality. Watson-Watt's team knew nothing about how to fit racks, brackets, and cable runs into aircraft, nor had they conducted the kind of shake tests and altitude tests required for any apparatus that was expected to endure the rigors of flight. During their first dozen flights, the airborne radar sets had burst into flames because of loose wires and malfunctioning power supplies. Yet Watson-Watt still refused to seek the assistance of industry experts.

This attitude partly reflected the typical British elite disdain for "trade." When Tizard's committee finally stepped in and instituted a sweeping reorganization of radar research and manufacture in early 1940, Watson-Watt erupted in a fury—not over being eased out of his position, but over being offered the new title of "Technical Adviser on Telecommunications." Watson-Watt insisted he should be called "Scientific Adviser." As he later huffily explained, "The difference between being a 'Scientific Adviser' and a 'Technical Adviser' was the difference between being a philosopher and a plumber."

By September 30, 1940, Fighter Command had still been unable to get more than a single Beaufighter in the air on any one night, and most of those missions had been aborted prematurely due to failure of the airborne intercept radar gear. During September and October, the Luftwaffe flew 11,000 night sorties over Britain; Fighter Command shot down a grand total of seven of the German planes. Though the Germans had proved that flak could be devastating to low-flying planes coming in against heavily defended battlefield targets, scoring a hit against high-flying bombers in the vastness of the night sky was a different proposition. In September London's antiaircraft defenses had only eleven gun-aiming radar units, and while the thunderous cacophony of the batteries cheered the populace, their haphazardly aimed shells did scant damage to the Germans.

Technically correct in rejecting ill-conceived stopgaps, but politically tone-deaf, Dowding rebuffed demand after insistent demand from the government that something drastic had to be done, and done fast, to remedy this dismal situation. The Air Ministry ordered him to send Hurricanes on random night patrols, much as had been done in the First World War; it was a genuinely bad idea, but instead of offering anything better Dowding scrawled huffy rejoinders in blue crayon on the memoranda, and implied that his authority was being undermined. In October Churchill made one last effort to underscore the urgency of the matter by creating a Night Air Defence Committee and naming himself as chairman; Dowding still did not seem to get the hint. Coming on top of his failure to resolve an increasingly bitter dispute between the commanders of No. 11 and No. 12 Groups over the proper employment of fighter squadrons north of London in meeting the German daylight raids, Dowding's unresponsiveness on the crucial matter of night defenses was the last straw. On November 13 the hero of the Battle of Britain was informed he was being relieved of his command.

It was a harsh blow, and many saw it as shabby treatment of the man who had built Fighter Command and seen it through its ultimate trial. But even some of Dowding's devoted supporters in high places agreed it was time for the fifty-eight-year-old air marshal to go. "It nearly broke my heart," Churchill told Marshal of the RAF John Salmond, one of those who had lobbied for Dowding's replacement; but, he conceded, "you were quite right." In his retirement, Dowding's bitterness combined with a quirky interest in spiritualism that dated back to that movement's heyday during the First World War, and the result was more than bizarre; Dowding began telling friends that he had been a Mongol chieftain in a past life and that a "Divine Will" had arranged for him to save the British nation in 1940. It was a singular end for a man whose singular contribution had been an unfailing appreciation of the importance of science and technology in defense.

Dowding's departure did nothing to immediately improve the night interception situation. The chief bottleneck was the slow production of Beaufighters and new Mark IV radar units and a lack of trained crews. The fall of 1940 saw another frustrating lag between a British scientific breakthrough and its practical implementation in improving night defenses. Throughout the spring of 1940, British Air Intelligence had been picking up clues that hinted at the existence of a German system that guided bombers to their targets with radio beams. A scrap of paper found in an He 111 that crashed in England in March mentioned a radio beacon called Knickebein. A decoded German radio message referred to establishing Knickebein at a given day at a particular latitude and longitude, which corresponded to a point over England. Two German prisoners taken from another crashed He 111 were interrogated about Knickebein and revealed nothing, but when they were alone one was overheard saying that no matter how hard their captors searched their planes they would not be able to find the apparatus.

R. V. Jones, a young Oxford-educated physicist, was the scientific officer on the Air Intelligence staff, and he began to put these tantalizing details together with mounting excitement. (The recommendation that the Air Ministry create such a post and that Jones fill it had come from the Tizard Committee, which had grown concerned at how little information seemed to be coming in about what the Germans were doing to apply science to warfare. Jones, brilliant, brash, confident, and wickedly funny, had leapt at the offer, exclaiming, "A man in that position could lose the war—I'll take it!") Jones immediately zeroed in on the prisoner's "they'll never find it" remark and realized that the only seemingly innocuous piece of equipment that had been found on the He 111s that could be used as a beam-guidance device was the Lorenz "blind-landing" receiver. This was a standard device that had been in use since the mid-1930s by commercial airlines and many air forces throughout the world. A radio transmitter at an airfield sent out two nearly parallel radio beams that slightly overlapped. The transmitter alternately sent a short pulse on one beam and a long pulse on the other. A pilot approaching the airfield who flew right down the middle of the two beams would receive the dots and dashes at equal intensity, and they would blend to form a steady tone. If he flew to one side, however, either the dots or dashes would grow louder and rise above the steady tone.

Jones saw that the same concept could work to guide an airplane on a preset course flying away from the beacon transmitter. He called the engineer who had examined the Heinkels and asked if he had noticed anything odd about the Lorenz receiver. "No," the engineer replied. "But now that you mention it, it is much more sensitive than they would ever need for blind landing."

By June 13 Jones was convinced and approached Lindemann, Churchill's scientific adviser and confidant; Lindemann penned a memorandum to the Prime Minister, who at once replied, in trademark fashion, "This seems most intriguing and I hope you will have it thoroughly examined."

Jones came into his office one morning a week later to find a note on his desk: "Squadron Leader Scott-Farnie telephoned and says will you go to the Cabinet Room at 10 Downing Street." An inveterate practical joker himself, Jones was sure he was the victim of retaliation from his office mates, but a few calls quickly assured him this was no joke. Leaping into a taxi, he arrived at

No. 10 twenty-five minutes after the meeting he had been hastily summoned to had begun. From the discussion taking place, Jones quickly became aware that none of the august personages assembled had quite grasped the situation, so when Churchill finally called on him to explain a technical point, Jones seized the moment and said, "Would it help, sir, if I told you the story right from the start?" Churchill was taken aback for a moment but replied, "Well yes it would!"

What followed was a moment that, in its style and consequences, would have been inconceivable in Hitler's or Göring's inner circles, but which owing to Churchill's naïve, though genuine, curiosity about all things scientific arguably changed the course of the war—and of aerial warfare to come. For twenty minutes Jones, a twenty-eight-year-old physicist, instructed the venerable leaders of Britain's war government on the arcana of radio navigation. Churchill would later vividly recall the scene of Jones "unrolling his chain of circumstantial evidence the like of which for its convincing fascination was never surpassed by tales of Sherlock Holmes or Monsieur Lecoq."

"When Mr. Jones was finished," Churchill also recalled, "there was a general air of incredulity." But the Prime Minister himself was convinced. He asked Jones what should be done; Jones said the first thing was to fly along the beams and confirm their existence. Churchill so ordered. If there was an official moment when electronic warfare was born, this was it.

The Prime Minister might have been convinced; the RAF was not. Tizard, his skepticism reinforced by his ongoing feud with Lindemann, dismissed it all as ridiculous. At a technical meeting at the Air Ministry that afternoon, another expert insisted that radio signals at the frequencies the Germans were thought to be using for their navigation beams could not reach England. The RAF's deputy director of signals said that in that case the beam-hunting flight scheduled for that evening should be canceled and Jones should stop wasting everyone's time. Jones, risking all, vehemently countered that he had heard the Prime Minister himself approve the flight, and he would see to it that the Prime Minister himself found out who had disobeyed his orders. The flight went on. The next day the smoking gun was in hand. "There is a narrow beam (approximately 400–500 yards wide) passing through a position 1 mile S. of Spalding, having dots to the south and dashes to the north," the pilot's report began.

At the previous day's technical meeting, the RAF's director of signals had asked, "What do we do if we find the beams?" Jones had whispered to a colleague next to him, "Go out and get tight!" The more serious answer was, first, to jam the signals. By later that summer, a special RAF wing set up to deal with the beams had devised a more subtle countermeasure. Code named "Aspirins" (the British had earlier given Knickebein the code name "Headache"), transmitters were set up to broadcast signals that sounded like a genuine Knickebein dash pattern. The result was that a German pilot correctly flying down the center of his beam would hear a superimposed dash and think he was steering too far to the left. (If it had been possible to synchronize the fake beacons with the real

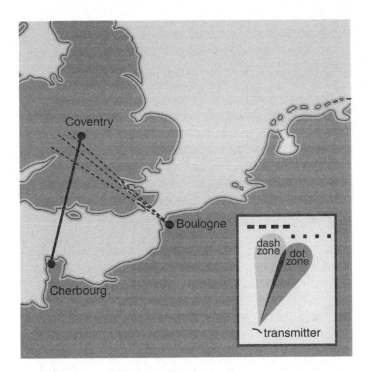

The X-Gerät beam-bombing system: Three cross beams marked
precise distances from the target. A separate director beam consisted
of two slightly overlapping beams alternately broadcasting dashes
and dots. When the dashes and dots merged into a single tone, the
pilot knew he was flying down the center line.

signals, the result would have been to "bend" the beams, but that was never
accomplished; the British had to be content with merely confusing the German
bomber pilots and making them distrust their apparatus.)

In early September, a Luftwaffe radio message encoded using the high-level
Enigma machine cipher was broken by the British cryptographers at Bletchley
Park, and this ushered in a new and deadlier stage in the battle of the beams. The
message referred to a beam no wider than twenty yards and mentioned that a
bomber of Kampfgruppe 100 was being fitted with a device called the X-Gerät,
the "X Apparatus." The X-Gerät turned out to be a far more sophisticated beam-
bombing device than Knickebein. Because of its complexity, it was carried only
by specially trained pathfinder bombers; this was to be the mission for Kampfgruppe 100, whose He 111s bore a distinctive Viking ship emblem.

Intersecting the director beam that established the bombers' flight path were
three cross beams. The first, set to cross the path at about fifty kilometers from
the target, simply warned the crew that it was approaching its destination. At
twenty kilometers, the second cross beam alerted the bombardier to press a but-

ton that started a clock in a small bomb-release calculator; on reaching the third cross beam, at five kilometers, the bombardier pressed the clock button again. That established the plane's ground speed over a distance of fifteen kilometers; combining this with altitude information that had been dialed in, the mechanism would then calculate the proper release point and automatically drop the bombs when it was reached.

In principle, jamming X-Gerät should not have been much harder than jamming Knickebein, but a series of technical blunders and RAF overconfidence caused delay upon delay in instituting effective countermeasures. Poor calibration of their frequency-measuring devices often led the British to jam on the wrong radio frequency. They had even mismeasured the audio tone of the X-Gerät signal, placing it at 1,500 cycles per second instead of the correct 2,000. A Kampfgruppe 100 Heinkel had crashed in Dorset on November 6, but through a farcical series of interservice squabbles, it was not until two weeks later that its equipment was finally recovered and examined: engineers discovered that filters in the receiver blocked any tones other than 2,000 cycles from getting through to the headphones or steering-indicator dial of the device.

In the meanwhile No. 80 Wing was still determinedly jamming the wrong audio tone when the Germans launched their horrific raid on Coventry. The story that Churchill had advance warning of the Coventry raid and chose to conceal the information to protect the secret of British success in breaking the Enigma traffic is a canard. The Luftwaffe's operational orders on the Continent traveled by landline and could not be intercepted. Enigma messages were picked up from time to time that referred to X-Gerät frequency settings for upcoming missions, and as the Blitz continued these would aid the British jamming operation. But they offered scant clues as to the intended target, and before the Coventry raid the traffic had seemed to refer to several possible targets that covered a wide swath of southern England. Churchill was convinced that the target was London, and was standing on the roof of the Air Ministry impatiently scanning the skies for the German bombers when the bombs began falling on Coventry on the night of November 14.

Although several important engine factories were among the designated targets for Coventry, Luftwaffe orders also instructed that, "by wiping out the most densely populated workers' settlements," the resumption of manufacturing in the industrial center of the city would be hindered. Many of the 450 German bombers carried parachute mines along with their loads of high-explosive and incendiary bombs, more than five hundred tons of ordnance in all. When the last bombers finally departed at 6 a.m., one-third of the city's factories, fifty thousand houses, and the fourteenth-century St. Michael's Cathedral were in ruins.

In the mirror-image psychology of strategic bombing, both sides found it to their advantage to play up the brutality of the Coventry attack. The German High Command issued a communiqué hailing the "utmost devastation" that had been inflicted on the enemy. The British press declared it another Guernica. In both languages a new verb appeared: *coventrate, koventrieren* in German, meaning to devastate a city from the air.

The German attack left an especially deep impression in two quarters. A senior RAF officer named Arthur Harris, the future commander of Bomber Command—"Bomber" Harris, as he would be known—was struck by the Germans' application of what he termed the principle of concentration: "the principle of starting so many fires at the same time that no fire services . . . could get them under control."

The other quarter where the echo of the German bombs loudly reverberated was thousands of miles away, across the Atlantic. The United States was still sitting out this war. But public opinion, once solidly isolationist, was rapidly swinging to Britain's side amid the British show of indomitable resolve under fire. Images of Coventry burning had appeared in newspapers across the country and were still vivid in the public mind when, a few weeks later, there arrived in the hands of the just-reelected President Roosevelt a personal letter from Winston Churchill—a letter that Churchill would call one of the most important he had written in his life.

The United States had been helping Britain within the constraints of its own cautious armaments buildup and the staunch Republican Party opposition to involvement in foreign wars. Under an amendment to the Neutrality Act that Congress reluctantly passed in 1939, Britain and France could buy surplus arms, paying cash in advance and transporting them on their own vessels. Roosevelt bent the rules as much as he could; when Hap Arnold, now Chief of the Air Corps, tried to blow the whistle and block the transfer of America's latest models of aircraft and aero-engines under the "cash and carry" rules, FDR none too subtly told him that there were places where officers who "didn't play ball" could be sent—like Guam.

But Britain's cash reserves had now almost run dry. Churchill told Roosevelt that only America's factories, her aircraft factories in particular, could lay the "foundation of victory." Churchill asked, simply, that the full industrial might of the mightiest industrial nation in the world be put behind the fight against Nazism.

Churchill's letter had caught up with Roosevelt in the Caribbean, where the President was taking a post-election holiday cruise aboard an American warship. A seaplane delivered it; Roosevelt read it with no comment and no obvious reaction. But his aide Harry Hopkins could see that he was turning it over in his mind in the ensuing days. "Then, one evening, he suddenly came out with it," Hopkins recalled, "the whole program." It was the program that would soon be known to the world as Lend-Lease. The U.S. government would provide loans to pay for, and ships to transport, the arms Britain needed to carry on the struggle. America, as FDR explained in one of his fireside chats a few weeks later, "must be the great arsenal of democracy." After an ugly political battle—one isolationist senator charged that Roosevelt's policy would "plow under every fourth American boy" and Roosevelt shot back that this was the most dastardly thing he had heard in a lifetime of politics—the Lend-Lease Act passed Congress on March 11, 1941. A few hours later, the White House sent British officials a list of military items available and asked Congress for $7 billion to cover the cost. And shortly thereafter, the White House decided that as long as the United States was not at war, the British would get all the new planes they could absorb from American factories.

In a war that was rapidly becoming a war of numbers in the air, this was a crucial turning point. Hitler and the Luftwaffe had counted on a short war. The German aircraft industry had made no plans to increase production, and so it did not: although output was scarcely sufficient to cover the Luftwaffe's operational losses throughout 1940 and 1941, German factories continued to operate on single eight-hour shifts. Machine tools looted from the occupied countries went into storage. Partly this was a matter of labor supply: where Britain mobilized women to work in factories, Nazi ideology demanded that women stay at home producing babies and tending the fireside. And so severe labor shortages hit all skilled industries in Germany as workers were drafted into military service.

But it was also a matter of organization, or rather lack thereof. Göring had placed a First World War fighter-ace buddy, Ernst Udet, in charge of the Luftwaffe's technical office, his major qualification being that he was a good Nazi and enough of a lightweight that he would pose no threat to Göring's authority. Udet had been a test pilot, a stunt flier, a film star, and was a celebrated socialite and ladies' man; he did not even pretend to know about industrial processes or aircraft engineering. Udet's attempts to increase aircraft production consisted chiefly in setting increased production targets, then revising the targets downward to match whatever the factories managed to produce.

The failure to increase aircraft output was exacerbated by Hitler's obsession with "wonder weapons," which grew more obsessive and disruptive just as the aircraft industry was finally beginning to shift from a strategy of quality to one of quantity. Udet had his own naïve technological obsession. Impressed by the potential of dive-bombing, which he had seen demonstrated on a prewar visit to the United States, he demanded that *every* German bomber be made into a dive-bomber. Udet had delayed the production of the Ju 88 bomber by two years by adding a dive-bombing requirement; now he insisted that the follow-on models to the Me 110 and the Do 17 likewise be redesigned. Even the planned four-engine Heinkel He 177 heavy bomber would have to be a dive-bomber. Trying to make a heavy bomber into a dive-bomber was an act of aerodynamic insanity; in vain Heinkel tried to talk Udet out of it. Heinkel's engineers had to completely redesign the wings to withstand the stress of diving. They attempted to reduce drag by mounting two engines front to back in a single streamlined engine compartment on each wing, both engines coupled to a single driveshaft turning a single propeller; it was an innovative notion, but heat built up so rapidly in the compartment that the engines frequently caught fire. Germany never would deploy a workable heavy bomber during the war.

By the time Udet was finally replaced in the summer of 1941, the damage had been done. A few months later Udet shot himself, leaving a message scrawled on the wall of his apartment accusing Göring of betraying him to "Jews" in the Air Ministry.

Udet's successor, Field Marshal Erhard Milch, immediately began to straighten out the mess and demanded that the German aircraft industry concentrate on quantity production of existing types. But by then the Anglo-American lead was insurmountable. By the second half of 1940, Britain's aircraft factories alone were outproducing Germany's. By the fourth quarter of 1941, combined British and American production was exceeding Germany's by 400 percent in fighters, 170 percent in twin-engine aircraft, and 4,000 percent in four-engine aircraft. The gap would only grow as the war went on.

Once, Göring had dismissed the possibility that American industry could pose any threat to German supremacy in the air; Americans, he said, "could only produce cars and refrigerators." In 1940, America's aircraft manufacturers would probably have agreed with that assessment, at least when it came to the seemingly fantastic notion of mass-producing airplanes on a factory assembly line, as if they were cars or refrigerators. Airplanes had never been built that way. America's aircraft makers had located in places such as southern California where the final assembly of large airplanes could be done outdoors. Countless parts had to be hand-fitted. Working with aluminum required special tools and jigs. But in May 1940 Roosevelt had issued an astonishing challenge to Congress: America, he declared, would produce fifty thousand airplanes a year. Henry Ford immediately told the press that he could produce a thousand a day. William Knudsen, the chairman of General Motors, who had been named by Roosevelt to take charge of production planning, offered the somewhat more realistic figure of a thousand warplanes a month from GM's plants. The airplane manufacturers for their part dismissed the notion of automakers building airplanes: "You cannot expect blacksmiths to learn how to make watches overnight," sniffed the president of North American Aviation.

But by the end of 1940 both the aircraft and automobile industries were mobilizing to do what the aircraft executives had declared impossible. By January 1, 1941, total factory floor space at plants producing airframes, engines, and propellers had doubled from just a year earlier. Hudson, Liberator, Boston, and Baltimore bombers, P 40 fighters, and thousands of Merlin engines for Spitfires and Hurricanes poured across to Britain. In April 1941 Ford broke ground at Willow Run near Detroit for a plant that became an instant symbol of America's gargantuan approach to war production. With seventy acres under one roof, the Willow Run plant enclosed a moving assembly line two-thirds of a mile long. A finished Liberator bomber, it was promised, would emerge every hour. This turned out to be harder than Henry Ford had supposed; a car had 15,000 parts, a B-24 had a million, and it was not until the end of 1944 that the plant produced a plane an hour. But the aircraft makers were swiftly adopting mass-production techniques themselves, and other huge factories—many of them more successful if less famous than Willow Run—were springing up across the vast American heartland: Chicago, Omaha, Cleveland, Indianapolis, Dallas, Tulsa, Oklahoma City. By the time the war was over, output per man-hour at aircraft plants would triple as automation and assembly-line methods replaced skilled handwork.

Mass production had one huge drawback, in that it threatened to exacerbate the inherent conflict between quality and quantity; to mass-produce a plane, its design had to be frozen. Even the smallest subsequent design change could play havoc with production schedules as lines shut down for retooling. The American strategy inevitably bet on quantity, but planners came up with an inventive solution to keep the lines rolling while incorporating at least some of the continual improvements. The factories would keep mass-producing according to the original design, but the finished planes would then go directly to "modification centers" that added the latest bells and whistles and sometimes even undertook structural modifications. Most of the modification work was done at the maintenance shops of commercial airlines, pressed into service for the task. In all, American factories would turn out 300,000 military aircraft and 800,000 aircraft engines during the sixty-two months from July 1940 to August 1945, more than meeting Roosevelt's call for 50,000 a year—this from an industry that, in 1939, had received a total domestic order of 435 airplanes from the U.S. Army Air Corps.

Spring returned to England in 1941 with a sense that a corner had been turned. America, though not yet in the war, was staunchly in Britain's camp, protecting her transatlantic lifeline and placing at her disposal the unstoppable dynamo of a gigantic industry now committed to producing military aircraft at a pace the world had never imagined possible. The Luftwaffe's night raids began to dwindle, and the Germans found themselves for the first time on the losing end of the British scientific counteroffensive. Beam jamming was yielding consistent results. By April and May, British jamming of a new, more sophisticated German beam system, the Y-Gerät, prevented the bombers from receiving their bomb-release signal well over half the time; on only two occasions in May did even 25 percent of the bombers receive the correct signal. Combined with the use of decoy fires, code-named "Starfish," the countermeasures frequently tricked the Germans into dropping their payloads on empty farm fields. The radar-equipped Beaufighters were now in service, too, and in May the British night fighters brought down some ninety Luftwaffe bombers.

But by then Hitler's attentions had already turned elsewhere: to the east, and disaster.

9

AIR VERSUS SEA

In 1941 the United States Navy stood on the verge of an intelligence revolution. The breaking of German U-boat signals, encrypted with the legendarily byzantine Enigma cipher machine, and the equally stunning solution of the main Japanese naval codes, would catapult intelligence from a neglected backwater, a career-ending dumping ground for hacks and has-beens, to the very center of Allied command and strategy. Before the war, intelligence had a bad reputation among naval commanders, and for good reason: it was almost worthless. Most intelligence reports consisted of gleanings from outdated foreign technical journals or newspapers, scraps of gossip picked up by bored attachés at diplomatic functions, mercenary concoctions of shady paid informants. Commanders who even bothered to sample this thin gruel would invariably pull out the few morsels that merely sustained their complacent prejudices about the enemy.

The revolution would come, but as of the first week of December 1941 complacency and prejudice maintained their unbroken reign. American intelligence officials had watched the rise of Japanese naval aviation and had seen what the Imperial Japanese Navy intended them to see, outmoded airplanes and unspectacular carrier operations. "Her personnel cannot send planes aloft or take them aboard as rapidly as American personnel," a former director of the Office of Naval Intelligence, Captain William D. Puleston, confidently asserted in 1941. Other American experts agreed that the Japanese as a race suffered from myopia, poor night vision, defects of the inner ear, mechanical ineptitude, and a want of individual spirit that made them poor pilots. "Every observer concurs in the opinion that the Japanese are daring but incompetent aviators," concluded Fletcher Pratt, a well-known American aviation writer.

There was more truth to another popular cliché about the Japanese: that they were copycats. The irony about this truth, though, was that it left fewer excuses for the West not to realize what sort of an air force and capability Japan was establishing. In the years immediately following the First World War, the Japanese went on a buying frenzy, acquiring foreign aircraft, machine tools, and patent licenses from American and European firms. A hundred Spad and Sopwith fighters arrived from Britain and France. Attachés and representatives of Japanese firms were busy in the United States and Europe snapping up manufacturing rights to Hispano-Suiza engines, Curtiss and Lockheed aircraft, and Hamilton Standard propellers. Japanese students began showing up to enroll in the newly established aeronautics programs at MIT, Stanford, and Caltech; Japanese industrial engineers and managers served apprenticeships at Douglas, Curtiss, Lockheed, Boeing, and Pratt and Whitney. Well into the late 1930s, the Japanese appetite for American aviation know-how was still on abundant display; the Japanese Navy ordered twenty two-seat versions of the P-35 fighter, an all-metal monoplane built by Seversky Aircraft, and Japanese firms secured licenses for North American's trainers and Douglas's DC-3 transports. Japanese factories would crank out these workhorse aircraft throughout the Second World War.

Along with foreign aviation technology and know-how came a flood of foreign experts, military and civilian. Soon after the end of the First World War, France sent a team of advisers to help the Japanese Army organize and train its air forces. The Japanese had asked for four experts. The French sent fifty-seven. Britain followed with a mission of its own, which provided extensive assistance and training for the Japanese Navy's fledgling aviation force. Thirty veteran British pilots of the First World War formed the entire founding faculty of the flight school established at the just-completed naval air station at Lake Kasumigaura, forty miles northeast of Tokyo. The school would subsequently train nearly every prominent Japanese naval aviator of the Second World War. The British experts also consulted extensively with the Japanese on the completion and initial flight operations of the world's first true aircraft carrier, *Hosho,* whose keel had been laid in 1919. Mitsubishi hired designer Herbert Smith of Sopwith to draw up plans for a family of first-generation carrier aircraft, including a fighter, a scout, and a torpedo bomber. Two other British firms, Gloster and Blackburn, designed Japan's second-generation carrier aircraft in the mid-1920s.

In 1931 two more British officers arrived to provide a select group of Japanese naval aviation officers instruction in advanced strategy and tactics. Among the topics of their seminars was the use of carriers in offensive operations. Among their eight students was a young and brilliant flier named Minoru Genda.

The British Royal Navy had from the start taken a progressive view of the role of the airplane in naval warfare. Britain had led the world in pioneering naval aviation advances during and even before the First World War. In 1911 the Royal Navy had already begun to carry out experiments in launching seaplanes and airplanes from the decks of cruisers and from platforms built over the gun turrets of battleships. In 1917 HMS *Furious,* a large light cruiser nearing completion, was commandeered for conversion to a dedicated aircraft carrier; a "flying off deck" was fitted on the forward end of the ship, and subsequently a separate "flying on deck" was added behind the superstructure, with small gangways connecting the forward and aft decks to shift airplanes between them. By the fall of 1918, twenty-two light cruisers had also been fitted with gun-turret platforms to help make up for the shortage of dedicated carriers. At the end of the war, the Royal Naval Air Service had 3,500 aircraft and 55,000 officers and men.

The Royal Navy's view of aviation was progressive; it was not revolutionary. Like all the world's navies, British sailors in the 1920s and 1930s still viewed the battleship as the final arbiter of power on the high seas. The job of aircraft at sea was to help the big guns do their job. Aircraft would find the enemy, slow him down with torpedo attacks, and then act as artillery spotters while the fourteen- and sixteen-inch guns of the battle fleet delivered the coup de grace. Given the mass of firepower that a battleship's guns represented, with a single shell weighing upward of a ton and a half, this was not on its face an unreasonable division of labor.

The officers of the Imperial Japanese Navy had their own persuasive argument for adhering to this progressive but limited view of naval aviation's role, one that went to the heart of Japan's well-established grand maritime strategy. The Japanese admirals had long been in the thrall of the theories of Alfred Thayer Mahan, the American "high priest of sea power," who saw the decisive clash of battle fleets as the key to securing free use of the seas while denying the same to the enemy. Mahan's writings had been translated into Japanese in the nineteenth century and were required reading at all naval schools and colleges. Japan's swift, stunning, and decisive defeat of the Russian fleet in the 1905 war only reinforced the Mahanian orthodoxy.

How to put Mahan's ideas to work against the far superior American fleet posed a thornier problem for Japan's naval contingency planners. The official solution, which dated back to as early as 1907, when tensions were rising between the two countries over America's treatment of Japanese immigrants, was that Japan would lure the American fleet across the Pacific, whittling it down with submarine attacks and night destroyer engagements, and then at last bring about the decisive clash in Japanese home waters. The advent of naval aviation fit smoothly into this overarching strategic concept: both long-range, land-based aircraft and carrier-based torpedo bombers would add their part to the whittling-down operation. The Washington Naval Treaty of 1922, which restricted Japan's warship tonnage to three-fifths of America's, gave further impetus to the notion that naval aviation would help Japan bridge the gap in capital ship strength, but it did not shake the orthodox belief that the capital ships would be decisive. No thought was given to the notion that carriers might operate independently.

Minoru Genda begged to differ, and the British lectures he attended in 1931 only confirmed his growing unorthodox convictions. There had always been one exception to the British view about the subordinate role of carriers in offensive operations. Since 1914 the Royal Navy had examined the idea of air raids on enemy ships in harbor, where they would be more vulnerable to a strike from the sky. (A two-hundred-plane raid on Wilhelmshaven had been in the works in

1918 when the war ended.) This was still a relatively conservative idea in terms of naval operational concepts; none of the British lecturers advocated using carriers independently or decisively in battles on the high seas. But attacking ships in harbor from the air would still represent a daring deviation from orthodox Japanese naval strategy.

In 1936 Genda was an officer student at the Japanese Naval War College when his class was given the assignment of working out suitable armament for an encounter between the navies of Japan and the United States. Genda turned in a paper that took the idea of independent carrier attacks to its logical, if astonishing, conclusion. "The main strength of a decisive battle should be air arms, while its auxiliary should be built mostly by submarines," he proposed. "Cruisers and destroyers will be employed as screens of carrier groups, while battleships will be put out of commission and tied up." Genda argued that carrier-based dive-bombers and torpedo planes would be able to outrange even the biggest guns; sea battles would be decided while the combatants were still separated by hundreds of miles, long before the battleships' big guns could be brought to bear. It was best to leave the big ships at home.

To say that his proposal was greeted coolly would be an understatement. "Not only was my idea attacked," Genda later recalled, "but even my mental soundness was doubted." Yet his belief in the "imperative" need for command of the air as a prerequisite to any military operation, on land or sea, only became firmer after he was assigned to combat duty in what Japan euphemistically called the "China Incident" in July 1937, and then even more so when he witnessed the aerial fighting in the Battle of Britain as an assistant attaché at Japan's embassy in London from March 1939 to September 1940.

Two months after Genda's return to Japan that fall, his British mentors offered another practical lesson in the growing dominance of air power at sea. Just before nine o'clock on November 11, 1940, twelve Swordfish torpedo bombers of the British carrier HMS *Illustrious* began to take off into the Mediterranean night. Their target was the ancient harbor of Taranto, situated at the arch of Italy's boot. Aerial reconnaissance sorties from Malta by a British Martin Maryland bomber the day before had confirmed that the entire pride of the Italian fleet lay at anchor in Taranto's outer harbor, the Mar Grande: six battleships, seven cruisers, numerous destroyers. More destroyers and light craft were moored in the inner harbor, the Mar Piccolo, which was also lined with fuel tanks and a seaplane base.

For months the British Commander-in-Chief-Mediterranean, Admiral Andrew B. Cunningham, had tried in vain to entice the Italians out of port for a decisive engagement that would remove the threat they continually posed to British convoys. Attacking the port itself from the sea was out of the question. Taranto's harbor was too shallow for submarines, and approaching the port within gun range of his cruisers and battleships would have exposed Cunningham's ships to attack by Italian land-based bombers.

But a British contingency plan for aerial attack on Taranto had been on the books since 1935, and it was just the sort of bold coup that appealed to Cunningham. A fighting admiral of the old school, certainly no intellectual or revolutionary, Cunningham nonetheless was never one to turn away a new idea if he thought it would win. He set the air attack for the night of October 21, the one hundred and thirty-fifth anniversary of the Battle of Trafalgar.

Taranto was probably the most heavily defended harbor in the world at that moment. Between shore batteries and ships, the Italians had close to a thousand antiaircraft guns in and around the harbor. Torpedo nets protected the battleships. Small white blobs that had shown up in the latest reconnaissance photos had much puzzled the intelligence officer on *Illustrious* until he flew to Cairo to seek the counsel of the RAF's photointerpretation expert; they turned out to be sixty barrage balloons whose steel cables posed an almost unseeable menace to low-flying planes at night.

The Swordfish torpedo bomber, however, was an anachronism that could never have survived the briefest daytime encounter with enemy fighters, so a night raid was the only possibility. Large, clunky, with a top speed of only about 140 miles per hour, the three-seat, open-cockpit biplane had the virtue of robustness but little else. Each could carry a single 1,600-pound torpedo or three 500-pound bombs. By throwing out one crew member and cramming an extra fuel tank into the backseat, its range could be extended. But it was in the midst of this refitting to prepare for the Taranto attack that a freak accident, triggered when a mechanic dropped a wrench and shorted an electrical contact, set off a blaze on *Illustrious* that destroyed two planes and damaged three more. Then three more were lost when contaminated fuel caused abrupt engine failures after takeoff. The mission was postponed. *Illustrious* was down to twenty-one serviceable aircraft.

Swordfish Mk I
carrier torpedo bomber
engine: 690-hp radial
top speed: 138 mph
armament: 1,600-lb.
    torpedo; two .303-in.
    machine guns

When the delayed raid finally began, late on the night of November 11, the Italians heard Cunningham's bombers coming from miles away. As the twelve Swordfish of the first wave approached, their crews saw the sky over Taranto teeming with a hail of exploding flak. Two Swordfish broke off and skimmed along the rim of the Mar Grande, dropping parachute flares from 7,000 feet to illuminate the big ships. The others swooped into the midst of the maelstrom.

First in was Lieutenant Commander N. W. Williamson; barely missing a barrage-balloon cable, skimming the waves at thirty feet, he released his torpedo at the first warship he saw, a destroyer. The torpedo missed the destroyer but exploded against a far more valuable target, the battleship *Conte di Cavour.* It was to be the most decisive hit of the entire raid. Williamson was in turn imme-

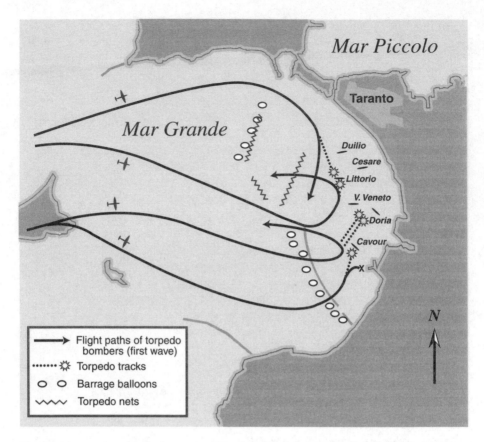

The British attack on the Italian port of Taranto, November 11, 1940. Japanese naval experts would closely study the raid while planning the Pearl Harbor attack the following year.

diately hit by antiaircraft fire and crashed into the water. Miraculously, both he and his crewman survived. (The Italian dockworkers who fished them out beat them up, but once in the hands of the Italian military authorities the British fliers were treated "almost like heroes" and plied with cigarettes; two nights later during an RAF bombing raid, they were serenaded by their Italian captors with a rousing chorus of "Tipperary" as they all huddled together in a shelter.)

By the time the second wave of Swordfish turned home, the Italians had fired 13,000 rounds of AA ammunition and shot down two of the twenty-one attackers; the British had crippled three battleships, severely damaged two destroyers, and set the seaplane base ablaze. Cunningham's carrier commander wasn't sure the raid had been a success and wanted to repeat it the next day. Cunningham wasn't sure that was a good idea, given what the aircrews had been through. "After all," one of the pilots said, "they only asked the Light Brigade to do it once." A storm that blew in later that day settled the matter, and the second raid was scrapped.

In fact, the British attack had been a stunning success, especially given that only five of the eleven torpedoes launched had both hit a target and exploded. The Italians hastened the surviving battleships north to Naples. All three of the damaged battleships would be out of commission for at least six months. *Cavour,* the most seriously damaged, settled to the bottom the next day after salvage crews gave up on towing her to shore.

In May of the following spring, a high-level Japanese delegation arrived in Italy with a long list of detailed technical questions about the Taranto raid. Two important things had happened in Japan in the meanwhile. In April the Japanese commander in chief of the Combined Fleet, Admiral Isoroku Yamamoto, had organized his carriers into a single, vast striking force. Named the First Air Fleet, it consisted of the carriers *Akagi, Kaga, Soryu, Hiryu,* and *Ryujo,* each accompanied by two destroyers. Genda was appointed air officer.

The other thing that happened was that Yamamoto had begun quietly preparing a plan for employing this unique force, a plan as audacious as the idea of organizing a fleet of carriers. "In the event of outbreak of war with the United States, there would be little prospect of our operations succeeding unless, at the very outset, we can deal a crushing blow to the main force of the American Fleet in Hawaiian waters using the full strength of the 1st and 2nd Air Squadrons against it," Yamamoto wrote in a letter that Genda was secretly shown in February. Genda was asked to draw up a preliminary plan.

He returned a week later with an outline. The attack would be difficult but not impossible. American carriers would be the primary target. Absolute secrecy and surprise were essential. If torpedoes could not be employed because of the shallow waters of Pearl Harbor, dive-bombers would have to be used instead.

In the end Yamamoto would have to threaten to resign to secure the approval of the conservative Naval General Staff for such a radical departure from orthodox strategy; even then the staff never abandoned the idea of the Great All-Out Battle but insisted that the Hawaiian Operation was only a preliminary to it, designed to even the balance of forces in the Pacific. Genda had pointed to the enemy carriers as the primary target, but in the final plan it was the American battleships that took the focus, holding their customary and paramount place in the worries of the Japanese naval planners.

Genda saw his plan as the first draft of a new rule book of naval warfare, in which the aircraft carrier replaced the battleship. The Naval General Staff admitted no such thing. They looked upon the Hawaiian Operation more as a sort of high-stakes commando raid, in which the gamble was that audacity and surprise would let them get away with breaking the rules just this once. The staff insisted that as soon as the carriers had completed their mission they were to turn home immediately and get back where they belonged, within the safety of the battleship fleet.

Yamamoto was certainly a gambler. From his time in the United States—he had studied at Harvard and served as a naval attaché in Washington—he had acquired a taste for poker, at which he was a habitual winner. He had also acquired a healthy respect for the might of American industry. In a letter to the Navy Minister in late October 1941, Yamamoto called the idea of going to war with the United States "risky and illogical."

But there was no doubt he would personally assume the risk if the decision to go to war was made. That was not just the gambler's instinct: for a senior admiral in the Japanese Navy, Yamamoto was unusually well versed in the technical details of naval aviation and was better able than most to appreciate the potential of this new weapon. In 1931 he had been appointed Chief of the Technical Division of the Navy's aviation bureau; the following year he was given responsibility for an ambitious "Aviation Technology Independence" program. The goal was to rapidly build up Japan's domestic aircraft industry and overcome America's quantitative and qualitative lead in naval aviation. In 1935 he was promoted to head of the entire aviation bureau and given the task of doubling aircraft production in two years.

In 1937 the bureau issued specifications for a new air-to-air fighter that would go "one step beyond" existing models in performance and endurance. The result was the famous Mitsubishi Zero, one of three superbly capable carrier airplanes with which Japan would enter the war. Light and highly maneuverable owing to its low wing loading, the Zero traded away a strong airframe and armor protection for the pilot in favor of performance. Despite a small engine—Japanese manufacturers never succeeded in producing engines of much over 1,000 horsepower during the war, their one big failing—the Zero could climb faster than a Spitfire and had a range twice as great, more than a thousand miles, with a top speed of 330 miles per hour. It was well matched by Japan's two other new carrier planes, the Stuka-like Aichi D3A dive-bomber (code-named "Val" by the Americans) and the Nakajima B5N torpedo bomber ("Kate"). They were both thoroughly modern, all-metal monoplanes that beat such relics as the Swordfish by a good hundred miles per hour.

A6M2 Zero
carrier fighter
engine: 950-hp radial
top speed: 331 mph
armament: two 7.7mm
    machine guns in nose,
    two 20mm cannon in
    wings

Other technical and tactical developments encouraged Yamamoto's confidence in Japan's emerging naval air weapon. Through extensive realistic testing, Japan developed its own fast and reliable airborne torpedoes in the 1930s. Powered by engines that burned kerosene and compressed pure oxygen, the Japanese torpedoes ran efficiently and, compared to torpedoes that used ordinary compressed air, produced minimal exhaust gases that could give away their track with a trail of bubbles on the surface.

Tests by both the Japanese and American navies in the 1930s also were proving the effectiveness of dive-bombing, especially against small and maneuvering targets of the kind that ships at sea presented. The U.S. Navy, having spent half a million dollars developing the Norden bombsight, largely gave up on high-altitude level bombing after exercises from 1932 on showed that no more than a few percent of bombs dropped from 10,000 feet could even hit a ship. Dive-bombing solved several problems at once. By pointing the nose of the attacking plane directly at the impact point, it greatly simplified the bomb trajectory calculation; the closer to 90 degrees the dive became, the less forward motion was imparted to the bomb upon its release and the less the pilot had to worry about the effects of altitude, speed, and wind. (Actually, from a purely theoretical and mathematical point of view, the dive-bombing problem was more complex and had more variables than level bombing. Carl Norden spent years scribbling equations and designing a bombsight that would automate the job of aiming for dive-

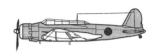

B5N "Kate"
carrier bomber/torpedo
bomber
engine: 1,000-hp radial
top speed: 235 mph
armament: 1,760-lb.
    bomb or torpedo load;
    7.7mm machine gun
    in rear cockpit

bomber pilots. But it proved so cumbersome to use that the Navy found pilots did a better job relying on their crosshair gun sight.) Releasing the bomb at a lower altitude of course inherently improved accuracy as well. In a 70-degree dive, average circular error dropped from about 400 feet for a bomb released at 10,000 feet to about 50 feet at 2,000 feet. Overall, in the same exercises that saw level bombers scoring hit rates of a few percent, dive-bombers consistently got about 20 percent of their bombs on target, even when the ships zigzagged or took other evasive action. The tests also confirmed that a rapid dive also made it difficult for antiaircraft gunners to follow their target.

All three methods—level bombing, dive-bombing, and aerial torpedoes—had their disadvantages, however. To withstand the aerodynamic stresses of the dive, dive-bombers had to give up some of their carrying capacity in favor of structural strength, and it was not clear that their small bombs could do much damage to the heavily armored deck of a battleship. Torpedoes were far more effective, striking ships at their most vulnerable spot below the waterline, but torpedo bombers had to make an almost suicidal approach, flying straight and level a few hundred feet above the water before releasing their weapon. Horizontal bombers could carry heavier loads but were forced to ever higher altitude by the growing effectiveness of AA guns.

Like the U.S. Navy, the Imperial Japanese Navy hedged and explored all three methods of attacking enemy ships from the air. For the attack on Pearl Harbor, Genda was convinced that the inherent technological limitations of all of the methods could largely be overcome through exhaustive training. In 1941 the

Imperial Navy already had what was probably the best-trained pilot force in the world. The carrier force was piloted by six hundred of the best of the best, men who averaged eight hundred flying hours. Washout rates in the Imperial Navy's flying schools were extraordinarily high. Out of fifteen hundred applicants to one class in 1937, seventy were chosen and twenty-five finished the course.

In earlier exercises, carrier aircraft crews had achieved impressive improvements in hit rates: 30 percent for dive-bombers, 10 percent for level bombers, close to 100 percent for torpedo bombers. Genda now pressed for intensive rehearsals of the Pearl Harbor attack to push those performances even higher. *Akagi*'s level bombers got their hit rate up to 40 percent. Because Pearl Harbor had an average depth of only 40 feet, the usual launch pattern for torpedo bombers could not be used (typically torpedoes would plunge to about 200 feet below the surface before rising back to their preset running depth). Genda had his airmen practice "stunt-like" flights in which they roared over treetops along the shore, then banked over the water, coming in at a mere 60 feet so their torpedoes would slice into the water without hitting bottom. At the end of September the Combined Fleet requested two huge, six-foot-square scale models of Oahu and Pearl Harbor. Genda set these up in his quarters on *Akagi* to work out in minute detail where each pilot would release his weapons during the attack. Japanese technicians that month also devised a last-minute fix: wooden fins that could be attached to the torpedoes to help keep them above a 36-foot depth.

In the years following the Pearl Harbor attack, there would be much recrimination in the United States over the "battleship admirals," the "gun club" that in its hidebound traditionalism had failed to recognize the potential of the airplane and the aircraft carrier, and thus had refused to believe that anything like December 7 could happen. But that was too simplistic. The U.S. Navy and the Imperial Japanese Navy were nearly identical when it came to aviation and naval strategy between the wars. For every American admiral like Charles McVay, head of the Bureau of Ordnance, who in 1921 declared that aircraft presented "no serious menace to the modern fighting vessel," there was one like William S. Sims, the naval commander in Europe in the First World War, who declared in 1925 that "a small, high-speed carrier alone can destroy or disable a battleship alone. . . . The fast carrier is the capital ship of the future."

American naval opinion generally fell between these extremes; on the whole it was solidly behind naval aviation in the crucial formative period in the 1920s and 1930s. Aviation was given an unequivocal endorsement in 1919 in a recommendation from the Navy General Board, which declared that "aircraft have become an essential arm of the fleet" and must be "capable of accompanying and operating with the fleet in all waters of the globe." The board recommended that every squadron of capital ships include one aircraft carrier. A statute enacted two years later required that all aviation units in the U.S. Navy be commanded by a naval aviator. Senior ships' officers who had no pilot training were given a quick course at Pensacola, Florida, where they could earn their wings. Young fliers joked about these "JCLs," the Johnny Come Latelys, but in fact the practice built a solid and informed constituency for aviation at the top of the command hierarchy. Among the many senior officers who learned to fly in their advanced years were Ernest King, the future wartime commander-in-chief of American naval forces, who earned his wings at age forty-seven, and William F. Halsey, Jr., who at age fifty-one received a waiver permitting him to fly wearing glasses (and who was the last in his flight class to solo).

Under the effective and politically astute leadership of Rear Admiral William Moffett, the Navy's Bureau of Aviation meanwhile funded industrial development and became a potent organizational force within the Navy advocating naval aviation. Moffett was a "master of publicity" who knew how to win favor with members of Congress and who arranged for Navy fliers to participate in air shows and in the 1932 Hollywood movie *Hell Divers*; he also was a superb administrator, seeing to it that aircraft and ships got built, pilots got trained, and aviation incorporated into command and fleet exercises.

Having someone with lots of stripes on his arm to run interference was always a blessing when the business at hand was to drag a military institution in a new direction, but it was especially so in the case of naval aviation; operating a carrier at sea required the mastery of technical and managerial skills that were utterly new to both fliers and sailors, and results would not come quickly. Building a carrier force was as much about building an institution as it was about building ships and planes. The manifest difficulty of landing an airplane on a tiny, heaving runway in the middle of an ocean was just the start of the new tasks to be learned. The planes needed to be armed and fueled and maintained all in the confines of a ship; thousands of bombs and torpedoes and hundreds of thousands of gallons of aviation gasoline had to be safely stored but rapidly brought on deck whenever needed; landings and takeoffs had to be coordinated with a precision unknown to landbound airport managers. Operating a carrier with a rhythm and purpose proved to be an art that could be perfected only through experience.

One of the most recondite skills was that possessed by the flight deck officer, whose job it was to locate the airplanes in the right place on the deck. On the American *Lexington* class ships, this meant a constant and perfectly choreographed reshuffling. Aircraft parked on deck had to be pushed forward to make room for other aircraft as they landed; they had to be pushed aft to make room for launching the next mission; they had to be pushed forward, where the gas lines were located, for refueling; they had to be pushed to midship, where the bomb and torpedo elevators were located, for rearming. It was like a chess game, figuring out which pieces to put where, all the while trying to think several moves ahead.

Billy Mitchell's stunts had reduced the question of air power at sea to a cartoon clash between planes and ships, bombs versus guns. In fact the real question resided at a higher level of operations and strategy. Few American naval officers doubted the importance of airplanes at sea, or the ability of air-delivered weapons to do serious damage to ships. The crucial question was the relative vulnerability of battleships and carriers. Battleships had heavily armored decks and layers of steel plate protecting their vital parts. They were accompanied by cruisers and destroyers that mounted multiple batteries of antiaircraft guns. Carriers by comparison were powder kegs, with thin flight decks and crammed with tons of explosive aviation gas and munitions. U.S. Navy exercises in 1934 suggested that if a carrier spotted an enemy carrier first, it could wipe it out in a single attack. Even with its own fighters, a carrier could not effectively defend itself in those pre-radar days; if a few enemy planes got through, the game was up. It was quite possible that *all* carriers might be lost in the "opening movements of a naval campaign," the commander in chief of the United States fleet concluded.

The only hope for such a weapon that could attack but not defend seemed to be either to deploy carriers in the midst of a battleship squadron or to employ them as a sort of stealth weapon, much like a submarine, that could stage hit-and-run raids. If the battleship was obsolete, as a few visionaries such as Admiral Sims and Captain Genda saw it, the aircraft carrier was as yet too immature to take its place as the queen of the seas.

Technological development and sheer desperation of a kind that only war can stimulate were about to change this. On November 23, 1941, the commander of the Japanese carrier fleet received his orders:

> THE CARRIER STRIKING TASK FORCE WILL PROCEED TO THE HAWAIIAN
> AREA WITH UTMOST SECRECY AND, AT THE OUTBREAK OF THE WAR,
> WILL LAUNCH A RESOLUTE SURPRISE ATTACK ON AND DEAL A FATAL
> BLOW TO THE ENEMY FLEET.

Two weeks later, at eight o'clock on a bright sleepy Sunday morning in Hawaii, the ship's band on the battleship *Nevada* was halfway through the national anthem for the morning raising of the colors when an aircraft roared over the deck and let loose a fusillade of machine-gun bullets. The stunned bandsmen and honor guard could see the unmistakable Rising Sun emblem on its wings. Within minutes all of the ships in the most exposed positions along Battleship Row had been struck by multiple torpedo hits. Antiaircraft gun crews struggled with locked ammunition boxes as Zeros swooped over and strafed the decks. Of the seven battleships tied up at Pearl Harbor that morning, *Nevada* alone got under way and fought off her attackers with a heavy barrage of AA fire—before running aground, damaged by a torpedo that had ripped a gaping hole in her bows. The six others were either sinking or immobilized between other sinking ships. The worst fate fell to *Arizona*; hit by an armor-piercing bomb that penetrated a forward magazine, she was torn in half by the ensuing explosion and swiftly took eleven hundred of her fourteen-hundred-man crew to the bottom with her.

A second wave of planes struck an hour later. The devastation they left behind was almost beyond comprehension: nineteen American ships hit, twenty-four hundred sailors dead, and some two hundred American planes out of action, most of them bombed on the ground as they stood lined up at their bases. Admiral Husband E. Kimmel, commander of the Pacific Fleet, was struck in the chest by a spent bullet during the attack. "Too bad it didn't kill me," he muttered.

A Japanese study the following summer confirmed the stunning success of the attack. Three hundred fifty planes had taken part. Of the forty torpedoes dropped, thirty-six hit, a 90 percent success rate. In sixty-five dive-bombing attacks, thirty-eight bombs had hit, an astonishing 59 percent. And of the seven squadrons of Kate torpedo bombers assigned to carry out high-altitude level bombing, six squadrons achieved at least one hit, for an estimated hit rate of 37 percent. Yamamoto modestly opined that such an increase in accuracy rate was "beyond the reach of human beings" and "the loyal fliers must have been inspired by the soul of the Emperor."

There were some disappointments. The study concluded that the 400-pound weapons carried by dive-bombers were insufficient to deliver a fatal blow even to a cruiser, much less a battleship. But the Emperor's inspiration apparently held sway for a few more days. On the night of December 10 the telephone at Winston Churchill's bedside rang. It was the First Sea Lord, Admiral Dudley Pound, and Churchill at first could not make out what he was saying. Finally Pound found his voice and blurted out the terrible news. "Prime Minister," he said, "I have to report to you that the *Prince of Wales* and the *Repulse* have both been sunk by the Japanese—we think by aircraft."

The ships, a battleship and battle cruiser, had been dispatched to Singapore to hold back the Japanese southward thrust; the carrier *Indomitable* had been meant to accompany them but had run aground during a training exercise. Believing he was outside the range of Japanese shore-based torpedo bombers, and keeping radio silence, the British admiral aboard *Prince of Wales* had not called for land-based fighters to provide cover as he steamed north in response to what later proved a false report of a Japanese landing at Kuantan. Spotted by a Japanese reconnaissance plane, the ships were attacked by dozens of Imperial Navy bombers flying four hundred miles from their base in Saigon. Each of the large ships was hit by multiple torpedo strikes. RAF Brewster Buffalo fighters belatedly dispatched from Singapore arrived three minutes after *Prince of Wales* had disappeared beneath the waves.

"As I turned over and twisted in bed the full horror of the news sank in upon me," Churchill recalled. "There were no British or American capital ships in the Indian Ocean or the Pacific except the American survivors of Pearl Harbor, who were hastening back to California. Over all this vast expanse of waters Japan was supreme, and we everywhere weak and naked."

There was another import to this shattering news. For the first time, a maneuvering battleship at sea had been sunk by air attack alone. As the historian Gordon Prange later observed, "It was really the loss of *Prince of Wales* and *Repulse* that clinched the conviction that naval airpower had come to stay."

The carriers of the U.S. Pacific Fleet were not at Pearl Harbor the morning the Japanese struck. *Lexington* and *Enterprise* were ferrying planes to Wake and Midway Islands; *Saratoga* was in repair. Now reinforced by *Yorktown* and *Hornet,* the American carriers became the nucleus of the resurrected Pacific Fleet, such as it was. Admiral Chester W. Nimitz, appointed to take command from the disgraced Admiral Kimmel, at once ordered a complete reorganization of his forces. The fleet had been built around a Battle Force and a Scouting Force. Nimitz now put the surviving battleships to work escorting convoys between Hawaii and the West Coast, and each of the operable four carriers was made the core of its own Task Force, accompanied by a protective screen of cruisers and destroyers. Carriers were still largely hit-and-run weapons of questionable survivability, but Nimitz believed it was essential to strike back as soon as possible, for American morale if nothing else, and February and March of 1942 saw a series of small Carrier Task Force raids against Japanese-held islands in the central Pacific.

April brought another token raid that had far more to do with boosting American morale than inflicting real harm on the Japanese. On the morning of April 18, sixteen B-25 medium bombers were catapulted off the deck of *Hornet*—something that had been rehearsed on a practice strip on dry land but never on a pitching aircraft carrier in heavy seas. At the controls of the first bomber was the unflappable test pilot of "blind-flying" fame, Lieutenant Colonel Jimmy Doolittle. Thirteen of the planes headed for Tokyo and environs, where they dropped four bombs apiece. The three others sought out targets in Nagoya, Kobe, and Osaka. The crews then were to continue eleven hundred miles toward China, where the plan was they would land in friendly territory under the control of Chiang Kai-shek. Instead they ran out of fuel and had to bail out or ditch; eight crewmen were captured by the Japanese and three were executed. Doolittle was among those rescued.

The damage done by the raid was slight. But the effect on Japanese morale was everything Douhet or Trenchard might have hoped for. The appearance of American bombers over Tokyo caught Japan completely by surprise; no one could explain where they had come from. (When asked by reporters, President Roosevelt grinned and said "Shangri-La.") Yamamoto started receiving hate mail for allowing the capital to be attacked. The admiral himself became racked with anxiety over the personal safety of the Emperor. Convinced that the bombers could only have come from Midway Island, America's westernmost remaining outpost in the Pacific since the fall of the Philippines, Guam, and Wake, Yamamoto now demanded that the General Staff agree to a plan he had been developing for months. It would be a final showdown with the Americans at this tiny island in the middle of the Pacific; by destroying the carriers that had escaped at Pearl Harbor, the American air threat would be neutralized once and for all and Japan's defensive perimeter secured thousands of miles from the homeland.

Yamamoto's plan, in the tradition of the Great All-Out Battle, was a hugely complex ambush that would lure the American fleet to its death. Following their familiar pattern, the Naval General Staff had balked and Yamamoto had threatened to resign. But after Doolittle's raid both the Navy and Army staff fell in solidly behind the plan.

Yamamoto's scheme would deploy eight carriers, but it envisioned delivering the decisive blow against the Americans with a huge concentration of conventional naval firepower, seven battleships organized into the "Main Body" of the Japanese assault. Four of the five heavy carriers assigned to the mission would form a separate Striking Force whose job was to support a troop landing on Midway, bombing the island's defenders and neutralizing its small air base. This would be the bait. "Although the enemy lacks the will to fight, it is likely that he will counterattack if our occupation operations progress satisfactorily," explained the Striking Force's commander, Admiral Chuichi Nagumo, who had also commanded the Pearl Harbor attack. When the American carriers sortied from Hawaii to relieve the island, the Main Body, which would hang back hundreds of miles to the west until the Americans had committed themselves, would spring forward for the kill. The air assault was fixed to begin June 4.

In the first week of May, a confused, chaotic action in the Coral Sea marked by mistakes and intelligence failures on both sides left the Japanese and the Americans each down one aircraft carrier. In many ways the battle confirmed the predictions of what would happen when two forces built around these inherently offensive and inherently defenseless weapons met on the high seas: a brutal exchange of long-distance blows in which landing the first punch counted for almost everything. Carrier warfare was high-risk, high-stakes conflict. The Japanese and American forces were 175 miles apart during the fiercest exchange of aerial attacks that took place on May 8; the ships themselves never caught sight of one another. "Nothing like it had ever been seen before," wrote Churchill later. "It was the first battle at sea in which surface ships never exchanged a shot. It also carried the chances and hazards of war to a new pitch."

In a contest in which accurate intelligence about enemy movements had suddenly assumed paramount importance, Nimitz possessed an incomparable advantage as the date of the Japanese attack upon Midway neared. On March 18, U.S. Navy code breakers in Honolulu had finally cracked the main Japanese fleet code, designated JN-25 by the Americans. For the first time they began reading current JN-25 traffic. On May 14 the code breakers deciphered a signal ordering an invasion force to move against an island designated AF, which had already been identified by the experts in Honolulu as the Japanese Navy's code name for Midway. Despite vehement objections from Navy headquarters in Washington, which in a vicious internal battle sought to discredit the Honolulu code breakers' interpretation, Nimitz stood by their conclusion and on May 17 ordered his three remaining carriers back from the South Pacific at once. Further JN-25 decrypts pinpointed the date of the Japanese attack. Nimitz's plan was, simply, to get there first and ambush the ambushers.

The initial Japanese air assault on Midway began as planned at dawn on June 4. Most of the American fighter force at Midway consisted of the slow and already obsolete Buffaloes. Twenty-six took off to repel the invading wave of bombers and Zeros; fifteen were shot down at once, and all but two of the remainder were destroyed on the ground when they landed in the middle of Japanese bombing runs over Midway's airfield. Two hours later, nine Army B-17s from Midway flying high over the carriers dropped seventy-two bombs and, despite wild claims of success that would subsequently be made, scored no hits at all.

But Midway's defenders were able to put up a strong antiaircraft barrage, and this took a heavy toll on the attackers. Of the 108 Japanese planes that formed the first wave, 67 were shot down or damaged beyond immediate repair. At 7:00 a.m. the leader of the attack radioed the carriers that a second strike was needed. Admiral Nagumo agreed, then hesitated. Ninety-three aircraft from *Akagi* and *Kaga* had been held back, fitted with torpedoes and armor-piercing bombs, ready to spring into action against any American ships that appeared. It would take an hour to change their armament to land-attack bombs. But there were no reports of American ships in the area. Nagumo was confident they were still back in Hawaii. He ordered that the changeover proceed.

At 7:28 a patrol plane reported ten enemy ships. It took some time for the message to be decoded and relayed to Nagumo; when he received it, he at once ordered the torpedoes and armor-piercing bombs loaded back on the planes. Forty minutes after the first report, the patrol plane sent another radio message, this time incorrectly identifying the American ships as cruisers and destroyers. Ten minutes later, Nagumo was hurriedly handed yet another report from the patrol plane: ENEMY FORCE ACCOMPANIED BY WHAT APPEARS TO BE AN AIRCRAFT CARRIER.

But at that moment the returning first wave of Japanese fighters and bombers was circling the carriers and running low on fuel; these planes would have to be recovered, refueled, and launched again before the second-wave bombers could be brought up to the flight deck. Nagumo could have launched a small attack at once against the American ships but apparently believed he still had plenty of time before his own ships would be in danger: due to a navigational error, the patrol plane had miscalculated the location of the American ships and placed them thirty miles farther away than they actually were—at some 200 miles distant, which, had it been true, meant the Americans either would have to send their bombers in without fighter escort, given the short combat radius of the F4F Wildcat fighter, or would have to wait at least another hour until the distance between the ships closed. Nàgumo decided to wait and launch a "grand-scale" attack.

What Nagumo did not know was that at that moment the first American aircraft were already on their way. Forewarned by the code breakers' discoveries, American patrol planes had been combing the likely area of the Japanese

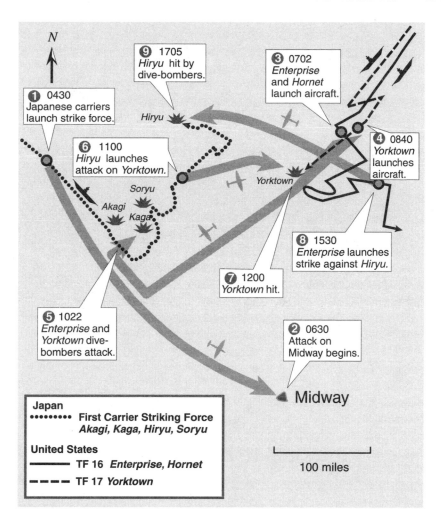

The Battle of Midway, June 4, 1942. Midway decisively established the supremacy of naval air power over the battleship fleets.

approach since before dawn. Shortly after 6:00 a.m., Admiral Raymond Spruance, approaching from the northeast in command of the carriers *Enterprise* and *Hornet,* knew where the four Japanese carriers of the Striking Force were. Spruance decided to attack as soon as possible, never mind the range; his pilots might be able to land at Midway, or they might just have to ditch on the way back, but getting in the first blow was worth almost any risk. At 7:02 a.m. his first planes were in the air.

The strike did not get off to a good start. Joined by a squadron launched from *Yorktown,* Spruance's two squadrons of TBD Devastator torpedo bombers immediately lost their fighter escort. When they began their runs against *Kaga* and *Akagi,* they were slaughtered by flak and Zeros; only four of the forty-one planes made it back.

But just as the melee was ending at about 10:30, forty-nine SBD Dauntless dive-bombers slipped in unnoticed at fourteen thousand feet. It was much to the credit of good luck and resourcefulness that they showed up at all; flying to the last reported position of the carriers, Lieutenant Commander Clarence Wade McClusky, the air group commander of *Enterprise,* looked down and saw nothing but empty ocean. At last he spotted the wake of a Japanese destroyer and decided to follow it. A few minutes later the two Japanese carriers were in full view. *Kaga,* her deck crowded with planes, fuel lines, and ammunition, took one direct hit. The carrier's communications officer hurried toward the bridge to urge the captain to move to safety; then came another explosion, and when he looked again the bridge was gone. Meanwhile, *Akagi*'s deck went up in a chain reaction of exploding fuel and weapons. Minoru Genda, who had been confined to bed with pneumonia and had dragged himself to the deck to watch the first air crews take off just a few hours earlier, now surveyed the scene of carnage and uttered a single word of ironic understatement: "Shimatta"—"We goofed." A third carrier, *Soryu,* was hit by the Dauntless squadron from *Yorktown* and erupted in flames. Only *Hiryu,* shrouded in mist, escaped the first wave to launch a counterattack. Her bombers sank *Yorktown; Enterprise* returned the blow and set her ablaze as the long day ended. Spruance then turned and steamed east into the night, keeping well out of range of the Japanese battleship guns that had been meant to decide the clash.

SBD Dauntless
carrier dive-bomber
engine: 1,000-hp radial
top speed: 245 mph
armament: 2,250-lb.
  bomb load; 2 to 4
  machine guns

In a mere seven hours, four of the six Japanese carriers that had attacked Pearl Harbor had been destroyed, along with three hundred aircraft and many experienced pilots. Yamamoto, who had taken personal command of the operation from the great battleship *Yamato,* slumped in his chair speechless when Nagumo at last dared to inform him of what had happened.

Within a year the fast-carrier force had consolidated its ascendancy over naval warfare. From early 1941 to mid-August 1942, six new American battleships had been commissioned and another six *Iowa* class and five huge *Montana* class battleships were on order. In the aftermath of Midway those plans were canceled. None of the *Montana* class ships would ever be built. At sea the battleship would become principally a platform for antiaircraft guns; for the rest of the war, its big guns would be employed not in fighting epic sea battles but to support amphibious landings by shelling enemy shore fortifications.

Congress hurriedly authorized the construction of ten *Essex* class carriers in place of the canceled battleships. These were veritable cities at sea: 27,000 tons, carrying a hundred planes apiece, manned by 150 officers and 2,550 men. More important, they were the first carriers to include an effective radar system integrated into a Combat Information Center (CIC)—a sort of seaborne version of the control rooms that were the secret of RAF Fighter Command's capability. By tracking friendly and enemy aircraft, the CIC was able to vector fighters and coordinate antiaircraft fire in a way that, for the first time, made it possible for a carrier to defend itself. In 1943 another development helped the U.S. Navy's carriers assume the lead in sustained combat operations at sea, by resolving a mundane but vital matter of logistics. Fast carriers needed fast tankers and supply ships that could keep up with the fleet. Japanese combat officers tended to give little attention to logistics, deeming it "boring." So it was, but America's construction of fast tankers proved to be the vital factor in sustaining this new, fast-paced kind of combat.

By 1943 America's industrial and technological mobilization, which Yamamoto so aptly feared, began to pay off in the areas of naval aircraft and armaments as well. Repeated failures of American torpedoes during the first two years of the war had been blamed by the Navy's Bureau of Ordnance on the incompetence of the men who fired them. The men knew better: time and again they saw torpedoes bounce off the sides of their targets without exploding, or run so deep under an enemy ship—even when set to run at a depth of zero—that their magnetic detonators would not trip. Finally, after the intervention of Admiral King, bureaucratic heads were knocked together hard enough to get the problem investigated and solved. The depth-setting mechanism was indeed defective, as were the detonators. By the end of 1943 America's fighting men could at last be reasonably certain that if they fired a torpedo it would go off when it hit something.

The Zero had outclassed the first wartime generation of American carrier-based fighters, but the advent of a new class of 2,000-horsepower American engines changed this. The state of the art was now such that bigger engines did not make for much faster planes; the circa-1940 fighter airframe had nearly reached the limits of the streamline revolution of the interwar years. Propellers were approaching a limit, too: beyond airspeeds of about 500 miles per hour, the airflow over the tips of a spinning propeller began to approach the speed of sound, causing turbulence and a sudden loss of thrust. But more powerful

F6F Hellcat
carrier fighter
engine: 2,000-hp radial
top speed: 380 mph
armament: six .50-in.
   machine guns in
   wings

engines, turning larger three- or four-bladed propellers, could generate greater thrust with each revolution, meaning that the same speed performance could be coaxed out of a much heavier plane, one that carried more guns, more fuel for longer range and endurance, and more armor protection for the pilot and for vulnerable spots such as fuel tanks. The Grumman F6F Hellcat and the Vought F4U Corsair that went into service in late 1942 and early 1943 outweighed the Zero by a factor of three but had significantly higher top speed and maximum ceiling;

moreover, the vulnerability of the lightly built Zero began to tell when American pilots discovered how easy it was to set its fuel tank on fire with a burst of incendiary bullets in the right spot.

In less than two years from start to finish, a revolution in doctrine had been launched and completed, with the tactics, logistics, weapons, intelligence, and strategy needed to realize it fully in place. Not only had the principle been established that command of the sea was impossible without command of the air, but the U.S. Navy had created the force to put that principle into effect. From Midway on, Japan was on the defensive. It was carrier-borne air power that was pressing her defensive perimeter inexorably back.

A half a world away, in the icy waters of the North Atlantic, air power and sea power were sorting out their new relationship in a like burst of experimentation born of necessity.

On August 27, 1941, a bizarre incident occurred that betokened the new balance of power at sea. The German submarine U-570, at sea barely three days on her maiden mission to strike at the merchant convoys supplying Great Britain, was running on the surface just south of Iceland. Her crew was green, figuratively and almost literally: high seas and inexperience had left many seasick. Suddenly an aircraft appeared and four depth charges exploded around the sub. That was enough for most of the crew, who promptly poured out of the conning tower in life jackets. The plane, a twin-engine Hudson bomber of RAF Coastal Command, turned and made a second pass, firing its machine guns as it roared a hundred feet overhead. *That* was enough for U-570's commander, who promptly ran up a bedsheet as a surrender flag and, exchanging lamp signals with the Hudson, agreed that he would not attempt to scuttle the U-boat if the Hudson ceased fire. For hours the plane circled its prisoner; a Catalina flying boat then arrived to take over guard duty; at last an armed trawler from Iceland reached the site, picked up the crew, and took the prize under tow. Although the U-boat's commander had managed to throw his Enigma machine and code books over the side, the capture of an intact German submarine was a major intelligence coup for the British.

U-boats would not make a habit of surrendering to airplanes, but the taking of U-570 did reflect a principle that British antisubmarine-warfare experts had established over the course of two years of battling the U-boat menace in the Atlantic: aircraft were among the most potent and feared weapons they could bring to bear in the fight. By the summer of 1941, a third of the damaging attacks on U-boats were being launched by aircraft. And while only two submarines had definitely been sunk in the 270 aerial attacks carried out through August 1941, considerable data already confirmed the powerful deterrent effect that air patrols and air escorts of convoys were having. The mere appearance of an aircraft was usually enough to convince a U-boat to dive. Running on the surface with its diesel engines engaged, a circa-1941 oceangoing German boat could make seventeen knots; submerged, it had to switch over to battery-powered electric motors that drove the boat at a maximum of eight knots, and then only as long as the batteries held out, which was not long—less than a day. Once the batteries were dead, the boat had to surface and run its diesel-driven generator to recharge them. Even a slow-moving merchant convoy could outrun a submerged U-boat. Keeping the boats down thus severely hampered their ability to find and catch up with their prey.

FW 200 Condor
maritime patrol/bomber
engines: 1,200-hp radial (4)
top speed: 240 mph
armament: 3,300-lb. bomb
 load; 20mm cannon, 5
 machine guns

Germany had tried and failed to challenge the Royal Navy for control of the high seas in the First World War. Far more successful had been its U-boat fleet, which in 1917 had nearly driven Allied shipping from the Atlantic and come perilously close to cutting off Britain's access to essential supplies. Germany started the Second World War with only about thirty oceangoing U-boats. But Admiral Karl Dönitz, commander of U boats, pressed Hitler to heed the lessons of the past and make the war on British merchant shipping the cornerstone of German naval strategy; by February of 1941 Hitler, impressed by the record tonnage sent to the bottom of the Atlantic each month by the U-boats and unimpressed by the accomplishments of his navy's capital ships, finally agreed. The next month, a half million tons of shipping, 139 ships, were sunk by German U-boats, mines, and aircraft. By June losses for the year had reached 2.8 million tons, more than 700 ships. Britain needed to bring in 31 million tons of supplies a year to keep running; for the first four months of 1941 actual imports were running at an annual rate of only 28 million tons.

Particularly alarming to Churchill was the disproportionate effect that German long-distance patrol aircraft were having. The German Navy had only a token naval aviation force and was neither well equipped nor well trained for maritime air operations. Dönitz had approached Hitler behind Göring's back to get a squadron of twelve of the Luftwaffe's long-range Focke-Wulf 200 Condors assigned to him for the fight against Allied shipping. Göring, who had spent years fighting off Navy claims to any piece of the sky ("Everything that flies belongs to me!" he once declared), at once tried to get them back; when he failed, he took his revenge by cutting off even the token cooperation he had until then given the Navy. Yet even with such small forces at its call, German naval aviation sank a half million tons of shipping during the months of March, April, and May 1941.

Summer finally brought a reprieve. The breaking of the naval Enigma cipher by the Bletchley Park cryptographers in June, the rapid buildup of armed escort vessels, and the somewhat desperate but effective measure of equipping fifty merchantmen with catapults that could launch a single Hurricane fighter if a

U-boat or German patrol plane was spotted gave the convoys the means to evade their pursuers or to fight back. (The Hurricanes of course had nowhere to land; the pilot was simply expected to ditch in the water, abandon his plane, and be fished out.) By July, sinkings of merchant ships by U-boats had dropped to 94 thousand tons; by August they were down to 80 thousand. The "happy time," as the U-boat captains had called it, was over.

The entry of the United States into the war opened up a new front in the U-boat war, and American ill-preparedness at once ushered in a "second happy time." Rushing to take what they rightly guessed would be easy pickings along the American coast, Dönitz's U-boats during two weeks in January 1942 sank thirty-five ships, more than 200 thousand tons of shipping, between Halifax and Cape Hatteras. Miami and other tourist towns along the coast resisted ordering a blackout, fearing it would be bad for the winter resort business, and night after night the U-boats surfaced to find their targets perfectly silhouetted against the bright city lights. The U.S. Navy resisted ordering merchant ships to move in coastal convoys and organizing a central intelligence and control system for the antisubmarine war, measures urged by the British as a result of their experience. "The situation is so serious that drastic action of some kind is necessary," Churchill warned in a letter in March to FDR's adviser and confidant Harry Hopkins. By May 1942, total monthly losses of Allied shipping to U-boats in the Atlantic exceeded 600 thousand tons, 114 ships. Tensions between the British and Americans flared; the commander of the Royal Navy's Submarine Tracking Room at one point furiously told the U.S. Navy's Deputy Chief of Staff, "We're not prepared to sacrifice men and ships to your bloody incompetence and obstinacy!" In June the U.S. Navy's own chief antisubmarine expert wrote a memorandum to Admiral King that declared, "The Battle of the Atlantic is being lost."

The crisis brought to a head another festering dispute within the antisubmarine war, and that was how best to employ aircraft in the fight. In both Britain and the United States, it was not the navies but the air forces that controlled coastal patrol aircraft, and this had resulted in debilitating organizational stand-offs. In April 1941 RAF Coastal Command was placed under the operational control of the Admiralty, but the RAF retained responsibility for training, equipment, and administration. A similar agreement in March 1942 placed the Army Air Forces' coastal defense unit, I Bomber Command, under the orders of the U.S. Navy commanders responsible for the various "sea frontiers" in the Western Atlantic, but again the admirals were powerless to exercise any control over how the units were equipped or trained. It was clear that neither the RAF nor the Army Air Forces put much stock in fighting submarines the way the navies wanted it done. There were tactical disputes over whether to undertake patrols to find and hunt down U-boats, which the airmen tended to prefer, or whether to stick to providing cover for the convoys, which the navies believed to be essential. But most of all there was a fundamental and increasingly bitter disagree-

ment over high-level strategy: whether air power wouldn't be better employed in another theater altogether.

The disagreement was sharpened by the fact that the types of aircraft that were proving essential to the U-boat war—large four-engine bombers with the range and endurance to mount long patrols and carry a lethal load of depth charges—were the same aircraft that both the American and British air forces needed to realize their grand vision of strategic bombing. Given a choice of taking the war to the German heartland or chasing U-boats around the Atlantic at the behest of admirals, the "bomber barons" left little doubt as to where their sympathies lay. "Bomber" Harris was especially indignant at the idea of frittering away his strategic aircraft on such a "purely defensive" task as convoy protection. Harris complained hyperbolically that if RAF Bomber Command were expected to supply any and every such request, "real or fancy," it would be reduced to the status of a "residuary legatee."

There was scant danger of that; in both Britain and the United States, the air force authorities still controlled the allocation of equipment, and the submarine-hunters of the RAF's Coastal Command and the AAF's I Bomber Command (soon to be renamed Army Air Forces Antisubmarine Command) were not high on the list of units designated to receive four-engine bombers. They barely made it onto the list at all. The head of Coastal Command had told the Air Staff in the summer of 1941 that he was 250 aircraft short of the 800 minimum required and that he also desperately needed to replace his short-endurance and obsolete planes, such as the Hudsons, with modern types that could patrol farther than a few hundred miles out to sea. The Air Staff replied that Bomber Command had first claim on all new production. A squadron's worth of long-awaited B-24 Liberators was delivered from the United States to Coastal Command in September; the following month, the RAF took away half of them for other duties.

At the end of May 1942 Admiral John Tovey, commander of the Home Fleet, fired a scathing broadside to the Admiralty protesting the "absolute priority" that had been given to Bomber Command as it began its increasingly heavy raids on German cities that spring. In a subsequent report, he summarized his arguments:

> The Navy could no longer carry out its much increased task without adequate air-support; that support had not been forthcoming. . . . Whatever the results of the bombing of cities might be, and this was the subject of keen controversy, it could not of itself win the war, whereas the failure of our sea communications would assuredly lose it. . . . It was difficult to believe that the population of Cologne would notice much difference between a raid of 1,000 bombers and one by 750.

Tovey urged the Admiralty board to threaten to resign en masse unless the necessary aircraft were supplied, so desperate had the situation become. The board was not prepared to go quite that far, but it kept pressing the case, and the "Battle of the Air"—as First Sea Lord Admiral Pound dubbed it—became increasingly acrimonious as 1942 wore on.

It was being fought with equal intensity on the other side of the Atlantic. The U.S. Navy had asked for two hundred B-24s and nine hundred B-25s and B-26s; General Arnold's headquarters replied that it would be impossible to meet the air force's own planned buildup of strategic bomber units if the Navy started making such demands on production. "There are no heavy or medium bombers available for diversion to the Navy," the AAF flatly declared.

Without these "VLR"—"very long range"—bombers, the Navy was staring at a five-hundred-mile-wide gap between Newfoundland and Iceland where the convoys were beyond the reach of air cover. American merchant sailors called it "Torpedo Junction." And it was there, in the late summer of 1942, that Dönitz began sending his wolf packs of U-boats to take up patrol lines. Their effect was deadly. Monthly losses of Allied ships surged upward once again.

As the urgency to counteract the U-boat menace grew through the fall of 1942, Allied air commanders were finally forced to respond with more than just stonewalling. On October 13 General Dwight D. Eisenhower, the commander of Allied forces preparing for the invasion of North Africa, told Eighth Air Force commander General Carl Spaatz, who was in charge of the still-embryonic American air force in Britain, that defeating the submarines is "one of the basic requirements to the winning of the war." Brigadier General Ira C. Eaker, Spaatz's bomber commander, responded with a plan that would show the Navy how God intended bombers to be used. Rather than "looking for a needle in a haystack—a submarine in the Atlantic" with endless patrols over open ocean, he proposed, American and British strategic bombers would strike the U-boats "where they are built and launched." It would be a classic demonstration of precision-bombing operations and strategic-bombing theory at work.

The U-boat pens that the Germans had constructed in the ports of occupied France along the Bay of Biscay were roofed with a dozen feet of reinforced concrete, impenetrable to conventional bombs. But surrounding the pens were locks, storehouses, rail yards, barracks; to Eaker's planners it was a classic choke point. Disrupting the Germans' precisely timed repair and resupply operations that kept the U-boat fleet at sea would strike at the heart of their war-making capacity.

From October 21, 1942, until just after the new year, American bombers dropped eight hundred tons of bombs in precision daylight attacks on the port facilities. It soon became clear that the precision was illusory, even when the bombers were sent in at the near-suicidal altitude of 7,500 feet. To the extent important facilities such as torpedo warehouses and workshops were being hit, it was as much a result of bombs missing a nearby assigned aim point as falling where intended. And despite some serious damage that was inflicted, the overall results were disappointing in the extreme. Saint-Nazaire was hit by five heavy raids in early November; two weeks later, according to agents' reports, the port was back in full operation.

The Eighth Air Force concluded that far heavier raids would be necessary, 250 sorties a week per base for eight weeks, which came to a staggering 10,000 bomber flights. The RAF argued that, given the nature of the task at hand, night area bombing made more sense: the best way to destroy the critical facilities in the port cities was to set fire to the cities themselves. In six raids in January and February 1943, Bomber Command got the job done. Thirty-five hundred of the five thousand buildings in Lorient went up in the flames set by British incendiary bombs; most of the remaining houses were rendered uninhabitable. Saint-Nazaire was next. By the end of May, Allied bombers had dropped three hundred tons of explosives and fifteen hundred tons of incendiaries on the city. The town itself was reported to be "virtually destroyed" in conflagrations that wiped out workers' houses, schools, churches, and civic buildings. "No dog or cat is left in these towns," Dönitz said at a meeting on May 4. "Nothing but the submarine shelters remain."

That was the rub: the submarine pens were indeed unscathed. Though they had received a dozen direct hits, the only damage that Allied experts could detect after poring over reconnaissance photos was a few barely visible pockmarks in their concrete roofs. Rail yards and power plants were damaged, but they were quickly repaired. Some critical shipyard facilities had been completely destroyed, but the Germans just moved the essential facilities inside the pens. Even the reduction in port capacity that this caused didn't make any difference. At the end of the raids, the Bay of Biscay ports could accommodate only 79 U-boats, down from 125. But that reduction almost exactly matched German U-boat losses at sea during the same period. The reduced capacity of the ports to service U-boats had almost no impact on the tempo of German submarine operations.

Those mounting losses at sea ironically owed much to the very-long-range B-24s that had finally been supplied to the antisubmarine operations over the Atlantic. It had taken political intervention at the highest level to make it happen—first from Churchill, who personally convened an "Anti-U-boat Committee" in November 1942 that ordered the diversion of British VLR aircraft to Coastal Command, and then from Roosevelt, who pressed his top military commanders to follow suit. "Every available weapon must be used at once to counteract the enemy submarine campaign," Roosevelt instructed his Army and Navy chiefs, General George C. Marshall and Admiral Ernest King. A few squadrons of VLR airplanes at last began to enter service in late 1942, and by May 1943 there were close to 200 in service with American, British, and Canadian forces. From the beginning of May to the first week of August, Coastal Command destroyed 41 U-boats, about 3 a week. By war's end, land-based aircraft would claim nearly a third of all U-boat kills, more than 300.

Small escort carriers also began sailing with convoys, bringing a substantial increase in air power to the battle. A variety of technical innovations during the spring of 1943—and the dramatic end, in March, of a long and frustrating black-out in decrypting the German naval Enigma signals—added their weight to the battle as well. The new ten-centimeter-wavelength radar carried by Allied patrol planes could not be detected by German warning receivers used by the U-boats. Enigma decrypts revealed that Dönitz suspected his mounting losses were due to the use of some new infrared detection device by the British. They weren't, but the British immediately fanned those suspicions with false double-agent reports. Dönitz responded by ordering that infrared-masking paint be applied to the conning towers, which happened to have the effect of *increasing* the boats' ten-centimeter radar reflection. When the Germans finally developed a new warning receiver that could detect centimeter-wave radar, the British again managed to spoof Dönitz, this time into believing that the receivers themselves were giving off stray radio emissions that the British planes were homing in on. Dönitz ordered the devices switched off—and to make sure they weren't used, U-boat captains were instructed to remove a critical part and lock it in the boat's safe.

At the end of May 1943, with his losses mounting and successes waning, a defeated Dönitz withdrew his U-boats from the North Atlantic. It was in no small measure a triumph of Allied tactical air power. Had the German Navy built a significant naval air force, the outcome might have been different. Yet the German interservice rivalry made the Allies' problems pale by comparison. In June 1942 Hitler ordered work stopped on the single German aircraft carrier then under construction and canceled the planned conversion of several merchantmen to small auxiliary carriers. The German Navy never mustered more than a token force in the air over the Atlantic.

As in the Pacific, however, the Allied command of the air in the Battle of the Atlantic was more than a triumph of numbers: it was also a triumph of a command system that could quickly put tactical intelligence in the hands of pilots and send them where the enemy could be found, a triumph, too, of an organization that could analyze scientific intelligence and operational experience and respond with the technological and tactical innovations needed to stay one step ahead of the enemy. Operational analysis time and again proved its value in applying mathematics and statistics to questions of strategy and tactics. The most important variable in determining how many Allied ships got sunk turned out to be how long a U-boat could stay out on each cruise. This in turn meant it was less important to sink the U-boats than to sink their resupply boats—the "milk cow" tanker submarines—and to make the U-boats lose time as they traveled to their Atlantic hunting grounds. The most effective way to do the latter was to concentrate air patrols over the Bay of Biscay, forcing the U-boats to submerge as they transited the 300-by-120-mile bottleneck to and from port. The Allies plied both tactics in earnest, and with devastating effect, in the summer of 1943.

Yet in the bitter fight over where to place the weight of the Allied air effort, the war against the U-boats provided an early warning of a much larger and deadlier controversy to come.

10

THE TEMPORARY TRIUMPH OF TACTICAL AVIATION

The job of writing the official history of the RAF's strategic-bombing campaign after the war fell to Air Chief Marshal Arthur Tedder's scientific adviser, Solly Zuckerman. An expert on primate anatomy and behavior, Zuckerman had come to the inner circles of Allied air strategy through a study he had been asked to carry out on the physiological effects of bomb blasts; this had led to work on a "bomb census" that kept track of every German bomb that fell on Britain and cataloged its effects; and that had led to a more direct role in the formulation of bombing policy.

Though not a statistician or physicist like many of the scientific advisers who would come to play an influential role in the Second World War, Zuckerman was a scientist par excellence, with an unshakable allegiance to objectivity and facts. The war had compelled an official and institutional recognition of the importance of scientific advice, and all of the services had quickly established operational research sections filled with young Ph.D.s. Whether the brass listened to these unmilitary outsiders was quite another matter, especially when they went beyond talking about the physics of radio waves or the chemistry of explosives and started offering advice about military strategy.

Zuckerman was on surer footing than most. He had a mixture of charm and integrity that allowed him to get along with generals and air marshals without ever being terribly impressed by them. When he set out to distill the conclusions about British bombing policy, he did not mince words. "In light of what we know now," he began his report, "it is clear that . . . both we and the enemy were ignorant, when hostilities broke out in 1939, of the scale of effort necessary for a decisive attack against a home front and its industries. The problem had no answer in past experience, and speculations were everywhere dominated by exaggerated ideas of what would be the effects of bombing on a centre of production and population."

Unsurprisingly, this was not a popular view in Bomber Command. The RAF suppressed the report, and it was not released until years after the war was over.

But to most of the top scientists in British defense circles, Zuckerman's conclusion had been obvious as early as the summer of 1941. Strategic bombing was not on its face pointless or impossible, but achieving the effects claimed for it, beyond satisfying the primal urge for retaliation, would require far more bombs and bombers than had been dreamed of, certainly more than the 150 or so airplanes that Bomber Command could put into the sky on any given night throughout most of 1941.

The RAF had begun the war with several sketchy "Western Air Plans" for attacking Germany. They reflected a hodgepodge of strategies. W.A. 1 targeted the aircraft industry. W.A. 4 focused on railroads. W.A. 5 aimed to destroy vital war industries in the Ruhr by striking the "common servicing system" they drew upon, such as electricity and coking plants. W.A. 6 aimed to knock out Germany's oil supplies. W.A. 8 was the classic Trenchard doctrine: widespread night bombing designed to cause dislocation and demoralization. By mid-1941 Bomber Command had tried several of these—especially attacks on oil, industry, and railroads—in addition to several more indiscriminate raids in retaliation for the Luftwaffe's bombings of London, Coventry, and other British cities and for its use of terror weapons such as naval mines.

The objective results of Bomber Command's attempts to strike specific targets were abysmal. Even if the British bomber crews had been able to find and hit targets in daylight, which they manifestly failed to do on several occasions, the night operations that had become almost exclusive practice since the RAF's heavy daylight losses of 1939 made precision out of the question. In August 1941 Lindemann, Churchill's science adviser and confidant, had a member of his staff, D. M. Butt, analyze 650 aerial photographs taken from the bombers during their night raids. What Butt found was so appalling that the air commanders reacted with literal disbelief. Only 22 percent of bomber crews who claimed to have hit their assigned target got so much as within five miles of it. In the more heavily defended and haze-bound areas of the Ruhr, the figure fell to 7 percent.

It didn't take a mathematical genius to grasp the implications of such numbers. "The war is not going to be won by night bombing," Henry Tizard advised the Air Ministry in late 1941, at least not until American aircraft production hugely increased the weight of bombs that could be delivered to make up for such woeful inaccuracy. Tizard was equally appalled by the losses that had been incurred to achieve such a dubious effect; in early 1942 he pointed out that in the preceding eight months Bomber Command had lost 728 aircraft, and even by the most primitive measure—Germans killed on the ground versus British bomber crews killed in the air—the campaign had been a losing proposition.

Tizard and other top scientific advisers to the government thus came down strongly on the Admiralty's side in their demand for long-range bombers to fight the U-boats, which made the Butt Report's findings even more of a political hot potato to the air leaders. Tizard one morning found a note on his desk from the Vice Chief of the Air Staff appended to a transcription of a German radio broadcast that had denounced the British bombing of German cities. "Like you," the note read, "the Germans are very anxious for us to stop raiding their towns." The commander in chief of Bomber Command at the time, Air Marshal Richard Peirse, dogmatically insisted that the figures Butt had come up with were impossible, for they "could not have produced the damage known to have been achieved." His successor, Air Marshal "Bomber" Harris, was equally glib in his dismissal of photographic evidence. He once told Churchill that "what actually occurs is much more than can be seen in any photograph."

To the scientists, the circular logic of such statements was flabbergasting. The historian Ralph Bennett, who had worked as an intelligence analyst at Bletchley Park during the war and knew well the obduracy the military mind was capable of, later wrote that Harris and other air commanders were so invested in strategic bombing as a policy that "they developed an abstract quasi-philosophical theory about it, to which ascertainable facts were almost irrelevant."

Almost, but not completely. Already in the summer of 1941 there were signs that the RAF was reverting to type, returning to the Trenchard doctrine of wide-area morale bombing in the face of evidence that trying to destroy specific targets was futile. The Air Staff began to talk up the "many signs" of a weakening in morale that had been detected in cities that had been bombed. And, as Harris would later explain, targets were now being chosen so that if the bombs "overshot or undershot" their intended aim point, they would still hit somewhere in "congested industrial areas . . . thereby affecting morale." This, in his view, was a sort of "halfway stage between area and precision bombing."

Harris's assumption of the leadership of Bomber Command in February 1942 would complete the journey. In retrospect, Zuckerman observed, the shift to a policy of area bombing was the tail wagging the dog. Strategy was being determined by tactics, when it was supposed to be the other way around: if the only thing bombers could do was scatter their bombs over large areas, then whatever targets existed that were spread over large areas—and this could only mean cities—would be the right targets to hit. And if it was illogical and circular, it was also politically inescapable. With the opening of a land front years off, bombing was the only way for Britain to strike at Germany, and the only way to convince Britain's new ally, Stalin, that it was doing something to help divert the Nazi onslaught against the Soviet Union.

On February 14, a week before Harris took command, the War Cabinet directed that a new air offensive would begin with the "primary objective . . . focussed on the morale of the enemy civil population and, in particular, of the industrial workers." The industrial cities of the Ruhr would be the primary targets. The Cabinet decision included in an appendix the mild suggestion that perhaps Bomber Command might continue to "consider the practicability" of attacking key oil, rubber, and power plants if some way to improve bombing accuracy could be found. Harris chose to ignore this. He had the marching orders he wanted, and he ran with them.

The new direction in bombing policy was given a vigorous boost a month later by Lindemann, who seems to have willfully, or at least recklessly, misconstrued a study that Zuckerman had just completed of the effects of German bombing on morale in the English cities of Hull and Birmingham. Zuckerman found that "in neither town was there any evidence of panic." The overall health and productivity of the workforce was unaffected. The only loss in factory output was the direct result of physical damage to plants.

Lindemann breezily declared that the study had shown just the opposite. On March 30 he circulated a memorandum to the Prime Minister stating that if the bomber force were directed against the centers of the fifty-eight largest German cities, one-third of the entire German population could be "turned out of their house and home." He continued:

> Investigation seems to show that having one's house demolished is most damaging to morale. People seem to mind it more than having their friends or even relatives killed. At Hull signs of strain were evident, though only one-tenth of the houses were demolished. On the above figures we should be able to do ten times as much harm to each of the 58 principal German towns. There seems to be little doubt this would break the spirit of the people.

Churchill was sold, and for a while at least he would speak enthusiastically of "dehousing" German workers. Although Churchill was never fully convinced that bombing alone would win the war, and often told his air commanders so (he warned them that they "must not spoil a good case by overstating it" and that he had no "unbounded confidence" in bombing), he acknowledged that it was currently the only way to keep fighting. "The only plan," he conceded, "is to persevere." In any case, the escalating chain of retaliation between Bomber Command and the Luftwaffe had settled the matter. Churchill was not about to let Hitler, or Stalin, believe that he would quail at matching brutality for brutality. "Morale is a military target," Churchill told Stalin a few months later in the spring of 1942. "We shall seek no mercy and we shall show none."

Harris was certainly the man to execute that directive. A veteran of the RAF's "air control" campaigns in Iraq and Palestine, the new leader of Bomber Command was in truth a skeptic, or at least an agnostic, on the question of targeting enemy morale. Morale, he thought, was a wild card. But he had no doubts about the efficaciousness of area bombing. His formula was hard, merciless, pragmatic. With his thick glasses and air of confident annoyance, Harris often looked and sounded like an irritated schoolmaster having to explain a simple lesson for the tenth time to a particularly dull class of boys. It was nothing but "panacea mongering," he said, to claim that bombers could demolish specific factories or other pinpoint targets. They couldn't. And even if they could, so what? The "panacea merchants," he snorted, wanted to "send a bomber to pull the plug out of Hitler's bath so he would die of pneumonia." The only way bombers could destroy *anything* was to destroy *everything*. Cities were *the* target—not factories, not morale, but the physical cities themselves. No modern society could last long with its metropolitan centers reduced to ashes.

The classic air-power theorists had never quailed at the notion of killing civilians. But there was more than that behind the unsentimental, ultralogical toughness that air commanders such as Harris so often displayed. Airmen had always seemed to feel a special burden of proving to their counterparts in the trenches that they were real warriors despite their cushy existence. In tales of great battlefield victories or defeats, legendary army generals might have a tear in their eye or a catch in their throat; great air commanders never do. Harris certainly never blinked. Even when he became the "willing fall guy," as the historian John Buckley observed, after Churchill and other politicians tried to dissociate themselves from the firebombing of Dresden in February 1945, Harris remained stubbornly, almost naïvely unapologetic. Enemy cities were the necessary target, he insisted again and again. Success was measured in the number of acres destroyed, later the number of square miles. The way to wipe them out was to husband one's resources into mass raids that would bring huge rains of incendiaries pouring down in the shortest possible time. As the incendiaries set the cities ablaze, high explosives in the bomb mix would harass firefighters, break water mains, and hinder access to the blazes. *That* was the recipe for success.

Harris raised the curtain on this Götterdämmerung policy on May 30, 1942. After a series of small experimental raids against Essen, Lübeck, Stuttgart, and other German cities in March and April, Harris pulled several hundred twin-engine bombers from training units to assemble a thousand-plane raid on Cologne. Fourteen hundred tons of bombs fell on the city in two and a half hours, two-thirds of the tonnage incendiaries. Six hundred acres were wiped out.

At the time, it is always worth remembering, Harris's blows against German cities were cheered throughout the Allied nations. Churchill later would praise Harris as a "vigorous" commander who, with his single-minded determination, whipped Bomber Command into an effective fighting force. Few qualms were raised about the morality of area bombing; the qualms raised by Tizard, Zuckerman, and others were solely about its effectiveness. Harris, however, was as impervious to evidence of the ineffectiveness of area bombing as he would later be to indictments of its immorality. The fact that even the routinely overoptimistic Ministry of Economic Warfare found the disruption to manufacturing following the Cologne raid to have been minimal seems to have made no impression on him whatsoever. "Victory, speedy and complete, awaits the side which first employs air power as it should be employed," Harris trumpeted in a memorandum to Churchill two weeks later.

The coming of the Yanks, however, presaged a showdown over how, exactly, air power "should be employed" that made the fights between Bomber Command and the scientists, or even Bomber Command and the Admiralty, look like

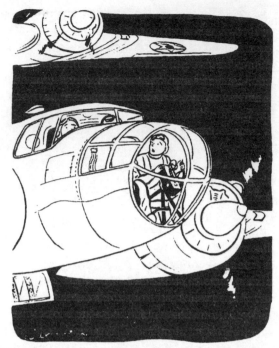

"Was that address 106 Leipzigerstrasse, or 107?"

At the start of the war, American confidence in precision bombing
was a matter of both official doctrine and popular pride. *(From*
Collier's, *September 26, 1942)*

sandbox squabbles by comparison. The RAF had concluded that daylight bombing was foolhardy, and that precision bombing was both impossible in practice
and pointless in theory. Now the Americans were insisting on doing what the
Brits had just declared foolhardy, impossible, and pointless.

Even before America's entry into the war, the British and American chiefs of
staff had held secret discussions in which they laid out a joint strategy "should
the United States be compelled to resort to war." Their final report, ABC-1, was
delivered on March 27, 1941, and it declared not only that Germany would be
the first target but that strategic air power would be the first weapon of choice,
with American bombardment forces operating "offensively in collaboration with
the Royal Air Force, primarily against German Military Power at its source."
This was vague enough to cover a multitude of differences.

In July President Roosevelt asked his Army and Navy chiefs to draw up a
specific estimate of the materiel they would need to do the job envisioned in
ABC-1. Arnold at once saw an opportunity to raise the flag of air force independence a notch higher; in the mad rush to get the report to the President's desk
in time, the Army General Staff was only too grateful when Arnold offered his

recently formed Air War Plans Division for the job of writing the air section. Working nonstop in the sweltering August heat on the top floor of the old Munitions Building (one of the "temporary" buildings that had been thrown up on the Mall in Washington during the First World War, the Munitions Building had stayed and stayed, a permanent eyesore), Arnold's planners were done in a week.

Arnold shrewdly suspected that the tight deadline meant there was not going to be much time for scrutiny or revision by the General Staff. The result was a document, AWPD/1, whose sheer chutzpah has rarely been equaled in military staff work. Ignoring the overall strategic conclusion of the report submitted to the President by the Army-Navy Joint Board, which stated that "only land armies can finally win wars," AWPD/1 took the attitude not only that the air force possessed a staff independent from and equal to the Army and Navy staffs, but that air power *could* very well win the war. And the way it could win was through an unwavering application of the theories so boldly elaborated over the years by the Air Corps Tactical School.

That AWPD/1 adhered so faithfully to these theories was hardly a surprise, for its four authors were all Tactical School instructors who had played a central role in developing them: Haywood Hansell, Laurence Kuter, Kenneth Walker, and Harold George. True to the "industrial web" concept, AWPD/1 took as its starting assumption that German war mobilization already posed "a very heavy drain on the social and economic structure of the state"; with the web stretched so taut, it would not take much to bring about total economic collapse, taking with it the "means of livelihood of the German people." The planners listed 124 targets whose destruction would finish off Germany: 50 electric power stations, 15 bridges, 15 marshaling yards, 17 inland waterway facilities, 27 oil plants.

Then they multiplied. Allowing for a wartime bombing accuracy 2.25 times worse than what was achieved in training, eleven hundred bomber sorties would be needed to destroy a 100-by-100-foot target. A six-month campaign destroying all 124 targets would therefore require a total of exactly 6,860 bombers, 98 groups. AWPD/1 was, as one military analyst later put it, a perfect demonstration of the "American propensity to see war as an engineering science."

The planners of course assumed that daylight precision bombing would be the only method of attack. And so as the first American bombers and air commanders began to arrive in England in the summer of 1942, they brought with them a philosophy of operations that ran smack into controversy with their allies even before they had gotten off the ground. That the buildup of American forces was much slower than expected only increased the British feeling that the Americans were hardly qualified to lay down the principles for prosecuting an air war. It would be another full year before the Eighth Air Force would begin to receive heavy bombers in any considerable number. As of the start of 1943, the Mighty Eighth had yet to drop a single bomb on German soil; even by the summer of 1943 it would have barely three hundred fully operational bombers in England. Articles in the British press began to belittle the American contribution, suggesting that the B-17 was a lemon, that American claims of precision bombing were empty boasts. The AAF retaliated with staff papers criticizing the British policy of "blitz bombing of large, city size areas" and suggesting that the only reason the RAF pilots flew at night was that they were afraid to fly in the day. The Anglo-American rift in bombing philosophy was growing deeper, and though it would repeatedly be papered over throughout the war, it would never truly be reconciled.

Where the RAF's Harris and the American commanders differed not a whit was in their conviction that strategic attack was the sine qua non of air power. Tactical aviation would play so negligible a role in the final outcome that for all intents and purposes it did not count. Throughout his memoirs Arnold would refer to tactical aviation, and even to major Allied ground operations such as the invasion of North Africa, as a sort of annoying distraction from the main event: "The direct strategic bombing of Germany by the RAF and ourselves remained, as I continued to state, the central road to Germany's defeat," Arnold said. And to be capable of "waging a decisive air offensive against the Axis powers," Arnold maintained, an air force "must consist predominantly of Heavy Bombers." That was exactly what AWPD/1 called for. Along with the ninety-eight groups of strategic bombers, a mere ten groups of fighters were to be deployed to Europe—not to escort the bombers on their missions, and certainly not to support ground troops, but only to help defend the bomber bases. As for battlefield attack, only thirteen groups each of light bombers and dive-bombers were called for in AWPD/1. These would be standing by "if it becomes necessary to invade the continent."

The attack groups were little more than a bone the air planners had thrown to the Army Chief of Staff, General Marshall, who they knew would not approve a plan that completely ignored ground support. Marshall approved AWPD/1, but not without private reservations; he thought the air staff "immature," too "busy taking stands" to be impartial, with far too much faith in strategic bombing and far too little grasp of overall strategy. The word *if* had made the planners' opinion clear: namely, that it would *not* be necessary to invade the continent. The AWPD/1 plan would work, according to its authors, "even in the event of a Russian collapse" on the battlefield. Germany would be crushed by six months of strategic bombardment, and that would be the end of that.

If there was a considerable irony in the Allied air forces' quickly becoming the world's leaders in the art and practice of supporting ground operations, a supremacy they would soon employ with crushing effect from the deserts of North Africa to the beaches of Normandy, there was a certain logic in this accomplishment's being worked by a man who had begun life as an outsider.

Arthur Coningham was born in Australia and grew up in New Zealand, a true self-made man. His father had had a brief career as a cricketer and a longer but luckless career as a con man. The family fled to New Zealand after his father's first elaborate con backfired: acting as his own attorney and aided by an impressive array of fabricated evidence (including his wife's suitably remorseful testimony on the witness stand), he staged a sensational divorce trial in which he accused his wife of having had an affair with a flamboyant local Catholic priest, from whom he sought five thousand pounds in damages. It might have succeeded had he not been duped in turn by an agent of the priest's attorney, who, posing as a disgruntled priest with a vendetta against the church, fed Coningham tainted evidence that the defense then crushingly rebutted in open court. In New Zealand, Coningham Senior landed in jail for six months when, again acting as his own attorney, he was convicted of defrauding his employer by collecting commissions on sales to customers who turned out not to exist.

Arthur Coningham had meanwhile won a scholarship to Wellington College, a local boys' school. When he was seventeen his parents divorced, in earnest this time, after his father had been caught having an affair, which prompted Arthur to write to his father with precocious self-assurance, "Look here, Coningham, although you are my father, I am ashamed of you." Arthur was not much of a scholar but was an excellent athlete, outdoorsman, horseman, and marksman, qualities prized alike by a British colonial boys' school and the British Army in the early 1900s. When the Great War broke out, he enlisted, served in Samoa and the Middle East, and then was discharged for ill health in 1916. But he soon recovered, made his way to England at his own expense, and probably using connections among those he had so favorably impressed at school and in the Army, was accepted into the RFC.

Coningham proved a natural and fearless fighter pilot. More than once he returned from missions covered in blood and with his aircraft nearly shot to pieces. ("From his own appearance and that of his machine, one might have imagined him fighting the whole German air force single-handed," commented a colleague on one of those occasions.) He won both the Distinguished Service Order and the Military Cross, and was quickly promoted to the rank of major.

He also proved a natural leader, whose dash, cheerfulness, and personal touch inspired his fellow fliers, while his skill at transforming himself "into a convincing English gentleman, with accent and tastes to match," as his biographer Vincent Orange put it, ingratiated him with the higher-ups in the RAF. After the war he took up yachting and polo and sports cars as he continued to rise quickly through the ranks. If he had some of his parents' skill at smoothly passing himself off as something he was not, he combined this with an unaffected personal integrity; he did not smoke, drank little, and disapproved of obscene talk.

Ambition and strong self-discipline are hardly uncommon in self-made men; easy assurance and charm, and a readiness to delegate authority and to calmly trust subordinates to do their jobs, are far rarer. Coningham by all accounts possessed these traits in abundance, and they came to the fore in the assignment that would prove the most important of his career, earning him the reputation as the architect of the modern system of tactical air support for ground operations.

Arriving in Cairo in July 1941 to take command of the British air forces battling the Axis in the Western Desert, Air Vice Marshal Coningham made it clear to his staff that he had two rules: he would not get involved in details or micromanage those under his command; but anyone whose mistakes cost lives would not get a second chance. "What a queer fellow he is!" exclaimed his administrative officer, Air Commodore Tommy Elmhirst, in a letter home that he wrote after working for Coningham six months. "Of the organisation, administration and supply of his force he knows nothing and never asks me or asks where or what I have been up to. But if anything went wrong and a squadron could not operate through failure or shortage of crews or supplies, he would know in a moment and be down on me like a ton of bricks."

The desert to the west of Cairo had become a major battlefield of the war through default, and Italian hubris. An Italian advance sixty miles into Egypt in September 1940 had been met by a sweeping British counterattack three months later; by February 1941 the Italians had lost half of their 250,000-man army to a British force one-eighth as large. Then a month later the tide swept back again: Hitler offered his Italian ally the services of General Erwin Rommel and two armored divisions, and by April Rommel had fully made good the Italians' loss of ground. Axis ground forces were once again threatening Egypt and Britain's foothold in the Middle East.

Coningham had been in England since the start of the war, serving as a group commander in Bomber Command, and when he arrived to take command of the Desert Air Force he found relations between the RAF and the Army in the field bordering on the dysfunctional. The pummeling that British ground troops had received from the Luftwaffe in France and Greece had convinced ground soldiers from privates to Chief of the Imperial General Staff Alan Brooke that the RAF had no intention of providing proper ground support for the Army. Brooke, who had commanded the withdrawal at Dunkirk, claimed that he hadn't once seen a British aircraft during the entire operation. British soldiers in Greece said that RAF stood for "Royal Absent Force." Brooke insisted that unless air-support units were placed under the direct control of ground commanders, they would never be there when they were needed.

After the disastrous performance in France, the Air Ministry and War Office had agreed to carry out joint ground-support exercises in Northern Ireland, and the RAF had agreed to establish an Army Cooperation Command in Britain that would take its place alongside Bomber, Fighter, and Coastal commands. But the ACC's existence was still mostly on paper; it often had more staff officers than aircraft and it conducted little or no training. Most RAF leaders, truth be told, agreed with Harris that there was little point in addressing such a remote contingency as providing air support for the British Army; as far as they were concerned, the Army had proved by its defeats in France, Greece, and Crete that its only role in the war would be to defend the homeland, and then to serve as an occupation force once Bomber Command had brought Germany to her knees.

The air units actually trying to fight in the Western Desert, meanwhile, were exasperated both by their difficulty in getting the RAF to send them enough planes and by the Army's demands for constant "umbrella patrols" to shield the troops against enemy air action. It didn't help, either, that during an abortive British attempt to lift the German siege of Tobruk in June 1941 the air and ground headquarters were located eighty miles apart, nor that radio links between air and ground units were inadequate, nor that the Army's own radio communications were so bad that it often did not even know with any precision where its forces were, and so could not tell close-support aircraft where to drop their bombs to avoid hitting their own troops.

That air power on the battlefront should be consolidated under a single air commander, not parceled out to individual ground units in "penny packets," as Coningham liked to put it, and that standing defensive patrols were a huge drain of resources that effectively conceded air superiority to the enemy while inviting defeat in detail—these, as far as air commanders saw it, were nothing more than conventional, well-established principles that went all the way back to the First World War. Yet the RAF's years of disdain for battlefield support of any kind (after all, many RAF leaders were insisting that the best way to support the ground battle was really to bomb a factory in Germany) had bred a poisonous atmosphere of mistrust. Any time an air commander refused a request for support, ground officers were more than ready to interpret it as a sign that the RAF simply didn't want to help them at all. Coningham saw that his mission was less to develop a new ground-support doctrine than to find a way to make the established doctrinal principles workable in practice, and palatable to the ground forces.

He received a welcome boost from the top almost immediately. With the Army-RAF breach jeopardizing the entire operation in the Middle East, Churchill weighed in solidly on the side of proven principle in a memorandum in September 1941:

> Never more must the ground troops expect, as a matter of course, to be protected against the air by aircraft. If this can be done, it must only be as a happy makeweight and a piece of good luck. Above all, the idea of keeping standing patrols of aircraft over moving columns should be abandoned. It is unsound to "distribute" aircraft in this way and no air superiority will stand any large application of such a mischievous practice.

But the Prime Minister's backing was no substitute for proving to resentful ground commanders that the RAF could deliver the goods. This depended largely upon detail of execution, not lofty principles. Coningham immediately began to demand results from his squadrons and to strike up cordial relations with ground officers. Air Marshal Tedder, the overall RAF commander in the Middle East, reported two months after Coningham's arrival that the air commander already had the soldiers "eating out of his hand" with his infectious good nature and enthusiasm. This in itself was something new for the RAF.

The specific improvements in operations Coningham pushed for in many ways echoed the tactics and command-and-control procedures that Wolfram von Richthofen and his close-support units had perfected years earlier in Spain, Poland, and France. Coningham stressed practice in escort flying and ground attack. He suggested that pilots hone their skills by firing at their airplane's shadow as they passed over the ground while an instructor flew overhead to observe. He ordered ground crews to spend hours drilling so they could get the planes refueled and rearmed faster. He set up his headquarters at Eighth Army's HQ and greatly improved the flow of communications through a network of radio-linked "tentacles" that reached directly to the battlefront. A joint Army-RAF Air Support Control center was located at each army corps or armored division to assess air-support requests and pass them on to air force headquarters for approval, but at the same time each control center was allowed to communicate directly with forward RAF controllers and rear airfields to let them know that a possible mission was in the offing. Once an air-strike request was approved, an RAF Forward Air Support Link located at a brigade could communicate directly with the attacking aircraft by radio to coordinate the strike. The average response time for air-support requests was cut from three hours to thirty-five minutes.

Coningham's breakthrough, though, was to make real to ground commanders the inherent potency of preserving air power's flexibility. "What a corps commander really wants," a tank officer told Coningham at lunch one day, "is a squadron at his disposal to come up on his call and bomb something in front of him." Coningham retorted: "Yes, and the whole lot would be immediately shot down by Me 109s because there would be no one central authority to ensure a fighter escort for them." Centralized command meant that aircraft could be massed to attack the enemy positions that were most critical for the overall ground operation, not frittered away against many small and unimportant targets; it also meant that when they were not urgently needed for close support, they could keep up a constant attack on Rommel's supply dumps and communications lines or conduct fighter sweeps and airfield attacks to keep the Luftwaffe on the defensive.

Unlike "Bomber" Harris, Coningham was an avid consumer of intelligence. Co-locating his headquarters with the commander in chief of the British Eighth Army ensured that he had a direct line to intelligence derived from German Enigma and Italian high-grade decrypts. In the offensive that began shortly after General Bernard L. Montgomery's arrival as the Eighth Army's new commander in August 1942, the flow of high-level intelligence was combined with Coningham's new air-support system with incisive effect. Since July, the Bletchley Park code breakers had been reading on an almost daily basis the German Panzer Army Operational Staff report, which included a detailed list of Rommel's supply situation. They had also been breaking Italian cipher messages that inventoried cargoes carried on ships traversing the Mediterranean to supply the Italian and German armies. British intelligence thus was able to calculate what materiel Rommel critically lacked on any given day—diesel fuel, gasoline, tires, food, 76.2mm antitank ammunition, 75mm tank shells—and pick out which vessels to strike.

"The convoy sets sail from Naples quietly and peacefully, escorted by three Italian torpedo boats," reported a German staff officer in Italy, "then just before it gets to Tripoli the British come flying over and of course it is only the German steamer that gets hit, and it carries a cargo for which Rommel is eagerly awaiting. Nothing happens to the Italian ships." The German officer was convinced that spies in the Italian ports were to blame.

By October 1942 nearly half the Axis shipping bound for Libya was being sunk. Singling out tankers, the British air attacks had gravely reduced Rommel's fuel supply. His Panzers were about to run out of gas.

Meanwhile, beginning with the Battle of Alam al Halfa in August, Coningham's fliers kept up relentless attacks on Rommel's road convoys, assembly areas, and supply dumps. In late October the Desert Air Force flew more than ten thousand sorties in a little over a week. It was the continuous pressure of these shuttle-bombing assaults—another leaf from von Richthofen's book—that, Rommel said, "pinned my army to the ground." Rommel later complained bitterly:

> Anyone who has to fight, even with the most modern weapons, against an enemy in complete command of the air fights like a savage against modern European troops, under the same handicaps and with the same chances of success. . . . The fact of British air superiority threw to the winds all the tactical rules which we had hitherto applied with such success. In every battle to come the strength of the Anglo-American air force was to be the deciding factor.

The Americans, however, still had some learning to do before their air force would be a deciding factor in ground combat. Landing on the west coast of North Africa on November 8, 1942, the U.S. Army proceeded to repeat every mistake in organizing ground-support aviation that Coningham had just spent a year correcting in the Western Desert. The U.S. Army field manual that governed the procedures, FM 31-35, "Aviation in Support of Ground Forces," had been issued by the War Department in April 1942, and it reflected the usual unworkable compromise that ensured army control and air force neglect. The Army corps commander called the shots, but the only aviation units organic to the Air Support Command under his direct control were observation squadrons. In January 1943, General Spaatz (who in addition to commanding the Eighth Air Force now was also the overall air commander for North Africa) approved adding fighter and bomber units to the XII Air Support Command in Tunisia. But the other half of the unworkable situation, Army control at the corps level, remained unchanged.

A few weeks later, Major General Lloyd R. Fredendall, commander of U.S. II Corps, refused to allow the XII ASC to respond to a request for support from a nearby Free French unit because he insisted they fly an umbrella patrol over his own ground forces. The French were hammered by a heavy German assault; the American regiment in action that day encountered no enemy air or ground resistance at all. When Spaatz protested that Fredendall should not be allowed to make any more "damned fool decisions" about the use of aircraft, the ground commander retorted that he had lost three hundred men to dive-bombers and he wanted American aircraft flying over his troops for forty-eight hours before every offensive: he "wanted his men to see some bombs dropped on the position immediately in front of them . . . so that their morale would be bolstered."

A little later, Fredendall's own Chief of Staff admitted to Spaatz that only a few men, nothing like three hundred (or seven hundred, as it became in one version that circulated), had been lost to enemy dive-bombers; he also acknowledged that a "defensive fear complex was being built up" in II Corps. The ensuing American rout at Kasserine Pass, which General Omar N. Bradley later called "probably the worst performance of U.S. Army troops in their whole proud history," only confirmed that judgment.

Once again, a political intervention at the highest level saved the day. Meeting at Casablanca in January 1943, Churchill and Roosevelt agreed to establish a unified British-American command that, among other things, placed all Allied tactical air forces under Coningham's control. Coningham took charge on February 17 and immediately began to institute his system. The next day he cabled all air commands to upbraid them for the completely defensive posture they had taken; even though targets were in evidence and bombers on call, he complained, no enemy positions had been attacked. Umbrella patrols were to be halted at once. The maximum offensive role would henceforth be assigned to each mission. He sent every ranking officer in Tunisia a pamphlet explaining how the Western Desert air system had worked.

Coningham's pamphlet, which almost verbatim became the basis for a sweeping new U.S. Army field manual on the use of air power that would be issued in July, emphasized in clear, simple language the fundamental principles of air command: The greatest asset of air power was its flexibility and ability to rapidly concentrate. Its command must therefore be concentrated. Air and ground staffs must work closely together, but "the Soldier commands the land forces, the Airman commands the air forces." The new U.S. Army manual, FM 100-20, echoed Coningham, too, in laying down the priorities of tactical air forces. First and foremost was to gain air superiority. Second was to isolate the battlefield by cutting off the enemy's supplies and reinforcements. Third was to carry out combined action with ground forces on the immediate front.

Ground-force officers sneered about "the Army Air Forces' Declaration of Independence"; yet it was and it wasn't. There was a strident insistence on the coequal status of land power and air power in FM 100-20. But the manual was also a pledge that the AAF would finally take battlefield air operations seriously, claiming it as one of its own fundamental missions. A new and powerful structure, the Tactical Air Forces, was established to carry out the job and given an institutional rank fully equal to that of the strategic bomber forces.

Still, many ground commanders refused to see the difference between independence and indifference. Their suspicion was persistently fed by the powerful psychological grip that aerial attack held on soldiers, as a famous incident that occurred several weeks after Coningham took charge in Tunisia underscored. On April 1, 1943, a flight of Ju 88s attacked an American position in Tunisia, killing three men, including an aide to Lieutenant General George S. Patton, Jr., who had succeeded Fredendall as II Corps commander. Patton filed a furious report: "Forward troops have been continuously bombed all morning. Total lack of air cover for our Units has allowed German Air Forces to operate almost at will." Coningham riposted with his own furious, and widely circulated, memorandum—he even sent a copy to the official historian at the Pentagon—pointing out that the Luftwaffe had flown only two missions over the II Corps front the entire day, versus 260 Allied sorties. Thus, Coningham concluded, the only possible interpretation of Patton's report was that it was an April Fool's joke. Either that or II Corps troops were "not battleworthy."

Patton erupted, saying he was "fed up with being treated like a moron by the British," but he was subsequently delighted when, during a meeting with Spaatz and Tedder the next day (the airmen had come to Patton's headquarters to smooth things over), a German aircraft interrupted the discussion by flying "right down the street, not fifty feet from the window," bombing and strafing, as Patton would later tell the story. "If I could find those sonsabitches I'd mail them each a medal," he gleefully claimed.

But no one was hurt, and in a pattern that would be repeated throughout the war, ground soldiers who would never expect immunity from artillery attacks reacted with anger and outrage, and not infrequently terror, if even a few enemy aircraft appeared overhead. Ragged enemy air strikes that caused few casualties and could not possibly alter the outcome of the fighting on even a small portion of the battlefront became fantastically exaggerated in the retelling. General Bradley, who had himself frequently berated the air force for not doing its job, was astonished when a report of "strong enemy air action," which had supposedly held up a ground advance during the fighting on the Cherbourg peninsula a week after D-Day, turned out to consist of two Me 109s that had strafed a regiment in bivouac. The sole casualty was the cook, who ended up with shrapnel in his buttocks. The incident opened Bradley's eyes to just how spoiled his troops had become, how they took total air superiority for granted. He marched off to the regimental commander and let him have it. "You cannot expect to *never* be under air attack, colonel," Bradley berated the officer. "This is a war, for God's sake!"

The Allied tactical air organization was born of necessity, but its chief weapon was born mostly by accident.

Throughout the interwar years the assumption of all the world's air forces had been that ground attack, if it were to be done at all, would be done either by general-purpose light and medium bombers or by specialized ground-attack machines such as dive-bombers. The Battle of France had exposed the suicidal vulnerability of regular bombers to flak when they ran in at low altitudes over the battlefield; the Battle of Britain had proved that even specialized ground-attack aircraft such as the Germans' fabled Stukas were liable to be shot down by enemy fighters, especially at the moment they began their dives. Another irony in a story replete with ironies was that the years of comparative neglect of ground-attack aviation by the RAF and AAF had saved the Anglo-American forces from committing themselves to these technological paths that would prove, just two years into the war, a dead end. When they finally began taking tactical air power seriously, a far superior tool for the job of ground attack was fortuitously at hand.

The idea of putting bombs on fighters went back to the First World War, when Camels and other fighters had been sent on ground-attack missions over the Western Front. Most of the world's air forces during the interwar years had made provisions for their fighters to carry bombs if needed. But the range and payload of fighters were always sharply limited compared to "real" bombers, and in the opening battles of the Second World War fighters had seen little service in the ground-attack role. It was the Germans who now revived the idea in earnest. After Fighter Command's dogged defense forced the German bombers hitting Britain to shift to night attacks in the late summer of 1940, the Luftwaffe found that Me 109s, each carrying a single 550-pound bomb or four 110-pound bombs, could still race across the Channel to carry out hit-and-run raids in daylight. The Germans christened these hybrids *Jagdbombers, Jabos* for short—literally, "fighter-bombers."

Fighter Command soon followed suit, fitting Hurricane fighters with two 250-pound bombs apiece and sending them back across the Channel on missions dubbed, in typically incomprehensible RAF banter, "Rhubarbs." The RAF had found that fighter sweeps over France meant to prod the German fighters into combat were often ignored by the Luftwaffe, but carrying bombs forced the enemy to take them more seriously. By fall 1941 Coningham was also using bomb-equipped Hurricanes, nicknamed "Hurribombers," in the Western Desert. These were soon joined by "Kittybombers," Curtiss Kittyhawks (the export version of the P-40 Warhawk) that each carried three 500-pound bombs.

Neither these aircraft nor any of the other fighter-bombers that would appear throughout the war had been specifically designed for ground attack. The Hawker Typhoon and the Lockheed P-38 Lightning were both intended to be interceptors. The most famous fighter-bomber of the Second World War, the Republic P-47 Thunderbolt, had also been conceived as an interceptor, then entered service as a short-range escort fighter with the Eighth Air Force in England, accompanying the strategic bombers on the first leg of their journey over France. The Thunderbolt found its role as a fighter-bomber even more by accident than its predecessors had; toward the end of 1943, P-47 pilots had begun strafing targets of opportunity in France with their leftover ammunition while returning from escort missions.

P-47D Thunderbolt
fighter-bomber
engine: 2,430-hp radial
top speed: 433 mph
armament: 2,500-lb.
bomb load or 10
rockets; eight .50-in.
machine guns in wings

Despite their chance origin, the fighter-bombers turned out to have formidable advantages over all other aircraft used for battlefield attack. Fighters had been designed to optimize speed and maneuverability over other attributes that, in the prewar period, were considered more important for ground attack, such as payload, extra gun turrets and extra crewmen to man them, and armor protection. But with the coming of the 2,000-horsepower engine, a fighter like the P-47 could carry the same 2,000-pound bomb load as a twin-engine attack bomber twice its size, such as the Douglas A-26 Invader, while suffering no impairment in its fighter-class performance. The P-47 had a top speed of 433 miles per hour, about 50 percent greater than any medium bomber and on a par with the best air-to-air fighters of either the Allies or the Axis at the time.

And speed, as it turned out, was what survival over the battlefield would depend on above all else. Fighter-bombers could roar in and out before antiaircraft could get a bead on them. Their small size and maneuverability made the AA gunners' job even harder. (The Thunderbolt had the added advantage of an air-cooled engine that could sustain considerable damage and keep running.) The dogfighting ability of the fighter-bombers meant they could defend themselves effectively against the air-to-air threat while on bombing missions. It also meant that the same aircraft could be used for all three of the essential tactical missions that Coningham, and FM 100-20, had delineated: air superiority, battlefield isolation, and close air support. Finally, fighter-bombers were easier to service and more responsive than bombers. They could fly from rough airstrips with narrow taxiways; and with a crew of one, they did not need the extensive crew briefing that bomber crews required for each new target. Given these advantages, fighter-bomber units typically flew three missions a day, compared to the two that were the most a medium-bomber unit could manage.

The only theater where specialized ground-attack aircraft continued to survive and perform well was on the Eastern Front. But the situation there was unique. By the time the Red Army had pieced itself together after the blow of the German invasion and the siege of Stalingrad and begun its slogging counteroffensive, Soviet aviation forces had amassed a two-to-one numerical superiority over the Luftwaffe. There was little finesse to the offensive espoused by Soviet military science in the year 1943. The sheer brutality of the war in the East, which would take the lives of six million Soviet soldiers by the time it was over, and the Soviet generals' knowledge that displeasing Stalin was not an option left little room for refinements. Tens of thousands of officers, including hundreds of generals, had been executed in Stalin's paranoid anti-Trotskyist purges before the war. The Red Army air force was particularly hard-hit, with 75 percent of its senior officers executed by 1939. Those who took their places were both inexperienced and chastened.

The Soviet Army's method of attack that emerged from this hell was grinding, methodical, and massive, with all the subtlety of knocking down a brick wall with a sledgehammer, one brick at a time. Breakthroughs against the Nazi lines were orchestrated by sheer concentration of firepower: thousands of tanks, artillery pieces, and attack aircraft, followed by human waves of troops. They would force a breach, move up, and then do it all again.

The star performer of Soviet attack aviation for these blunderbuss assaults was the Ilyushin Il-2 Shturmovik, or "armored attacker." The Shturmovik represented the apotheosis of the specialized attack airplane. It was armed with two 20mm cannons and two .30-caliber machine guns and could also carry four 220-pound bombs and eight rockets. The crew, engine, and fuel tank were protected from ground fire by 1,500 pounds of armor plating; in later models, this was beefed up to more than 2,000 pounds. A shell of anything less than 20mm caliber would bounce off the Shturmovik like a BB.

Il-2 Shturmovik
ground attack
engine: 1,600-hp V-12
top speed: 251 mph
armament: four 220-lb.
    bombs and 8 rockets;
    two 20mm cannon,
    two 7.62mm machine
    guns in wings

By the time of the Battle of Kursk in the summer of 1943, the Soviets could throw more than a thousand such ground-attack aircraft against the German front. Ground attacks invariably were preceded by a massive artillery barrage and waves of hundreds of attack aircraft striking enemy artillery and mortar positions, clearing the way along the planned axis of advance. There was absolutely no surprise about it.

Shturmovik pilots often flew in formations as low as thirty feet off the ground, sending their cannon shells ripping into enemy tanks and trucks. "A very effective and unpleasant" weapon, complained one German officer, who noted its invulnerability to light flak and the impossibility of getting a bead on the incoming planes with heavier flak guns when they employed their hedgehopping tactics.

The Shturmovik had some of the characteristics of a fighter-bomber; its top speed of 290 miles per hour at ground level was better than that of medium and light bombers. But it was still markedly inferior to what the best fighters could do. When jumped from the rear, the Shturmoviks were extremely vulnerable and easily shot down. They proved an effective weapon on the Eastern Front not only because of the Soviets' massive air superiority by 1943 but also because of the Soviets' acceptance of massive human losses as the cost of reclaiming the Motherland.

By the spring of 1944, as the Normandy landing approached, the fighter-bomber had become the workhorse of Allied tactical air power. The huge tactical Ninth Air Force in Britain commanded thirteen groups of P-47s, some sixteen hundred aircraft. But nothing had changed the essential convictions of the bomber commanders. Deep down, many of the top Allied air commanders remained unpersuaded that tactical aviation was anything more than a sideshow. Harris went so far as to assert that using air power to strike German military targets in France in advance of the invasion, as Eisenhower now called for, "might give the specious appearance" of helping the ground troops, but in reality would "be the greatest disservice we could do them." The best way to help the ground troops was to keep bombing strategic targets. When Germany was thus forced to surrender, further ground fighting would be unnecessary. (General Brooke, Chief of the Imperial General Staff, sarcastically summarized the position Harris had presented at one meeting with Eisenhower: "Harris told us how well he might have won the war had it not been for the handicap imposed by the existence of the other two services.")

Spaatz thought the Normandy landing was not merely unnecessary; he thought it doomed to failure. At times he seemed genuinely mystified as to why anyone would even want to attack the Germans on the ground. "Why undertake a highly dubious operation," he wrote his fellow AAF general Hoyt S. Vandenberg, "when there is a sure way to do it?"

With less than three months to D-Day, the wrangling over the right way for air power to contribute to the success of ground operations had left the air plans for the invasion in a state of chaos. The disputes ran the gamut from high theory to low political intrigue; the question of what to do was hopelessly entangled with the questions of who would do it and who would get to supervise. It didn't help that the Americans didn't want to be commanded by the British, or that the strategic-bomber forces didn't want to be commanded by anyone.

Paralleling Eisenhower's Allied Expeditionary Forces Command, with its painstakingly balanced distribution of key commands between American and British officers, were national chains of air command that utterly confused the lines of authority within air units. British Air Chief Marshal Trafford Leigh-Mallory was Eisenhower's commander of the Allied Expeditionary Air Force, under which fell all tactical air forces, including the U.S. Ninth Air Force. But the Ninth Air Force was at the same time administratively subordinate to Spaatz's newly created U.S. Strategic Air Forces in Europe, or USSTAF, which encompassed the American air forces in both England and Italy. And Spaatz in turn had always considered himself answerable only to Hap Arnold, the AAF's commanding general in Washington. Arnold in fact had created USSTAF explicitly to give an American bomber commander a status fully equal and "parallel" to that of Harris and Eisenhower. As Arnold explained in a letter to Spaatz, this was so the American strategic-bombing effort would be ensured of proper "credit" for the defeat of Germany.

Meanwhile, Spaatz and Harris both deftly sidestepped the efforts of Leigh-Mallory—whom they disdainfully considered a lightweight—to begin directing the pre-invasion air campaign, and in particular his efforts to have the heavy-bomber force help with the job. By March 22, 1944, a despairing Eisenhower wrote George Marshall, threatening to resign: "The air problem has been one requiring a great deal of patience and negotiations. Unless the matter is settled at once I will request relief from this command."

Eisenhower's planners clearly saw that the crux of the problem for the invasion was not *getting* ashore but *staying* ashore. They thus wanted pre-invasion air operations to focus on paralyzing the Germans' ability to mobilize a quick counterattack. The plan favored by Leigh-Mallory and Tedder (who now held the number-two position at Eisenhower's headquarters) targeted the French rail system. The strategic-bombing and railroad experts that Leigh-Mallory had consulted calculated that 45,000 tons of bombs would be needed to do the job; this was clearly a task that only heavy bombers were capable of. On March 25 Eisenhower held a conference with all of his air leaders and forced a decision; the next day he formally approved the "Transportation Plan." On April 13 control of the strategic-bomber forces shifted to Eisenhower's headquarters and the bombing of the rail centers began in earnest.

Spaatz privately made it clear, however, that he was giving the Army generals the tactical air support they demanded for D-Day only because he feared that if he didn't, the air force would become the scapegoat for the debacle that was sure to follow. "This ---- invasion can't succeed, and I don't want any part of the blame," he stormed at one meeting. "After it fails, we can show them how we can win by bombing." Spaatz had also been on edge because he was afraid Harris would be able to slip out from under Leigh-Mallory's control using the excuse that the RAF's night-bombing tactics were ill suited to hitting transportation targets. "Harris is being allowed to get off scot-free," a vinous Spaatz complained to Solly Zuckerman after dinner as his guest at an Oxford college High Table one night that spring. "He'll go on bombing Germany and will be given the chance of defeating her before the invasion while I am put under Leigh-Mallory's command."

Inevitably, even the forced involvement of the strategic bombers brought a very strategic kind of thinking with it. The Transportation Plan was a form of battlefield isolation, but an extremely strategic form, what would come to be called "strategic interdiction" as opposed to "battlefield air interdiction." The idea was to cut off the supply system at its source rather than to cut off the battlefield. (It was while the debate over the Transportation Plan was raging that Zuckerman, one of its chief architects and advocates, first heard the term *interdiction* used this way; he wrote to one of the editors of the *Oxford English Dictionary* to ask if this was proper usage. "To my astonishment," Zuckerman recalled, the editor's only reaction was one of delight, "since, as far as he was concerned, it meant he now had a new entry.")

As formulated by Zuckerman and Leigh-Mallory's bombing and railroad experts, the Transportation Plan bore more than a passing resemblance to the surgical Tactical School theories of strategic attack on critical nodes and resulting systemic collapse. The plan called for hitting eighty rail centers, not near the landing area in Normandy where the battles would be fought but hundreds of miles away in northern France and Belgium. The chief targets at these centers were to be the repair shops that supposedly constituted the weak link in the entire rail system. Rather than trying to cut lines or blow up switches or demolish locomotives, which could only do local damage, the attacks on repair shops would destroy the facilities that the entire network relied upon to keep running on a daily basis.

By the end of April, 33,000 tons of bombs had been dropped on rail centers. The damage was enormous. But German troop and supply trains, which had amounted to only 40 percent of French rail traffic at the outset, were still getting through without any serious delays. The Germans had simply bumped civilian traffic to keep the military trains moving. By mid-May Eisenhower's intelligence staff had concluded that the bombing—which would reach 71,000 tons by D-Day, and also kill 4,750 French civilians in the process—was a failure. There was simply too much redundancy in the rail system to try to squeeze German supply movements this way. The railroads still had four times as many rail cars and eight times as many locomotives as the entire German Army in France needed to meet its essential needs.

The strategic bombing experts had, from the start, dismissed the idea that any effort should be expended trying to destroy railroad bridges. But in the face of such dismal results, Montgomery, who was to command the British ground forces that would cross on D-Day, began pushing for just that. The experts replied that it would take 1,200 tons of bombs per bridge. Leigh-Mallory ruled that it would be a waste of effort.

That would have been that, but for the commander of the American IX Fighter Command. Major General Elwood R. "Pete" Quesada had served under Coningham in Tunisia and was a true believer in the fighter-bomber. Itching to show what his planes could do, Quesada had already disobeyed orders by sending his pilots racing along rail lines to shoot up trains. Now he began importuning Leigh-Mallory to allow his fighters a crack at the bridges. The proposition that a P-47 carrying one ton of bombs could have any effect on a target that the experts had just declared to be able to resist a thousand-ton bombardment was greeted with incredulity at Leigh-Mallory's headquarters. But Quesada, skating perilously close to insubordination, kept insisting and finally was given permission to try.

On May 7 eight Thunderbolts roared down the Seine toward the rail bridge at Vernon, thirty-five miles from Paris. Flying below the height of the bridge, each plane released a pair of 1,000-pound bombs as it pulled up at the last second, sending the bombs hurtling directly toward the bridge abutments. The sixteen bombs, a mere eight tons, cut the bridge in half. On May 10 Leigh-Mallory approved an all-out attack against the Seine bridges. On May 20 he also approved wide-scale fighter-bomber sweeps against moving trains.

The next day nearly eight hundred fighters swept over northern France. The results were swift and spectacular. French train crews deserted en masse, especially after the fighters began setting stalled trains ablaze by dropping their external fuel tanks on them and then torching the gasoline with a burst of incendiary bullets on a return pass. By May 26 rail traffic was barely moving during daylight hours. By June 6, D-Day, 475 locomotives had been put out of action and every bridge across the Seine south of Paris had been dropped. The number of trains crossing the Seine per week had fallen from 200 in mid-April, and 150 in early May, to 8 in the first week of June. Eisenhower's intelligence section had calculated that for the Germans to concentrate a counterattacking force in Normandy and keep it supplied they would require 175 trains a week.

Where strategic interdiction against the entire rail network had manifestly failed, the much more focused tactical interdiction campaign to isolate the battlefield had manifestly triumphed—not just because fighter-bombers could deliver bombs with greater accuracy than could heavy bombers but also because they were going after the right targets. The destruction of the bridges had taken only 4,400 tons of bombs in total, one-fifth the amount the strategic bombing experts had predicted.

After the war, the U.S. Strategic Bombing Survey would cite the Transportation Plan among the accomplishments of the strategic-bomber offensive. Assigning credit for military successes is always a tricky business, but giving credit for a success to a command that not only did not accomplish it but indeed opposed it as the wrong objective carried out by the wrong force against the wrong target using the wrong strategy and the wrong tactics—this was taking credit to an extreme.

The overwhelming air superiority that the Allies had achieved by D-Day was another accomplishment of the tactical air forces that the Survey appropriated to strategic air power in its conclusion that "Allied air power was decisive in the war in western Europe" and "made possible the success of the invasion." At a pre–D-Day briefing on May 15 at Montgomery's headquarters attended by all of the brass, from division and corps commanders through to Churchill and even

King George, it was the forty-year-old fighter chief Quesada who was called upon to explain the tactical air plan. Major General J. Lawton Collins—"Lightning Joe" Collins, commander of U.S. VII Corps—asked the crucial question: "Pete, how are you going to keep the German Air Force from preventing our landing?" Quesada didn't bat an eye. "There is not going to be any German Air Force there," he replied.

As Quesada recalled, his answer was met first with snickers, then with Winston Churchill's skeptical growl: "Ahhhh, young man, how can you be so sure?" Quesada repeated his assurance: "Mr. Prime Minister, because we won't let them be there. I am sure of it. There will be no German Air Force over the Normandy invasion area." He was basically right. On D-Day, the Allies were able to put up ten thousand planes, including five thousand fighters; the Germans, three hundred.

That was what "made possible the success of the invasion."

Taking a page from the Luftwaffe's occupation of forward airfields during the invasion of Poland and France, Allied fighter units were standing ready to jump to France as soon as the first airfields were secured within the Normandy beachhead. One British officer recalled the "memorable sight" at RAF Hurn, outside of Bournemouth on the Channel coast, where hundreds of fighter pilots were staging for the cross-Channel deployment. Since they "only had what they stood up in," after a few days' waiting "they began to acquire a rather unshaven and disreputable appearance"—an impression reinforced by the pistols and knives most of the fliers had armed themselves with in case they were shot down over enemy lines. Soon the mess began to look like a scene out of a Wild West film, "hundreds of whiskery unkempt ruffians, festooned with pistols, sitting round small tables drinking beer and playing cards." Except for one "strange and incongruous detail": from "this barbarous and wild-looking mob there arose a babble of conversation, but all in the purest accents of the English Public School."

Throughout the Allied breakout from Normandy and the drive across France, Quesada seized every chance he could to show the ground forces what his mob of fighter pilots could do for them. "Nothing conventional about Quesada," marveled an aide to Bradley after the two generals had spent the day of June 17 together touring the battle zone. "When he talks power, he means everything but the kitchen sink." Quesada struck up a close working relationship with Bradley, who, as General of the U.S. First Army, was the principal American ground-force commander in the Normandy invasion. "Bradley liked me," Quesada said simply, "and I liked him a hell of a lot." Quesada even tossed aside some of the air force's hard-won principles of independence in battlefield operations, agreeing to put up standing patrols over Allied armored columns. An air force pilot riding in a lead tank talked directly to pilots in the air over a VHF radio to coordinate air and ground fire.

The idea of using qualified pilots as ground spotters had been Quesada's, and he proposed it to Bradley in mid-July; Quesada explained that he wanted to make sure that "the direction from the ground will be in the language the fighter boy will understand." Bradley liked the idea at once and agreed to send two tanks over to IX Tactical Air Command headquarters to have them fitted with radios. There followed a small farce that spoke volumes about the institutional barriers Quesada and Bradley were breaking down. The First Army's ordnance officer assumed the order was a mistake and sent the two tanks to the 9th Infantry Division. Eventually that mix-up got straightened out and the tanks were redirected to the right place, but when they arrived there an airman shooed them away. "Get the hell out of here!" he shouted to the drivers. "This is the air force!"

When the radio-equipped tanks finally entered the battle, the results were immediate. The partnership of aircraft and tanks produced a deadly armored juggernaut as the planes blew aside German antitank gun positions ahead of the advancing tank columns. The system was quickly adopted throughout the First Army, and then the entire American 12th Army Group. Once again, Field Marshal Erwin Rommel, who now commanded the two German armies trying to stem the Allied invasion, would rue the Allies' air superiority: "There's simply no answer to it," he wrote in a letter to his wife in the midst of the fighting. Fittingly, it was tactical air power that ended Rommel's own war: on July 17, his staff car was strafed by two RAF Spitfires, killing his driver and sending the car out of control; Rommel suffered a fractured skull when he was thrown out of the vehicle. Sent home to Germany for treatment, he poisoned himself in October after being implicated in the failed plot to assassinate Hitler.

Despite Quesada's unbounded personal enthusiasm for close air support in the battle zone—a far cry from what ground officers had come to expect from airmen—these operations never delivered as much bang for the buck as air superiority and battlefield air interdiction missions, which with good reason remained the top two priorities for tactical air employment. In their interdiction role, the fighter-bombers continued to hammer railroads, German supply convoys, and airfields with overwhelming effect. In the days immediately following the Normandy landings, fighter-bombers had repeatedly bombed and strafed trains carrying German reinforcements to the battle zone, cut rail lines, and shot up anything that moved along the roads. Kampfgruppe Heintz, sent from Brittany on five trains on the morning of June 6, finally had to set out on foot, taking a week to make the 120-mile trip. Other reinforcing divisions suffered a similar fate. The 2nd SS Panzer Division, "Das Reich," a fully motorized armored division, was forced to take circuitous detours and arrived from Toulouse only on June 13; again and again the tank crews had been forced to leap from their vehicles and take cover beneath their hulls when Allied planes attacked, and the division soon abandoned any attempt to move during daylight hours. A full two weeks after D-Day, the German Seventh Army defending Normandy had received only five divisions of reinforcements; German plans had called for shifting seventeen divisions within a few days of an Allied landing. By the time Hitler committed suicide in his bunker and the final collapse came, the Luftwaffe had ceased to exist, and 100 bridges, 4,000 locomotives, and 28,000 rail cars lay in ruins across western Europe as a result of attacks by the Allied tactical air forces.

For close air support, survivability remained the basic problem. Even if the fighter-bomber had, through its speed and tactics, reduced the attrition rates suffered by dive-bombers and other special-purpose attack aircraft when going after troops in the field, it was still a manifestly dangerous business. In a typical two-week period during the breakout from Normandy, IX Tactical Air Command lost eighty fighter-bombers, half to flak and a quarter to small-arms fire. Even when they weren't shot down, their effectiveness under fire dropped dramatically. A study of fighter-bomber attacks on the bridges over the Savio River in Italy during the spring of 1944 found a sharp drop-off in bombing accuracy directly correlated with the intensity of flak fire. With no flak, P-47s could put half of their bombs within 180 feet of their target and required 30 bombs to score one hit. With medium flak, accuracy dropped to 300 feet, requiring 84 bombs per hit; with heavy flak, it was 420 feet and 164 bombs.

And even under the best of circumstances, no air-to-ground weapon seemed to be accurate enough to be an effective—much less a cost-effective—tank killer. Rockets helped, but not much; though somewhat more accurate than bombs, they had no near-miss value given their much smaller warhead. Later research found that it took an average of 3,500 bombs or 800 rockets to get a single tank hit. The hundreds of tanks that fighter-bomber pilots claimed to have destroyed turned out to have been smoke and bangs that had often done no damage at all. After the Normandy fighting, the Operational Research Section of Montgomery's 21st Army Group combed the battlefield and examined 301 tanks and self-propelled guns left by the retreating Germans. Only 10 were found to have been hit by air-to-ground rockets. On close examination, many of the vehicles proved undamaged: they had simply been abandoned, their gas tanks empty. Interdicting fuel supplies to the battlefield had been vastly more effective than trying to blow up the tanks themselves.

A few attempts at improving the effectiveness of close air support by substituting sheer mass of aerial firepower for accuracy, however, proved utter fiascos. Heavy bombers reduced the hilltop monastery of Monte Cassino in Italy to piles of rubble; the only result was to create obstacles for the advancing Allied troops and cover for the German defenders. Heavy-bomber attacks on the beach defenses on D-Day had the limited effect of demoralizing the German gun crews but left their concrete-reinforced emplacements almost unscathed. Most disastrous of all was a series of strikes by Bomber Command, the U.S. Eighth Air Force, and the medium bombers of the Ninth Air Force against German troops just ahead of Allied positions during the breakout from Normandy. The usual problems of accurately marking and communicating friendly troop positions were compounded by the

12,000-foot bombing altitudes the heavies had to employ to avoid the worst of the enemy flak, and by the rigidity of mission planning and execution that was always entailed in operating large airplanes from distant bases and flying them in formation. The bomber commanders insisted that ground troops had to withdraw 3,000 yards from the front to clear a safety zone, which of course meant giving up ground they had just bitterly won from the Germans.

But even with such precautions and a series of increasingly stringent procedures to avoid mistakes, mistakes kept happening. One of the worst occurred on July 25 at Saint-Lô, when thirty-five of the fifteen hundred attacking American bombers bombed "short," wounding 463 American troops and killing 101, including Lieutenant General Lesley J. McNair, the former commander of United States ground forces. American troops began to make sardonic references to the "8th and 9th Luftwaffe." A few weeks later the same thing happened again, and this time some Allied troops opened fire with antiaircraft guns on the errant B-17s that were bombing their position. Cheers were heard when some of the planes were hit. After another "short bombing" disaster, this one the work of RAF Bomber Command, killed 112 Allied troops on August 14, the further use of heavies in close air support was halted.

Neither heavy bombers with their mass nor fighter-bombers with their agility could bring close air support to its ultimate goal of delivering a knockout blow against enemy troops and armor engaged on the battlefield. The intimately linked problems of survivability and accuracy of fire remained the twin bugbears of close operations.

But by the summer of 1944, as Allied armor and aircraft pressed on across Europe, military aviation had seen three harbingers of technological revolutions that much later would unleash tactical air power on the battlefield with the omnipotence that the air-power theorists had always imagined air power would possess everywhere *but* on the battlefield.

On May 6, 1941, outside his factory at Stratford, Connecticut, Igor Sikorsky, wearing his customary fedora and coat and tie, had climbed into the open seat perched at the forward end of a slender girder-frame aircraft, took off straight up, and hovered in the air for one hour thirty-two and a half minutes before settling gently back to the ground. As a young man in Russia, Sikorsky had tried to build a helicopter in his first attempt at aircraft construction. Now, thirty years later, Sikorsky's VS-300 was not the first helicopter to fly, but it had set a world record. More important, it proved that the single-rotor configuration, which Sikorsky had patented in 1931, could solve the fundamental problem of precise control and smooth transition from vertical to horizontal flight, which had plagued all other attempts to build a practical helicopter.

It was, said the inventor, "a wonderful chance to relive one's own life, to design and construct a new type of flying machine without really knowing how to do it, and then climb into the pilot's seat and try to fly it." The key to Sikorsky's invention was a brilliant mechanical system that alternately increased and decreased the pitch of the rotor blades as they turned through each revolution. By moving a control stick, the pilot could change the point in the cycle where the pitch of each blade was greatest: to the side in order to make the plane bank and turn, or directly fore or aft to make the plane fly forward or backward. By the end of the war, Sikorsky had sold 425 two-seat helicopters to the U.S. Army Air Forces, Navy, and Coast Guard.

On September 9, 1943, the Italian battleship *Roma* became the first victim of a technology whose implications for a revolution in aerial warfare, when they were finally recognized and embraced five decades later, would be even greater. Flying at 22,000 feet, twelve Dornier Do 217 bombers of the Luftwaffe's special experimental bomber group KGr 100 had caught up with *Roma* and her sister ships *Italia* and *Vittorio Veneto* as they were steaming for Malta. With the collapse of the Italian government, Admiral Carlo Bergamini had made a fateful decision to defect to the Allies. Unfortunately, the Germans had deciphered his radio messages and knew his intentions. Each of the Dorniers carried a new weapon: a 3,000-pound radio-steered bomb called the Fritz X. One struck *Roma* squarely amidships, punching a hole through her hull and detonating below. A few minutes later, a second bomb hit just forward of the bridge, starting a raging fire that the crew unsuccessfully battled until it reached the forward magazine, setting off a huge explosion. The battleship sank quickly, taking 1,254 officers and men to their death, including Bergamini. *Italia* was hit, too, but escaped serious damage.

The Germans would deploy and use several hundred guided bombs of several different types. The Fritz X was the simplest and most effective. The bombardier would visually observe the bomb's fall, aided by a flare in the tail of the weapon. Using a joystick, he operated two radio-controlled tail fins to adjust the bomb's course as it fell, deflecting it up to 500 meters from its uncorrected impact point in range, 350 meters side to side. By the time the Luftwaffe had been neutralized in 1944, German remotely piloted bombs had sunk five more Allied ships and damaged ten others, including the British battleship *Warspite* and the American cruiser *Savannah*.

And on April 21, 1944, a British scientist reported during a top-secret meeting at the Royal Aircraft Establishment that American scientists had just revealed work they had been doing for the past two years at MIT's Radiation Laboratory and the DuPont chemical company. Using rolls of foil-backed materials and special paints, they treated the metal surfaces of aircraft so that they would either absorb radar energy or generate two reflections of an incoming radar signal, each a quarter wavelength out of phase with the other so that the two reflections would cancel each other out. If the technique could be perfected, it would make an airplane disappear from an enemy radar screen.

11

THE ALLIED BOMBER OFFENSIVE

"The Romance of the Air," said the novelist E. M. Forster in a letter written to his fellow writer Christopher Isherwood toward the end of the fighting in Europe, is "war's last beauty parlor." No one could retain any illusions that war was a chivalrous business; if any remnant of this myth had managed to crawl bleeding from the trenches at the end of the First World War it had been exterminated by the grim totality of the six years' fight against the Axis. No one wrote poems about the beauty of dying for one's country amid the stunned horrors of Gestapo executions, Japanese death marches, hillsides of mud and corpses in the Pacific, or the tonnage of death raining down upon the home front. When the poets of the Great War had turned bitter, they addressed their most searing anger to the smug complacency of the civilians at home who, unable to imagine the horrors of war, mouthed platitudes about gallantry and sacrifice. Now the mutilated corpses were not just in far-off Flanders but on the streets of London. The home front was the war front; the real war was the first-hand experience of civilians and soldiers alike.

And yet the romance of the air still beckoned, and young men still flocked to it by the millions. When America entered the war, the AAF was besieged by applicants desperate to become pilots; by early 1943 it had a surplus pool of close to 100,000 men awaiting flight training. The AAF was deemed by its rival services to have so unfair an advantage in attracting recruits that for a time the Army forbade the air force from advertising on the radio.

The glamour of the air rubbed off on everything associated with flying; even enlisted slots for ground duty were hugely oversubscribed. In November 1942 the AAF's Aviation Cadet Branch reported that its supply of enlisted candidates for various technical positions was backlogged from eight to twenty-four months. The following month voluntary enlistments were terminated.

In the single combat of fighter pilots there was still a ghost of legitimacy to the glamorous image; probably no other combatant in any service was quite so able to defy the mass anonymity of modern warfare, to retain so much of a command over his own destiny. But only a fraction of British and American air force recruits were destined to become fighter pilots. What the Allies needed in the hundreds of thousands were the crews to man the strategic bombers.

As the bomber crews quickly discovered, their front in the sky bore no more resemblance to gladiatorial single combat than did the trenches of Flanders. In its sheer, overwhelming scale, the strategic-bomber offensive was eerily akin to the grinding, anonymous, assembly-line war of attrition that had dehumanized, or at least deromanticized, the ground soldier in the last war. A bomber crew was a mere cog in an immense machine that spanned half the globe, a machine that scooped men up and hurtled them into the air with all the finesse of a catapult smashing boulders against the stone walls of an enemy castle.

In Britain the construction of sixty-seven bases for the American Eighth Air Force was under way at a cost of £1 million per base; it was the largest civil engineering program ever undertaken in the history of the United Kingdom. Across East Anglia, amid hayricks and flocks of grazing sheep, scenes of rural England straight from a Thomas Hardy novel, concrete tarmacs and cities of Quonset huts sprang up. Back home a gargantuan bureaucracy was feeding recruits into the system that would train 193,000 American pilots, 100,000 bombardiers and navigators, 300,000 aerial gunners, 300,000 radio specialists, and 670,000 aircraft-maintenance specialists by the time it was all over.

There was no way to handle this flood of men without the chaos, screwups, and hurry-up-and-wait inanities that have ever plagued military bureaucracies and driven soldiers to distraction. A U.S. Air Force historical study later concluded that "an ill-fated combination of all-out recruitment, inadequate facilities for handling the men recruited, and precipitate ups and downs in aircrew requirements" led to "masses" of applicants and recruits cooling their heels at every stage of the process, enduring endless boredom and often difficult physical conditions.

For many new recruits their first glimpse of the romance of the air looked an awful lot like mud: "Floorless, stoveless tents pitched on marshy ground . . . throttling dust . . . rain and muck . . . extremes of heat and cold." George Hull, an RAF flight sergeant posted to RAF Wigsley, wrote home to a friend that "washing facilities are confined to a few dozen filthy bowls and two sets of showers an inch deep in mud and water." Nor were airmen exempt from the standard army fate of dreadful food: Flight Lieutenant D. G. Hornsey recalled his mess hall's "combined tea and coca urn" that produced cocoa-flavored tea at supper and tea-flavored cocoa at breakfast, the latter meal invariably featuring pig's liver as the staple item.

Both the AAF and the RAF had well-developed systems for training pilots and assessing their fitness, based now on decades of experience, but it was inevitable that the enormous expansion of training programs would create strains that no amount of peacetime preparation could cope with. In the United States, many of the new instructors were civilian teachers who had been lured by "overzealous AAF recruiters" eager to outbid the Navy with extravagant prom-ises of the rank and promotions they could expect. Once in the service, those assigned to the Laredo Central School for Flexible Gunnery in Texas, for exam-ple, found they were destined to an eternal purgatory as privates with no hope of promotion but a full allotment of KP duty. They suffered the humiliation of hav-ing to deliver a rigidly prescribed curriculum that tolerated no individual initia-tive. In a final twist of the knife they saw their students, automatically promoted on graduation, walk out the door wearing sergeant's stripes.

With demoralized instructors giving canned lectures at a breakneck schedule, the doses of physics and aviation theory that were force-fed the trainees in ground school were by most accounts a colossal waste. Bert Stiles recalled the "attempt at education" he was subjected to as a pilot-to-be as "the saddest, poorest, most incomplete I ever ran into," eight classes a day "giving the maximum of predi-gested information in the minimum time." Other parts of the curriculum, such as the daily hours of Morse code practice insisted upon by AAF headquarters, seemed little more than pointless hazing of the kind armies have always used to dispel any lingering illusions recruits might harbor about their individuality.

The AAF's particular devotion to standardized psychological and aptitude testing did nothing to dissuade recruits from the conviction that they were anonymous pieces of machinery being hustled along a rapidly moving assembly line, with little say in their own fate. Officer applicants were hit with a barrage of two full days of physical exams, psychomotor evaluations, and multiple-choice, machine-scored academic tests designed to sort out assignments for pilot, navi-gator, and bombardier courses. The tests actually had a decent record of predict-ing those who would ultimately pass each course, but they only added to the impersonal, and arbitrary, aura of the whole process. The joke was that if you were fast and got the right answers on the exams they made you a pilot; if you were slow and got the right answers they made you a navigator; if you were fast and got the wrong answers they made you a bombardier.

There was still the fascination of flying that awaited those who made it that far. Jaded as we are today about air travel, which has become about as exciting as riding a bus, we forget that flying an airplane was something only a tiny fraction of mortals had ever done in 1942. Even the crews of the big bombers—the gun-ners who crammed themselves into the improbably tiny ball turret that hung out into space below the ship's belly or who crouched in the narrow waist of the rear fuselage, knocking icicles off their oxygen masks in the forty-below of the unheated and unpressurized cabin, risking frostbite if they took off a glove to unjam their machine guns or if a wire in their notoriously unreliable electrically heated suits broke—even they could find compensation in the intricacy of the expensive gadgets entrusted to them by the government and in the thrill of flying itself, of looking down on cloud and earth from thirty thousand feet. Flight Sergeant Hull, who served as a navigator on a Lancaster bomber, was so fed up that he took to crossing out the RAF crest on his stationery when he wrote home;

"I hate the R.A.F.," he declared, "this camp in particular, I hate the job we do, not for myself but for those who are lost in the gamble." But then he added: "I do like to fly, it fascinates me."

Most Allied bomber crews did not hate the air force, at least not so overtly, but by the time they were dropping bombs on Germany they could nurture few illusions about the glory or heroism of the air. It was the most experienced crews that had the fewest illusions. Sergeant J. P. Dobson, pilot on an RAF Wellington bomber of 218 Squadron, left an account of one of the rare acts of outright rebellion by a bomber crew; it is revealing precisely because it shows how no one ever "got used" to the job.

On the day of the mission Dobson is awakened by a cheerily sardonic aircraftsman: "You're to be Queen of the May and it's just exactly 0600 so you've got plenty of time to get your party frock on." In the ops room everyone is sullen. The wing commander enters and delivers the bad news: It is a "special mission" over enemy territory. "The mission is directed against a factory situated twenty miles southeast of the German capital. This target must be completely demolished at all costs and no aircraft captain is to deviate from his bombing run for any reason whatsoever. I am to inform you that for all practical purposes your chances of return are nil as weight of fuel must give way to bomb load." The crews file in to lunch "like condemned men"; no one can force down a bite. When Dobson wanders out to look over the plane, his navigator calls him over. "We're not going," he abruptly tells Dobson. "We're all married and some of us have kids and so you can tell old ——— that he can do without us for tonight. Anyhow we've all done '30' [their mission quota] and so really we are entitled to a rest and certainly not going to chuck our lives away on this damned death-but-no-glory stunt."

Dobson informs the commanding officer of what's happened and is given a completely new crew with zero experience. To keep the new men from seeing the sure-to-be even more morose scene in the mess hall at supper, he tells them it is a tradition that men on their first missions wait in their rooms. They take off into the darkening sky, and once over the target area the bomber is shot down; Dobson ends up spending the rest of the war as a POW.

Even on less "special" operations, everything about a bomber mission conspired against the notion that crews were anything but pawns in a vast scheme over which they had no control. The principal American bomber, the B-17, was a huge, complex machine that took ten men to operate: pilot, copilot, bombardier, navigator, radio operator, flight engineer, ball-turret gunner, two waist gunners, tail gunner. Pilot and copilot sat before an array of a hundred and fifty dials and switches and lights, reviewing a checklist of fifty-six items before each takeoff. Once in the air, the plane became part of an even larger machine. Each group, usually eighteen to twenty-one bombers, would array itself into a tight "combat box," with each plane assigned a spot in a geometric pattern that staggered the planes in altitude and position. From the top and side views, they formed a V-shaped wedge, with one squadron above and to the right and another below and to the left of the lead squadron, but it was more intricate than that: each plane had to keep a precise position within the formation so it could fire all its guns with minimal risk of hitting another bomber. These groups in turn were part of a still larger juggernaut, the combat wing: three combat boxes, sixty-three planes, packed into a space 7,000 feet wide by 3,000 feet high by 1,800 feet deep setting out across hundreds of miles of European sky.

The idea was both to interlock fire and to mass forward fire into an impenetrable wall against enemy fighters that tried to attack head-on. The psychology was that of a nineteenth-century infantry or cavalry charge: safety in numbers, in moving unquestioningly in step with the man to the left and the man to the right. It was the stragglers, those who broke formation, that were the first to be picked off by the German fighters.

The dangers and the need for discipline mounted as the bombers reached the

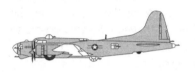

B-17F
heavy bomber
engines: 1,200-hp radial (4)
top speed: 314 mph
armament: 4,000-lb. bomb
load; 10 to 13 guns

I.P., the initial point of the bomb run; from this moment until bomb release about ten minutes later, the formation would have to fly straight and steady if the Norden bombsight could do its job. Even pilots became passive observers as the bombsight took over the steering of the plane. Fighter pilots lived or died by their aerobatic skills in executing evasive maneuvers in the heat of combat. At the live-or-die moment in bomber combat the pilots were merely passengers along for the ride. Opening the bomb-bay doors slowed the planes down and

hindered their maneuverability in any case. "The most important characteristic of a lead bombardier is determination that nothing shall deter him from placing the bombs of his formation on the assigned target, and that nothing the enemy may do will distract him," was the law laid down by Major General Curtis E. LeMay to the crews in his group of the Eighth Air Force. It was as robotic as a First World War infantry attack.

The odds of surviving were about as good, too. General Arnold's statistical control unit tried to convince bomber crews that they had a 60 percent chance of completing their tour unscathed, and that even if they were shot down their chances of survival were 50 percent; their overall odds of making it were thus 80 percent. In fact, a careful postwar study of 2,051 crewmen who began the required cycle of twenty-five missions in six American bomber groups in England found that by the end of the war 63 percent of the men had been killed or were missing and 10 percent had been wounded. The average Eighth Air Force bomber crewman could expect to complete only fifteen missions before fate caught up with him. Statistics from RAF Bomber Command were much the same: half of all the crewmen who flew on bombers during the war were killed, a quarter were wounded or taken prisoner; only a quarter survived the war unscathed. They might as well have been at the Somme or Pickett's charge.

The fate that awaited those who managed to safely bail out from flaming wrecks going down over enemy territory did not inspire much cheeriness, either. In the First World War, escaping from the enemy had almost resembled a chivalric game. A 1918 RFC lecture for fliers offered this hearty advice if taken prisoner:

> Reasons for attempting escape, useful even when unsuccessful.
> (a) As an occupation
> (b) As an annoyance to the Germans

Escaping from the Nazis had quickly ceased being a game. Squadron Leader S. A. Booker, the navigator of a Halifax bomber shot down over occupied France just before D-Day, left a harrowing account of his torture by the Gestapo and the SS after he was betrayed by a member of the French Resistance while trying to evade capture. He was accused of being a British spy posing as an RAF officer; his torturers knocked out his teeth, kicked him in the groin, and beat him repeatedly on the knee with a rubber truncheon. He was eventually transferred to the Nazi camp at Buchenwald. Some downed airmen were simply lynched by local Nazi Party henchmen or other civilians, encouraged by broad hints from Heinrich Himmler, the SS chief; Himmler had stated that the German police could not be expected to protect airmen from the justifiable fury of the citizenry.

Forced to face danger that arbitrarily packed some off to oblivion and capriciously chose to spare others, bomber crews coped by means of fatalism, superstition, and only rarely outright anger, hysteria, or rebellion. "All Eighth Air Force fliers were highly superstitious," said John Morris, a B-17 waist gunner in the 91st Bomb Group. Morris slept through the 3:00 a.m. wake-up call for his first mission; then he got lost in the dark and couldn't find the briefing hut; having missed the crew briefing, he stumbled over to the mess hall to find it shut down and deserted, too. He only caught up with the rest of his crew as they were already on the way out to their airplane. Despite this inauspicious start, he survived his first mission unscathed. And, he recalled, "I was so impressed with my survival that I resolved to repeat the preparatory routine in all future missions. And I did: up late, circuitous route, no briefing or breakfast."

Relatively few men cracked up completely under the strain; only 1.5 percent of American bomber crewmen had to be grounded permanently for stress or other non-physical conditions; another 3 percent had to be temporarily relieved while they recovered their shattered nerves and pulled themselves back together for another go. In a few units where morale fell apart, crews would regularly "discover" mechanical troubles early in the mission and head for home; others became known as "fringe merchants" for their practice of cutting their bomb runs short and ditching their load far from the aim point to avoid the worst of the flak.

But if the overwhelming majority stuck to it and did their job, the strain still took its toll in countless ways. Sixty-eight percent of Eighth Air Force flight surgeons reported cases of crew weight loss once bombing operations began to intensify in the summer of 1943. The first ten missions were the worst, the flight surgeons reported, with many crewmen suffering extreme anxiety and a sense of hopelessness. Those who made it through this initial period tended to cope better, as familiarity and team spirit helped counteract battle anxiety. But as crews approached the end of their tour, anxieties shot up again, with dangerous levels of fatigue, nervousness, insomnia, and loss of appetite common.

Everyone talked about being "flak happy," and the creative coping mechanisms that cropped up among all bomber crews certainly suggested that everyone was skating pretty close to the edge. A B-17 navigator in the 381st Bomb Group had an armorer make him a steel-plate jockstrap, which he wore religiously on every mission. Chuck Halper, a pilot in the 385th Bomb Group, recounted a ritual observed aboard his plane. Every few missions the navigator's bowels would give way in midflight:

> He carried big paper containers with lids which he picked up from the mess hall. And despite layers of clothing, parachute harness, Mae West, heated suit, trousers, long johns, etc., he would make his contribution in the containers, slap the lid on and get dressed. When the bombardier took over on the bomb run, the navigator would hook up to a portable oxygen bottle, pick up the carton and go back and stand in the bomb bay. When the first bomb fell free he would throw his container out. We were all somewhat flak happy and each of us accepted this behavior as perfectly normal.

The crew gave the navigator's personal contribution to the strategic bombing of the enemy the ersatz German designation *flieger sheit.*

Because the Eighth Air Force was not out in some hellhole, slogging forward toward an objective, cut off from the "normal" world but rather was based in a place that seemed not unlike home—with beer, dances, big bands, and an abundance of unattached young girls in male-depopulated Britain—it was all the more strange that crewmen were required to go off and risk their lives every few days. "Nobody expected to live through it," said Burrell Ellison of the 392nd Bomb Group. "I know the completion of tour of duty was a surprise to me. It was akin to being born again."

Flight surgeons urged the Eighth Air Force to set the twenty-five mission requirement to give the men a concrete goal in a battle that offered no milestones of success, no definition of when the task had been completed. (The inducement backfired, though, when crews later in the war reached twenty-five missions only to learn that the quota had been raised to thirty, then thirty-five missions, and morale plummeted.) The AAF also tried to keep up morale by shoveling out truckloads of medals and relentlessly working the public-relations angle to ensure proper "credit" for its fliers. Anyone who completed five missions in the Eighth Air Force got the Air Medal; anyone who completed his tour got the Distinguished Flying Cross; mere survival, rather than heroic deeds, became the currency of accolades in this war. But this attempt at morale-boosting had a tendency to backfire, too. The more medals that were distributed, the less they were valued by the men; but when standards for earning a medal were tightened, those who no longer got medals complained about the unfairness of the new rules.

For the first and last time in history, the U.S. Army and the RAF had managed to turn what had been an elite corps of airmen into a mass army of citizen soldiers, with all the sardonic, cannon-fodder mores traditional to ground troops. The songs and poems of the First World War airmen had been full of black humor about the bodily mayhem they were destined to suffer, but they were infused with cheerful braggadocio. The airmen of the Second World War adopted a far more obscene, fatalistic, and deeply cynical idiom. They were cynical about the commanders, cynical about the point of it all, and in this RAF song (to the tune of "John Brown's Body"), cynical about the "twats" in the Operations Room who planned the missions:

> We joined the fucking Air Force 'cos we thought it fucking right,
> We don't care if we fucking fly or if we fucking fight,
> But what we do object to are the fucking Ops Room twats,
> Who sit there sewing stripes on at the rate of fucking knots.
> (Chorus)   Ain't the Air Force fucking awful?
>   Ain't the Air Force fucking awful?

A ditty that American aviators sang begins conventionally enough as a black-humor account of a pilot smashed to the point that "all they could find were spots," but then it takes a sardonic turn unimaginable during the First World War. Commenting on the $10,000 Government Service Life Insurance policy each airman carried, the chorus goes (to the tune of "Casey Jones"):

> Ten thousand dollars going home to the folks.
> Ten thousand dollars going home to the folks.
> Oh won't they be delighted! Oh won't they be excited!
> Think of all the things they can buy!

Arnold and Spaatz worried constantly about the state of morale in the bomber crews, even conducting anonymous surveys to gauge the mood. One such survey of three thousand officers and men in the Eighth Air Force revealed that a con-siderable majority had "occasionally" or "quite often" felt that a particular mis-sion was not worth the cost. "Men who have flown a large number of combat missions," the survey reported, "are more likely . . . to have some doubts about the selection of targets and the value of particular missions." Among crews with fewer than ten missions, 50 percent had such doubts; it rose to 65 percent for crews with ten to nineteen missions and 73 percent for those with twenty or more.

As they began their first major operations against Germany in the summer of 1943, the American bomber forces remained committed as ever to the policy of daylight precision bombing. But policy was one thing and reality another; the main reason crews were growing skeptical was that they knew their bombs were mostly falling wide of the mark. Even bombing in daylight, the Americans, just as the British before them, were frequently missing their targets by miles. Bomber crews began making jokes about "killing sheep" or conducting "a major assault on German agriculture."

The American commanders steadfastly resisted pressure from British air leaders to abandon daylight bombing and join forces with the British in night area bombing, but both the theory and the practice of precision bombing were getting their sharp edges blunted in the realities of war and alliance politics. At the Casablanca conference in January 1943, Eaker, who had by then succeeded Spaatz as commander of the Eighth Air Force, had visited Churchill to sell him on the American policy. While the result was a victory of sorts, it inevitably, like all political compromises, placed comity ahead of clear enunciation of principle. In fact, it fudged strategic policy altogether. Eaker, who had a degree in journal-ism, had cannily written a one-page pitch that he handed to the great orator, and Churchill was obviously taken by the rhetorical flourish with which Eaker closed: "If the R.A.F. continues night bombing and we bomb by day, we shall bomb them round the clock and the devil shall get no rest."

Churchill himself began using Eaker's phrase about the devil, and the direc-tive subsequently approved by the American and British Chiefs of Staff at Casablanca duly authorized a "combined" bomber offensive of the two nations against Germany. But it was combined only in the sense that it encompassed two uncompromisingly divergent strategies and operational approaches under a kitchen-sink statement of objectives. "Your primary objective," the chiefs ordered, as if what they were enunciating really were a single, unified plan, "will be the progressive destruction and dislocation of the German military, industrial and economic system, and the undermining of the morale of the German people to a point where their capacity for armed resistance is fatally weakened."

In reality there was little if any coordination of RAF and AAF operations. Harris took the Casablanca directive as a clear warrant to continue his program of reducing German cities to ashes. On the night of July 27, 1943, he sent close to eight hundred bombers over Hamburg in a raid that dropped 1,200 tons of incendiaries, most of them falling on the densely built-up residential areas in the eastern section of the city. The weather was warm and dry, and the blazes that started merged into a single vast inferno, what would come to be called a "firestorm." Temperatures rose possibly as high as 1,000 degrees Celsius as the blaze created its own vortex, sucking the oxygen from the air and setting up vast wind currents to keep the flames fed. Some 40,000 civilians were incinerated or asphyxiated. Hamburg's mayor estimated that thirteen square miles had been destroyed, which nearly doubled Harris's running tally of destruction of the fifty-eight principal German cities. By March of 1944, the tally would reach thirty-six square miles; by March of 1945, sixty-seven square miles. By the time Harris was done, about 50 percent of the urban area of Germany was destroyed. Only the cities in the extreme eastern reaches of the country would escape the RAF's area bombing.

Harris in the meanwhile never ceased his efforts to get the Americans to abandon "panacea" bombing and join the RAF campaign. "We can wreck Berlin from end to end if the U.S.A.A.F. will come into it," Harris insisted. "It will cost us 400–500 aircraft. It will cost Germany the war." The Combined Bomber Offensive issued a progress report in November 1943 insisting that the night attacks on morale were so effective that they might cause Germany to surrender even before the American program of precision assault on the German economy began to bite:

> The maintenance of morale is the greatest single problem confronting the German authorities. . . . The increasing death roll is the most important factor and coupled with military failures, the general attitude is approaching one of "peace at any price" and the avoidance of the wholesale destruction of further cities in Germany. The housing situation and the general morale are both so bad that either might cause a collapse before industry became unable to sustain the war effort.

The British Assistant Chief of Air Staff for Intelligence weighed in at the same time with a memorandum asserting, "We are convinced that Bomber Command's attacks are doing more towards shortening the war than any other offensive, including the Russian."

With or without the Americans, Harris was going to try. From November 1943 to March 1944, Bomber Command launched sixteen major raids against Berlin. But even Harris admitted they did not succeed in reproducing the results of the Hamburg attack. Berlin was a more modern city, its buildings more spread out and less prone to catch fire, and the incendiary raids there never managed to get a firestorm going. The attacks, Harris later conceded, "did not appear to be an overwhelming success."

German morale unquestionably did suffer. Absenteeism rates in German factories regularly reached 20 to 25 percent by 1944. Postwar surveys found that

91 percent of German civilians said the bombing was the most difficult hardship they had had to endure. Eleven percent of the entire population of Germany, and perhaps as much as 50 percent of urban dwellers, were left homeless by the attacks on cities. Yet Harris turned out to be simply wrong in his assumption that killing cities was the way to kill industry and the economy. As the U.S. Strategic Bombing Survey concluded after the war, "Allied bombing widely and seriously depressed German civilians, but depressed and discouraged workers were not necessarily unproductive workers." And urban dwellers turned out not to be hysterical mobs of lumpen proletariat unable to fend for themselves. Within days of the Hamburg raid, mail service, transportation links, and electric power had been restored. People learned to cope and improvise. Meanwhile, the German war industry was buffered by a considerable excess capacity. Contrary to the nearly universal assumption of Allied strategists, the German economy was not at all stretched taut at the start of the war: only 49 percent of the German GNP was going to war production in 1941. Germany had a huge surplus of machine tools and factory space; in fact, German factories as of 1943 had twice as many machine tools per worker as did England, and 90 percent of German industry continued to operate on a single shift throughout the war. When Albert Speer was brought in by Hitler to take charge of the Reich's armaments production, he was able to triple arms output by 1944 in spite of the devastation caused by Harris's bombers.

Few factories, indeed, were located in city centers: another of Harris's miscalculations. British experts later estimated that the overall loss in German war production attributable to town-area attacks in 1944 and 1945 amounted to no more than 1 percent.

Nor did loss of morale, as crushing as it was to many German civilians who had lost homes and family members, foment revolt or even resistance against the Nazi regime. Harris, to be sure, had never put much stock in this aspect of the traditional theories, but it was at least a tacit assumption in the talk of an impending German "collapse." Yet, if anything, the Allied bombing and the government-supplied relief to those who had lost their homes in it solidified the Nazi hold over a demoralized citizenry. As the Strategic Bombing Survey found, "However dissatisfied they were with the war, the German people lacked either the will or the means to make their dissatisfaction evident. . . . Their morale, their belief in ultimate victory or satisfactory compromise, and their confidence in their leaders declined, but they continued to work efficiently as long as the physical means of production remained. The power of a police state over its people cannot be underestimated."

While the Americans continued to resist the British strategy of area bombing on principle, in practice they were not accomplishing anything very different by the winter of 1943–44. Flying the combat-box formation meant that individual bombers could not steer precisely to the target; instead, the lead bombardier guided the entire group and the rest of the planes dropped their bombs when the leader dropped his. "Dropping-on-the-leader" reduced the subordinate bombardiers to "toggliers" whose complex training was now employed in nothing more than flipping a toggle switch on cue. The system blanketed the target area and vicinity with a pattern of bombs that more or less matched the shape and size of the bomber formation. Using this technique, American bombers were now getting 25 percent of their bombs within 1,000 feet of the aim point, which was an improvement, but it still meant that 75 percent of bombs were doing nothing but "killing sheep," or killing civilians who lived in the vicinity of a precision target.

As bad weather set in toward the end of 1943, the increasing American use of several "blind-bombing" methods developed by British scientists further erased the distinction between American daylight precision bombing and British night area bombing. One set of systems—Gee, Oboe, and several other variations—used radio signals transmitted to or from fixed ground stations to determine the bomber's location. H2X used ground-mapping radar carried aboard the bomber; the radar display in theory took the place of visual observations through the Norden bombsight when the target was obscured by clouds. But interpreting the cluttered images on the radar tube took considerable skill and experience, and generally the only well-defined ground features that emerged from the fuzzy green glow of the CRT display were the sharp boundaries between land and water.

"Blind bombing" proved an apt description. (Spaatz in fact tried to banish the term, asking Arnold to substitute such euphemisms as "overcast bombing technique" when issuing press releases about the missions.) Even discounting "gross

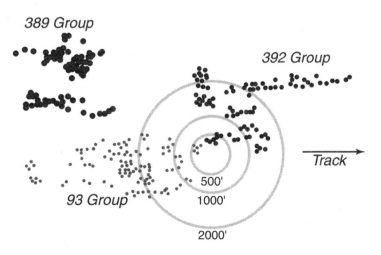

A typical blind-bombing pattern, from the raid on Siracourt,
January 31, 1944. Usually, the aim point was hit only by "drenching"
the entire target area with bombs. *(Library of Congress)*

errors," in which bombs fell more than 5 miles from their aim point, H2X bombing yielded an average circular error of 1.1 miles under five-tenths cloud cover; it increased to 2.5 miles under ten-tenths cover. And ten-tenths was the condition under which the majority of bombs were dropped during the northern European winter. "Gross errors" were supposedly those attributable only to equipment malfunction or outright human blunders, but they accounted for a whopping 42 percent of all bombs dropped using H2X under ten-tenths cover. The fact was that bombardiers generally didn't have a clue what they were seeing on their radar screens without the occasional glance through a break in the clouds to help orient themselves.

As its official history later acknowledged, the only way the AAF was hitting its intended targets was by "drenching" the entire target area with bombs. Spaatz at one point ordered that Operational Research Section reports that rated groups "purely on the basis of percentage of bombs dropped within a radius of a particular aiming point" no longer be circulated because they were "poor for morale." Even after a full year of experience with blind-bombing methods, their effectiveness improved little; in the last quarter of 1944 some 80 percent of all Eighth Air Force missions used blind-bombing devices, and more than half were judged by the operational analysts to be total or near-total failures.

The AAF's insistence that it nonetheless intended to hit specific targets in such operations was not, however, merely window dressing. For one thing there was enormous pressure to simply keep trying in the hope that more precise results would come with experience—and also a fear that to let up now would give the enemy a propaganda victory; bad bombing was better than no bombing. And with enough effort, some precision targets would still be hit, no matter how imprecise the aim. Even as inaccuracy eroded the neat distinctions between the American and British approaches, a few objective differences remained, notably the proportion of incendiaries used by each force.

For another thing, at this point in the war, there was little need for window dressing, since most Americans thought that killing German, and Japanese, civilians was a perfectly good idea. In 1938, a Gallup Poll had found 91 percent of Americans agreeing with the statement "all nations should agree not to bomb civilians in cities in wartime." Three days after Pearl Harbor, 67 percent said they favored unqualified and indiscriminate bombing of enemy cities, with only 10 percent expressing unqualified opposition. The only criticism of American bombing policy that appeared in the many articles on the subject in popular magazines was the frequent complaint that America was not hitting German and Japanese civilians hard enough. In September 1942, *Time* called for destroying thirty-one German cities to shorten the war. An article in *Harper's* in January 1943 advocated burning the Japanese out of their homes with aerial attacks. In early 1944, three-quarters of Americans surveyed expressed approval of bombing even historic buildings and religious shrines if military leaders believed such attacks were necessary.

When the *New York Times* reported on its front page of March 11, 1944, that twenty-eight noted clergymen, educators, and professional people had signed a protest against the American bombing of German civilians, and had called upon "Christian people . . . to examine themselves and their participation in this carnival of death," the story provoked a storm of letters that ran fifty-to-one against the idea that being good Christians had anything to do with the matter. "Socking the rapacious German nation with every pound of high explosives available" (as one writer urged) was the course advocated by the overwhelming majority of those who responded. One writer denounced the signers for "softheartedness" and giving "a great lift to enemy morale." An evangelical pastor declared, "God has given us the weapons, let us use them."

Hap Arnold had summed up his own views on what would later come to be called "collateral damage" in a stern memorandum sent to his air staff in early 1943. In response to a message that had circulated to theater commanders raising questions about "the bombing and machine gunning of civilians in connection with bombing raids," Arnold reminded the staff that it was a brutal war and that no reduction in effort could be tolerated. "This does not mean that we are making civilians or civilian institutions a war objective," he stated, "but we cannot 'pull our punches' because some of them may get killed."

But the moral certainties became less absolute as the British firebombing raids intensified. Churchill began to question his own earlier enthusiasm for unrestricted city bombing. "Are we beasts? Are we taking this too far?" he exclaimed, jumping up from his chair during a showing of a film of British bombing in July 1943. In August 1944 Spaatz had written Arnold that he was being "subjected to some pressure" from the British Air Ministry to "join with them in morale bombing." He urged that the United States keep its hands clean, at least as far as its declared policy went: "I personally believe that any deviation from our present policy, even for an exceptional case, will be unfortunate. There is no doubt in my mind that the RAF want very much to have the U.S. Air Forces tarred with the morale bombing aftermath which we feel will be terrific." Eaker also counseled against joining in explicit targeting of cities, writing Spaatz on January 1, 1945: "We should never allow the history of this war to convict us of throwing the strategic bomber at the man in the street."

Yet American strategic theory, going back to AWPD/1 and the Tactical School lectures, had always left open the possibility of a final, morale-breaking attack on civilians to swiftly conclude a war once the enemy was near defeat. And in the final months of the war the AAF began to do exactly that—even as it simultaneously began to grow increasingly worried about a public backlash. The result was an increasing attempt by American air commanders to rationalize what were, in reality, deliberate attacks on civilians as a continuation of precision-bombing policy. In so doing, they would forever muddy the waters and leave the AAF's motives and intentions throughout the war subject to suspicion and doubt. When the Eighth Air Force joined in Bomber Command's obliteration of Dresden, it was still ostensibly carrying out precision daylight strikes on rail yards, but the American bombers carried the same 40 percent proportion of incendiaries that the British bombers did for their night raids. Other "transportation" targets across Germany selected for attack in the final months were, as Arnold readily acknowledged in private, chosen not for any military objective but "to give every [German] citizen an opportunity to see proof positive of Allied air power."

When an Associated Press wire story reported that the Dresden raid marked "a long-awaited decision to adopt deliberate terror bombing of the great German population centers as a ruthless expedient to hasten Hitler's doom," Arnold's information chief hurriedly cabled Spaatz: "What do we say? This is certain to have a nation-wide serious effect on the Air Forces as we have steadfastly preached the gospel of precision bombing against military and industrial targets." The AAF immediately put out the word that the AP story "was erroneously passed by the censor," and that there had been "no change" in American bombing policy. As always, it was directed solely against "military objectives."

It was ironic that the American air war against Germany ended with an apparent abandonment of the policy of precision bombing of industrial targets, for despite the miscalculations of theory and the failures of execution, the system had finally begun to work by the last few months of the war.

A major refinement in American precision-targeting plans and theory had occurred shortly after the Casablanca directive in early 1943. In preparation for the first American bombing assaults on Germany later in the year, Arnold had brought in a group of prominent civilian experts, including the Wall Street banker Elihu Root, Jr., and a Princeton diplomatic and military historian, Edward M. Earle, to serve on an Advisory Committee on Bombardment; it was later called, a bit inaccurately, the Committee of Operations Analysts. Although the committee fully embraced the "scientific method" of strategic warfare as espoused by the Tactical School and AWPD/1, it did so with an important difference: its members actually knew something about industry and economics. The COA set up subcommittees to scour the data on each industrial sector in Germany and to answer the critical questions the AWPD/1 planners had never asked: How did German war requirements relate to the supply of various basic materials? And how much would those supplies be reduced by the destruction of specific targets?

The COA approach also differed from AWPD/1 in a more fundamental conceptual respect. Gone was the abstract, almost magical tenet of the Tactical School tradition: the belief that snipping away a few vital strands would cause the *entire* civilian economy and social structure to collapse, leading to swift capitulation. The COA's main tenet was, by contrast, pure strategic interdiction: the experts sought to identify exactly what materiel the German military in the field needed in order to keep fighting, and then cut it off at its source. The effect would hardly be instantaneous, but in time—and assuming the German military was forced to keep using up its war materiel through actual fighting in the meanwhile—the enemy war machine would have to grind to a halt.

The primary industrial target that the COA first identified seemed to fit the bill perfectly. Ball bearings were an indispensable component in all high-speed machinery, notably aircraft. The German ball-bearing industry was highly concentrated, with half of all production coming from plants in a single location, Schweinfurt, in central Germany. Based on their experience, American and British experts believed that stockpiling of ball bearings was infeasible. On August 17, 1943, the Eighth Air Force sent 200 B-17s to attack Schweinfurt. The raid did extensive damage to the factories. September production fell 65 percent. On October 14, 228 B-17s struck the plants again. "Now we have got Schweinfurt," Arnold triumphantly told reporters.

The only trouble was they hadn't. The damage to the factory roofs was much more serious than the damage to machine tools within, and in a matter of months production was back to pre-raid levels as repairs were made and tools dispersed to new plants. In spite of "the enormous devastation wreaked" on German factories in 1943 and 1944, the U.S. Strategic Bombing Survey concluded after the war, only 6.5 percent of machine tools were damaged or destroyed, and Germany "never experienced a general shortage of machinery." It also came out after the war, when Erhard Milch and other German officials in charge of aircraft production were interrogated, that Germany *had* in fact laid in large stockpiles of bearings. Although a "brief shortage" had occurred in early 1944, this was quickly made good by increasing Swiss and Swedish imports and by redesigning many critical aircraft components to eliminate bearings or use substitutes, such as sleeve bearings, which reduced the consumption of ball bearings in the aircraft industries by 60 percent. "As it was," Speer later wrote, "not a tank, plane, or any other piece of weaponry failed to be produced because of a lack of ball bearings."

This was far from the only trouble. The raids had been incredibly costly: 36 B-17s had been shot down by German flak and fighters in the first raid and 62 in the second; 138 had been damaged, many beyond repair. One American navigator was puzzled by the scores of burning haystacks in the fields below until he realized they were B-17s. No matter how much stomach American bomber commanders had for taking casualties, losing 27 percent of the attacking force on a single mission was not a sustainable attrition rate.

All of the American targeting plans, from AWPD/1 on, had listed as a primary "intermediate" objective the destruction of the German air force. But true to strategic-bombing gospel, the method envisioned for achieving air superiority was to combine attacks on aircraft factories—cutting off the enemy air force at its source—with the unstoppable defensive fire of unescorted bombers in the air.

The fact that the first mission flown by the Eighth Air Force, in August 1942, had seen 108 RAF Spitfires accompanying the twelve American B-17s to Rouen and back had done nothing to alter the faith of AAF commanders that escorts would be unnecessary once massed formations of B-17s began fighting their way through the Germans.

That the American air leaders were still inhabiting a fantasyland in the early months of bombing operations was evident in Spaatz's assurance to Arnold that "bombing accuracy does not diminish under fire, but rather increases" and in Eaker's unquestioning acceptance of outlandish claims by B-17 gunners that they had shot down vast numbers of German fighters. On one mission in October 1942, American gunners claimed to have destroyed 102 enemy planes. The true number turned out to be 1.

Despite Arnold's confident words to the press, the Schweinfurt raids had plunged him into a crisis of confidence about the B-17. Arnold's nickname, "Hap," was short for "happy," and he always had a twinkle in his eye and a smile on his face—so much so that one of his close aides was firmly convinced that his boss was "really dumb"—but the image was profoundly misleading. He was, to be sure, no intellectual: he had finished in the bottom half of his class at West Point, earning the sarcastic observation of his classmates that "by diligent efforts, he has overcome any hankering for work." But Arnold was in fact an intensely serious man, often humorless, and in his years since West Point he had developed the habit of long hours and hard work if nothing else; so much so that the strains of 1943 had already brought on two heart attacks.

Now, as Assistant Secretary of War Robert A. Lovett would later recall, Arnold was "having a hell of a time hanging on." It seemed that the theory of daylight precision bombing, on which he had staked so much, was a dead end, given the numbers of B-17s that were blown out of the sky on each mission. Arnold vented his mounting frustrations at Eaker. In a letter to Arnold, the Eighth Air Force commander was reduced to defending himself against the accusation that his command was "lazy, cowardly or ineffective." Arnold in turn impatiently rejected excuses—about the weather, a shortage of crews, or a lack of long-range escorts—for the failure to get results.

After a visit to England in October, however, Arnold became convinced that escorts were the essential missing element. The Schweinfurt raids had finally proven that bombers could not survive without them, especially as the Luftwaffe's fighter tactics had grown increasingly difficult to counter through the spring and summer of 1943. Almost from the start, German pilots had adopted the psychologically unnerving tactic of flying parallel to the bombers for some time, just out of the bombers' gun range, then gradually pulling ahead several miles and turning in for a head-on attack. By April 1943 they had discovered through trial and error exactly where the B-17 was most vulnerable; during a B-17 raid on Bremen that month, the German fighters had waited until the moment the bombers opened their bomb-bay doors, then swung into position at ten o'clock and two o'clock, just outside the limit of traverse of the B-17 nose guns.

The August 17 raid on Schweinfurt had brought confirmation of a new and deadlier tactic, and a new weapon to go with it. A pack of Focke-Wulf 190 fighters, their 20mm cannons blazing, approached a B-17 formation when all of a sudden "a very large flash burst from the center of each enemy aircraft, obliterating it from view," according to a report from one of the bomber crewmen. "These aircraft then dived under our formation." A few seconds later, the air-to-air rockets—for that was what the new weapons were—exploded, crippling two B-17s. During the second Schweinfurt raid, the Germans' technique was further refined: first came a wave of single-engine fighters attacking head-on with cannons; then a wave of twin-engine fighters firing rockets; then the single-engine fighters, having refueled in the meanwhile, attacking again.

As early as April 1942, Spaatz had acknowledged that it might be a good idea to outfit fighters with jettisonable extra fuel tanks so they could provide escort on long-range missions. But he had been more keen to try out another idea: strip a B-17 of its bomb load, add extra guns and ammunition and a ton of armor plating, and place one of these "battle cruisers" in each combat box to beef up its defensive firepower. Twenty-five of these strange hybrids, designated the YB-40, were built. When they finally arrived in England in May 1943, they proved a flop in combat, in part because they were simply too heavy to keep up with their formations.

Had drop tanks for fighters been made a top priority, they could easily have been manufactured and delivered back in 1942. There was not much to them. But the program proceeded at a lackadaisical pace. Eaker—who in June 1943 called a proposal to study the possible use of the P-51 Mustang as a long-range escort "premature," and as late as October 1943 thought that, with a big enough formation, escorts wouldn't be needed at all—assigned the project only a number-four priority. The British and American production authorities kept bouncing responsibility back and forth. The Americans thought it would be faster and easier to have the tanks made in England; the British were sure it would be better to have them made in the United States. When drop tanks for the Eighth Air Force's P-47s finally began to trickle in during the summer of 1943, Eaker still hesitated to make aggressive use of fighter escorts. Even with its extra fuel tanks, the P-47 did not have the range to escort missions deep into Germany, including Schweinfurt. Only with the arrival of the P-51 at the end of 1943 would it be possible for fighters to accompany bombers all the way to their targets and back. And only with Arnold's decision to remove Eaker as commander of the Eighth Air Force in December would there be the determination to use this new tool, which by now all of the bomber commanders, but Eaker, had finally recognized as essential. The commanders who had for years clung to the belief in the self-supporting bomber were now the fighter escort's chief advocates.

The shock that even the P-47s had given the Luftwaffe when they appeared over German territory had been immense. General Adolf Galland, commander of German fighter forces, went to see both Hitler and Göring personally to inform them of the development. Göring declared that he did not believe it. ("What's the idea of telling the Führer that American fighters have penetrated the territory of the Reich?" Göring angrily demanded of Galland. "I herewith give you an official order that they weren't there!") Accepting the truth was difficult for the simple and painful reason that— as Göring later acknowledged when interrogated by Spaatz and other American air commanders immediately following the German surrender—the arrival of the long-range escort fighter could only mean one thing: the beginning of the end of the Luftwaffe.

P-51D Mustang
escort fighter
engine: 1,695-hp V-12
top speed: 437 mph
armament: six .50-in.
    machine guns plus 10
    rockets

At first the escorting fighters were ordered to stick close to their charges. But slowly it dawned on some of the American strategic commanders that they were seeing the problem backward. If the primary mission was to defeat the German air force, then bombing raids were less about bombing than about provoking the German fighters into aerial combat. It was the bombers that were playing the supporting role in this final showdown between the Allied air forces and the Luftwaffe. "The first duty of the Eighth Air Force fighters is to bring the bombers back alive," read a sign that Jimmy Doolittle, the new commander of the Eighth, saw on the wall of his fighter commander's office in January 1944. He ordered him to take it down. A new sign took its place: "The first duty of the Eighth Air Force fighters is to destroy German fighters."

It might even more candidly have read "to kill German fighter pilots." The next five months brought to a climax the war of attrition—and attrition of trained fliers in particular—that had begun with the Battle of Britain. Doolittle ordered his fighters to begin flying much more aggressively, breaking away from the bombers and chasing the enemy whenever he showed himself. In January, aerial fighting wiped out 30 percent of the entire German fighter force. In February, another 30 percent. In March, 56 percent. The Luftwaffe's combat squadrons in Europe had begun 1944 with only 2,395 single-engine fighter pilots, only 1,495 of them fully combat ready. By June, 2,262 fighter pilots had been killed or put out of action by injury, an attrition rate of close to 100 percent. The Luftwaffe had begun the war with pilots who had flown close to 250 hours in training, versus about 200 for British pilots. By summer 1944 British and American pilots were getting about 350 flying hours in training while the Germans were down to 110. This was a war of attrition the Luftwaffe could not win. Machines could be replaced; men could not.

American bomber crews complained about being little more than "fighter bait" in these operations, but in truth they were for once accomplishing something. The effort to destroy the Luftwaffe on the ground by bombing airframe

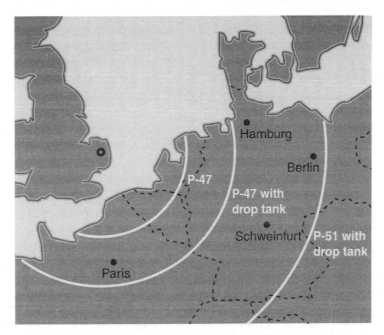

Approximate range of fighter escorts from their bases in England. The
arrival of long-range P-51s at the end of 1943 proved a decisive
turning point in the war of attrition against the Luftwaffe.

factories had by comparison largely proved another dud. From January to mid-
April 1944, the American strategic air forces dropped 27,000 tons of bombs on
every known aircraft plant in Germany. In the fall of 1944, German aircraft production reached an all-time peak. As had happened following the attacks on the
Schweinfurt plants, the Germans had responded by dispersing production, many
of the machine tools having survived unscathed throughout the bombing. The
attacks merely postponed the Germans' planned increases in aircraft production:
Speer's goal of building two thousand fighters a month by February 1944 was
not reached until June. When interrogated after the war, both Galland and Milch
"expressed the opinion that the attacks on airframe factories were not of decisive
importance." Once again, the American industrial experts had underestimated
the redundancy in the German economy and its ability to improvise and adapt.
What had looked on paper like a choke point was at best a minor pressure point.

Germany's oil industry had originally been number three on the list of priorities drawn up by the Committee of Operations Analysts. But in June 1944 a
revised priority list from the committee moved it ahead of ball bearings. Oil was
a more promising target than ball bearings on several scores. For one thing,
refineries were large and complex installations that could not be picked up and
moved elsewhere very easily. For another, fuel was one thing armies definitely
needed to keep running: There was no substitute for it. And Germany, like Japan,

had always depended heavily on imports; before the war, 60 percent of Germany's oil came from countries outside of Europe.

Almost immediately, the attacks on oil began to yield huge dividends. Refineries, oil depots, and the vast hydrogenation and Fischer-Tropsch plants that converted coal to synthetic oil were all targeted. Most oil installations were big targets; moreover, American daylight visual-bombing accuracy had been slowly and steadily improving, and by the summer of 1944 the Eighth Air Force was managing to drop 70 percent of its bombs within 2,000 feet of the aim point, 40 percent within 1,000 feet. On July 20 a message sent to Tokyo by the Japanese ambassador in Berlin, Baron Hiroshi Oshima, was decrypted by Allied code breakers. Oshima reported that the attacks on synthetic-fuel plants, which were by this point supplying more than half of Germany's demand for finished oil products, were "a source of very great concern" to the German authorities, so much so that "the question arises whether large-scale operations by the German Army may not be affected." On August 18 another message from Oshima was broken; in this he reported on a conversation he had just had with Speer, who told him that, with its concentration on oil targets, Allied bombing was "for the first time" in a position to "deal a fatal blow to Germany." Three days later, a signal from the chief of Japan's naval mission in Berlin reported that production had dropped 50 percent as a direct result of recent attacks and that Germany's stocks of oil would run out in six months.

The impact on aviation gasoline was especially acute. By the end of the year, production would fall to almost zero, with reserve stocks down to less than a month's supply. During the last German counteroffensive in the West, the Battle of the Bulge in December, plans called for German motorized units to capture Allied fuel stocks in order to keep running. In the east, the Germans assembled

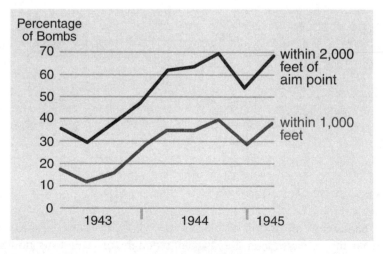

Bombing accuracy attained by the U.S. Eighth Air Force against Axis targets.

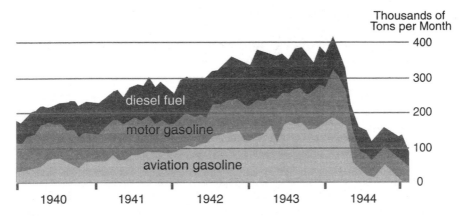

German production of transportation fuels: The belated shift to the oil target by Allied bombers in summer 1944 had an immediate and dramatic effect.

1,200 tanks to make a last stand against the Russians at the Vistula; when they ran out of gas, they were immobilized and overrun.

The British radio-guidance system called Oboe had meanwhile made it possible to hit something smaller than a city in night bombing, and in September 1944 Harris had been ordered to join the American effort, making oil his number-one priority as well. A glance at the target statistics compiled after the war shows that Harris ignored the order. From mid-September to the end of the year, his bombers dropped 24,000 tons of bombs, 11 percent of their total, on oil plants versus 102,000 tons, or 48 percent, on towns. The proportion directed at the oil target was virtually unchanged from the previous period. When challenged, Harris blamed the weather and German defenses, but Air Chief Marshal Charles Portal, the Chief of Air Staff, saw that these were just excuses; it was nothing but "the magnetism of the remaining German cities," he said, that was drawing Harris away from his orders.

The inclusive vagueness of the Casablanca directive had at last come back to haunt the air commanders, for Harris was able to cite it to justify his actions: "I have always held and still maintain," he asserted in March 1945, "that my derectif which was 'the progressive destruction and dislocation of the German military, industrial and economic systems' could be carried out only by the elimination of German industrial cities and not merely by attacks on individual factories, however important they might be in themselves."

In the end, only 12 percent of the entire two million tons of bombs dropped by Allied strategic bombers in Europe during the war were directed against oil targets, and almost all of this from June 1944 on. If it was painful to contemplate what might have been had the entire might of Allied air power been directed against Germany's oil supply from the start, most of the victorious British and American air leaders chose not to contemplate it. As far as they were concerned, air power had been vindicated as an institution and that was all that mattered.

"We won the war," Spaatz declared. And as for the postmortem of the Strategic Bombing Survey, he said he was "never that interested enough to read it." In one sense he had a point. Aside from destroying the German oil supplies, the other real accomplishments of the Allied strategic-bombing campaign had little to do with either the choice of targets or how well they were hit. Protecting against the bombers had forced Germany to divert two million soldiers and civilians to ground antiaircraft defense, along with 30 percent of the heavy guns, 20 percent of the ammunition, 50 percent of the electronics, and a third of the optical instruments produced by German industry. This was materiel and manpower that could have been going to the front lines instead. The decisive showdown against the Luftwaffe in the spring of 1944 had also demonstrated the relative unimportance of what it was the bombers managed to hit: what mattered was provoking the fighters to respond and get shot down. In retrospect, the 600,000 German civilians killed in the process seemed almost incidental to what the air war had been about.

Strategic bombing had turned out to be largely a war of attrition, and American resources and manpower had won that war—though at prodigious cost to victor and vanquished. At its peak in July 1944, the AAF had 2,400,000 men, 31 percent of the entire U.S. Army. The American strategic air forces in Europe lost 8,237 bombers in the course of the war; total American bomber-crew casualties in the combat in Europe were 73,000, including 29,000 killed. That was nearly twice the number of American soldiers killed in all of the Normandy campaign, more than the entire number of U.S. Marines killed in all of the brutal fighting in the Pacific. RAF Bomber Command lost 8,325 bombers with 64,000 aircrew casualties, including 47,000 killed.

It was "Bomber" Harris who, back in 1937, had made the observation that air warfare was infinitely preferable to "morons volunteering to get hung in the wire and shot in the stomach in the mud of Flanders." One of the great ironies of the Second World War was that the bombers had delivered no escape from the horrors epitomized by the trenches of the First World War—not even for the soldiers who flew them, much less for the civilians on the ground below.

The first Allied vehicles to enter Paris on the morning of August 25, 1944, were five tanks of Charles de Gaulle's Free French forces, who had been given the honor of leading the liberation of their capital city from four years of German occupation. The sixth vehicle was an American jeep. Dodging sniper fire from the German rear guard, the jeep cut down backstreets and eventually found its way to rue Pierre Curie, home to France's famed Radium Institute.

The Americans were on a scientific-intelligence mission, code name Alsos, of pressing importance. The chief scientist of the mission was Samuel Goudsmit, a Dutch theoretical physicist who had been working at MIT on radar research.

Summoned to Washington in the fall of 1943, he was interviewed by a screening committee, which explained that his job would be to follow immediately in the wake of American troops after the Allied landing and collect whatever information he could on German scientific developments.

Goudsmit naturally assumed this meant he was to concentrate on German radar. He was astonished when, immediately after the interview, an Army major pulled him aside and said, "You understand, of course, that what you really are going to do is look into the atomic bomb development."

In December 1938, two German physicists, Otto Hahn and Fritz Strassman, had published in the open scientific literature their discovery that the uranium atom, when struck by a neutron, would split in two, releasing prodigious quantities of energy in the process. The implications were not lost on any of the world's physicists. "Within a week," recalled Philip Morrison, a student of J. Robert Oppenheimer's at the University of California at Berkeley, "there was on the blackboard in Robert Oppenheimer's office a drawing—a very bad, an execrable drawing—of a bomb."

The big *if* was whether, in the process of splitting, the uranium atom would release neutrons that could in turn cause further uranium atoms to split, thereby initiating a self-sustaining chain reaction. Within a few months it looked as though the answer was yes, and simultaneously in Germany, Britain, and the United States the lid was slammed down on further open publication.

But still there were many uncertainties. The American and British physicists had at first thought that perhaps tons of uranium would be needed to sustain a chain reaction, making the idea of a weapon impractical. But by February 1940, experiments had succeeded in extracting tiny amounts of the more reactive of the uranium isotopes, U-235, which made up only about 1 percent of natural uranium. Theoretical calculations now revealed that only a few kilograms of U-235 might be enough to make a bomb.

This critical discovery was made by Otto Frisch, a German émigré in England; he quickly calculated, too, that while the costs of extracting enough U-235 would be huge, in principle it could be done by cascading thousands of separation units of a kind that had already been shown to work in the laboratory. There seemed little doubt now that an atomic bomb could be made—a bomb, Frisch calculated, that would be equivalent in its explosive force to several thousand tons of TNT.

"I have often been asked," Frisch wrote many years later, "why I didn't abandon the project there and then, saying nothing to anybody. Why start on a project which, if it were successful, would end with the production of a weapon of unparalleled violence, a weapon of mass destruction such as the world had never seen? The answer was very simple. We were at war, and the idea was reasonably obvious; very probably some German scientists had had the same idea and were working on it." Finding out just how far the Germans had progressed was Goudsmit's job.

The American-British bomb project, now directed by Oppenheimer and an Army Corps of Engineers brigadier general, Leslie R. Groves, had proceeded on a gargantuan scale. A complex of laboratories, an entire secret city, arose on a secluded and tightly guarded mountaintop in Los Alamos, New Mexico. Hundreds of millions of dollars in construction contracts, and orders for staggering amounts of raw materials and machinery, were approved on Groves's say-so. When a shortage of copper wire threatened to delay production of the tons of electromagnets needed for the huge U-235 separation plant going up in the backwoods of eastern Tennessee, Groves sent his assistant to see if the Treasury Department could lend silver from its reserves as a substitute. Groves's aide asked for ten thousand tons. "Colonel," came the icy reply, "in the Treasury we do not speak of tons of silver; our unit is the troy ounce." The Treasury ended up turning over 395,000,000 troy ounces—13,540 tons, about 200 million dollars' worth.

There was no reason to discount the possibility that Germany might be at work on a comparable scale. In 1941, the Norwegian resistance reported that the Germans were ordering vast quantities of heavy water from a plant in Norway, and the only conceivable use of heavy water in large quantities was to act as a moderator in a nuclear reactor designed to transform natural uranium into plutonium, a second route to the production of bomb-grade fissionable material. Despite the escape to the West of many Jewish physicists from German universities, Germany still claimed many of the brightest stars in the world of nuclear physics, chief among them the great theoretician Werner Heisenberg, director of the Kaiser Wilhelm Institute for Physics.

The specter of a German bomb program hung over all of the American physicists as they struggled to solve the myriad interconnected problems of distilling the primal power of the universe into a weapon of warfare. Hitler's boasts of "secret weapons" had mostly proved empty so far. The atomic bomb in Hitler's hands would be no empty boast. To those indoctrinated in the secrets of atomic fission, it was a terrifying thought.

In Paris, Goudsmit at first found few leads. But a visit to a French optical firm revealed that one well-known German physicist was now a professor at the University of Strasbourg. Other records showed that the university had ordered new equipment for nuclear research. In November Patton's troops reached the city. The German physicist and his laboratory staff were found occupying a wing of the university hospital, trying to pass as medics. Taken into custody, they would tell Goudsmit nothing. But a further search at the university turned up a trove of interesting-looking papers, and Goudsmit and a colleague carried them back to their makeshift quarters to study them and prepare translations.

"It was a rough evening," Goudsmit recalled:

> The Germans were shelling the city from across the river; our guns were answering. Air raids and air battles raged overhead. We had no

light but a few candles and a compressed gas lamp. In the center of the room, our soldiers were playing cards. Fred and I sat in a corner on easy chairs and began to scan the German files.

We both let out a yell at the same moment, for we had both found papers that suddenly raised the curtain of secrecy.

Their shout had sent the soldiers grabbing for their rifles, but what they had found—including copies of letters to Heisenberg that discussed German efforts to separate U-235—had fully justified an outcry. "The conclusions were unmistakable," Goudsmit said. "Germany had no atom bomb and was not likely to have one in a reasonable time." German efforts to separate U-235 had been a failure. Their primitive nuclear reactor had yet to produce a sustained chain reaction. "In short," concluded Goudsmit, "they were about as far as we were in 1940."

After the war, Heisenberg would insist that the German bomb program had been deliberately sidetracked by the German scientists, who refused on principle to develop such a terrible weapon. Otto Hahn, whose original discovery of fission had made the bomb possible, would have none of that. Hahn, Heisenberg, and eight other German physicists captured by the Allies were being held in a farmhouse outside Cambridge, England, when news of the atomic bombing of Japan came. British intelligence secretly recorded their conversation. Carl von Weizsäcker, a leading German theoretician who had worked at Strasbourg, seconded Heisenberg's claim that the German bomb program had failed because the scientists had deliberately dragged their feet.

"If we had all wanted Germany to win the war," von Weizsäcker announced, "we would have succeeded."

"I don't believe that," Hahn retorted.

Hahn's explanation for why Germany had lost the race to build the bomb was much simpler: "If the Americans have a uranium bomb," he told Heisenberg at one point, "then you're all second-raters. Poor old Heisenberg . . . you may as well pack it up."

There was never any fear that Japan could build a bomb. And so with Goudsmit's discoveries in a candlelit tent in Strasbourg, the nightmare prospect of facing an atomic-armed enemy was dispelled.

But the Manhattan Project was by now a force unto itself. A few of the project scientists tried to raise the question of whether it was justifiable to use the weapon on Japan—or wise even to test it, thus letting the world know of its existence. Leo Szilard, a Hungarian-born theoretical physicist, one of the original "men from Mars," had been the main driving force in launching the Manhattan Project; in 1939 he had approached Albert Einstein to have him write to President Roosevelt warning of the possible danger of a German atomic bomb and urging that American research on atomic fission be accelerated. Now, as the project neared its momentous conclusion, Szilard went to Einstein again and got a letter of introduction to the President. After several false starts, the White House finally sent him to see Secretary of State Jimmy Byrnes.

The meeting was a disaster. Szilard had thought deeply about the possible implications of atomic weapons at a time when most people could not even have conceived of their existence; it was easy to dismiss him as a visionary. Byrnes was immediately irritated, too, by Szilard's insistence that only scientists "actively engaged" in the atomic-bomb project were qualified to assess the future role atomic weapons would play in the world. The chief point Szilard tried to make to the Secretary of State was that, although the United States for the moment had a monopoly on atomic weapons, this would surely not last once their existence was revealed. Nuclear weapons in the hands of other nations would neutralize the great military advantage that the United States now commanded through her vast industrial might. The "most immediate danger" of using or testing the bomb, Szilard told Byrnes, was that it "will precipitate a race in the production of these devices between the United States and Russia."

Byrnes was an old politician. He proceeded to give Szilard a lesson in reality. "He said," Szilard recalled, "we had spent two billion dollars on developing the bomb, and Congress would want to know what we had got for the money" if the bomb were not used. And so, as Byrnes's assistant Benjamin Cohen would later put it, the handling of the atomic bomb became "somewhat a Greek tragedy." The bomb would be built and tested, and "short of complete and incontrovertible evidence that the Japanese were through," it would be used on Japan, simply because it had become impossible for anything else to happen.

The Los Alamos scientists had designed two types of bombs. The more complex version had at its center two small hemispheres of plutonium, a mere fourteen pounds in all, surrounded by a layer of TNT molded into an intricate geometric pattern. Its shape was the result of months of exhaustive theoretical calculations—and about twenty thousand experimental detonations that had echoed through the canyons surrounding the mountaintop laboratory. The idea was to focus the explosive force of the TNT, much as a lens focuses light, so that the resulting shock wave formed an inward-traveling sphere that would squeeze the plutonium core with the same force in every direction. When the compressed plutonium reached the critical density at which enough fission-produced neutrons would strike other plutonium atoms before escaping, the self-sustaining chain reaction would be triggered and an atomic detonation would occur.

The other bomb design was much simpler: a "bullet" of uranium would be shot by an explosive charge into a larger "target" chunk of uranium to form the critical mass. This "gun-assembly" weapon was almost certain to work, and the scientists decided no test of it was necessary. The plutonium "implosion" bomb was far less certain, and by the summer of 1945 the Los Alamos scientists were sweating over a thousand technical details as they prepared a make-or-break test of the device. Contractors and a contingent of soldiers erected two towers at a desolate spot in the New Mexico desert; it was so far from anywhere that the

Army unit stationed there won an award for the lowest VD rate in the entire U.S. Army. As a rehearsal for their test procedures, the scientists stacked a hundred tons of high explosive on one of the towers and blew it up. On July 14, a Saturday, the "gadget," as it was called with extravagant understatement, was hoisted into position on the remaining tower. Two days later, just before dawn, the final arming switches were thrown. A B-29 came over at 30,000 feet to simulate a bomb run; the air force wanted to know what the effect of an atomic blast would be on the plane that would drop the real thing.

The ball of fire came first: silent, the brightest light anyone had ever seen. At the shelter five miles away where the project's top scientists were watching, it took half a minute for the sound of the explosion to arrive. Then came the wind. A hundred thousand photographs would record the explosion in exhaustive detail, along with a phalanx of other instruments. But Enrico Fermi, the Nobel Prize winner who had fled fascist Italy in 1938, had prepared his own homely test, dropping pieces of paper from six feet off the ground and measuring how far they were blown by the wind to estimate the bomb's explosive force. He put it at ten thousand tons of TNT. Later measurements would pin it down at 18.6 kilotons, about four times what the builders had predicted.

Kenneth Bainbridge, the experimental physicist in charge of the test, congratulated everyone. Then he said to Oppenheimer, "Now we are all sons of bitches."

Four hours later, the gun-assembly bomb, dubbed Little Boy in its completed form, was on its way out of San Francisco Bay aboard the cruiser *Indianapolis,* headed for the AAF's B-29 base at Tinian in the South Pacific. Ten days later, two C-54 cargo planes and three B-29 bombers followed, carrying the components of an implosion bomb, Fat Man.

It would prove convenient for decades afterward, convenient to both the defeated Japanese and the victorious Americans, to adhere to the story that Emperor Hirohito had been a remote, passive figurehead who bore no responsibility and no guilt for Japan's military actions during the war. The truth was different. As his biographer Herbert P. Bix found from an exhaustive examination of Japanese documents that became available only after Hirohito's death in 1989, the Emperor asserted strong control over both overall war policy and specific military operations—often with disastrous consequences for Japanese troops and Japanese civilians alike. Far from agonizing over his people's mounting suffering as American forces pressed relentlessly on toward the Japanese homeland, far from advocating surrender once Japan's defeat had become a near certainty in the fall of 1944, the Emperor exhorted his commanders again and again to fight to the death. He saw his people's willingness to sacrifice their lives in the tens or hundreds of thousands, millions even, as proof of what he called "our imperial destiny." As far as the Emperor was concerned, dying for the Emperor was the greatest virtue that his subjects could aspire to.

Many of them agreed. On New Year's Day 1945, the Emperor inspected the special ration packets being prepared as the last meals for the "special attack forces," young men who had volunteered to fly their gasoline- and explosive-laden planes into Allied ships and troops and bombers. There was no shortage of enthusiastic volunteers. The name given to the suicide forces, *kamikaze,* meant "divine wind," an allusion to the miraculous typhoons that, in Japanese legend, twice in the thirteenth century arose to repel the invading Mongol fleet. Suicide attacks were a weapon the American, British, and Australian troops fighting the Japanese could not understand. To Hirohito they were the supreme manifestation of the superiority of the Japanese spirit of sacrifice and honor. During a briefing on the battles raging near the Philippines in early January, the Emperor astonished his military aide by twice rising and bowing deeply when his aide mentioned one of these "special pilots."

The kamikaze attacks were deeply unnerving to the Allies; they were also extremely effective. During the invasion of Okinawa, nearly two thousand kamikazes struck the Allied fleet. They managed to get past the radar pickets, the hornet's nests of fighters mounting standing air patrols, and the antiaircraft guns enough times to sink thirty-six American ships and damage hundreds of others, including dozens of battleships and aircraft carriers. Five thousand American sailors were killed in the attacks. Most of the kamikazes piloted conventional dive-bombers or light bombers, but several hundred specially built suicide planes were also unleashed. Little more than a huge bomb with small wings and a cockpit welded on top, the Ohka ("Cherry Blossom") was dropped from a bomber and then propelled into its target by three small solid-rocket motors. The Ohkas could reach 575 miles per hour in a dive, and some were used to attack American bomber formations with an unanswerable effect. The constant fear of such attacks left bomber crews shaken and jumpy.

Blaming his ministers for the continuing military failures, Hirohito brought down the Japanese government in April. A new cabinet agreed to put out feelers to the Russians for mediation to help end the war. But the new cabinet's vision of "peace" was so far from the Allied terms of unconditional surrender as to border on the delusional. The new Army Minister, Korechika Anami, insisted that since Japan was still holding territory it had conquered, Japan had therefore won the war and any peace treaty must acknowledge that fact. "Concurrently," Bix notes, "the controlled press waged a daily die-for-the-emperor campaign."

The articles in the popular American press that had been urging the United States to give Japan "everything it had" invariably called attention to the flammability of Japanese cities. In February 1944, the AAF built four "Little Tokyos" at Eglin Field in Florida to test what effect incendiary bombs would have on the densely packed wood-and-paper houses that typified Japanese urban construction. The results were encouraging, particularly when a 100 percent load of incendiaries was employed. Adding high explosives in the mix tended to reduce the effect; the houses were flattened by the blast and did not burn nearly as well as when they were left standing.

Haywood Hansell, who now commanded American heavy bombers in the Far East, was a dedicated believer in precision bombing. He also had a new tool, the first "very heavy" bomber, the B-29. The "Superfortress" was twice as big as the B-17 in almost every way measurable: twice the weight, twice the range, twice the bomb load, twice the engine horsepower, and twice the cost—a staggering $600,000 apiece. The test model had not yet flown when Arnold in September 1942 approved the first order for 1,644 of them. Thousands more would be ordered before their first combat mission. The "three-billion-dollar gamble," they called it.

The B-29 had a sleek, shining metallic, streamlined exterior, none of the bumps and bulges of the drab-painted B-17. It had other advanced features: a pressurized cabin that freed the crew from oxygen masks, and remotely controlled gun turrets. Originally justified as a way to hit Germany from bases in Egypt or North Africa when bad weather hindered operations from Britain, the B-29 inevitably came under the covetous eyes of commanders in the Far East. With the B-29 the Japanese mainland could be hit from bases in China and, once American troops had secured them, the Mariana Islands. Roosevelt was convinced that using bombers

B-29
heavy bomber
engines: 2,200-hp
    radial (4)
top speed: 357 mph
armament: 20,000-lb.
    bomb load; ten to
    twelve .50-in. machine
    guns, 20mm cannon

against Japan would offer a great morale boost to Chiang Kai-shek's attempts, thus far unsuccessful, to unify the Chinese under his leadership and fight off the Japanese on the ground.

By the time the bomber offensive in the East finally got under way, the political expectations that were riding on it allowed no time for patience with Hansell's dedication to precision bombing of Japanese industry. Adding to the pressures, the B-29 was beset with bugs that caused about 20 percent of missions to be aborted, notably when its 2,200-horsepower air-cooled radial engines kept catching fire. Tailwinds over the target areas pushed ground speeds close to 600 miles per hour, beyond what the Norden bombsights were calibrated for. At the beginning of 1945, Arnold relieved Hansell of his command and brought in a man with a reputation as a charger, Major General Curtis LeMay. He took command of XXI Bomber Command on January 20.

While commanding his B-17 group in England, LeMay had invented the combat-box formation, which he now brought to the Far East along with the ruthless perfectionism for which he was both hated and respected. A self-described "tough guy," LeMay had frequently led dangerous missions himself.

Leaving nothing to chance, he immersed himself in technical details and insisted that the survival and effectiveness of his aircrews depended on high standards and harsh methods. Arnold had made it clear that if LeMay failed to get results with the B-29s in Japan, he would be fired. Failure was the one thing LeMay never tolerated, in himself or anyone else. The oldest of seven children, LeMay had grown up in ignominious poverty that produced in him an "angry stimulation" to succeed. To pay his way through college, he had worked from 5:00 p.m. to 3:00 a.m. six nights a week at a steel foundry. LeMay suffered from Bell's palsy, a nerve paralysis that made his face droop, and he took to keeping a cigar clenched in his teeth to hold his facial muscles in place. It only made him look more grimly determined.

LeMay pulled gun turrets, armor shielding, ammunition, and gunners out of the B-29s to increase their bomb load. Then he started pulling fuel out, too, until he determined the minimum the planes needed to get there and back. Each pound of fuel less meant one pound of bombs more.

At first, precision attacks on Japanese aircraft plants remained LeMay's official top priority. By March, LeMay's results looked not much better than Hansell's. "This outfit has been getting a hell of a lot of publicity," LeMay grumbled to his public-relations officer on March 6, "without having really accomplished a hell of a lot in bombing." This was about to change. Three nights later, LeMay sent 334 B-29s at altitudes as low as 5,000 feet over the Japanese capital. The weather was ideal for firebombing: a high wind, dry, clear. A pathfinder squadron marked the target area for each group with 70-pound bombs filled with napalm, jellied gasoline that instantly started large blazes. The other planes followed with hundreds of thousands of 6-pound oil bombs. The entire 1,667-ton bomb load of the attacking force was incendiaries. Crews were briefed to drop bombs individually within their assigned target areas in spots where fires were not already burning in order to spread the blaze as far as possible.

As in Dresden, the AAF adamantly denied that it had abandoned its policy of precision bombing. LeMay publicly insisted that the Tokyo raid had aimed only at industrial targets. After the war he was less circumspect. Although LeMay said the aim had not been "slaughtering civilians for the mere sake of slaughtering," he also acknowledged the slaughter for what it was. "We knew we were going to kill a lot of women and kids when we burned that town," LeMay said. "Had to be done." On another occasion he remarked: "Killing Japanese didn't bother me very much at that time. It was getting the war over that bothered me. So I wasn't worried particularly about how many people we killed getting the job done." By the time the Tokyo fires were extinguished, 15.8 square miles of the city were gone, probably 100,000 civilians were dead.

Five months later, on August 6, the gun-assembly uranium bomb, Little Boy, was dropped on Hiroshima. It destroyed 4.7 square miles and killed probably 70,000 to 80,000 outright. The second atomic bomb, Fat Man, dropped on

Nagasaki three days later, destroyed 1.45 square miles and killed an estimated 35,000 in the initial blast.

To Americans, to the world, the atomic bomb was a weapon so powerful that no sane person could believe further resistance was possible. Yet the atomic bombs killed considerably fewer people than had LeMay's months of conventional firebombing. The new weapons had fallen on two provincial cities far from Tokyo; indeed, the principal reason Hiroshima and Nagasaki were chosen as targets was that almost all of Japan's other major cities had been burned to the ground already. Making charred debris bounce, it was thought, would constitute an inadequate demonstration of the new weapon's awesome destructive power to the Japanese people. But the end result was rather the same, since most Japanese had not even heard of the atomic bombings by the time of the Japanese surrender, much less were they pressing their government to capitulate because of them. If the Emperor had told them to fight on, the Japanese people would no doubt have fought on.

The Japanese Army certainly still wanted to fight on. A cabinet meeting summoned to discuss the Hiroshima bombing the following day was canceled because the Japanese Army representatives didn't show up: they claimed to have "more pressing business." Hirohito himself vacillated and procrastinated on issuing a surrender statement to the very end, even after the second bomb was dropped on August 9. "Obviously," noted Bix, "Hirohito sought to justify his decision to surrender by citing the dropping of the atomic bombs. . . . Whether the emperor and his advisers ever really believed that, however, is unlikely."

Years later, George Marshall would be one of the first to recognize what had happened: "There is one point that was missed and that frankly we missed in making our plans," Marshall admitted. "And that was the effect the bomb would have in so shocking the Japanese that they could surrender without losing face." At least since spring, and probably for a full year, Japan's leaders had known that, militarily, their situation was hopeless. Douhet, Mitchell, and the Tactical School theorists had never conceived of the possibility that an enemy might continue to fight on in a militarily hopeless situation, its armies facing defeat after defeat, its cities laid waste night after night. If destroying an entire major city from the air in a single night were enough to make a country surrender, Japan would by now have surrendered many times over. The timing of the Japanese surrender in the wake of the atomic bombings said more about the thralldom of Japan, and its leaders in particular, to a nationalistic ideology of honor and supreme self-sacrifice than it said about the logic of victory through air power.

Indeed, to the extent that LeMay's B-29s contributed to the ultimate victory, their spectacular leveling of Japanese cities was probably a lesser factor than another mission that to this day remains "one of the least-known B-29 operations and yet arguably the most effective," in the words of the aviation historian Kenneth P. Werrell. After repeated urgings from Admiral Nimitz and the Committee of Operations Analysts, Arnold reluctantly agreed to allow B-29s to be used to drop naval mines in the waters off the Japanese homeland beginning in April 1945. LeMay didn't like the idea, either—not only for the usual airman's reason that it was a wasteful diversion from the real mission of strategic bombing, but more particularly because he thought mine-laying sorties were bad for morale: the operations were classified top secret and so his crews could not get publicity for them, nor could they get the immediate satisfaction of witnessing the "results" of their missions. Mine-laying never amounted to more than a small fraction of B-29 sorties. Yet the results were, as Nimitz later exulted, "phenomenal." In the last five months of the war, 800,000 tons of Japanese merchant shipping, nearly 300 vessels, were sunk or crippled by air-dropped mines, accounting for some 60 percent of all Japanese shipping losses for the period. All of the ports opening on the Pacific were shut down. Traffic through the Shimonoseki Strait, the critical waterway between Honshu and Kyushu, was cut in half. Japan depended upon imports carried by merchant ships for 80 percent of its oil and 90 percent of its iron ore, and now those imports had nearly ground to a halt. A ton of mines secretly dropped at sea probably did infinitely more to harm the Japanese war effort than a ton of incendiaries dropped with far more visible effect on Japan's population centers.

To the American air leaders in August 1945, however, any such subtleties in explaining the Japanese surrender would have seemed absurd, laughable. On August 10, LeMay and Spaatz were widely quoted as stating that the air-delivered atomic bomb had made armies unnecessary in the future; Marshall sent them a curt cable rapping their knuckles; Spaatz said he had been misquoted. But four days later, Spaatz sent public-relations guidance to AAF headquarters that brimmed with the same Air Force triumphalism: "It can be said that air power emerged from its trial completely vindicated and mature in stature. It was the B-29 Superfortress which effectively used this revolutionary weapon in a decisive manner. No better instrument has been found to utilize the military might inherent in the atomic bomb." Then Spaatz added a word of discreet advice: "It would be inconsistent with A.A.F. dignity and restraint to make these statements boldly and brazenly. In an overall strategic story the implication can be skillfully woven into the piece so no reader can fail to draw the unmistakable inference that air power was the outstanding factor in our victory."

Robert Lovett, the Assistant Secretary of War for Air, wholeheartedly agreed. "The decisive factor in the sudden collapse seems to have been air power," he wrote Spaatz, "and the old myth of not being able to defeat a major country with its armies intact except by invasion seems to have been shattered once and for all."

But the air leaders were worried that people might see the atomic bomb as a special case, separate from air-power theory and doctrine. They wanted to be sure that the atomic bombings were understood to be a vindication of those theories, not a substitute for them. "While I am naturally feeling very good about peace being effected with Japan," Arnold confided to Spaatz, "as far as the Army Air Forces are concerned it is, shall I say, unfortunate that we were never able to launch the full power of our bombing attack with the B-29s. The power of these attacks would certainly have convinced any doubting Thomases as to the capabilities of a modern Air Force." As the Strategic Bombing Survey pointed out, a conventional raid of 210 B-29s would have produced the same effect as the Hiroshima bomb. The "inference" Spaatz wanted people to be sure to draw was that since the atomic bombing of Hiroshima had forced the Japanese surrender, conventional B-29 attacks could have done so, too. The fact that they had manifestly failed to do so for month after month does not seem to have posed any logical problem to him.

Another public-relations gambit poignantly reflected the air force's determination to make the outcome of the war fit America's prewar theories of air power. The day after the Hiroshima bomb fell, AAF headquarters in Washington cabled Spaatz. "It is understood that the Secretary of War in his press conference tomorrow will release a map or photostat of Hiroshima showing the aiming point and the general area of damage," the cable stated. "It is believed here that the accuracy with which this bomb was placed may counter a thought that the CENTERBOARD project [the atomic bombings] involves wanton, indiscriminate bombing."

Against all odds, American faith in the virtues of the doctrine of precision aerial bombardment had survived the dawn of the nuclear age.

PART FOUR

OMAHA TO BAGHDAD, 1946-2003

12

STRATEGIC AIR COMMAND

If the reaction of Americans to the Japanese surrender could have been reduced to a single word, it was relief rather than exultation. There was, noted the historian Eric F. Goldman, something strangely perfunctory about the celebration of the crowds that poured into New York's Times Square on V-J Day. Everyone talked about "the end of the war," not "victory." The urge to just forget it all, to turn inward and homeward, back to the mundane joys and mundane worries of families and jobs and cars and lawns and fishing and golf, was everywhere. *U.S. News & World Report* greeted the peace with an issue that spoke not a word of the war or the international situation but instead merely expressed concern over what impact peace would have on the economy. Millions of young men, who had put off college and marriage for the war, were eager to get home and pick up the strands of their prematurely abandoned youth; millions of workers, men and women alike, were itching to spend some of the $100 billion that had been saved during the war as unemployment vanished, incomes doubled, and the conversion of American factory production from consumer goods to war materiel left them nothing to buy.

In Manila, Paris, and Frankfurt, GIs chanting, "We wanna go home!" rioted. The *New York Times* editorialized its dismay over the effect that this "breakdown of Army discipline" would have on foreign observers and worried over what it said about America's ability to meet her new "international obligations." Members of Congress, knowing a political tidal wave when they saw one, sided with the mutineers. Politicians fell over one another to demand that the Army speed up demobilization. "Remember," solemnly announced one congressman, "there is no place like home."

What followed was not even demobilization, President Harry S Truman later recalled: "it was disintegration." There was no stopping it. To generations of Americans, Eric Goldman said, "concern over international matters was to be confined to unfortunate periods of war. Foreign policy was something you had, like measles, and got over with as quickly as possible." That seemed to suit many Europeans just fine, too. No sooner was the war over than reports from every European country, former enemy and former ally alike, began to include a new word: anti-Americanism. "From England to Germany they have had enough of us," wrote an American Army chaplain in an article in *Christian Century* that was widely quoted. The American soldier in Europe, he said, "is not very clear in his own mind about why he fought, or what victory means. . . . There he stands in his bulging clothes, fat, overfed, lonely, a bit wistful, seeing little, understanding less—the Conqueror, with a chocolate bar in one pocket and a package of cigarets in the other. . . . The chocolate bar and the cigarets are about all that he, the Conqueror, has to give to the conquered."

Intellectuals fretted over the implications of the atomic bomb; most Americans did not. Popular articles treated the bomb more as a scientific marvel than as a pivotal component of the changing world order and America's new role in it. Just as they had in 1900 with their forecasts of technological marvels in the new century to come, magazines exuded predictions: push-button meals, personal helicopters, picture telephones, and countless other gadgets would make life easier and more amusing in the age of limitless scientific progress that lay ahead in a world where man had conquered even the atom.

Truman was the man who had unflinchingly ordered the use of the bomb against the Japanese and who had been appalled when Robert Oppenheimer later came to his office in a state of mawkish anguish, declaring that as a scientist he had blood on his hands. "The blood is on my hands," Truman snapped back. "Let me worry about that." He told his aides he never wanted to see that "cry-baby" again. Now the President replaced the toy cannon on his desk with a toy plow, ordered the Presidential Seal redesigned with the eagle's head facing the olive branch of peace and away from the arrows of war, and turned his attention, like everyone else, resolutely homeward. Putting the economy on a sound basis, Truman believed, would itself now be the most important step the United States could take for national and international security. The defense budget was slashed from 40 percent of the GNP to 4 percent. To his Secretary of Commerce, Henry A. Wallace, the President confessed one worry he did have about the newly unleashed power of the atom: the limitless bounty of atomic energy might so reduce the workweek that people would "get into mischief."

"It would be ironical," said Robert A. Taft, the unreconstructed isolationist Republican senator from Ohio who now found the nation once again in sync with his views, "if this Congress, which really has its heart set on straightening out domestic affairs, would end up being besieged by foreign problems."

There was nothing in the least ironical about Joseph Stalin's paranoia and expansionist ambitions—except to those who still fostered illusions about America's erstwhile ally, or to those who thought the world might go away if they only shut their eyes tight enough. Stalin quickly set to work setting them straight, reneging on promised elections in Poland, scorning an idealistic American proposal for international control of atomic weapons, brazenly staging a Communist coup in Czechoslovakia. In a rare public speech in Moscow in early 1946, Stalin had declared that war between Communism and capitalism was inevitable; accordingly, he said, nothing must be permitted to stand in the way of increased Soviet military production.

It was not right-wing Red-baiters who called the United States back to her duty, and definitely not right-wing isolationists such as Taft; it was liberal internationalists, scholar-diplomats such as George F. Kennan, who had studied Russia and knew her well, who now began to press for a new American policy that would face up to the threat the Soviet Union posed, and to the responsibilities the United States had now acquired in a world she had helped to free from the yoke of tyranny. From the U.S. Embassy in Moscow, Kennan sent an eight-thousand-word cable in February 1946 warning that Americans had not even begun to grasp the Kremlin's "neurotic view" of the world. Grafted to age-old roots of Russian insecurity and paranoia, Bolshevik revolutionary ideology had sent forth a tangle of new shoots but in the same stunted and twisted pattern. The Soviet leadership nurtured an almost religious faith in the impossibility of peaceful coexistence with the West. In a subsequent article in the journal *Foreign Affairs,* written under the byline "X," Kennan repeated his arguments to a wider audience and introduced a new word to the lexicon of war and diplomacy. The only alternative to the conundrum of facing World War III on the one hand or capitulation to relentless Soviet expansionism on the other, Kennan proposed, was what he called "containment"—a policy of steadfastly resisting Soviet advances throughout the world by building strong democracies that would possess the economic stability and political will to stand up to Communist expansion. Bernard Baruch, who as America's United Nations representative had developed the proposal to place atomic arms under international control, added another new term. "Let us not be deceived," Baruch declared. "Today we are in the midst of a cold war." The phrase immediately caught on.

Kennan, Baruch, and other internationalists advocating American engagement in Europe saw economic aid as the chief weapon in this new kind of war. "Global New Dealism," the Taft isolationists sneered, but Congress quickly approved Truman's emergency requests for aid to Greece and Turkey, and then the breathtaking sum of $17 billion to fund George Marshall's plan to extend aid to all of war-torn Europe.

Yet it was becoming increasingly impossible to ignore the purely military dimension of the emerging conflict. As the historian William Manchester observed with grim irony, there were no "Wanna Go Home" riots in the Red Army. The U.S. Army was down to one armored division; Russia had thirty. Millions of Soviet troops continued to occupy Eastern Europe. Even short of a World War III, such a preponderance of force spoke loudly. On June 24, 1948,

Stalin exercised that voice with a direct challenge to the West: Soviet troops cut off Berlin.

Truman four days later ordered a massive airlift to keep Berlin supplied. It would become, in Truman biographer David McCullough's words, "one of the most brilliant American achievements of the postwar era," though this was far from apparent at the time. It would also become one of the most imaginative applications of air power under the strange new rules of cold-war military confrontation, though this, too, was far from immediately apparent—especially to the Army Air Forces planners who, since the Japanese surrender, had turned their attention to building an independent, atomic-armed Air Force for the postwar era.

Shortly after the end of the war, Arnold had appointed Spaatz to head a board that would examine the implications of the atomic bomb for air strategy and Air Force organization. The board presented its conclusions in October 1945. Although the bomb "has not altered our basic concept of the strategic air offensive," Spaatz stated—the atomic bomb was just "an additional weapon," and no change in basic doctrine, operations, or force structure in the postwar Air Force was needed to deal with it—atomic weapons had nonetheless made the strategic air offensive a more potent, and more certain, war-winning strategy than ever. "No nation will risk war with the United States without themselves possessing this weapon or one of equal capabilities," the board confidently declared.

Arnold echoed these conclusions in his report to the Secretary of War the following month: "It must be apparent to a potential aggressor that an attack on the United States would be immediately followed by an immensely devastating air atomic attack on him. The atomic weapon thus makes offensive and defensive airpower in a state of immediate readiness a primary requisite of national survival." The atomic bomb, he said, "had made Air Power all-important."

In these and many other statements, public and private, the air leaders left no doubt that they saw the atomic air offensive as the cornerstone of postwar air strategy. Spaatz, succeeding the retiring Arnold as Army Air Forces commanding general, announced that in making his budget allocations for 1946 he had no choice but to give first priority to "the backbone of our Air Force—the long-range bomber groups and their protective long-range fighter groups organized in our Strategic Air Command."

The Strategic Air Command—SAC—had been created in early 1946 as the practical embodiment of the absolutist air-power concept that had gained swift and nearly unchallenged ascendancy in the postwar AAF. SAC's stated mission was to "conduct long range offensive missions in any part of the world." The impossibility of America's matching Soviet might in troops and tanks led even ground commanders such as Omar Bradley, chairman of the newly created Joint Chiefs of Staff, to agree that the atomic-strike mission would necessarily be the backbone not just of the Air Force but of America's entire military posture. A Joint Staff committee in March 1948 concluded:

Atomic bombs will be used by the U.S.

Strategic concept: To destroy the capacity and will of the enemy to continue hostilities. Initially to launch attacks designed to exploit the power of atomic weapons against the war-making capacity of the enemy.

The overwhelming physical and psychological power of the atomic bomb did nothing to encourage subtle reexamination of the old dogmas of strategic-bombing theory. Writing in the *Air University Review* in 1948, Lieutenant Colonel Joseph L. Dickman clearly spoke for many of his Air Force colleagues when he proclaimed that the atomic bomb had now fully vindicated Douhet's principles. In late 1947 the Air Force Directorate of Intelligence undertook a study of Soviet targets for the atomic bomb, and the result was a report grandiloquently titled "To Kill a Nation."

Explaining precisely how such strikes on enemy targets would translate into immediate victory seemed to matter less than ever. "To Kill a Nation" spoke vaguely of destroying Soviet government control, industrial capacity, and "support" by using atomic weapons against the large industrial centers clustered in and around seventy Soviet cities. But reading between the lines, it was clear that the authors thought it didn't matter very much where the bombs fell, given the awesome destructive power and psychological impact of atomic explosives. At a conference in late 1948 at which senior Air Force leaders met to receive and scrutinize the plans of all commands, the Director of Intelligence, Major General Charles P. Cabell, asserted, "If a shooting war should come, we know that atomic destruction must be delivered to the heartland of the Soviet Union. Only *there* can Russia be stopped! The Russians know this." When SAC's turn came, its representative melodramatically dimmed the lights and, in what one member of the audience would later recall as a "voice of doom," unveiled SAC's plan to deliver the entire atomic stockpile against Russian targets in a single mission of crushing psychological effect.

In words that consciously echoed the classic and magnificently vague phrases of Douhet and the Air Corps Tactical School theorists, official Air Force doctrine and plans throughout the 1950s would speak of attacking the enemy's "will to fight," its "national structure," "vital centers," the "enemy heartland." *These* were the targets for the atomic bomb. The Air Force's first official doctrine handbook issued in the early 1950s, Manual 1-2, *Basic Doctrine,* asserted simply, "No nation can long survive unlimited exploitation by enemy air forces utilizing weapons of mass destruction." Or, as Manual 1-8, *Strategic Air Operations,* explained in best Tactical School manner:

Somewhere within the structure of the hostile nation exist sensitive elements, the destruction or neutralization of which will best create the breakdown and loss of will of that nation to further resist. The fabric of modern nations is such a complete interweaving of major single elements that the elimination of one element can create widespread influence upon the whole.

Attacking these "sensitive elements" with atomic weapons, Manual 1-8 explained, would thus bring about a swift "collapse of the national structure."

It did not at all diminish the airmen's enthusiasm for the strategic atomic offensive that this mission also offered the best political argument in favor of their long-sought goal of Air Force independence, with a considerably more than equal slice of the shrinking postwar defense budget to go with it. W. Stuart Symington, the first Secretary of the Air Force, went about giving speeches in which he openly ridiculed the idea of an equal division of the defense budget among the three services. This so-called principle of "balanced forces," he said, was being advocated only by "ax-grinders dedicated to obsolete methods" of warfare.

It was all just common sense, airmen believed. "Air leaders realized that here was an opportunity to put warfare on an economical, sensible, reasonable basis," Ira Eaker insisted. "The atomic bomb made it possible to fight wars cheaply." Armies and navies were, by comparison, hugely expensive, an anachronistic luxury in an age when a force of one hundred bombers could, in a single mission, deliver the equivalent tonnage of all the bombs dropped in the entire Second World War. Airmen became less and less guarded in publicly declaring their conviction that only "token" armies and navies would be needed in the future. The nation has only "one real defense," Spaatz declared: "a planned and ready air offensive."

This point was not lost on the Navy, which quickly became the most vehement opponent both of an independent Air Force and of strategic bombing as a basis for American war strategy. An enduring effect of the Navy's clumsy political maneuvering was that, for years to come, many in the Air Force would dismiss any criticism of its strategic bombing strategies and plans as just so much interservice politics in disguise.

Both Marshall and Eisenhower strongly supported an independent Air Force as part of a postwar reorganization that would include a centralized Department of Defense and permanent Joint Chiefs of Staff to coordinate the policies and operations of all three services. Eisenhower was convinced that the future of war lay in joint operations that rested on what he called the "three-legged stool" of air, land, and sea. He had been reassured by the practical arrangements worked out during the war to coordinate air and ground operations; these, he felt, had finally laid to rest the old fear that the Army would be left in the lurch unless aviators were kept under its direct control. Eisenhower thought "no sane person" could oppose the change. Naval leaders may or may not have been sane, but oppose it they did. Even after defense unification and the creation of the independent Air Force were a done deal in September 1947, the Navy continued a backroom fight to keep the upstart service from poaching on what it saw as its traditional preserve as the preeminent guarantor of the nation's security.

The backroom fight burst into the open in an incident that would become famous as "The Revolt of the Admirals." One observer captured the pettiness of the affair more aptly in dubbing its opening skirmish "The Battle of the Mimeograph Machines." The immediate precipitating event was Secretary of Defense Louis A. Johnson's decision in April 1949 to cancel the Navy's planned $188 million supercarrier, USS *United States*. At almost the same time, senior Navy officials learned that the President had approved an Air Force request to purchase additional B-36 strategic bombers. They had to read about it in the newspapers, they said.

The Navy retaliated with a story tailor-made to get the attention of newspaper columnists and publicity-hungry congressmen. With the help of a former Hollywood scriptwriter, and supplied with rumors eagerly dished by the Glenn L. Martin Company, whose B-48 jet bomber had been canceled in favor of the additional B-36 purchases, the Navy produced and secretly circulated an anonymous document accusing Symington of having awarded the B-36 contract as part of a deal in which he would become president of the bomber's manufacturer, Consolidated Vultee, once he left office. The newspaper gossip columns and congressional hearings followed like clockwork. At one hearing, Symington dared a congressman who questioned him about the kickback accusation to repeat the charge off the floor. He declined to do so. The origins of the "Anonymous Document" were eventually exposed, and the corruption charges vanished.

But the Navy had obtained the platform it had been seeking for a much broader attack, and now the big guns opened up. Vice Admiral Arthur W. Radford testified that in his view the Air Force's atomic air offensive amounted to "war of annihilation." He called the huge, six-engine, propeller-driven, $3.7-million-apiece B-36 "a billion dollar blunder" that was slow and vulnerable. Just as the "battleship admirals" had clung to an obsolete weapon to protect their turf, Radford said, so the "bomber generals" were now "fighting to preserve the obsolete heavy bomber—the battleship of the air."

Rear Admiral R. A. Ofstie followed with a sweeping assault on the notion of strategic bombing, arguing that it did nothing to address America's true national security challenges: the defense of Western Europe, the command of the sea, protection of forward bases, and the early reduction of enemy combat power. "Must the Italian Douhet continue as our prophet because certain zealots grasped his false doctrines many years ago and refused to relinquish his discredited theory in the face of vast, costly experience?" Ofstie asked the House Armed Services Committee. "Must we translate the historical mistake of World War II into a permanent concept merely to avoid clouding the prestige of those who led us down this wrong road in the past?"

The naval commanders had raised some fundamental and valid points, but they proceeded to shoot themselves in the foot by arguing in effect, and simultaneously, that atomic bombing was immoral, that it was ineffective, and that the Navy could do it much better. (During one hearing, the head of the Naval Ordnance Branch asserted, "You could stand in the open at one end of the north-south runway at the Washington National Airport, with no more protection than the clothes you now have on, and have an atom bomb explode at the other end of the runway without serious injury to you." He soon retracted that assertion.) That the Navy was pressing to have carrier-based aircraft armed with atomic weapons certainly seemed to undermine any pretense that a serious debate over strategy and principle was taking place. The skirmish became so political and acrimonious that, in the end, the Chief of Naval Operations was fired and two senior Navy captains—including a future CNO, Arleigh A. Burke—were rebuked for having publicly attacked Secretary of Defense Johnson and having challenged administration policy.

Few Americans in 1949, inside the halls of power or out, heard the much more serious arguments against placing so much reliance on the atomic air offensive as the basis of national security. One argument was so shocking that only a handful of people were privy to the information; even the President of the United States was in the dark about it for a while. But the fact was that for several years after the end of the war the United States was scarcely capable of delivering an atomic weapon at any target. During the Berlin crisis, B-29s were dispatched to Europe; everyone knew that the B-29 was the plane that had dropped the A-bomb on Japan, but what they didn't know was that only a handful of the planes had been modified to be atomic-capable, and the planes that went to Europe were not among them. As of June 1947, the United States had only thirteen atomic bombs in its stockpile; a year later, the number was probably about fifty, still a tiny fraction of the number needed to execute SAC's emergency war plan. SAC had no fighters with the range to escort the bombers on a mission over the Soviet Union. Intelligence about Russian targets was poor to nil, based mostly on captured German photographs and records covering areas that had been occupied by the Nazis. In-air refueling was in its infancy; SAC's strike plans relied on staging first to forward bases in the Arctic, but operating under constant subfreezing conditions was proving a huge technical challenge. During one exercise, a pilot trying to land had to break the cockpit window with the crash-ax when it became covered with thick ice that the B-29's deicing equipment could not handle. Moreover, most crews had practiced only the mechanics of flying and had scant training in combat tactics or launching a coordinated attack. The atomic air force was in many ways a hollow threat.

Largely shielded from the public, too, were the grave doubts that at least a few top civilian leaders entertained about the viability of a strategy that hinged on treating the atomic bomb as if it were just another weapon. A fundamental problem was that the threshold for ordering such a devastating response was so high, or at least ought to be, that SAC was little more than a last-ditch insurance policy against a Pearl Harbor–style surprise attack. Strategic atomic attack was, in truth, less a strategy that fit the new policy of engagement than a throwback to the heyday of American isolationism, when the Great White Fleet shielded the country from a hostile world. "Air power romanticism," as Walt Rostow would later observe, was the natural American successor to the naval romanticism of half a century earlier; it was a way to act like a world power without getting too involved.

Truman had profound misgivings about whether the Air Force could be trusted to understand the difference. During the Berlin crisis, Symington proposed that custody of atomic weapons be transferred from the civilian Atomic Energy Commission to the Air Force. "Our fellas need to get used to handling it," he breezily explained to Truman at a White House meeting called to discuss the matter. Truman made it clear that he was not about to cede the basic principle of civilian control of atomic weapons. "You have got to understand that this is not a military weapon," he told Symington. "It is used to wipe out women and children and unarmed people, and not for military uses. So we have got to treat this differently from rifles and cannon and ordinary things like that." Symington cluelessly repeated his line about "our fellas," and Truman abruptly ended the discussion saying, "This is no time to be juggling an atom bomb around." That night David E. Lilienthal, the chairman of the AEC, recorded in his diary, "If what worried the President . . . was whether he could trust these terrible forces in the hands of the military establishment, the performance these men gave certainly could not have been reassuring."

George Kennan, for his part, voiced skepticism about whether even carrying out an atomic strike on Russia would have the effect the Air Force planners and Joint Chiefs imagined. "If you drop atomic bombs on Moscow, Leningrad, and the rest," he reportedly said when he was briefed on the plan to strike Russian cities, "you will simply convince the Russians that you are barbarians trying to destroy their very society and they will rise up and wage an indeterminate guerilla war against the West." A high-level military review of SAC plans in April 1949 had actually raised similar doubts, cautioning that raining atomic bombs on seventy Soviet cities would "not per se bring about capitulation, destroy the roots of Communism or critically weaken the power of Soviet leadership to dominate the people"; on the contrary, it might actually "validate Soviet propaganda . . . unify the people . . . and increase their will to fight."

Still, the report concluded, the logic and advantages of early use of atomic weapons in a conflict would be "transcending." This was a curious kind of logic, and it would become even more curious as a result of a development just five months later that stunned even the scientific experts, who had confidently predicted that America's monopoly on atomic weaponry would last for years. On September 3, 1949, a B-29 bomber modified as a "weather reconnaissance" aircraft collected an air sample while flying east of Kamchatka in the Soviet Far East. Three weeks later, analysis confirmed the presence of radioactive isotopes, for which there was only one possible explanation. Truman released a short statement: "We have evidence that within recent weeks an atomic explosion occurred in the U.S.S.R."

Thanks to an entirely unintended consequence of the Allied strategic bombing of Germany, the Soviets also had the means to deliver the bomb. Driven to relocate as far east as possible to escape the Allied attacks, four-fifths of the German aircraft industry wound up in areas under Soviet control at the end of the war. Airframes, jet engines, radar-guided missiles, V-2 rockets, and several hundred thousand skilled workers, engineers, and designers fell into Soviet hands. Two American B-29s that had to force-land at Soviet fields in the Far East during the war proved to be another windfall; they were promptly dismantled and copied to become the Soviet Tu-4 bomber, which entered service in 1947.

Harold C. Urey, the Nobel Prize winner who had been a leader in America's bomb project, was asked by reporters for a comment on the news of the Soviet atomic explosion. "There is only one thing worse than one nation having the atomic bomb," he replied. "That's two nations having it."

During the B-36 hearings, Defense Secretary Johnson had suggested that any technical questions about the effectiveness and performance of the Air Force's new bomber be left to the Pentagon's Weapons Systems Evaluation Group, a new interservice body created just to settle such disputes and provide the Joint Chiefs with independent assessments. In January 1950 the group's director briefed Truman and Johnson on its findings. The B-36 had its origins in a 1941 requirement for an intercontinental-range bomber that would allow the United States to continue to fight against Germany even in the event Britain fell. When it became clear Britain would not fall, the plans were put on hold; the first prototype did not fly until a year after the war was over, in August 1946. According to the evaluation group's analysis, the B-36's intercontinental range, some 10,000 miles, did give it a logistical advantage over the Air Force's other strategic bombers, which depended on forward basing or the as-yet-unproved technology of aerial refueling to reach targets in the Soviet Union.

But all of these propeller-driven heavy bombers would face a withering attrition rate were they to attempt an attack on the Soviet Union, with 30 to 50 percent shot down. The B-36 could fly higher than the B-29, with a service ceiling of over 40,000 feet, but it was an elephant of a plane, with the aerodynamic grace to match. At 160 tons, it was so heavy that only three runways in the entire country had concrete thick enough to support its weight. Consolidated's engineers eventually had to redesign the landing gear, replacing the single huge tire on each gear with four smaller ones to spread out the load. The B-36 had a maximum speed of only 345 miles per hour, slower than the B-29, and the evaluation group's study concluded that its attrition rate would likely be every bit as high as that of the planes it was to replace.

At the end of the briefing, Johnson smilingly declared that the study's findings had vindicated his decision to proceed with the B-36 program. Truman retorted that, as far as he could tell, it showed exactly the opposite. But even the most stalwart defenders of the B-36 in the Pentagon were keenly aware that it was at best a stopgap. Back in 1947 the Air Force's own Heavy Bombardment Committee had concluded that only a bomber of radically new design, with speeds in the range of 500 miles per hour or greater, could meet the emerging mission requirement. That meant jets.

The jet engine was an excellent demonstration of the principle that the more revolutionary the invention, the more obstacles it faces. The first obstacle had been the war itself. The basic idea for propelling an airplane without a propeller had been around throughout the 1920s and 1930s, and in both Britain and Germany jet aircraft were under construction when the war began. But production demands to meet immediate war needs had forced a decision to freeze existing types rather than pursue radically new technologies. Nearly all major aircraft that saw service during the Second World War, fighters and bombers alike, were designed before the war began; the new aircraft that did appear were largely improvements on time-tested concepts rather than true innovations. Just as had happened during the First World War, some advances in speed and performance were achieved, but only by pushing the existing paradigm to its limits—mainly by pouring more engine horsepower into the same basic airplanes. Were it not for the war, the jet airplane might actually have gotten off the ground sooner. No one was about to gamble the war effort on an unproven idea.

But the jet faced other obstacles, too: institutional, technological, even cultural. Chief was a circular trap in which radically new technologies often find themselves. Aircraft designers and engine experts in the 1920s and 1930s routinely dismissed the jet on the grounds that jet engines are far less efficient than piston engines, except at speeds exceeding 400 or 500 miles per hour—and airplanes didn't go that fast. They magnificently missed the point that the jet might *make* airplanes go that fast. Technological pessimism became a self-fulfilling prophecy.

In fact, high-speed flight had scarcely been explored by aerodynamicists precisely because there seemed to be no way to reach such speeds. The National Advisory Committee for Aeronautics, for all of its forward thinking in the 1930s, rebuffed Theodore von Kármán when he urged construction of a supersonic wind tunnel. Ironically, NACA's director, George Lewis, had cited the very limitation of propeller propulsion in arguing to von Kármán that there was no need to investigate aerodynamic phenomena at higher speeds. The only part of an airplane that ever encountered high-speed air flows was the rapidly whirring tips of the propeller blade. But as everyone knew, the efficiency of propellers began to decline sharply as the air flow over the tips approached the speed of sound. So what would be the point of building a wind tunnel that generated such high speeds?

It was surely no coincidence that the man who would invent the jet engine had broken out of this circular thinking before he ever sketched an engine design. Frank Whittle was a flight cadet at the RAF College in Cranwell in 1928 when he wrote a term paper titled "Future Developments in Aircraft Design." Whittle, the son of a foreman in a machine-tool factory, had acquired much of his early engineering education on his own. Forced to leave college when his father's attempt to start his own business failed, Whittle spent hours in the local library reading about gas and steam engines. He tried to enlist in the RAF, was rejected for being under height and weight, but finally got in on his third try. After two years he was accepted for one of the highly competitive cadetships that allowed enlisted men to train as pilots. He quickly gained a reputation as a daredevil pilot and a naturally gifted scientist and engineer.

In his term paper, Whittle went to the heart of the matter. He pointed out that speeds of 500 miles per hour or greater would be attainable only at very high altitudes, where drag was minimized due to low air density. And since neither propellers nor piston engines work well in thin air, some other form of propulsion would be necessary to fly at those speeds and altitudes.

The idea of the jet engine was not new, and its simplicity of operation had long been recognized as a virtue. A thermal power jet has only three stages: air is compressed; fuel is injected, causing the mixture to ignite; and the resulting explosive expansion of gases, directed outward, provides thrust. The main problem was how to compress the air in the first stage. The only ideas that had been tried, and these only in the lab, were to use the cylinders of a conventional piston engine to compress the air or to have a piston engine turn a rotary air compressor. Neither approach had proved practical.

Even before the First World War, efforts had been made to develop another appealingly simple aero-engine, the gas turbine. Like the standard piston engine, the gas turbine produces mechanical power by turning a shaft; but instead of driving pistons up and down that then turn a crankshaft, the hot combustion gases pass through a turbine—basically a rotating fan—that turns the shaft directly. Experiments had been disappointing, however; calculations suggested that trying to drive a propeller with a turbine was inherently inefficient compared with using a conventional piston engine.

Whittle's great innovation, which he hit on only a year and a half after writing his cadet term paper, was to combine the two concepts in a single engine: the "turbojet." A portion of the energy in the exhaust gases formed in the jet's third stage would be used to turn a turbine; the turbine shaft would then drive the compressor that fed the air in the first stage. Because most of the energy in the gases would still be available to generate thrust, the mechanical losses inherent in driving a turbine would be minimal. The brilliant part of the concept was that in the high-altitude, high-speed regime that the turbojet itself made possible, the engine would become more efficient. At high altitudes, low air temperatures would increase the thermal efficiency of the engine's work cycle; and at high speeds, a portion of the first-stage compression would be obtained by the "ram" effect as air entered the engine at high velocity. In fact, fuel consumption, although high, would be almost independent of forward speed in a turbojet. In January 1930, Whittle filed for a basic patent on the concept.

Getting the authorities interested would prove a formidable challenge, however. The Air Ministry promptly dismissed Whittle's idea as impractical, and a few years later let his patent lapse when it refused to pay the £5 renewal fee. It was not until 1934, when the RAF sent him to Cambridge University to further his education as an engineering officer, that Whittle was able to resume work on his ideas, file for new patents, and, with the encouragement of his professor—since Cambridge had no lab facilities suitable for this kind of work—seek private financial support to begin building an experimental model. In March 1936 Power Jets, Ltd., was formed with a subscribed capital of £2,000. As a serving officer, Whittle was required to have his own shares held in trust for him by the Air Ministry.

Work was interrupted for several weeks in June while Whittle studied for his final exams. Whittle later recalled that he managed to turn his worst disadvantages into an advantage by keeping the design as simple as possible; the lack of facilities, funding, and time made it impossible to do anything else. On April 12, 1937, the engine was ready for testing. It worked from the first. By May it was running at speeds of up to 12,000 rpm.

In the meanwhile, Whittle had also managed to secure the interest of the Air Ministry, largely thanks to the intervention of Henry Tizard, whom Whittle had met at a Cambridge University Air Squadron dinner the same month he had founded his company. Tizard was at once impressed by Whittle and his ideas; by the fall of 1936, he had persuaded the Ministry's Directorate of Scientific Research to have a leading expert on turbines make a full report, which strongly endorsed Whittle's approach (though it cautioned that jet propulsion would be practical only for "special purposes, such as the attainment of high speed or high altitude for a short time").

The Air Ministry at last said it would provide financial support, but months of wrangling over terms ensued, on lines uncannily reminiscent of the Wright brothers' failed negotiations with the British government three decades earlier. Power Jets wanted to sell the invention to the government outright for £10,000. The Air Ministry objected that Whittle was trying to finance his future work by making an "exceptional profit" from this sale, and pointed out that the usual research contract for developers of new engines was based on running time of the engine, at a fixed rate of £25 per hour. The Ministry said it would make an exception in his case and make it £200 an hour for twenty hours of increasing running speeds, plus £1,000 for a report already completed. Whittle pointed out that this was still half of what he was expecting, and that by imposing the Official Secrets Act and claiming the right to his patents as Crown user, the government had left him with no other possible customers.

Even after the government began to provide more direct support for Whittle's efforts to perfect his engine, the muddling continued, and it was only in the spring of 1940 that the Air Ministry was ready to fund production. Rover was selected to manufacture the engines and Gloster to build the airplane that would incorporate the revolutionary new propulsion system, but both were chosen less because of any obvious expertise than simply because they were not too busy with other vital war work. Gloster rose to the occasion and quickly designed an effective airframe, but Rover immediately caused trouble by refusing to admit the validity of Power Jets' patents. The government ordered Whittle to turn over the design of his new W.2 engine, citing wartime necessity. Rover then spent two years fiddling with the design, raising suspicions that it was trying to change just enough so it could continue to manufacture the engines after the war without running afoul of Power Jets' claims to the invention.

On May 15, 1941, a small experimental Gloster fighter powered by one of Whittle's W.1 engines made its first flight. Tizard again pressed the case with higher authorities—he had called the jet fighter a potentially "war winning gamble if it comes off"—and Churchill sent forth one of his famous "Action This Day" memoranda suggesting that the project should be speeded up at once. But it was not until November 1942 that work got under way, when the job of building the engines was handed over to Rolls-Royce. (In 1944 the government nationalized Power Jets, but Whittle himself received nothing from the sale; only in 1948 was his great contribution recognized with a knighthood and a £100,000 ex gratia payment.)

Although both Gloster Meteor and German Me 262 jet fighters saw combat in the final months of the war—the Meteors shot down a dozen V-1 flying bombs and the Me 262 took a toll on Allied bombers over Germany—both came too late to make a significant difference. An experimental German jet had first flown two years before Whittle's, but the Me 262's debut had been delayed by shortages of chromium and nickel needed to make heat-resistant steel for the turbine blades, and the engines were dangerously unreliable; early models could run for only about ten hours before the blades became hopelessly distorted. Still, when it flew, the Me 262 outperformed any aircraft in the world, with a top speed of 540 miles per hour.

American engine makers had missed the boat completely. The Air Corps' procurement system during the 1920s and 1930s allowed engine makers to recoup their research-and-development costs as overhead when they won a production contract. An unintended result of this generosity was that it tended to discourage risk-taking, since only incremental innovations that were likely to pay off quickly in a successful production model were usually considered worth pursuing by management. In 1941 Hap Arnold had been in England and had witnessed the first Gloster jet flights; he quickly secured British agreement to turn over Whittle's engine design to General Electric, which had experience with gas turbines. Within a year an American prototype jet fighter, the Bell Airacomet P-59, was flying. But no American jet would see service during the war. The

Right: A Bf 110 flies over Poland in the opening battle of the Second World War. *(Courtesy James S. Corum)*

Below: German bombing run in Poland. *(Library of Congress)*

German Stukas over Norway in 1940.
(Courtesy James S. Corum)

Newly promoted Field Marshal Erhard Milch (left) and General Wolfram von Richthofen in France, shortly after the French surrender in 1940.
(Courtesy James S. Corum)

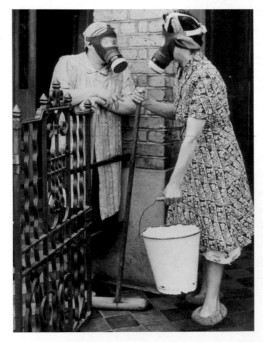

British women wearing gas masks during a civil defense drill. Widespread fear that the war would begin with German gas attacks on London soon gave way to apathy; by spring 1940, fewer than 1 percent of Londoners were obeying instructions to carry their masks with them at all times.
(Imperial War Museum/Copyright Fox Photos)

Above: Fighter Command chief Hugh Dowding
(Imperial War Museum)

Right: Luftwaffe commander Hermann Göring
(Imperial War Museum)

Luftwaffe officers look across the Channel to the white cliffs of the English coast.
(Imperial War Museum)

British fighter pilots—
in a typical insouciant
pose.
(Imperial War Museum)

Above: Aircraft spotter on a
London rooftop. St. Paul's
Cathedral is in the background.
(National Archives)

Reports of RAF successes, though
eagerly accepted at the time,
proved greatly exaggerated.
(Imperial War Museum)

Scenes from the Blitz
(*clockwise from left*):
firemen on a London street
after a night raid; crowds
sheltering in an Underground
station; Churchill visiting the
bombed-out cathedral at
Coventry; Anderson shelter
amid debris from German
attack on Norwich.
*(top: National Archives;
others: Imperial War Museum)*

A P-51 Mustang crated for shipment to Britain, October 1942.
(Library of Congress)

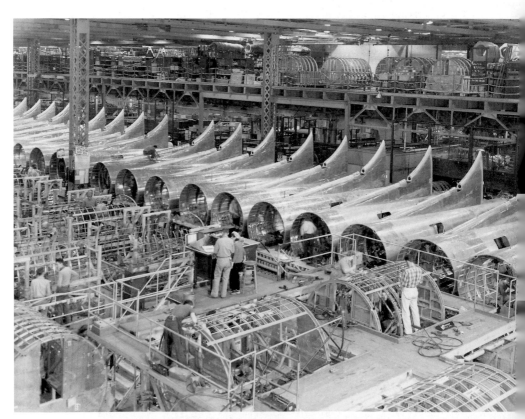

B-17s under construction at Boeing's Seattle plant.
(Library of Congress)

Above: Battleship *West Virginia* in flames after the Japanese attack on Pearl Harbor. *(National Archives)*

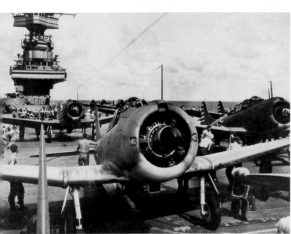

SBD Dauntless dive bombers aboard the U.S. carrier *Yorktown*. *(U.S. Naval Historical Center)*

A Japanese plane shot down near the Mariana Islands, June 1944. *(National Archives)*

Control room at Uxbridge,
coordinating tactical air operations
for D-Day.
(Library of Congress)

By D-Day, fighter-
bombers had destroyed
every bridge across the
Seine south of Paris.
(Library of Congress)

P-47 Thunderbolts over northern
Italy.
(Library of Congress)

Monte Cassino, after the futile attack by heavy bombers.
(Public Record Office)

Dresden, after the Allied fire bombing.
(Library of Congress)

B-29s dropping incendiaries
over Japan, June 1945.
(Library of Congress)

GEN. H.H. ARNOLD

Clockwise from above left:
"Bomber barons" Curtis LeMay,
Carl Spaatz, and "Hap" Arnold.
(Library of Congress)

USS *Bunker Hill* hit by two kamikazes, May 11, 1945.
(National Archives)

Atomic bomb explodes over Nagasaki, August 9, 1945.
(Library of Congress)

Robert T. Jones, the self-taught American aerodynamicist and inventor of the swept-back wing. *(NASA)*

Frank Whittle's first experimental jet engine (*right*), and the Gloster jet-fighter prototype powered by a later version of the engine.
(right: Public Record Office)

F-86 jet fighters in Korea, June 1951. *(National Archives)*

The multibillion-dollar SAGE system combined radars and vacuum-tube-powered digital computers to track Soviet bombers and to guide fighters to intercept them.
(William J. Cook)

The entrance to SAC headquarters in Omaha, early 1960s.
(U.S. Air Force/National Air and Space Museum)

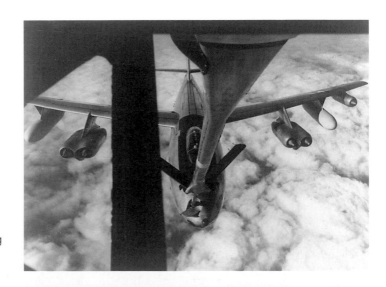

A B-47 jet-bomber refueling in midair.
(Library of Congress)

Helicopters approaching a target in
Vietnam.
(National Archives)

Though woefully ill-suited to ground attack,
F-105s flew 75 percent of Rolling Thunder
bombing missions in Vietnam. An EB-66
jammer leads a formation, June 14, 1966.
(National Archives)

One reason interdiction failed: Communist
porters carrying supplies up a mountain trail
in South Vietnam, 1965. Total Communist
supply needs amounted to a mere seven
2½-ton truckloads per day during the period
of guerilla warfare in the south.
(Library of Congress)

Precision and stealth in Desert Storm
(*clockwise from above*): the launching
of a Tomahawk cruise missile toward
Iraq; reinforced hangars struck by
precision weapons; an F-117 stealth
fighter; a bridge and trucks destroyed
in the Euphrates River valley.
(*Department of Defense*)

A B-52 takes off from Diego Garcia to strike targets during the war in Afghanistan in 2001. Real-time intelligence and satellite-guided JDAM bombs (*below*) allowed large bombers to fly effective close support missions for the first time in history.
(*U.S. Air Force*)

An RQ-1 Predator drone, armed with Hellfire missiles.
(*U.S. Air Force*)

established engine companies, Pratt & Whitney and Wright, were deliberately kept out of the program lest it interfere with their war production of piston engines, a decision that sharply limited American production of turbojets not only during the war but for several years after.

If the United States had been caught napping by the turbojet, it was nonetheless at the forefront of two breakthroughs in aerodynamic science that would turn out to be every bit as fundamental in ushering in the postwar jet age. One came as a consequence of exploring the limitations on faster propeller speeds that were to doom piston engines to obsolescence. Aerodynamicists had long recognized that the equations describing airflow become mathematically manageable only if the simplifying assumption is made that air is incompressible, that its density is unaffected by pressure or speed. This is a valid approximation when working with speeds up to about 350 miles per hour, and so in designing wings and airframes in the postwar era aerodynamicists could routinely make use of these simpler equations. Propeller-tip speeds were the one place where aircraft designers were routinely likely to encounter compressible flow, however, and in the 1920s NACA had begun a series of experimental studies in a small high-speed wind tunnel to determine some rule-of-thumb corrections that could be applied to the standard airflow equations when designing propellers. John Stack, the NACA aerodynamicist who pioneered much of this work, also began to study the phenomena that cause drag and turbulence to sharply rise as airflows approach the speed of sound—what Stack termed the "compressibility burble."

As a result, Stack was the only aerodynamicist able to explain the sudden and violent crashes that had occurred during tests of Lockheed's P-38 Lightning fighter in 1941. Twice, test pilots lost control after putting the plane in a dive; the tail surfaces began shaking violently, the controls seemed to freeze, and the plane plunged with a will of its own into an ever steeper dive. The first time this happened, the tail was ripped from the plane and Lockheed test pilot Ralph Virden was killed. Ben Kelsey was at the controls the second time it happened; he managed to get the canopy open to bail out, and "suddenly the plane was gone," so violently was he thrown from the cockpit.

Stack was able to show that, at speeds exceeding 450 miles per hour, compressibility effects in the air moving over the P-38's wing caused a sudden loss of lift, setting off a chain reaction of events. The loss of lift caused the angle of downwash leaving the trailing edge of the wing to be reduced, which in turn altered the angle of attack of the airstream that encountered the tail surface in a way that tended to increase net lift from the tail. That forced the nose of the plane to "tuck under" into a dive.

It would not be until the advent of supercomputers in the 1970s that the equations for compressible flow could be solved, but Stack's explorations of the "transonic" region provided an empirical basis for the design of workable airframes as jet aircraft crept up to and beyond the sound barrier in the 1950s and 1960s.

It was another NACA scientist who provided the other American break-through for the jet age. Robert T. Jones was, much like Frank Whittle, a largely self-taught scientist with a natural bent for his subject. Jones had dropped out of college after a year at the University of Missouri, worked for a while as a crew-man in a flying circus, and, when the Depression hit, wound up as an elevator operator in Washington, D.C. In 1933 he began to take night classes in aeronau-tics at Catholic University taught by the legendary Max Munk, and with this connection his fortunes began to change. Within a year, Jones had been hired for a temporary position at NACA, where his talents were immediately recognized. By 1936 the laboratory wanted to hire him permanently, but found that civil-service regulations required a college degree for a starting position at the lowest engineering grade. The regulations did not, however, specifically repeat this requirement in spelling out the qualifications for the next higher grade, so NACA was able to exploit the loophole and start Jones at that level.

In 1944 Jones had worked out some equations for the aerodynamics of short delta-shaped wings as part of research on a proposed glide bomb being devel-oped for the AAF. Thinking his theory was "so crude" that "nobody would be interested in it," he shoved the results into his desk. A year later, Jones was work-ing on the theory of supersonic flow and was surprised to find that the equations he obtained for a delta wing at supersonic speeds looked much like the low-speed equations he had derived a year before. Wondering how this could be, he recalled an observation Munk had made years earlier: The aerodynamic characteris-tics of a wing depend mostly on the compo-nent of the airflow moving in a direction perpendicular to the leading edge. With a highly swept wing, that component of air velocity could remain well in the subsonic range even at speeds that approached or exceeded the speed of sound. This realization had stunning implications, for it meant the problems caused by turbulence, shock waves, and compressibility effects as airplanes approached the speed of sound could be avoided if wings were simply swept back.

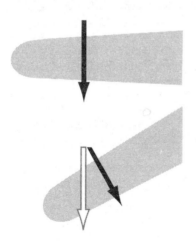

Swept wings reduce the component of airspeed perpendicular to the leading edge, avoiding the sharp rise in turbulence and drag that occurs at near-supersonic speeds.

Jones's superiors at NACA initially refused to accept his finding; it seemed too preposterous, he was told, and he was ordered to cut out the "hocus pocus" about swept wings before publishing his report. But wind-tunnel experiments soon confirmed that a swept-back wing would reduce drag by a factor of four in supersonic flight. By June

1945, NACA finally agreed to circulate to the aircraft industry a confidential report on Jones's findings. His credibility was almost immediately bolstered by the discovery that German scientists had independently arrived at the same conclusion. In fact, the German aerodynamicist Adolf Busemann had publicly presented a paper on the swept-back wing at an international scientific conference in 1935, but it went largely unnoticed. (During one dinner at the meeting, the conference chairman had presented Busemann with a facetious sketch he had done on the back of the menu card; "Busemann's airplane of the future," it was labeled, and it depicted an airplane with not only swept wings but swept tail and swept propeller as well.)

The impact on the American airplane industry was immediate and revolutionary. Boeing's B-47 jet bomber, then under design in a straight-wing configuration, was quickly redesigned with swept wings. It would become the progenitor of the famous Boeing 707 and every other jet transport to come. A few months later, North American Aviation hastened to redesign its XP-86 jet fighter with swept wings, too; on April 26, 1948, it became the first combat aircraft to go supersonic.

And by the end of 1948, Boeing unveiled a completely overhauled concept for its proposed B-52, the bomber that was to take the Air Force into the next generation. Originally conceived as yet another propeller-driven giant in the line of succession from the B-17, the B-52 in its new iteration was a radical departure: swept wings, eight jets, a range of 8,000 miles, top speed of 570 miles per hour. As great as the delays had been in bringing the jet engine and the science of high-speed flight to practical fruition, the elements now all swiftly fell into

B-47
bomber
engines: 7,200-lb. thrust
turbojet (6)
top speed: 610 mph
armament: nuclear or 10,000
lbs. conventional bombs;
two 20mm cannon in tail

place as a fully mature technology. This maturity proved itself in the remarkable fact that B-52s would still be flying as the mainstay of America's bomber forces a half century later.

SAC's dubious ability to execute the atomic offensive that the Air Force had by 1948 embraced as its prime mission was in part due to a lack of equipment. The other part of the problem was that SAC was operating the way the peacetime American military had always operated. Curtis LeMay took charge of SAC on October 19, 1948, determined to fix both problems. LeMay's meteoric rise, and the reputation for take-charge efficiency that had propelled it, had only continued after the war. He was now America's youngest four-star general since Ulysses S. Grant. LeMay's aim, which would verge on obsession, was to put SAC on a war footing and keep it there. SAC was to be "combat ready," in the

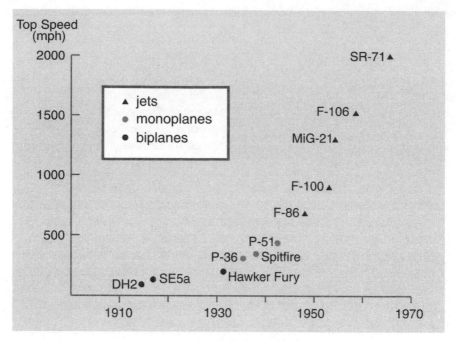

The effect of the monoplane and jet revolutions on aircraft speed.

phrase LeMay would use again and again. "We had to be ready to go to war not next week, not tomorrow, but this afternoon," he insisted. "We had to operate every day as if we *were* at war." In the past, the peacetime military had largely been a caretaker force; to fight a war required mobilization, training, months if not years of preparation. In this new kind of war there would be no time for that.

LeMay zeroed in on pernicious examples of peacetime attitudes that had to be stamped out: he was apalled to note that SAC's crews had been ordered to keep their planes at low altitudes to avoid causing equipment malfunctions and that practice targets were rigged up with radar reflectors so they would show up well on the planes' radar scopes. In January 1949 LeMay ordered every bomb group in SAC to take part in a mock bombing run on Dayton, Ohio. This time there would be no phony shortcuts or cheat sheets. The missions would be flown at 30,000 feet. Crews were given 1938 aerial photographs of Dayton to work from; this was more up-to-date information than was available for most of Russia. The crews would have to find their targets using radar images without any artificial gimmicks. LeMay's intention was that the groups would fail so he could give the entire command a much-needed kick in the pants. He was not disappointed. Many planes aborted with engine or radar failures. Those that did make it over the target missed their aim point by an average of two miles. "Not one airplane finished that mission as briefed. Not one," LeMay said. He had made his point.

Just as he had during the war, LeMay drove his crews relentlessly, left nothing to chance, and demanded perfection. "I couldn't afford to differentiate between the incompetent and the unfortunate," he said of his practice of firing people who didn't perform to standards. LeMay concluded that the most experienced pilots and crewmen were actually the most likely to cut corners, and thus to have accidents or make mistakes; he ordered detailed manuals and checklists—Standing Operating Procedures—prepared for every conceivable task, insisting that everyone follow them to the letter. Crews were each assigned specific targets in the Soviet Union and expected to know them inside out. Photographs were simulated to show how each target would look on a radar screen, and crew members had to be able to produce a drawing of their target from memory.

At the same time, LeMay used his influence in the Air Force and in the public arena to build up SAC as the Air Force's elite unit, with the perks to match. LeMay was granted personal authority to confer "spot" promotions on his men, and SAC consistently had the highest promotion rates in the entire Air Force. At SAC's new headquarters in Omaha, Nebraska, and at other SAC bases, LeMay ordered that the cooks be sent to the area's best hotels as unpaid apprentices to improve the quality of food in the mess halls. He replaced open-bay barracks with dormitory-style housing that had two beds to a room, saying this was a necessity for men who were on call for round-the-clock operations. At one point he even solemnly insisted in testimony to Congress that "the lack of adequate housing on or near Strategic Air Command bases has direct, immediate, and inevitable bearings on the ability of this nation to survive a major nuclear war." No one complained that he was overdoing it.

Within just a few years, the relentless training and lobbying began to pay off. Radar bombing accuracy improved to an average error of 1,400 feet. Aerial refueling was perfected, and by October 1952 there were seventeen air-refueling squadrons in operation, up from one in 1948. The number of aircraft and personnel in SAC had tripled. LeMay boasted that SAC had run mock bombing attacks on every single city in the United States with a population of more than 25,000; San Francisco, he said, had been "bombed" six hundred times in one month.

SAC had also become a household word, thanks both to LeMay's reputation and to an unabashed marketing campaign. A 1955 movie starring Jimmy Stewart, titled simply *Strategic Air Command,* had done much to publicize SAC's catchphrase that only by being constantly vigilant and combat-ready could it keep the peace. PEACE IS OUR PROFESSION, proclaimed billboards that appeared at SAC bases throughout the country.

As a basic principle, the idea of maintaining a war footing to prevent war was not hard for most people to grasp. But when atomic bombs were involved, the logic of deterrence began to lead down odd and mind-twisting paths. On the one hand, there was widespread agreement with the point Truman had made to Symington: the bomb wasn't a military weapon at all. This became all the more apparent after the United States and then the Soviet Union exploded the first thermonuclear bombs, whose force was measured not in kilotons but in megatons. Under such circumstances, nuclear war could only mean mutual annihilation. "There is today not the slightest chance of anyone winning a war," John Slessor, the RAF's recently retired Chief of Air Staff, wrote in *Air Force Magazine* in 1954. President Eisenhower's Secretary of Defense, Charles Wilson, echoed the thought: "I assure you that in my opinion everybody is going to lose in the next war."

On the other hand, deterrence demanded that in order to prevent the other side from ever starting a nuclear war, the threat of retaliation had to be believable and ever-present. That left SAC in the position of relentlessly practicing to do something that no sane person would possibly do, of spending billions of dollars on airplanes whose sole reason for existing was that they should never be used for the one job they were designed for, of placing on a hair trigger a weapon that, if the trigger were ever actually pulled, would cause instant mutual suicide.

The result was a paradoxical blend of make-believe and dead earnest. Practicing flying their airplanes was one thing, but it cut close to home when tens of thousands of SAC personnel were sent to "survival school," where they learned how to hide in caves and live off the land for weeks while awaiting rescue, in case they were forced to ditch after launching World War III.

The inherent absurdity of these contradictions provided the setup for the extremely black humor of Stanley Kubrick's satirical film *Dr. Strangelove.* Along with its black humor it actually did a brilliant job of capturing the way American nuclear strategists at think tanks such as the RAND Corporation grappled with the perverse logic of nuclear deterrence. One of the film's best moments comes early on, when the B-52 crew receives the order to carry out a nuclear attack on the Soviet Union. The crew members react with sheer incredulity to the idea that they could possibly be ordered to do what they have been training every day to do. The radio operator calls the pilot on the intercom.

"Major Kong," he diffidently begins, "I know you'll think this is crazy but I just got a message from base that decodes as Wing Attack Plan R. R for Romeo."

The major, played with perfect humorous incongruity by the rodeo and cowboy-movie star Slim Pickens, irritatedly tells him off: "How many times have I told you guys I don't want no horsing around on the airplane?"

"I'm not horsing around, sir," the radioman insists. "That's how it decodes."

"Well, I've been to one World Fair, a picnic, and a rodeo," the major erupts, "and that's the stupidest thing I ever heard come over a set of earphones."

The attack order has been given to an entire B-52 wing by a SAC base commander who has lost his marbles, and for the rest of the film, as the President and the Pentagon desperately try to recall the attacking bombers, Kong's crew (accompanied in the background by the strains of "When Johnny Comes Marching Home") bravely overcomes one challenge after another, dodging antiaircraft missiles, extinguishing fires, running dangerously low on fuel, and finally jury-rigging the plane's wiring when the bomb-bay doors fail to open. The audience, of course, hopes at every step of the way that they will fail, thus averting World War III. Instead, of course, they heroically succeed, thus triggering doomsday.

To many the movie's most memorable scene, however, is Peter Sellers's appearance as the wheelchair-bound title character. A top civilian scientist steeped in the mysteries of nuclear combat, Dr. Strangelove epitomizes the arcane high priesthood of deterrence theory, consulting his circular slide rule and invoking studies by "The Bland Corporation" while glibly tossing about predictions of the millions who will be annihilated on both sides.

RAND had been created in 1945 as a continuation of the civilian scientific work that the AAF had come to rely so heavily upon for operational analysis during the war. Donald Douglas, the president of Douglas Aircraft, had approached Hap Arnold with the idea; Arnold was immediately enthusiastic, and as soon as the war was over came up with $30 million, left over from his wartime research budget, to launch the venture. RAND stood for "Research and Development," although LeMay later grumbled that it ought to stand for "Research and No Development," since all it did was churn out paper. One of its major tasks would be to try to make sense out of America's nuclear-war-fighting plans and strategy.

Despite their Strangelovian reputation, the RAND experts and other civilian consultants who scrutinized the Air Force's nuclear-targeting plans were from the start appalled by the military's shallow thinking and casual assumptions about "winning" a nuclear war. In 1950, Bernard Brodie, a Yale University political scientist who had done extensive research on air targeting theory, had looked at SAC's plans and concluded that all the talk about inducing Soviet "collapse," or "killing a nation," or delivering a "Sunday punch" (one of LeMay's pet phrases) was simply nonsense. For one thing, almost no thought had gone into how striking particular targets would even achieve a particular effect. But more to the point, nuclear weapons required thinking about targeting strategy in a way entirely different from what generals were used to. In fact, Brodie argued, the only real value of nuclear weapons was in the threat they posed, and this value would be completely lost once they were used. It was like killing the hostage, he said. SAC's plans to deliver the entire stockpile in one blow fit conventional Air Force doctrines about concentration of force, but in a crisis between two nuclear-armed states the important goal—arguably the *only* important goal—was to continue to hold a threat over the enemy so that he would not simply retaliate by blowing up every American city.

A key implication of this thinking was that in a first strike it would be better to target Soviet nuclear forces than Soviet cities and industry. But many SAC commanders apparently could not grasp such subtlety, and indeed many seemed to have faced down the paradox of getting ready for a war that could never be fought by deciding that it could, too, be fought.

The Air Force was the most technical of the services, but the entry of thousands of aviation cadets during the Second World War had saddled the postwar

Air Force with a large cadre of pilot-officers who had never finished college. Many Air Force leaders in the 1950s and 1960s were certainly smart, but they typically were not well educated, articulate, or given to studying the nuances of geopolitics. "New hardware was welcomed with more enthusiasm than were new ideas in the realms of strategy, concepts, and doctrine," Air Force Secretary Eugene M. Zuckert recalled in 1965. The Air Force of the 1950s, as the military historian (and Air Force F-16 pilot) Colonel Mike Worden put it, "continued to value experience over education, action over contemplation." Only 44 percent of Air Force officers had a college degree in the mid-1950s; in the Army and Navy the percentages were double that.

There was certainly more than a little brash, I-don't-give-a-damn anti-intellectualism in the attitude of SAC commanders toward the advice of civilian strategists. In his 1965 memoirs, LeMay glibly suggested that even after the Soviet Union acquired nuclear weapons the United States might have delivered an ultimatum "to behave your damn selves" or face preemptive nuclear attack. General Thomas S. Power, who was LeMay's handpicked successor as SAC commander despite being the only Second World War–generation Air Force general who had no education beyond the high-school level, exploded during a briefing in December 1960 when a RAND consultant first presented the "counterforce" concept to him. RAND's analysis had concluded that if SAC limited its initial attacks to hitting Soviet nuclear forces in the event, say, of a Soviet conventional attack on Western Europe, the Soviets would still be deterred from retaliating against American cities, since their own cities would remain at risk. The war could be brought to an end with perhaps a few million casualties on both sides, versus hundreds of millions dead if SAC executed its preferred "maximum effort" attack plan. The key concept was "restraint," but this word sent Power into a rage.

"Why do we want to restrain ourselves?" Power shouted. "Restraint! Why are you so concerned with saving *their* lives? The whole *idea* is to kill the bastards! Look, at the end of the war, if there are two Americans and one Russian, we win!"

Air Force opposition to counterforce weakened, however, when it became clear that the Navy's submarine-launched Polaris missiles lacked the accuracy for anything but city-busting attacks; under counterforce targeting, the more precise manned bombers of the Air Force would still be needed to hit thousands of "designated ground zeroes" at Soviet air bases and missile complexes.

This was the flip side of the ultrarealistic training that LeMay had made SAC policy; the mission itself was so fantastic and hypothetical that it was hard to tell whether SAC's stated requirements reflected legitimate national security needs or merely the Air Force's institutional ambitions. The nuclear mission had been the Air Force's unquestioned ticket to institutional security and growth. From 1952 to 1960, the Air Force had received an average of 46 percent of the three services' total annual budget. Yet, as a few dissenters noted, SAC's dominance had ossified Air Force thinking. One Air Staff planner complained that SAC suffered from the "curse of bigness"; it had become dogmatic, status quo, interested only in "more of the same," especially more bombers that flew higher, faster, and farther.

In the nuclear competition, where weapons were less weapons than they were symbols of national technological superiority, it was especially easy to fall into the mind-set of technology for technology's sake. Hap Arnold had often frustrated his more technically competent subordinates through his ignorance of basic science and mathematics (he didn't know how to use a slide rule, one recalled, and would become manifestly impatient with technical explanations), but by the Second World War he had acquired a sort of layman's admiration for scientific wizardry. Even before setting up RAND, he had convened a panel under von Kármán with the charge to look as far as possible into the future and make the "boldest" predictions about advances that might affect air power in decades to come.

Such speculation was only prudent, but inevitably this sort of thinking led the Air Force to push some profoundly crazy ideas—and, given the mood of the times, for Congress and the public to support them wholeheartedly. One of the craziest ideas was the atomic-powered airplane. No one could come up with an explanation of why this might be a good idea. Oppenheimer dismissed it as "hogwash." But LeMay pushed it, and senators and congressmen were if anything more enthusiastic. Republican Senator Homer Ferguson of Michigan, explaining his support for the project, solemnly declared that "we have really got to keep out in front in science." There were rumors of Russian experiments on nuclear-powered flight, and Democratic Senator Henry M. Jackson of Washington State warned ominously that "if Russia beats us in the race" for atomic-powered bombers "our security will be seriously endangered." In the end, the Air Force would spend a billion dollars on the idea, flying an operating nuclear reactor around in the bomb bay of a B-36 for no discernible reason (the plane was powered by conventional engines). Engineers never came close to figuring out how to convert nuclear power into a practical means of propulsion that could lift an airplane off the ground, much less how to shield the crew from radiation. Nor did they specify what would happen if such a plane crashed.

Of course, even crazy predictions occasionally turn out to be right. One of the boldest predictions made by von Kármán's committee was that intercontinental ballistic missiles might replace manned bombers. Vannevar Bush, for one, had been scornful of such ideas, telling the Senate Atomic Energy Committee in December 1945, "we have plenty enough to think about that is very definite and very realistic" without "fantastic" talk about intercontinental rockets armed with atomic warheads. Experts who wrote "these things that annoy me," Bush huffed, were imagining something that was "impossible today and will be impossible for many years."

The announcement from Moscow on October 4, 1957, that Sputnik, the first artificial satellite, had been successfully launched into earth orbit had been a shock almost greater than the Soviet atomic bomb. Everyone at once understood that a rocket that could put a payload in orbit could also power an ICBM. Another intrusion of reality occurred on May 1, 1960, when a Soviet surface-to-air missile shot down an American U-2 reconnaissance aircraft flying 70,000 feet over the Russian city of Sverdlovsk. Both were cruel blows to SAC's abiding faith that America's security was to be found in ever-higher-performance manned bombers. The Air Force Chief of Staff continued to grumble that ICBM crews would develop a "static, nondynamic," "Maginot Line" mind-set if they sat in bunkers for hours on end, but within a few years the growing capabilities of both ICBMs and SAMs would dramatically alter SAC's posture. By 1964 the number of ICBMs exceeded the number of bombers in SAC's arsenal. In the end, it would be the missileers who would have the last word. One missile crew at Whiteman Air Force Base in Missouri posted in their control bunker a sarcastic epitaph for the manned nuclear bomber. "We will continue to deter potential aggressors with our hardened and dispersed ICBMs," it read, "and at the same time afford the remainder of the Air Force the opportunity to indulge in their favorite hobby—FLYING."

One might have thought that the post-Sputnik decline in the fortunes of the manned bomber would have been accompanied by a rise in the fortunes of the Air Force's nonstrategic forces. One would have thought wrong. The tactical air forces had been on the short end of the stick for so long, and had lost so much institutional clout to SAC, that no news was ever good news for the Tactical Air Command. TAC shrank from forty-one wings in 1957 to twenty-three in 1960, while SAC's budget (already seven times TAC's in 1957) grew by a third to meet the costs of developing and deploying the new ICBMs. By 1962, the American nuclear arsenal, the vast majority of it still under SAC, had doubled from pre-Sputnik levels, reaching 26,500 warheads.

TAC had been established as a separate, permanent command right after the war; it had been one of Eisenhower's essential conditions for his support of an independent Air Force. Pete Quesada was named its first commander, a choice that seemed to promise further Air Force commitment to the air-ground cooperation Quesada had done so much to establish during the war. But almost immediately the atmosphere soured. An Air Staff historical study, officially approved by Spaatz in June 1946, gave the Ninth Air Force and other wartime tactical air forces a pat on the head—they had performed "marvelously the tasks assigned to them"—and then declared them obsolete in the new atomic age. Quesada privately expressed a worry that he was becoming an unintentional "conspirator in an ugly mistake"; as he put it years later, "We were forgetting the teachings of the war in Europe." He told young pilots that if they had any ambitions they had better try to get into SAC, otherwise they'd never make it past colonel. On

December 1, 1948, General Vandenberg, the Chief of Staff, ordered all fighter aircraft reassigned from TAC to a new Continental Air Command, which would handle both tactical and air-defense missions. Its new commander then announced that training for close air support would take time away from the much more important air-defense mission; the B-29 and B-36 could provide all the ground support the Army needed. Quesada was left with a planning staff of sixty-six and told to write manuals and plan training exercises. He asked to be reassigned, and by the following year he had left the Air Force.

General O. P. Weyland, who commanded TAC from 1954 to 1960, recalled that "people in the tactical business were outnumbered and outvoted" again and again on policy matters. The reign of the bomber generals only tightened after LeMay became Vice Chief of Staff of the Air Force in 1957 and then Chief of Staff in 1961. By the time he was through, three-quarters of the highest-ranking officers at Air Force headquarters had come directly from SAC.

Remarkably, even the American military's initial debacle in Korea had done almost nothing to shake the convictions of LeMay and his fellow true believers that, as LeMay told an Air Force commanders' conference shortly after becoming Vice Chief, the United States "could no longer afford the luxury" of devoting significant air resources to supporting ground forces. America's new jet fighters had proved so ill suited to ground attack when the fighting in Korea broke out that Second World War–era Mustangs were hurriedly pulled out of storage and rushed in to fill the gap. The F-80 jet fighter, designed as a high-altitude air-defense interceptor, didn't even have racks under its wings to carry bombs. With their short range and high takeoff speeds, jets required forward deployment and long runways, and tactical air engineering units were woefully underequipped, undermanned, and undertrained for the tasks of airfield construction. The procedures for air-ground coordination so painstakingly worked out in North Africa and Europe during the Second World War had been almost completely forgotten; as the U.S. Air Force official history of the Korean War noted, the command situation for tactical air support during the early fighting was "fantastically confused." Most of the tactical aircraft were based in Japan, more than 150 miles away, and timely intelligence was rarely available. In his report after the war, General Weyland, who took command of the American air units in Korea in June 1951, commented bluntly, "An astounding facet of the Korean War was the number of old lessons that had to be relearned."

Even more astounding were the new lessons that went unlearned. Most notable among these was the fact that a massive Second World War–style strategic air interdiction campaign had almost completely failed to affect the course of the fighting in Korea. American planners, and it would not be for the last time, simply did not grasp how minuscule were the logistics needs of an Asian army, nor how much could be supplied by coolie labor conscripted on a massive scale. More than 500,000 North Korean laborers worked to repair rail lines and bridges damaged by interdiction strikes; every night, an army of oxcarts, horse-drawn wagons, pack animals, and coolies would emerge from hiding and take to the roads, often removing a section of a bridge span before dawn to make it look as though the damage were unrepaired. By late 1951, the Communist forces had work crews stationed every four miles along rail lines and at every rail junction; rail cuts were repaired in two to six hours. They had also tunneled extensive caves near the front to conceal stockpiles of food and supplies accumulated between offensives. General Lemuel C. Shepherd, Commandant of the Marine Corps, bluntly called the air interdiction effort a "fizzle." But most American air leaders insisted that interdiction, combined with strategic bombing of North Korean dams, industrial centers, and hydroelectric plants, which altogether accounted for more than half of all sorties during the war, had indeed worked and were what ultimately forced the North Koreans to settle.

American airmen were equally quick to insist, however, that there was probably little point in trying to draw any real lessons from Korea, since the war had been an "aberration," a "special case" that would never be repeated. "Certainly, any attempt to build an air force from the model of the Korean requirements could be fatal to the United States," the report of the U.S. Far East Air Force declared. As far as airmen were concerned, Korea had been a complete misuse of air power. General Nathan F. Twining, the Chief of Staff, went so far as to say that "limited war" was nothing but a "fad," a "convenient excuse for deferring hard decisions and sweeping fundamental principles under the rug." He thought air power should have been used in Korea as basic Air Force doctrine dictated: in an all-out effort to achieve complete victory over the seat of enemy power— namely, the Soviet Union. Instead air power had been squandered (in LeMay's words) as mere "flying artillery."

LeMay's position toward Korea, and that of other air leaders, was simple if not exactly logically consistent: nuclear weapons and SAC's plans for using them had made wars like that in Korea impossible, and the fact that the Korean War had nonetheless occurred did not alter this principle; on the contrary, the war would not have happened in the first place if only the politicians had understood how to fight a war the Air Force way. On the one hand, the Air Force argued, "small wars fought in remote areas with obsolete explosives and outmoded concepts are no threat whatsoever to the survival of the United States." On the other hand, LeMay declared, only if the United States were lacking in "intelligence and resolution" would it be possible for the Soviets to start a localized war and not have it immediately become a general war.

In other words, if America's war aim in Korea was less than total victory over Communism, then it was simply fighting the wrong war. Lieutenant General George E. Stratemeyer, Weyland's predecessor as air commander in Korea, would be the first though hardly the last U.S. Air Force commander to complain that his "hands were tied" by politicians who weren't willing to win a war they had started. "That is not American," Stratemeyer fumed. Shortly thereafter, Stratemeyer began issuing deranged pronouncements about a Jewish conspiracy that was sapping American strength, but his views on the failure to unleash the full force of the air weapon in Korea were widely shared by airmen. As Air Force historian Robert F. Futrell observed, "The emphasis of air planners was in making war fit a weapon—nuclear air power—rather than making the weapon fit a war."

Blaming the politicians was an understandable reaction from commanders whose baptism of fire had been the Second World War. But it also reflected a remarkable strategic naïveté. What Stratemeyer, LeMay, and the other air leaders overlooked was that the constraints imposed by the "politicians"—who notably included the Joint Chiefs of Staff—precisely reflected America's limited war aims in Korea: repel the North Korean attack, stop the fighting, and do so without starting World War III in the process. In ruling out air strikes on Chinese territory, the Joint Chiefs had pointedly stated, "It would be militarily foolhardy to embark on a course that would require full-scale hostilities against great land armies controlled by the Peking regime"—this especially at a moment when Communist forces continued to pose a threat all around the world, and in Western Europe most notably.

Blaming political constraints also deflected attention from the fact that, within the considerable leeway the airmen *were* given to select and attack targets in Korea, they had often favored the wrong targets or lacked the means to identify and attack the right targets. Airmen complained loudly that they had not been permitted to bomb the bridges over the Yalu River that allowed 200,000 Chinese troops to flood into North Korea; even when this restriction was lifted in November 1950, they were still barred from crossing into Chinese airspace and so could only approach the bridges perpendicularly, flying along the river rather than along the length of the span—which greatly reduced the odds of setting a bomb on the target. But more significant was a woeful lack of aerial reconnaissance capability, and a misuse of what was available, that incredibly had allowed the Chinese infiltration to go undetected for weeks in the first place. That failing was then compounded by a chronic shortage of effective fighter-bombers, an insufficient capacity to suppress enemy flak, and an almost total inability to carry out successful night attacks or deliver munitions with precision. B-29s were sent in at 21,000 feet to avoid the flak; at those heights, hitting a bridge was close to impossible. By the time even half of the seventeen bridges were finally destroyed, the Chinese army was already across the Yalu and had seized the initiative.

Along with the Chinese ground forces came Chinese MiG-15 jet fighters, 445 of them by mid-1951, that began to harry the B-29s and F-80s. Rushing fifty new F-86 fighters into operation and scoring a thirteen-to-one kill ratio with them against the MiGs, the American air forces were able to stem this aerial juggernaut but never completely halt it; with the advantage of sheer numbers, the Chinese repeatedly were able to send "trains" of sixty to eighty MiGs southward to bounce American bombers on interdiction missions.

All of this pointed to intrinsic weaknesses in Air Force doctrine, equipment, and preparations for conventional war. By LeMay's reasoning, though, conventional wars were a thing of the past. A force properly designed to defeat the Soviet Union in a total war would create a "strategic umbrella," he said, that would deter limited wars as well. Or as Air Force Manual 1-2 put it, "The best preparation for limited war is proper preparation for general war." (It was during the late 1950s that Earl Long, the erratic and flamboyant governor of Louisiana, uttered a pithy statement that suggested how all-encompassing the nuclear-deterrent umbrella had become in the American mind. After federal courts ordered the integration of schools, Long confronted the leading segregationist politician in the state, Leander Perez. "What you going to do now, Leander?" he asked. "The Feds have got the atom bomb.")

MiG–15
single-seat fighter
engine: 6,000-lb. thrust
   turbojet
top speed: 670 mph
armament: 37mm plus
   two 23mm cannons;
   rockets or 2,000-lb.
   bomb load

Within a few years it was as if Korea had never happened. LeMay in 1957 told the Air Force Science Advisory Board that, although "token" surface forces would still be required to "show the flag" and to "insure that Soviet violation of Free World territory automatically becomes a clear-cut act of war" (and, of course, to serve as "clean up and policing forces" once air power had delivered the victory) he simply could not "conceive of major protracted engagements by surface forces which have been previously disembodied by a properly executed nuclear air offensive." At a conference that same year, senior air commanders agreed that the Air Force should make no further increases in its inventory of high-explosive ordnance and set a goal of eliminating it altogether as soon as national policy permitted. LeMay said the Army didn't need the Air Force to provide air support, in any case; the Army could henceforth handle that job itself with short-range surface-to-surface missiles.

F–86A
single-seat fighter
engine: 5,200-lb. thrust
   turbojet
top speed: 685 mph
armament: six .50-in.
   machine guns and 8
   rockets

The immediate effect on TAC of such thinking was not only to keep it on a shoestring but increasingly to co-opt it into the nuclear-strike mission. Seeing which way the wind was blowing, TAC commanders, Quesada included, had from the start realized their best survival strategy was to make TAC indispensable even if its traditional mission was dead. A board that Quesada headed just before he left the Air Force in 1949 argued that TAC's primary mission would be as an appendage to the strategic air campaign; fighters would go after targets whose destruction would "tend to accumulate the initial effect of the atomic bomb." (In an abject bit of revisionism, the board even disavowed any claims that the tactical air forces in the Second World War had made their greatest contribution by attacking German troops and tanks; rather, it was "in the augmentation and acceleration of the strategic air campaign" that their efforts had really paid off.)

The advent of small thermonuclear weapons that could be carried by fighters, and the huge increase in the number of targets to be covered by SAC's strike plan as Soviet ICBMs began to be deployed, only increased the pressures on TAC to become a subcontractor of SAC. Fighter pilots joked about having been "SACumcized." The result was reflected in the airplanes that emerged in the late 1950s and early 1960s. A plane did not have to be able to dogfight or withstand ground fire over the battlefield to drop a nuclear bomb on the Soviet Union as part of SAC's total-war plan; it just needed to go fast.

So, too, did fighters intended for TAC's other total-war mission: defending the United States in turn against a Soviet nuclear attack. By 1957 two-thirds of the 100 fighter squadrons in the Air Force were permanently assigned to the Air Defense Command. Fighters built for this job, such as the F-102, were little more than flying darts, designed to accelerate to supersonic speeds and shoot down Soviet bombers with nuclear-armed air-to-air rockets or missiles. The weapon itself would be fired automatically by the aircraft's radar system once the plane was in the right position. And to get the planes in the right position, the Air Defense Command was spending billions of dollars on a system called SAGE ("semi-automatic ground environment"): Two dozen huge digital computers, each containing 50,000 vacuum tubes (and a three-megawatt power supply), were built by IBM and installed at ground-control centers across the country. Linked by telephone lines, they translated radar data into a continually updated video map display so controllers could follow the air battle in real time. The F-102's successor, the F-106 Delta Dart, could even be flown automatically by SAGE from the moment of takeoff. Fighter-interceptors such as the F-102 and F-106 were conceived from the start not so much as fighters but as components of a vast electronic fire-control system to defend the United States during World War III. The planes' capability for dogfighting, let alone ground attack, was nil.

A few voices in the wilderness had predicted early on that one important consequence of mutual nuclear deterrence would be to make the world once again safe for conventional war. In April 1950 Vannevar Bush wrote to Omar Bradley warning that it was a dangerous and unrealistic strategy to put so much faith in massive atomic retaliation, and to neglect conventional land forces and air support, now that the Soviets had the bomb:

> We cannot win a war and emerge in sound condition without adequate armies of our own. . . . We have no tactical air force worthy of

the name, nor have our Allies. Our enemy has always placed great
weight on tactical air and is doing so now. We cannot allow our
armies or those of our Allies to fight without such support or they
will be overwhelmed. We had better get at it.

Air Force planners, Bush concluded, "have been drawn down a single line of
reasoning much too long." Bush particularly urged that work begin at once on
advanced air-launched antiarmor weapons to counter Soviet tank strength.

General Matthew B. Ridgway, the Army Chief of Staff, raised a similar concern in 1954, cautioning that reliance on massive retaliation could leave the
United States in the impossible position of having to choose between resorting to
atomic weapons or acquiescing in Soviet conventional gains. But the most influential and trenchant criticism came in 1957 in a brilliant analysis of the shortcomings of United States nuclear strategy by Henry A. Kissinger, then an
obscure instructor in the government department at Harvard University.

Nuclear Weapons and Foreign Policy presented a sweeping, merciless, rigor-
ously argued indictment. Kissinger began by distilling the dilemma of nuclear
weapons to its essence: "The more powerful the weapons . . . the greater
becomes the reluctance to use them." The bomb might be a credible deterrent
against a surprise attack on the United States, but only because any nation so
attacked could be counted on to respond "in a fit of righteous indignation" with
a war of "maximum destructiveness." The United States had failed to develop a
strategy for translating its military superiority into political advantage for circumstances short of total war. Notably, even America's complete monopoly on
the bomb in the immediate postwar years had failed to deter the Soviets from
expanding their control over Eastern Europe.

The failing was at its roots a conceptual one, Kissinger argued. American
military planners had looked upon the atomic bomb "merely as another tool in a
concept of warfare which knew no goal save total victory and no mode of war
except all-out war." Now, in the face of a growing Soviet capability to inflict devastating damage in return, "our reluctance to engage in an all-out war is certain
to increase." The growing potential destructiveness of war had also exacerbated
a traditional American tendency to view war and diplomacy as polar opposites:
the object of war was total victory through unconditional surrender; the object of
diplomacy was an equally absolute *prevention* of war, at all costs. Yet, especially
against a foe such as the Soviet Union, the ability to convincingly threaten to use
force was an indispensable component *of* diplomacy. And the only way to wield
this threat in an age of nuclear stalemate was to build a credible capability for
limited war. Indeed, given the impossibility of waging a decisive all-out conflict,
such "intermediate" applications of power were the best way to bring about
larger strategic changes favorable to the United States.

As Korea had shown, fighting a limited war required "radically" different
weapons, planning, concepts of operation, and training from what it took to pre-

pare for SAC's all-out nuclear strike plan. In particular, limited war required high mobility and flexibility, since targets were not knowable in advance. Kissinger believed that both battlefield nuclear weapons and improved conventional forces were required to restore credibility to America's "extended deterrence" over Western Europe and to halt the Soviets' relentless, piecemeal erosion of America's position throughout the world.

Other voices slowly swelled the chorus. A RAND study noted that the U.S. Air Force was not equipped to "perform the tasks which may be required of air power in local war." An Air University study headed by Bernard Brodie concluded that the traditional strategic doctrine of striking directly at "sources of national power" as opposed to enemy military forces would not be applicable in limited conflict. TAC's General Weyland made so bold as to opine—it was "obvious" to him, he said—that "we must have adequate tactical air forces in being that are capable of serving as a deterrent to the brush-fire type of war, just as SAC is the main deterrent to global war."

But most senior Air Force officers who agreed with such views—and these men were rare the late 1950s—had the wisdom to keep it to themselves. Shortly before his retirement as Chief of Staff in 1965, LeMay made it clear that he had not changed his thinking one iota. "All conventional forces do is delay the inevitable nuclear confrontation," he told an interviewer. If anything, LeMay insisted, "we have gone too far with our conventional capability."

13

HARD KNOCKS

Lyndon Johnson once called Vietnam "a raggedy-ass little fourth-rate country." That pretty well summed up the confidence with which America's leaders in the fall of 1964, military and civilian alike, contemplated a show of force that would convince the Communist insurgents in that far-off raggedy-ass land that they didn't stand a chance against the mightiest and most technologically advanced nation on earth.

The confidence ran the gamut from the cosmic to the trivial. It was in the very air that Americans breathed: What would seem fatuous, shamelessly hubristic, a mere decade later when the very word "Vietnam" had become an emblem of American defeat and humiliation seemed just common sense then. American technological prowess was unquestionable. The United States' triumph over Nazi Germany and Imperial Japan and her subsequent resolute stand against Communist expansionism in Korea and Europe had given Americans a confidence not only in their ability to resist the mightiest of foes but in the manifest rightness and goodness of their cause. From the President's inner circle to the Special Forces "advisers" who began to arrive in South Vietnam in the thousands in the early 1960s, the idea that the United States of America could lose a tiny war in a primitive country—or even that Americans might be seen by the local inhabitants as colonialists rather than as champions of freedom and liberty that they so obviously were—would have seemed incredible. "Probably the only people who have the historical sense of inevitable victory," commented the British historian Denis Brogan, "are the Americans."

Robert S. McNamara, President Kennedy's and now President Johnson's Secretary of Defense, was the personification of another facet of that very American self-confidence. McNamara had been a young teacher at Harvard Business School at the beginning of the Second World War when Robert Lovett, the Assistant Secretary of War for Air, had approached the school for help in training the officers who would manage and organize the huge buildup of hundreds of thousands of pilots and airplanes. McNamara was the standout in the program; he had an awesome ability with facts, figures, and logical analysis, and he was soon tapped to manage the myriad complexities of the B-29 bomber program. After the war, McNamara and the other Harvard "Whiz Kids" convinced Henry Ford II that the modern, quantitative management methods that they had used to transform the Air Force into a vast industrial enterprise, a supercompany really, could revolutionize and rationalize the auto business, too. McNamara rose quickly through the ranks at Ford, eventually becoming the company's president.

As Secretary of Defense, McNamara apparently believed that one could run a war the same way one ran an auto plant. He wanted to see the numbers on everything: troop supplies, ratios of enemy killed, ordnance expended. On his first visit to Vietnam, in May 1962, he spent two days filling his notebook with statistics, then announced to reporters that the Americans and South Vietnamese had the Communists on the run. Neil Sheehan, a UPI reporter, intercepted the Secretary as he was getting into his car to dash back to the airport and asked him, off the record, how he could possibly be so confident "about a war we had barely begun to fight." McNamara replied with his trademark steely stare. "Every quantitative measurement we have," he said, "shows that we're winning this war."

The Second World War had proved the importance of operational analysis in modern warfare, but McNamara's version was so extreme that at times it seemed like a parody of operational analysis. At Ford, and now at Defense, associates and aides discovered that the avalanche of statistics that McNamara let forth in meetings was often simply a weapon he wielded to beat aside opposition to his ideas and plans; it was power masquerading as scientific objectivity. They learned that the way to stay in his good graces was to tell him what he wanted to hear, and in the form he liked best, statistics.

General Paul D. Harkins, head of the U.S. Military Assistance Command Vietnam, certainly got the message. He made it clear to his staff that they were to report only good news and good numbers. Harkins titled his weekly report to the Pentagon the "Headway Report." Intelligence from the field was edited by his operations staff, which systematically reduced the size of the enemy, inflated the numbers killed, discounted the many signs of the Communist cadres' fanatical commitment to their cause, and glossed over the pathetic performance of the South Vietnamese Army troops. (It turned out that South Vietnam's strongman Ngo Dinh Diem, fearing that his Army might not be around to protect him from coups in Saigon, had forbidden any offensive actions that might result in casualties.)

Anyone who looked very closely might have wondered from the start if the Americans were backing a losing horse. The South Vietnamese government had been established after Ho Chi Minh's Communist-led rebellion forced the French to withdraw in 1954 and Ho declared the Democratic Republic of Vietnam on the territory his forces held in the north; a million refugees, two-thirds of them Catholics and most of the rest the families of soldiers and police officers who had served in the French colonial government, fled south. Diem's South

Vietnamese Army was commanded entirely by former Vietnamese officers and noncoms of the French colonial army. To many Vietnamese, the regime in the South would remain fatally tainted by these associations; their fight against the regime was but a continuation of the anticolonial struggle. American reporters noted that villagers did not welcome Americans the way the South Koreans had. One of the more reliable American intelligence reports called attention to the unsettling fact that 80 to 90 percent of the guerrilla fighters in the South were locally recruited. American officials in Saigon were afraid to drive even a few miles south of the city without the protection of a military convoy. Across vast stretches of the country, the Communist guerrillas operated with near impunity, assassinating thousands of local government officials, setting ambushes by night, melting back into the countryside by day. Harkins ignored it all and explained that American willpower and American firepower would carry the day.

All of the confident American self-delusions came together in late 1964 as the situation on the ground rapidly deteriorated and planners began to cast about for a decisive military move that would bring the Communist insurgency to a halt. Operation Rolling Thunder, which would drop 643,000 tons of bombs on North Vietnam in the longest sustained bombing campaign in history, was the product of all of these self-delusions, none more than the determination of air planners and civilian strategists alike to make a supremely unconventional war conform to conventional American beliefs.

As the historian Mark Clodfelter observed, the growing American sense by the fall of 1964 that an air campaign could break the impasse in Vietnam was more a "mood than a belief," a product of "frustration more than conviction." It was in any case a perfect reflection of the desire to force the chaotic jungle fight against a shadowy enemy into a conventional form that the United States military thought it could master. Many airmen, as one Air Force analyst put it many years later, "hoped escalation would clarify and simplify the conflict."

Overshadowing the doubts and skepticism that were occasionally voiced was the pervading sense that Vietnam was becoming, as McNamara put it in a key policy memorandum approved by Johnson in March 1964, "a test case of U.S. capacity to help a nation meet a Communist 'war of national liberation.' " In such a test of wills, McNamara and the other top civilian advisers believed, what really mattered was standing firm and showing American resolve. Many of Johnson's civilian advisers were strongly influenced by the theories of nuclear conflict and the notion that an enemy could be coerced to surrender by demonstrating the certain destruction that awaited if he continued to resist. As McNamara argued to Johnson, the pressure on North Vietnam at any given moment "depends not upon the *current* level of bombing but rather upon the credible threat of *future* destruction which can be avoided by agreeing to negotiate or agreeing to some settlement in negotiations." After all, no one could doubt America's theoretical ability to incinerate any adversary on earth if she so chose; the whole point of bombing the North would thus be to show that the United States was serious, that the threat of unlimited further destruction was "credible." In November an interagency working group chaired by Assistant Secretary of State William P. Bundy accordingly proposed a graduated series of air strikes against the North "designed to signal U.S. determination, to boost morale in the South and to increase the costs and strains upon the North."

By its very nature that meant holding back, not just to prevent the war from escalating into World War III with the Chinese and the Soviets, a risk Johnson constantly worried about, but also, as National Security Adviser McGeorge Bundy argued, so the North Vietnamese would realize that "there is always a prospect of worse to come." As in the nuclear-war scenarios, the idea was to keep the hostage alive.

LeMay and other air leaders advocated a much more concentrated series of air strikes and prepared a list of ninety-four targets in North Vietnam to be struck in a sixteen-day blitz. It was once again the classic expression of the Air Force strategic-bombing credo: Strike at industry and transport in the heart of the enemy's territory, and you will destroy his basic war-making capacity and interrupt the flow of supplies to the field. As an added bonus, the resulting disruption to the civilian economy would demoralize the government and populace and (in the words of the operational commander of Rolling Thunder, Admiral U. S. Grant Sharp) engender "general frustration, anxiety, and fear." LeMay would later describe his plan with the unforgettable phrase "bomb them back into the Stone Age," insisting that if he had only been permitted to carry it out, he could have won the war "in any two-week period you want to mention."

The inflexibility, lack of imagination, and political incomprehension underlying these proposals, which air commanders and the Joint Chiefs would advocate again and again with only minor variations, heightened Johnson's mistrust of his generals and convinced him he had to keep tight personal control over the bombing once it began. McGeorge Bundy sent Johnson a memo warning that airmen would always fudge on the issue of civilian casualties; it was a lesson, he said, he had learned from former Secretary of War Henry L. Stimson, who always felt he had been "hornswoggled" by Hap Arnold over the firebombing attacks on Japanese cities. (Stimson, fearing that the United States might "get the reputation of outdoing Hitler in atrocities," had sternly questioned Arnold over the bombing policy in the final months of the war, but Arnold had glibly reassured him that the AAF was doing its best to stick to precision targeting of Japanese industry.)

All that generals know how to do, Johnson said, was bomb and spend. On one occasion he dressed down the Army Chief of Staff, General Harold K. Johnson, right in front of the general's staff: "Bomb, bomb, bomb, that's all you know. Well, I want to know why there's nothing else. You generals have all been educated at taxpayer expense and you're not giving me any ideas and any solutions for this damn little pissant country. Now, I don't need ten generals to come in here ten times and tell me to bomb."

Johnson would later boast that the way he ran things, "they can't even bomb an outhouse without my approval." He wasn't joking. Every Tuesday Johnson would hold a White House luncheon meeting with McNamara, McGeorge Bundy, Secretary of State Dean Rusk, and presidential Press Secretary Bill Moyers. Until October 1967, there was not even a military representative in attendance. The group would approve "packages" of targets that could be struck during the following one to two weeks. They also set the number of sorties that would be allocated to each target. The North Vietnamese capital of Hanoi and its major port, Haiphong, were off-limits altogether; no targets could be struck within a fixed radius of either city, not even in response to antiaircraft fire, without direct approval from the President. Also off-limits was a buffer zone along

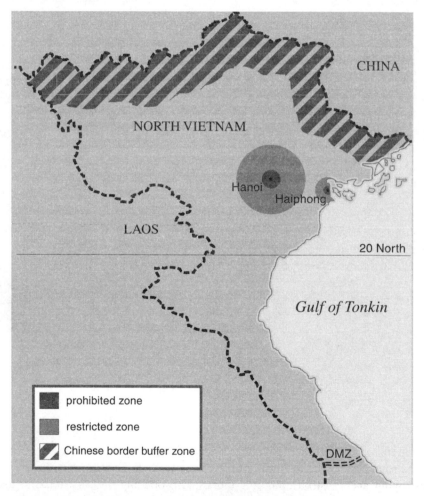

Off-limits areas during Rolling Thunder. Prohibited zones could be attacked only with direct authorization of the President.

the Chinese border, as well as airfields and SAM sites, lest Chinese or Soviet construction workers be injured or killed.

The Rolling Thunder plan as it unfolded was a carefully graduated series of escalations with periodic pauses to allow the North Vietnamese to cry uncle. Attacks began mainly south of the nineteenth parallel; in May 1965 they moved up to the twentieth parallel; by July isolated targets north of the twentieth parallel were struck. The goal, the planners explained, was to lead up to "a situation where the enemy must realize that the Hanoi and Haiphong areas will be the next logical targets." The plan was a fantastic delusion. The North Vietnamese were as mystified as the American pilots who had the job of carrying it out. (If you want to send a message, went the sarcastic saying in the Air Force, call Western Union.)

The White House micromanagement led to some absurd consequences familiar to students of the perverse incentives of bureaucracies. Commanders were always sure to fly their full budgeted allotment of sorties, whether or not they were needed to destroy the assigned target, because if they didn't use them all up they wouldn't get as many next time. At one point when American forces were suffering from a shortage of munitions, strike aircraft were each sent off with less than a full load of bombs to avoid having to cut back on the number of sorties.

The rules and limits provided rich material for military men who would later complain that their teeth had been drawn by meddling civilians. (Admiral Sharp complained of all the "pussy-footing around" with bombing pauses and target restrictions.) But by the time Johnson announced a halt to the bombing on March 31, 1968—at the same time stunning the country with the announcement that he would not run for reelection—almost all of the restraints had been released. Not only had the original ninety-four targets whose bombing LeMay said would reduce North Vietnam to the Stone Age been destroyed, but nearly all save a half dozen or so of an expanded list of 242 targets advocated by the Joint Chiefs had been taken out as well. By April 1967, General Earle G. Wheeler, Chairman of the Joint Chiefs, commented at a White House meeting that "the bombing campaign is reaching the point when we will have struck all worthwhile fixed targets except the ports."

Mark Clodfelter, an Air Force major and a history professor at the Air Force Academy, was among the first from within the Air Force to challenge the conventional military wisdom that civilian meddling lost Vietnam. He concluded that Rolling Thunder could *never* have worked. In the words of another Air Force skeptic, Colonel Dennis Drew, Vietnam had exposed a "doctrinal void," a stagnation in Air Force thinking that led leaders to keep trying to drive a square peg into a round hole. The most glaring defect in Rolling Thunder was that, even if destroying an industrial country's means of producing and distributing goods were the magic formula for victory, North Vietnam was simply not an industrial country. McGeorge Bundy had once made an offhand retort to LeMay's suggestion of bombing North Vietnam back to the Stone Age by suggesting that maybe North Vietnam was already in the Stone Age. It might have been a good idea if he had followed this thought to its logical conclusion. Industry accounted for a mere 12 percent of the country's minuscule $1.6 billion GNP. North Vietnam had a single steel plant. When the Joint Chiefs drew up their initial target list, intelligence analysts could find only eight industrial installations in the whole country worth listing.

Making a virtue out of necessity, American air planners tried to argue that the North's tiny industrial base made what it did have that much more vital, and thus the county was *more* vulnerable to strategic bombing than an industrialized nation would be. This might have made sense if North Vietnam had been populated by Americans who couldn't imagine how to get along without electricity or the corner gas station, but it was ridiculous given North Vietnam's reality as a subsistence agricultural economy. A country of peasants used to doing without modern conveniences had little difficulty adapting to the loss of the few they did have.

In December 1966, the North Vietnamese government permitted a sympathetic American newsman, Harrison Salisbury of the *New York Times,* to inspect the effects of the American bombing. Although he was duped into repeating some wildly exaggerated tales of civilian casualties, he accurately observed and reported the improvisations by which the North had swiftly overcome the loss of bridges, rail lines, and oil-storage depots. If a bridge was down or a track cut, for example, "bicycle brigades" were called up:

> Five hundred men and women and their bicycles would be sent to the scene of the break. They would unload the stalled freight train, putting the cargo on the bikes. Each bicycle would handle a six-hundred pound load, balanced across the frame with a bar. The bicycles would be wheeled, not ridden, over a pontoon bridge, and on the other side of the break a second train would be drawn up. The cargo would be reloaded and moved on south.

By 1968 an estimated 60 percent of all oil-storage facilities had been wiped out; the North Vietnamese responded by filling thousands upon thousands of fifty-five-gallon drums with fuel and dispersing them along roadways, in villages, rice paddies, everywhere. A study by top American civilian scientists in December 1967 reported to McNamara that as a result of measures North Vietnam had taken in response to the American bombing, it had actually *increased* its capacity to transport war materiel.

The other problem with Rolling Thunder was that, while it was true that the southern insurgency was being helped and supplied by the North, it was also true that even as late as mid-1967 North Vietnam had only 55,000 regular troops in the South; the rest of the 300,000 or so Communist fighters were indigenous guerrillas who were supplied locally and who fought an average of only one day in thirty. Overall Communist supply needs were so small, Mark Clodfelter estimated, that they could be met by a grand total of seven 2½-ton trucks per day. That amounted to less than 1 percent of what the North itself imported from abroad every day, less than 0.2 percent of what its ports and roads and railroads were capable of handling had the North chosen to operate them at full capacity. A Pentagon analysis in February 1967 calculated that a completely "unrestrained" air campaign, even backed by the mining of Haiphong harbor, would reduce the North's import capacity to 7,200 tons a day, which was still about 200 times what it needed to keep the war going.

Johnson was politically astute in fearing that reports of civilian casualties would arouse hostility against the American war effort even among America's allies; even with his elaborate restraints and refusal to approve targets that might result in "collateral damage," the inevitable inaccuracies and outright mistakes took their toll and triggered protests from European leaders. At the start of Rolling Thunder, the average circular error of bombs dropped by U.S. Air Force fighter-bombers was 750 feet; it took several years to cut that even by half. Ironically, the years of cant about American precision-bombing ability had easily convinced Salisbury and other observers that civilian targets such as housing and irrigation dikes that had been hit must have been hit deliberately.

The even greater irony was that Johnson's attempts at political damage control through cautious target selection were in the end almost for naught. Salisbury's front-page reports in the *New York Times* received huge attention, and it proved extremely difficult for the Pentagon to rebut them. Salisbury had obviously been manipulated by the North Vietnamese, and his naïve conclusion that the U.S. Air Force was deliberately "dropping an enormous weight of explosives on purely civilian targets" was clearly false (as the author Tom Wolfe would later observe, the North Vietnamese had played the *Times* man like an "ocarina"), but in fact the Pentagon's estimates of civilian deaths caused unintentionally by the bombing were not far from what Salisbury was reporting. A RAND analysis concluded in 1966 that, as far as its impact on North Vietnamese morale was concerned, the bombing had provided Hanoi "with a near-ideal mix of intended restraint and accidental gore." It was never enough to hurt militarily, but quite enough to keep the North supplied with a quantity of outrages to display to visiting journalists and peace activists.

The other air war would never make the same headlines. While American aircraft were dropping half a million tons of explosives on the North to dubious effect throughout the four years of Rolling Thunder, four times that weight of bombs were falling in the South in support of American and South Vietnamese ground forces. By the time the United States pulled out in 1973, more than six million sorties had been flown and eight million tons of bombs dropped, with more than 95 percent of the sorties and 90 percent of the bomb tonnage directed against targets in the South and nearby Laos and Cambodia.

A vast proportion of the air effort in the South would be expended in an attempt to cut off the Ho Chi Minh Trail, the network of jungle paths and roads that were the North's major supply and infiltration route. This was another attempt to conventionalize the war, applying another model of Second World War air power—interdiction—to a situation for which it was manifestly ill suited. At the strategic level, the interdiction campaign suffered from the same miscalculations as Rolling Thunder, grossly overestimating the insurgency's dependence on materiel and grossly underestimating the enemy talent for making do.

The contrast between American technological wizardry and Vietnamese improvisation was never more starkly displayed than in the interdiction campaign. At an "Infiltration Surveillance Center" in Thailand, American commanders stood by each night as IBM 360/65 computers analyzed signals relayed from a phalanx of tiny air-dropped sensors that could detect sounds, vibrations, electrical emissions from engine ignition systems, even human urine. In time twenty thousand of the sensors, cleverly camouflaged to resemble vegetation, would be dropped, at a cost of billions of dollars. (One type of sensor was made to look like dog excrement until someone realized that there weren't any dogs in the area. It was redesigned to look like a piece of wood.) When the computers had pinpointed the general location of an enemy movement, strike aircraft would be sent in to saturate the area with bombs. "McNamara's Wall," they called it. It was a technological marvel, just the thing to excite the technocratic enthusiasm of the Defense Secretary. At the peak of the interdiction campaign, hundreds of fighter-bombers and dozens of B-52s were carrying out strikes each day, destroying tens of thousands of trucks and wiping out as much as 90 percent of all the supplies the North attempted to move south.

The truth, the almost inconceivable truth, was that it didn't matter. Supplies never fell below what American analysts calculated to be the Communist forces' minimum requirements. The interdiction campaign certainly did not prevent the North Vietnamese from building up sufficient forces, including several hundred tanks, to launch a major conventional offensive in the South on Easter 1972. To conceal their movements, the North Vietnamese tied together the branches of the overhanging forest canopy, carved out new routes, built hidden underwater bridges across rivers using sandbags, supplemented trucks with oxcarts, boats, and porters. Reconnaissance photos would capture an image that looked as though it had come straight from the moon, almost nothing but bomb craters, and there, snaking along the few bits of intact earth, would be a new-plowed road.

To the public, the press, and many politicians and civilian strategists, it probably would have sounded like the old joke about the operation being a success even though the patient died to have pointed out that American forces achieved some remarkable technological and tactical successes even while losing the war. But in fact the interdiction and jungle campaign in the South was a proving ground for several key innovations that would be of enduring, even revolutionary, importance in the application of air power to the battlefield in future wars. Partly this was because, whatever the failings of doctrine and strategy at higher levels of military and civilian command in Vietnam, improvisation and experimentation always occur at a tactical level when the necessities of war replace the by-the-book practices of peacetime.

One of the most unconventional innovations to emerge from Vietnam—for it defied almost all accepted Air Force notions of aircraft performance, roles, and tactics—was the side-firing gunship. The basic idea had been around for decades, though it had never gone anywhere. A gun aimed out the side of an airplane that is flying in a slow, steady circle will always remain fixed on a point of ground at the circle's center. Rather than taking a single sweeping pass, a gunship orbiting around a target could thus bring a sustained, highly accurate fire to bear. And if the gunship were a big transport such as the DC-3, it could carry some very big guns indeed: machine guns, Gatling guns, even howitzers.

Flying slow, obsolete transports at low altitudes was not the Air Force's idea of a tactical air mission. When several Air Force officers tried to push the concept in the early 1960s, arguing that it was just what was needed in Vietnam, they ran into heavy resistance. Tactical Air Command leaders thought the idea would be a failure. Even worse, it might be a success, which would set an extremely awkward precedent. However valuable gunships might be in a counterinsurgency war against troops lacking sophisticated surface-to-air weapons, TAC, like everyone else, was convinced that counterinsurgency wars would never be a major part of its mission, and did not want to be stuck with a weapon that would be a sitting duck in a high-intensity, high-threat environment such as Europe. It would be hard to make a case for buying expensive high-performance jets if cheap and slow prop planes could do the job just as well or better. Moreover, TAC had been fighting tooth and nail against the Army's recent attempts to build up its own force of armed helicopters to provide battlefield air support; one of the major arguments it had been using to counter the Army was that helicopters were too slow to survive on the battlefield. If gunships proved successful it would be rather hard to sustain that argument.

Kept alive by a few enthusiasts, the gunship idea bounced from TAC to the Air Force psychological warfare laboratory to the Aeronautical Systems Division and finally won approval for a small test program in mid-1963. The impressive results quickly convinced some of the skeptics, and on the night of December 23, 1964, a military DC-3 carrying three 7.62mm Gatling guns (the aircraft was now officially designated the AC-47, A for attack) went into action for the first time, repelling a Communist attack on a hamlet in the South. American troops promptly dubbed the machine "Puff the Magic Dragon," which was both a reference to the puffs of smoke that filled the air as its guns blasted out their six thousand rounds a minute and an ironic allusion to the nicey-nice children's song popularized at the time by the saccharine singing group Peter, Paul, and Mary.

Subsequent gunship models were equipped with advanced infrared night-vision sensors, digital targeting computers, and, in the definitive AC-130 version, much increased firepower: four 20mm Gatling guns at first, later a devastating 105mm howitzer as well. From January 1968 to April 1969, four AC-130s flew just 4 percent of the interdiction missions against the Ho Chi Minh Trail in Laos but accounted for 29 percent of the trucks destroyed and damaged. General Creighton Abrams, the last American commander in Vietnam, commented that the AC-130 had been one of three weapons that had proved unqualified successes during the war. (The other two were the wire-guided anti-tank missile and the laser-guided bomb.) AC-130s would play crucial roles in subsequent conflicts, conventional and unconventional, in Grenada, Panama, Iraq, Somalia, Bosnia, and Afghanistan.

The Vietnam War would also be the making of the helicopter in battlefield operations. For all of the Army's enthusiasm for the idea, the helicopter-borne air assault and the armed attack helicopter were merely speculative concepts at the start of American involvement in Vietnam. Small helicopters had been used for medical evacuation in Korea, and the increasing lift capacity of models that appeared throughout the 1950s had fired the imaginations of Army planners who saw bigger possibilities. Among these possibilities was reclaiming part of the tactical air support mission that had been ceded to the Air Force. Many in the Army continued to doubt the Air Force's commitment to the mission; only by having aviation units directly answerable to ground commanders, they felt, could the Army count on getting the air support it needed, when it needed it.

The 1947 act creating the Air Force had permitted the Army to retain aviation units "organic" to it. Subsequent interservice agreements and Defense Department directives restricted the definition of *organic* to missions such as observation and liaison, and explicitly assigned tactical airlift and close air support to the Air Force. Worried that the Army would always try to stretch the definition, however, the Air Force wanted to have the physical characteristics of permitted Army aircraft spelled out as well, and a 1952 agreement did just that, requiring that the Army's fixed-wing planes not exceed 5,000 pounds. But perhaps because the Air Force thought helicopters would never amount to much, it agreed that the Army could operate any helicopters it wanted to. This was the loophole the Army now began to enthusiastically exploit.

Besides providing aerial firepower under direct Army control, the helicopter could, in the view of its proponents, solve a problem that had been nagging ground commanders since the Second World War. In April 1954, *Harper's* published an article by Major General James M. Gavin, who had commanded the famous 82nd Airborne Division during the war. Titled "Cavalry, and I Don't Mean Horses," Gavin's article was subsequently reprinted in many military journals and had a profound impact on ideas about army aviation. Gavin pointed out that, as important as tanks and armored divisions had become to modern ground warfare, they lacked the mobility to carry out missions that traditionally had been assigned to cavalry: screening movements, deep reconnaissance, raids in the enemy's rear areas. The mobility of troops was in fact currently no greater than it had been during the war. Helicopters, which were capable of carrying a dozen combat-equipped men apiece, could fill this gap by serving as the mounts of a new "Air Cavalry."

Much of the official Army rationale for air mobility and armed helicopters was couched in terms of the Soviet threat in Europe. Gavin argued that the threat of battlefield atomic weapons would require armies to disperse over much wider areas than had been necessary in the past; air mobility would thus be required not only to maintain contact and reconnaissance over these greater distances but also so that troops could be quickly dispersed for protection against atomic attack and then just as quickly concentrated to launch a counterattack. By 1955, the Army was testing antitank weapons to see if they could be fired safely and effectively from helicopters and had started to work out doctrine, techniques, and tactics for air-mobile units.

This focus on a conventional "big war" scenario to some extent merely reflected the Army's wish for helicopters; at the moment, this seemed the best justification for them. But as a consequence precious little thought was given to using air mobility and armed helicopters in counterinsurgency warfare, so little that when General Johnson first asked his aviation experts if any study had been made of how air mobility might work in Vietnam, they were astonished at the general's question.

There was a relevant precedent, however, and it became increasingly relevant as the first American helicopter companies began arriving to assist South Vietnamese troops in December 1961. Six years earlier the French, fighting rebels in Algeria, had been the first to employ helicopter air mobility in war, and a counterinsurgency war at that. Although the French lost, the introduction of the helicopter had dramatically evened the score against guerrillas who had been operating in small groups with near impunity and considerable advantages in mobility, surprise, and knowledge of the local terrain. Transporting an assault party by helicopter allowed the French forces to quickly assemble for action before a rebel force could be tipped off that they were coming. Within minutes, an entire battalion of 480 men could be ferried to a landing zone on a couple of dozen helicopters shuttling between multiple pickup points. At the landing zone, the first troops would be out the door when the helicopter was still six feet off the ground, the last man out before it even touched down.

When the Algerian rebels began to acquire machine guns and shoot down helicopters approaching their landing zones (rebel marksmen also, quite effectively, began simply picking off the pilots), the French began to precede each landing with a series of bombing and strafing runs by fixed-wing attack aircraft.

But the timing had to be precise, or the rebels would still be able to get back into position in time to fire on the arriving helicopters. This became a compelling argument for arming a portion of the helicopter force, so that the landing party would arrive with its own fire support and protection. By the time the war ended in 1962, the French had experimented with mounting a variety of weapons on helicopters, including machine guns, rockets, and 20mm cannons.

Vietnam catalyzed a rapid transformation of experimental U.S. Army units that had been set up to test the air-mobility concept into actual operational forces. In July 1965 the 1st Cavalry Division (Airmobile) was activated as a combat-ready unit, and in September it was ordered to Vietnam. The 1st Cav had 428 helicopters, five times the number of a regular infantry division, but it had only about half the number of jeeps and other vehicles. Tactical and technical lessons the French had learned in Algeria were quickly relearned and put into practice in Vietnam, most notably the need to beef up the firepower of attack helicopters. The first helicopters sent to Vietnam were unarmed, but the Communist forces soon began setting ambushes at likely landing zones, even rigging up bows and arrows that would be triggered automatically by the helicopters' downwash. Within months, in October 1962, UH-1 Huey helicopters armed with rockets and machine guns were being deployed to Vietnam to escort the transport helicopters and clear landing zones. From May 1966, heavier ACH-47 Chinook gunships, equipped with grenade and rocket launchers, 7.62mm Gatling guns, and 20mm cannons that could be swept through a 360-degree arc, were also brought into service, to be followed in turn by the first dedicated attack helicopter, the AH-1 Cobra.

Although armed helicopter gunships had evolved directly from the tactical demands of air-mobile operations, inevitably the Army began to rely on this source of ready firepower under its direct control in other situations. The Air Force began muttering about the Army's "overenthusiasm" for its latest toy, noting that from 1960 to 1965 the number of helicopters operated by the Army had doubled to five thousand. To circumvent the official restriction against supplying its own close air support, the Army came up with a tortured semantic definition that categorized attack helicopters as part of "a family of ground firepower systems" and explained that when helicopters were called in to blast enemy positions with rockets and guns, what they were doing was not close air support but rather "direct aerial fire support." In reality the only difference between the two was whether the mission was being flown by the Air Force or the Army.

The Air Force argued that the substitution of helicopters for more survivable fighter-bombers would result in an inevitable loss of combat power, and for proof it could point to situations such as the bloody fight at Ap Bac on January 2, 1963, where several helicopters were shot down approaching a landing zone and the situation was only retrieved when fighter-bombers appeared on the scene. An even worse debacle occurred during Operation Lam Son 719 in March 1971, when American helicopters airlifted South Vietnamese troops into action against a well-prepared force of 25,000 regular North Vietnamese holding positions along the demilitarized zone and in the Laotian panhandle. Two hundred of the 600 helicopters involved in the operation were shot down. The U.S. Air Force, by comparison, flew 8,000 tactical sorties in support of the assault with a loss of only seven aircraft.

A further source of Air Force antagonism was the Army's refusal to place attack helicopters under the joint tactical air-support system that was designated to coordinate and allocate ground-attack missions under a single air manager. (The command system in Vietnam was so chaotic and fragmented, however, with the Air Force, Navy, and Marines each directing its own strike missions, that this otherwise valid complaint lost some of its force.)

By the end of the war, the Army had lost a total of 2,400 helicopters to anti-aircraft and small-arms fire. Like the fixed-wing gunship, the attack helicopter had proved to be a devastating and very flexible weapon under the right conditions but a sitting duck in a high-threat environment. Its clear advantages were its ability to operate from close to the front, often under poor weather conditions that kept fixed-wing aircraft grounded, and its ability to unleash a sustained stream of fire at a fixed point on the ground and continue to do so until verifying that the target had been destroyed.

The addition of highly accurate, guided antitank armament in future attack helicopters would make them a potent weapon in the standard arsenal of armored forces throughout the world in decades to come. Survivability remained the crucial factor, however. In Vietnam, helicopter pilots found that as long as the enemy had only light weapons, they could operate in full view of enemy positions and simply stay high enough to remain out of range. But when facing heavy machine guns or bigger weapons, the only solution was to keep out of sight altogether, by flying at very low level, at night, under poor visibility, or following circuitous routes. Future attack helicopter tactics would emphasize "nap of the earth" flying and quickly "popping up" from behind terrain features to minimize exposure to enemy fire.

Air mobility was something of a mixed bag as well. Air assault by helicopter clearly had advantages over the Second World War–style parachute drop in speed, flexibility, and the ability to concentrate troops at the landing zone. But it would emerge from Vietnam as more of a specialized mission to be used for small raids or special-forces penetrations than as a staple of ground-maneuver warfare against a well-prepared enemy.

Few of the official nicknames dreamed up for military aircraft by the Air Force or by industry admen in the years after the Second World War ever caught on with the pilots who flew them. The names were a bit too self-consciously slick and phony. And so the B-52 was always just the B-52, never the "Stratofortress," and no one ever called the F-104 the "Starfighter," at least not with a straight face. Planes that had particularly unpleasant performance characteristics, however, readily acquired unofficial nicknames. Officially, the F-105, which would fly 75 percent of the Rolling Thunder air strikes against North Vietnamese targets, was the "Thunderchief." Unofficially, it was the "Thud" or the "Lead Sled."

Though designated a fighter, the Thud was no such thing; it was a supersonic nuclear bomber designed for what would in all probability be a one-way dash into the Soviet Union. The F-105 weighed as much as a B-17 and could carry twice the B-17's normal bomb load. To reduce drag and increase speed, the plane had been designed with short, stubby wings, which was fine for high-speed runs at very low altitudes but made it woefully unmaneuverable, especially at higher altitudes. Drag considerations had likewise led designers to do away with the bubble cockpit that afforded the pilot an unobstructed view to the rear, a feature that the aerial duels of the Second World War had proved to be a matter of life and death in any dogfight.

F-105D
nuclear strike fighter
engine: 24,500-lb. thrust
    turbojet
top speed: 1,390 mph
armament: 14,000-lb.
    bomb load; 20mm
    cannon

Early models of the plane had another problem: they broke. The Air Force "Thunderbirds" precision-flying team quickly abandoned the F-105 and went back to the older F-100 after one of their planes, apparently unable to take the stress of a routine vertical-turn maneuver, split apart and exploded in midair, killing a team pilot in May 1964. Extensive modifications were ordered on all F-105s to strengthen the fuselage. There was no way the F-105 was ever going to be a dogfighter.

Pressing the F-105 into service as a ground-attack aircraft revealed still more flaws and limitations. Little consideration had been given to survivability or ruggedness in an airplane that was going to fly a single mission in World War III. Because it lacked self-sealing fuel tanks—another feature that had been standard equipment for fighters in 1940—even small-caliber rounds could cause fuel leaks and explosions. Standard aircraft-design practice called for connecting the flight controls to the rudder, elevator, and ailerons with a pair of hydraulic lines so that in case one failed a backup would keep the plane flyable. But in the F-105 both lines ran side by side through the fuselage, so when one was hit usually both would go. When this happened, the elevator would lock in the full up position. A jury-rigged fix was hastily contrived in mid-1965 that provided the pilot with a mechanical lock to hold the elevator in the right position if the hydraulics were lost; this at least would increase the odds of keeping the plane flying until it was over friendly territory and the pilot could eject. Self-sealing fuel tanks and remotely controlled fire extinguishers were retrofitted as well. Still, by the time the last F-105 was replaced by the F-4 Phantoms in 1970, nearly 400 had been lost in Vietnam, more than half the total number built.

The F-4 was a marked improvement, but that was in part dumb luck rather than intentional design. The Phantom still had plenty of problems trying to play the part of an air-to-air fighter in a shooting war. Like other Air Force fighters of the period, the F-4 had not been built for fighting at all; it was in fact originally a Navy plane intended to defend aircraft carriers against slow maritime patrol bombers, while also doubling as a fighter-bomber. McNamara pressed the Air Force to acquire it as a multirole tactical aircraft in his drive for cost efficiency and commonality. The F-4 weighed just as much as the F-105, it lacked self-sealing fuel tanks like the F-105, and it had poor cockpit visibility like the F-105, but it did have two engines and bigger wings, which improved survivability and maneuverability. Though not optimal as either an air-to-air fighter or an attack jet, it was better at either role than a nuke carrier like the F-105, or than the F-105 successor that the Air Force had wanted to buy before McNamara's intervention in behalf of the F-4. Its engines, though, were notorious for producing a heavy black smoke trail that made it easily visible to an enemy fighter at a range of fifteen to twenty-five miles, robbing it of the advantage of surprise, which experience in wars from the First World War on had proved to be of decisive consequence in aerial combat.

The F-4 also notably lacked a gun, which might have seemed a basic failing in a fighter. Instead, the F-4 carried only missiles. It was a notion that was not so much wrong in principle as woefully premature in practice as of 1965. Guided air-to-air missiles had been under development by both the Navy and the Air Force since the early 1950s, and great things were expected of them. By homing in on the infrared plume of an enemy plane's engines (as the Navy Sidewinder and the Air Force Falcon did) or riding a radar beam reflected off the target (as the Navy Sparrow did), missiles promised a one-shot, one-kill capability that would eliminate all of the arduous calculations of deflection shooting or the need to get close on the enemy's tail.

F-4C
fighter-bomber
engines: 17,000-lb. thrust
    turbojet (2)
top speed: 1,400 mph
armament: 16,000-lb.
    bomb and missile load

The trouble was that all three missiles had been conceived for use against slow-moving and slow-maneuvering Soviet bombers. The Sparrow had a hard time picking up the radar reflection of the small MiG-17s and MiG-21s flown by the North Vietnamese; the Falcon was handicapped by a small warhead and the lack of a proximity fuse, which required it to achieve a direct hit to have any effect; and even the Sidewinder, which would prove by far the most effective of the three (58 percent of missile kills for 40 percent of missile launches), could be outmaneuvered by the MiGs, causing the missile to "go stupid," as pilots put it. The Falcon, though, was so bad that when the new F-4D models of the Phantom arrived in Vietnam in 1967 and pilots found that among its new and improved features was that it could fire only Falcons and not Sidewinders, some pilots made their own surreptitious retrofits to the plane's wiring and kept using the Navy missiles.

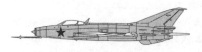

MiG-21
single-seat fighter
engine: 12,675-lb. thrust
    turbojet
top speed: 1,300 mph
armament: 30mm cannon plus
    2 missiles

Even the Sidewinder hit its target only about one time in seven. And time and again F-4 pilots had found themselves too close to a MiG to use their missiles at all. So, after much lobbying by pilots, the F-4 was retrofitted with a 20mm cannon carried in a pod mounted under the fuselage. The pod was not an ideal solution, since it increased drag and tended to flex as the gun was fired, scattering the bullets, but it was better than nothing. During the month of May 1967, F-4s and F-105s shot down twenty-one MiGs, a third of them with gunfire. The F-4E, which entered service in 1968, was finally equipped with a built-in gun, and scored half of its kills using it.

The tactics, training, air-to-ground weapons, and electronic-warfare capabilities needed to fight a war in hostile territory had been equally slighted in TAC during the Cold War years. At one point during the Vietnam War the Air Force, incredibly, ran out of bombs and had to buy back five thousand that had been sold to West Germany for scrap. (The scrap price was $8,500; the Air Force paid $105,000 to get them back.) Despite the extensive use that had been made of radar countermeasures in the Second World War and Korea—including jamming transmitters and "chaff," strips of metal foil that filled radar screens with a cloud of reflected signals—only in 1961 did the Air Force begin a research-and-development program to equip fighters with such antiradar devices. U.S. Air Force fighters in Vietnam at first did not even have simple receivers that could detect when they were being tracked by enemy radars.

The most underestimated threat of all was that posed by SAMs. Despite the manifest success of a Soviet SAM in the 1960 U-2 shoot-down, the general sense was that fighters did not have to worry about them. Radar countermeasures would just add unnecessary weight, sacrificing speed, which was the fighter's best defense. Again, what might have been true for a one-time nuclear penetration mission was not true for a sustained campaign when pilots flew over the same ground day after day. There followed, in the words of one Air Force historian, "an agonizing relearning process and a hurried adaptation of weapon systems back into an arena thought to have been eliminated" from aerial combat in the nuclear age.

The threat posed by the North Vietnamese air defenses had been underestimated by civilian leaders, too, though for other reasons. On a visit to Saigon in early 1965, Assistant Secretary of Defense John T. McNaughton had dismissed a request from Harkins's successor, General William C. Westmoreland, to attack the SA-2 sites that had been spotted under construction around Hanoi; McNaughton was sure that the Russian-built sites were just part of the same game of signaling and brinksmanship that dominated the Cold War calculus. Just as American civilian leaders were not going to risk a real superpower confrontation by striking North Vietnamese SAM sites or airfields, which might cause Chinese or Russian casualties, McNaughton was sure the North Vietnamese would never be allowed to fire the SAMs at American planes. "You don't think the North Vietnamese are going to use them!" he exclaimed. "Putting them in is just a political ploy by the Russians."

On July 24, 1965, an F-4 was shot down by one of these "political ploys." Johnson gave permission for a strike on two SAM sites thirty miles west of Hanoi that appeared to be the start of an outer ring of defenses around the capital. The U.S. Air Force immediately received another agonizing lesson in what it was going to take to deal with modern air defenses. Three days later, a huge force, fifty-four F-105s, was sent in against the two outer ring sites. To avoid being tracked by the SAM radars, they were instructed to fly below 500 feet. They ran right into a flak trap: dummy missiles had replaced the real SAMs, and the North Vietnamese had moved in a phalanx of antiaircraft artillery. Four planes were shot down and two more crashed after colliding in midair.

Countering enemy air defenses would prove to require a complex blend of new tactics, specialized weapons, and vastly improved electronic spoofing devices and command, control, and intelligence. All of this the Air Force now proceeded to learn on the job. TAC rushed into service an experimental radar-surveillance plane code-named Rivet Top, a forerunner of the well-known Airborne Warning and Control System, or AWACS, that allowed air commanders to track friendly and hostile fighters deep in enemy territory, beyond the limited range of ground- or ship-based radars. Rivet Top also could compare SAM radar signals with a database of launch sites to locate SAM threats and warn pilots. (An earlier radar plane, supplied by Air Defense Command, had been designed to track only incoming Soviet bombers over water; its radar screens simply filled with clutter when it was used to look down over land. In Vietnam the best it could normally do was determine that an enemy plane was somewhere in a thirty-by-thirty-mile sector, which hardly helped American fighter pilots. "Forty-five thousand cubic miles of sky," complained one pilot, "and you tell me there's a MiG in there.")

Also rushed into action were the first of a series of specialized radar-hunting aircraft, the Wild Weasels. Using a two-seat trainer version of the F-100 (later the F-105 and then the F-4 would take over the job), the Wild Weasels carried an array of radar-location equipment, a backseat operator to handle them, and a Navy radar-seeking missile, the Shrike, that could be fired when a SAM radar was detected.

Yet none of this was ever more than a partial and temporary solution to the growing SAM threat, which grew more formidable with every passing month of the conflict. The North Vietnamese moved their SAM launchers frequently and began locating radars away from the missiles themselves, which made it increasingly difficult to destroy the missiles and launchers. The Wild Weasels fired thousands of Shrikes, but it was becoming clear that their intimidation value was their greatest asset; the main effort of the Wild Weasels was to keep the SAM radars off the air for the duration of a raid. Generally, the radars came back on as soon as the last Wild Weasel had departed. More menacingly, the North Vietnamese system was becoming an interlocking network; tracking data from early-warning radars or airborne MiGs could be "handed off" to a SAM site, allowing the SAM radars to come on the air only at the last minute before launching.

That American tactics and technology remained far from a complete solution to the problem of enemy air defenses became abundantly clear in the renewal of the bombing of North Vietnam in 1972. Seeking to force the North Vietnamese to sign a peace treaty that would allow the United States to withdraw from the mire that Vietnam had become, President Richard M. Nixon in December 1972 ordered the most intensive bombing of the North yet. B-52s were sent over Hanoi, Haiphong, and other targets north of the twentieth parallel, dropping 15,237 tons of bombs in twelve days; another 5,000 tons were dropped by fighters. It was the first time that B-52s had flown over the enemy capital, and they were meant to deliver a message along with their 30-ton bomb loads. As Nixon's National Security Adviser, Henry Kissinger, had explained in a memorandum earlier in the year advocating the strikes, the use of the very heavy bombers would be "a warning that things might get out of hand."

B-52D
bomber
engines: 10,500-lb. thrust
    turbojet (8)
top speed: 640 mph
armament: 43,000-lb. bomb
    load; four .50-in. machine
    guns in tail

The damage done in the offensive, code-named Linebacker II, was nonetheless real enough. Rail lines, storage depots, power plants, oil facilities, and bridges were hit in some 2,000 sorties.

Yet despite intense electronic jamming and strikes against a dozen SAM sites, fifteen B-52s were shot down and ten more damaged by SAMs during the operation. It didn't help that SAC headquarters in Omaha had insisted on planning the missions and had directed the bombers to make a turn over the target that happened to leave them particularly vulnerable to the North Vietnamese air defenses. But even discounting that blunder, ground-based air-defense systems had proved to be a formidable threat. Radar aiming and rapid firing mechanisms had made antiaircraft artillery a weapon of newly menacing capabilities. Indeed, pilots in Vietnam found that flying below 3,000 feet was nearly suicidal given the lethality of North Vietnamese antiaircraft artillery—AAA or "triple A" as pilots called it. In all, some 1,400

U.S. Air Force planes were shot down by ground fire and 110 by SAMs during the course of the war, versus 67 in air-to-air combat.

On December 29 Hanoi indicated its readiness to resume negotiations, and a month later, January 27, 1973, an agreement was signed in Paris. It wasn't much of an agreement. It got the United States out—Nixon declared that "peace with honor" had been achieved—but the North Vietnamese gave almost nothing in return, except a face-saving concession to, in effect, slightly postpone their conquest of the South. The North was not required to remove its troops from their current positions in South Vietnam, or for that matter acknowledge that there was such a thing as South Vietnam. The last United States troops left two months later. On April 30, 1975, the last South Vietnamese government troops defending Saigon surrendered to Communist forces.

A week earlier, President Gerald Ford had already relegated Vietnam to history. Americans, he said, should not keep "refighting a war that is finished."

Of course Americans would keep refighting it, especially over who was to blame for it all. Even two decades later there would remain, as Mark Clodfelter noted, an "almost universal Air Force perception" that (as an article in *Air Force Magazine* put it in 1986) "timorous military amateurs who were setting policy in Washington" had prevented the Air Force from winning the war in 1965. Many Air Force officers set the apparent success of a mere week and a half of no-holds-barred bombing in Linebacker II against the failure of four years of Rolling Thunder's gradualism to draw the conclusion that while the Vietnam War might have left American prestige, military morale, and social cohesion in tatters, it hadn't laid a finger on air-power theory and doctrine. Several air leaders wrote self-exculpatory memoirs with titles such as *Strategy for Defeat* reiterating the theme. General William W. Momyer, who had commanded the Seventh Air Force in Vietnam, went so far as to insist that the United States might have lost the war but the Air Force won it: Like "the surrender of Japan in the 1940s," Momyer said, Linebacker II was "additional evidence" of the fundamental rightness of traditional Air Force concepts.

There were so many flaws in this argument it was hard to know where to start in debunking it. For one thing, as Clodfelter pointed out, in 1965 the Communist insurgency was still just that, an unconventional guerrilla war that did not depend upon supplies from the North: B-52s could have bombed every railroad line, power plant, oil tank, and airfield in North Vietnam and it wouldn't have made any difference. By 1972 North Vietnam was engaged in a conventional, armored ground offensive that was highly susceptible to interdiction by air: American air strikes against the North were for the first time aligned with the enemy's strategy. And Linebacker II had followed months of attacks earlier in the year that had hit North Vietnam where it was now vulnerable. Having spent eight years chasing a shadowy enemy in the jungle, the American forces now were finding and attacking convoys of a hundred or more trucks, tanks, and artillery moving in daylight under clear skies on open roads.

In addition, Nixon's far more modest political aims for Linebacker II—securing from Hanoi what was little more than a fig leaf to cover America's abandonment of a war it was heartily sick of—stood in marked contrast to Johnson's attempt to end the war on American terms by bombing the North in Rolling Thunder. "Neither the war nor the American objectives were the same in 1972 as they were in 1965," Clodfelter concluded from his analyses.

In a sense, though, many in the Air Force did heed President Ford's advice; their blanket insistence on the old verities mainly revealed a desire to put Vietnam out of their minds, to pack it away as a bad dream best forgotten. The 1979 edition of the Air Force's basic doctrine manual, AFM 1-1, did not even mention the Vietnam War by name, much less seek to draw any lessons from it; it referred to the war only elliptically as an "unpopular conflict that was not and is not yet clearly understood." On the one hand, the post-Vietnam doctrine manuals sought refuge in the tried-and-true formulas. The strategic nuclear mission reappeared as the "highest defense priority." The proper aim for strategic attack was, as ever, "devastating bases or industrial centers behind enemy lines." SAC commanders had complained bitterly that Vietnam was "ruining SAC" by taking B-52s off nuclear alert and forcing them to drop iron bombs over battlefields; now they were more than happy to return to the placid make-believe world of practicing for a war that would never happen and dreaming up justifications for a new, fabulously expensive manned supersonic strategic bomber, the B-1. Air Force doctrine manuals and commanders were careful to always equate "strategic" with "nuclear" once again; as far as the Air Force was concerned, said one observer, having the B-1 fly a conventional mission would be like hitching a Thoroughbred to a milk wagon. "There was no damn way we were going to risk losing a $100 million strategic asset in some conventional shoot-out," one general was quoted as saying.

But on the other hand, as a few younger Air Force officers now dared to point out, there was something painfully hollow about these echoes of traditional certitudes. Colonel Dennis M. Drew observed in an article published in the Air Force's *Air University Review* in 1986, that, before Vietnam, the Air Force knew what it stood for, even if it was naïve and simplistic. But now, despite the bold front, the confidence was gone. The Air Force was wandering in an "air power wilderness." The 1979 doctrine manual was the nadir. It was a "muddle," "wallowed in generalities," contained "almost nothing about the nature of war, the art of war, or the employment of air power." Airmen dubbed it the "comic book"; it had a lot of pictures and big quotes, a "triumph of form over substance," Drew said, but there was hardly anything in it that would help airmen run a war. Vietnam had left a vacuum in thinking that the Air Force leaders, still dominated by bomber generals who had entered the service in the Second World War, sought to fill with the old platitudes.

The major alternative to the old platitudes, however, was a set of new platitudes that found wide voice in the press and the public, and that was equally anti-thetical to confronting the detailed tactical and doctrinal lessons of the Vietnam War. This was what would be called the "Vietnam syndrome," the conviction that "force had lost its utility" altogether, as the military analyst Joshua Muravchik described it: "Even small countries, it seemed, could find the means to thwart large ones. This was held to be especially true for America because of the pecu-liar ineptness of our armed forces. . . . And even in the unlikely event that Amer-ica could employ force successfully, we would so alienate other people that the victory was bound to be pyrrhic."

In particular, there was remarkably ready acceptance of the uncritical, sweep-ing judgment that Vietnam had exposed fundamental flaws in the American tech-nological approach to warfare. With all of its firepower and high-tech wizardry, the United States had been revealed as a muscle-bound giant. This was true for air power above all: "Air power cannot win wars" was the mantra repeated in article after article whenever the question of use of force arose. It was perhaps an inevitable reaction to the technological hubris with which the United States had embarked upon the war, yet it further postponed the examination of what pre-cisely had occurred in Vietnam and what lessons this held for future wars: What was the point, after all, in trying to fix the failures and build on the successes if by definition the use of air power was futile?

In the end, what forced the Air Force to confront the defects in doctrine, tech-nology, and training that Vietnam had so painfully exposed—and equally, to invest in the tools that Vietnam had shown to be effective—was the rise to power of the generation that had fought there and learned its lessons the hard way. The Vietnam-era pilots were overwhelmingly fighter pilots; they were better edu-cated, far less parochial, accustomed to the idea that air power belonged on the battlefield and not just in the strategic-bomber force, and comfortable working with other services.

Even before the Vietnam generation assumed the top leadership posts of the Air Force (that would not happen until the 1990s), the bomber generals were on their way out. The Second World War generation reached retirement age in the late 1970s, and in its place came officers who had entered the service in the Korean War; like the Vietnam generation, they, too, were mostly fighter pilots. The consequences were nothing short of a cultural revolution. The change from bomber to fighter culture came practically overnight: in 1975 bomber generals outnumbered fighter generals on the Air Staff two to one; by 1982 there wasn't a single bomber general left in a key staff job. Even SAC itself, by 1985, was commanded by a fighter pilot. (By the early 1990s every four-star general in the U.S. Air Force would be a fighter pilot, as would more than two-thirds of the commanders of major commands. And by 1993 SAC would be gone, having been merged with TAC into a single Air Combat Command that combined bombers, fighters, tankers, and reconnaissance aircraft into unified "composite" wings. It was the fighter pilots' ultimate revenge.)

What would grow into an avalanche of post-Vietnam reforms began with a modest and long-overdue reemphasis on realistic training for tactical air operations. For almost two decades, instruction and practice in air-combat tactics had taken a backseat to the nuclear-strike mission in TAC. In Vietnam much of the training that fighter pilots got was on the job; a historical review by Air Force analysts found that the first ten missions were when most pilots got killed. In 1975 the Air Force instituted a training program called Red Flag, which five times a year staged exercises over the Nevada desert to give beginner pilots highly realistic experience. Fighter squadrons had always practiced with one-on-one mock combats, but Red Flag was a virtual air war. Attack forces flew in large strike packages, just as they would in battle, encountering "Aggressor" squadrons that imitated MiG tactics and facing mock SAMs that were fired at them. (The mock missile batteries could send up a smoke trail that simulated a real SAM trajectory.) Electronic sensors kept track of weapon "firings" and scored "kills." Within a few years, Red Flag had inspired the Army and Marine Corps to adopt similar combat exercises.

The new guard began to institute other changes. Air Force fighter pilots participated in joint large-force training exercises with the Army as part of what the commander of TAC called "an unprecedented cooperative effort to develop concepts, procedures and tactics" for battlefield attack—rather than wait for the next war to break down the parochialism and bureaucratic self-interest that had always relegated interservice cooperation and planning for the use of tactical air power to the bottom of Air Force priorities in peacetime. General Lew Allen, Jr., the Chief of Staff, initiated a program in 1982 with the unmistakable name of Project Warrior to encourage officers to study the history of warfare and the role of air power in combat. Suddenly Air Force journals were filled with serious, analytical articles—many by younger officers who were clearly irked by the Air Force's 1979 "comic book" manual—that sharply challenged received wisdom about air doctrine and called for new concepts in the application of air power to the battlefield. By the mid-1980s, as Air Force Lieutenant General Merrill A. McPeak observed, the Army and the Air Force were on the threshold of an era of "greatly increased joint effectiveness on the tactical battlefield."

The changing of the guard came at a critical moment for the future of the fighter aircraft. Since the mid-1960s, a small coterie of Pentagon mavericks had been questioning the entire technological course that fighter development had been taking. The "Fighter Mafia," these men came to be called, and their godfather was a pugnacious, cocky, profane, perennially insubordinate Korean War fighter pilot, Major John R. Boyd. As Boyd himself was the first to admit, people thought he was "bright but screwy." Everyone agreed he was a man of unshakable convictions and integrity; most would also have agreed that, like most people of unshakable convictions and integrity, he could also be insufferable, "a 24 karat pain in the ass" as one fed-up general called him. In his later years, Boyd developed a thirteen-hour briefing on the principles of warfare that began with the Battle of Marathon and ended with twentieth-century guerrilla warfare, which he would eagerly inflict upon anyone willing to listen.

Boyd had begun thinking about basic principles of fighter performance while flying in Korea, and had early on concluded that the conventional measures of how good a fighter was, factors such as speed and ability to hold a tight turn, did not fully explain why the F-86 so consistently outperformed the MiG-15. The MiGs in fact could turn tighter than the F-86. But the F-86 was able to transition from one maneuver to another more quickly: it had greater agility.

By 1960 Boyd had written what would become the standard fighter tactics manual and pulled a clearly insubordinate bureaucratic trick to get it adopted over the objections of his commander. In the mid-1960s he decided that, to develop his ideas further, he needed to go back to school and get a second bachelor's degree, this time in engineering (his first was in economics); he pulled another bureaucratic stunt to get the Air Force to send him to Georgia Tech. It was there that he worked out a brilliantly simple theory that for the first time related, in rigorous scientific terms, the characteristics of an aircraft design to the factors that mattered in air-to-air combat. Boyd's Energy Maneuverability Theory basically compared how the key performance characteristics of two fighters varied as a function of speed and altitude—which together determined a plane's instantaneous energy state—and then examined how an energy advantage could be translated via maneuvers into a favorable firing position.

The inescapable conclusion of Boyd's theory was that much of the conventional wisdom guiding fighter development was flat wrong. To the faster-bigger-heavier dynamic that had been at work since the Second World War, McNamara had added a well-intended but misguided enthusiasm for efficiency and commonality that sought to combine every possible mission—air superiority, close air support, nuclear strike, all-weather ground attack, all-weather interception—into a single plane. The result was a series of ever costlier, more complex, and less agile fighters in the "Lead Sled" vein. McNamara's decreed successor to the F-4 was the F-111, a complex, 100,000-pound, electronics-laden plane with swing wings that could be positioned straight out for maneuvering or swept back for supersonic flight. In Vietnam the F-111s had been plagued with problems; four had promptly flown into the ground when they were deployed for two months in 1968, leading to their hasty withdrawal from combat. (Some of the crashes appeared to have been caused by problems with the terrain-following radar that was designed to allow the plane to fly at very low altitudes. A message was posted on the squadron bulletin board: "Flak effectiveness is 5%—missile effectiveness is 8%—ground effectiveness is 100%—AVOID GROUND.")

Now the F-111's successor was in the works, the "F-X" experimental fighter, and it was being conceived as a sort of super F-111. It would be another swing-wing high-tech marvel, capable of Mach 2.7 speeds. But to Boyd it looked like another beast as far as the all-important factor of agility was concerned: "overweight, under-winged, overly expensive, overly complex, ineffective," he con-

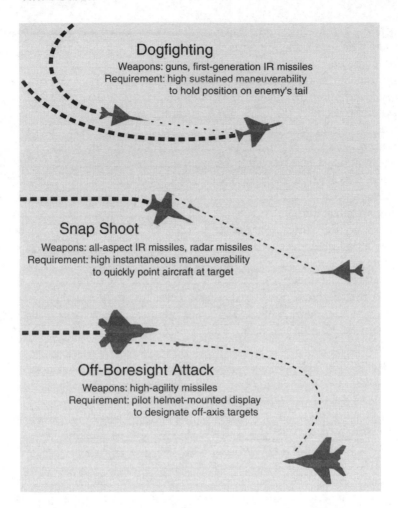

Dogfighting
Weapons: guns, first-generation IR missiles
Requirement: high sustained maneuverability
to hold position on enemy's tail

Snap Shoot
Weapons: all-aspect IR missiles, radar missiles
Requirement: high instantaneous maneuverability
to quickly point aircraft at target

Off-Boresight Attack
Weapons: high-agility missiles
Requirement: pilot helmet-mounted display
to designate off-axis targets

The evolution of fighter requirements, as new weapons systems dictated new tactics.

cluded. (Boyd commented to his superiors that though he had never designed an airplane, "I could fuck up and do better than this.") One of the major implications of energy maneuverability was that a high top speed, which would rarely be employed in combat anyway, was far less important than a high thrust-to-weight ratio. High thrust to weight meant that an aircraft could nimbly change attitude; a light enough plane with big enough jets could even have a thrust-to-weight ratio of greater than 1, meaning it could go straight up like a rocket. The F-105 had a thrust-to-weight ratio of about 0.5; the F-4's was 0.6; the F-X concept, when Boyd was asked to critique it, was at 0.75. Boyd proposed cutting the design weight by a third and getting the ratio up almost to 1. In the end, the F-15, the fighter that would emerge from the original F-X concept thanks to the

Fighter Mafia's relentless efforts, had a thrust-to-weight ratio of 1.07. "Not a pound for air to ground," read a huge banner that hung in the F-15 program office. The Fighter Mafia was determined that for once this would be a real fighter, unencumbered with other requirements.

Boyd and company began pushing for an even lighter fighter, which they jocosely termed the "F-X-X," the plane that would become the F-16. When even some of the fighter generals balked, fearing that a second fighter program would undermine funding and support for the F-15, Boyd pulled another typical act of insubordination. Just before a briefing of the Air Staff that was to decide the plane's fate (he had been tipped off it would be killed), Boyd went to Defense Secretary James Schlesinger and got the project approved. Boyd walked into the briefing room and announced, "Gentlemen, I am authorized by the Secretary of Defense to inform you that this briefing is for information only. The decision has been made to procure the F-16."

Boyd and the Fighter Mafia did not win all their battles; one they notably lost was their effort to strip fighters of advanced electronics equipment. Their argument, which resonated throughout the "military reform" movement of the 1970s and 1980s, was that weapons systems had become too complex, too elaborate, too "gold-plated" with gee-whiz gizmos that added weight, drove up costs, and drove down reliability, and which were predicated in the first place on a McNamara-esque view of warfare as a sort of glorified management problem, when what really counted in warfare was flexibility, surprise, and "intangibles" that could never be

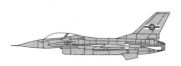

F–16
single-seat fighter
engine: 25,000-lb. thrust
    turbojet
top speed: 1,500 mph
armament: 20mm cannon;
    12,000-lb. bomb and
    missile load

anticipated. They had a point: as a result of increasingly complex radars, avionics, controls, engines, and weapons systems, the number of maintenance hours per flying hour of military aircraft had tripled or quadrupled since the Second World War; by the 1960s it was up to forty hours of maintenance on the ground for every hour in the air. Something on an F-111 would, on average, fail every twelve minutes. And costs were rising so quickly—by a factor of 100 since 1945, with no slowing down in sight—that, as the aerospace executive Norman Augustine observed, if current trends continued it would not be long before the Air Force, Navy, and Marines would have to share a single airplane.

Yet, as noted by the Air Force's historian, Richard P. Hallion, there was a simplistic nostalgia in some of the arguments of the reformers, who would often point to famous warbirds of the past such as the F-86 or the P-51 as exemplars of the "small, austere" fighter, while ignoring their high cost and sophistication for their time, their clear dominance over their less sophisticated rivals of the day, and for that matter their often serious shortcomings compared to modern aircraft, notably their high accident rates. (F-86s crashed so often in accidents, an average of one per week in the early 1950s, that they were known as "lieutenant eaters.") The reformers also failed to anticipate the impact the solid-state electronics revolution would have in dramatically lowering costs and increasing reliability of avionics. Compared to the previous generation of fighters, the F-15 and F-16 required only half as much maintenance time per flying hour. Computer technology also reversed the inevitable connection between complexity of an aircraft's systems and complexity of operating them: in place of the bewildering "switchology" of the F-4—which, the reformers had rightly pointed out, was so complex that to operate radar and missiles the crew frequently had to look down and lose sight of their target—the new fighters had computer-generated "head-up" displays projected on a clear screen in the pilot's line of sight, with key functions operated by buttons on the throttle and stick.

But most of all, the reformers had greatly underestimated the necessity of technological systems for countering the problem of enemy air defense. The reformers had looked at the chilling effectiveness of SAMs and AAA in Vietnam and concluded that the answer was lots of cheap airplanes because so many of them were sure to be shot down in a war. This was not a prospect that sat well with most of the men who flew airplanes, however; nor did it sit well with what was, in the aftermath of Vietnam, a strengthened American belief that if wars must be fought, technology should substitute for blood.

More to the point, some hardheaded calculations were beginning to show that, unless some revolutionary means could be devised to defeat ground-based defenses, tactical air power would scarcely be able to operate over the battlefield at all, so heavily had the balance shifted between offense and defense since the early 1960s. A standard Soviet field army now deployed an array of antiaircraft missiles and guns that could effectively deny its enemy the use of the entire airspace from 1,000 feet to 16,000 feet, and for dozens of miles beyond its front. There were fewer SAM threats at higher altitudes, but they were not absent altogether, and higher altitudes made it harder to find and hit ground targets. That left only one choice: to fly at very low altitudes and take advantage of terrain to hide from air-defense radars. NATO pilots now trained relentlessly for this mission, but it wasn't a happy solution. Flying at 500 miles per hour a couple of hundred feet off the ground left little margin for error and little time to spot and hit a target. It was not much of an answer to AAA, either. Lots of cheap fighters or no, the extremely high kill rates calculated for the new air-defense systems suggested that the half-life of any United States tactical air forces that faced them might be as little as nine days. Whether the attacking force was large or small in absolute numbers did not alter the prediction that 97 percent would be shot down in six weeks. Sending a larger army running against a brick wall would be unlikely to do anything about the wall.

The Bekaa Valley would not be anyone's obvious choice as a place for one of the great strategic or technological showdowns of the late twentieth century. Hemmed in between two mountain ranges in Lebanon's northeast corner, the Bekaa Valley consisted of grain fields, the occasional opium and hashish plot, and a scattering of archaeological ruins. To the outside world, its main claim to notice was the town of Baalbek, whose Roman temples stood on the site of an even more ancient temple hill that had once been a shrine to the Canaanite god Baal, one of the idols whom the Israelites of the Bible raged against.

The Lebanese civil war had dramatically altered the strategic significance of what went on north of Israel's border. In 1977 Palestine Liberation Organization fighters, escaping Christian militias and Syrian forces that had taken up positions outside Beirut, fled south. From their new positions in southern Lebanon they began launching artillery and rocket attacks and infiltration raids on northern Israeli towns. As the PLO attacks intensified, so did Israeli retaliation, and the Israeli Air Force presence over Lebanon became nearly constant as Israeli jets struck PLO positions and conducted reconnaissance flights. By 1981, several clashes between Israeli and Syrian aircraft had taken place. Then in April 1981, following the downing of two Syrian helicopters by Israeli F-16s, Syria began to deploy SAM brigades in the Bekaa Valley. It was a move that immediately challenged Israeli air superiority over all of southern Lebanon.

Before the 1973 Yom Kippur War, Israeli air commanders had regarded the SAM threat with almost smug disdain; now they viewed the Syrian move as deadly serious. It was a lesson they had learned the hard way. Despite the warning the Vietnam War had provided about the growing effectiveness of Soviet-built air-defense systems, the Israelis were reassured by their own swift and overwhelming defeat of the Syrian and Egyptian air defenses in the opening minutes of the 1967 Six-Day War. Throwing practically its entire air force into a single surprise attack—only twelve jets were left to protect the country—Israel had destroyed 204 aircraft, half of Egypt's air force, in little more than a half an hour. All but nine of the Egyptian planes were destroyed on the ground. Hundreds of French-made Durandal runway-attack weapons were dropped on Egyptian airfields. This secretly developed device released a 180-pound bomb on a parachute; when it was just over the ground, a rocket would fire to drive into the pavement before it detonated, heaving up a huge crater. Several of the twenty-seven SA-2 missile sites the Soviets had provided the Egyptians were also hit.

The Israeli strike force encountered almost no ground fire. That was in part because they flew below the SAM radars. But it was also for a reason that might have made the Israelis far less confident in dismissing the capability of the weapons: Field Marshal Abdel Hakim Amer, the Egyptian commander, was flying back to Cairo from an inspection of his frontline troops in the Sinai and had issued orders that no antiaircraft weapons be fired lest his plane be hit by mistake. A major in Cairo commanding one antiaircraft artillery battery was the only one to disobey orders and try to shoot back at the Israeli fighters. ("I thought I'd be court-martialed for it," he later admitted.)

Israel continued to score successes in taking out Egyptian SAM sites during the war of attrition along the Suez Canal that dragged on until a cease-fire in August 1970. In an extraordinary display of cockiness, Israeli aircraft hit every Egyptian air-defense site along the entire length of the Canal on Christmas Day 1969 and that night landed a team of commandos who seized an advanced Soviet-built surveillance radar station near Ras Gharib, dismantled the equipment and antenna, and flew it out on two heavy-lift Sikorsky CH-53D helicopters.

In the months that followed, however, Egypt began a discreet but massive buildup that quietly began to change the equation. Among the latest Soviet hardware that Egypt now received were two formidable antiaircraft weapons, the SA-6 missile system and the ZSU-23 gun. Unlike the fixed SA-2, both were highly mobile, mounted on light-tank chassis, and both could be aimed by either radar or optical tracking. The SA-6 had a range of seventeen miles and could also engage low-flying aircraft. This had been a gap in the older SAM systems, which were ineffective below about 3,000 feet. The Egyptians at the same time greatly expanded their network of fixed SAMs and began hardening the missile emplacements with concrete and sand and heavily protecting them with guns.

The result came as a stunning blow to the Israeli Air Force in the 1973 war. In the first forty-eight hours, some forty Israeli aircraft were shot down, about 14 percent of the Israeli Air Force's entire frontline combat strength. In all, the Egyptians and Syrian forces would fire thousands of missiles in the course of the war; on average it took fifty-five SA-6s to kill one aircraft, which was little better than how the SA-2 had performed in Vietnam. But the greatly increased density of SAMs, and the severe limitation on freedom of action they had imposed on the Israeli Air Force, threatened the very premise that fighters would in the future be able to secure air superiority in the face of sophisticated, interlocking defenses. To General David Ivry, the fighter pilot who took command of the Israeli Air Force in 1975, the lesson of the war was simple and inescapable: "We have to find an answer to the SAM batteries."

By the summer of 1982, the Syrians had to move nineteen batteries into the Bekaa Valley. The Israeli operation to take them out that began at 2 p.m. on June 9 had been meticulously rehearsed. Strategic surprise was "impossible under the circumstances," as one Israeli Air Force colonel noted. "The Syrians were waiting for us."

For more than a year beforehand, the Israelis had been sending unmanned drones over the Syrian sites, equipped with cameras and radio receivers to locate the batteries and measure the radar frequencies they operated on. For the attack, the drones were sent over again, this time with transmitters that emitted signals to mimic the signature of a much larger fighter jet. The aim was to provoke the Syrians to turn on their targeting radars. As soon as they did, F-4s that had been standing off fired radar-seeking missiles that had been programmed to the known Syrian radar frequencies. Simultaneously, long-range artillery and surface-to-surface missiles were brought to bear, their fire adjusted by other drones that sent back instantaneous television images of the targets. Meanwhile, still more drones were monitoring three major Syrian airfields to provide immediate warning when Syrian MiGs were scrambled.

In the huge aerial battle that ensued as Syrian aircraft rose to challenge the assault on their missile sites, Israeli F-15s and F-16s were fed in a highly controlled sequence into the battle, vectored by an airborne warning-and-control plane and also aided by the long-range airborne radars of the F-15s themselves, which were able to serve almost as mini-AWACS to fill in radar coverage. At one point, ninety Israeli and sixty Syrian fighters were in the air; this was probably the largest aerial clash since the Second World War. Israeli claims of having downed eighty-five Syrian jets for a loss of none over two days of engagements were at first dismissed as exaggeration, but were later confirmed by U.S. Air Force officials: forty had been shot down by F-15s, forty-four by F-16s, and one by an F-4. Although all of the air-to-air engagements occurred within visual range, 93 percent of the kills were made with missiles rather than guns. A significant factor in the Israeli success was extensive jamming of Syrian ground-to-air communications, which cut pilots off from ground controllers. Israeli fighters, by contrast, remained in constant radio communication with commanders, who received continually updated intelligence.

The air-to-air successes would receive much of the attention, but it was the destruction of all nineteen SAM sites within the first two hours of the operation that was to prove the battle's real historical significance. In a single afternoon, the Israeli Air Force regained complete air superiority. It is no exaggeration to say it had also gained a new lease on life for air power. The Bekaa Valley operation proved that even a modern air-defense system could be defeated through innovative tactics—tactics that relied far more on high technology than brute force or attrition. Above all, it depended upon the high technology required to win what was increasingly being termed the "fourth dimension" of military conflict, the electronic spectrum.

The Bekaa Valley operation also reawakened interest in the U.S. Air Force in drones. Although they had been used in Vietnam, support had languished in the years since. Advocates of unmanned aircraft frequently blamed this on the fact that fighter pilots saw drones as a threat to their job ("All the customers wear wings," grumbled one contractor about his frustration in trying to sell drones to the Air Force), but in fairness there had also been legitimate doubts about their capabilities, especially compared to the costs in an era of tight budgets. But here was proof of their utility that was hard to ignore: Drones were less a threat to the mission of the fighter pilot than a weapon that would protect fighter pilots against what had for more than a decade seemed a threat to the very premise of air operations.

14

PRECISION, AT LAST

Everyone always called them "smart bombs," which was an unfortunate bit of exaggeration. The term made them sound like the sort of high-tech fantasy weapons that defense reformers complained about: a bomb with its own built-in brain, a bomb that thought for itself, a bomb that could eliminate human beings from the equation altogether. This was certainly how many of the reformers portrayed them: just another gold-plated "magic weapon" being pushed along by the Pentagon's "high-tech juggernaut," which ignored the realities of combat, especially combat against a low-tech opponent of the kind the United States had been humiliated by in Vietnam.

Yet there wasn't anything terribly complex or high-tech about them. When the Second World War ended, American weapons scientists were working on two dozen different guided-bomb projects. The basic idea was not even, as the saying went, rocket science: Put a set of fins on a bomb so it can be steered rather than fall at the whim of gravity, wind, and bad aiming from 30,000 feet up, and bombing accuracy could be improved from thousands of feet to dozens of feet or less. The laser and the navigation satellite had not been invented in 1945, but every other means that would eventually be used to provide the steering control for a smart bomb was in the works. There were projects to develop bombs that would be guided by a bombardier over a radio link, bombs that would home in on a heat source, bombs that would search for a visual contrast such as a ship's waterline, bombs that would follow a radar signal reflected from the target, bombs that would transmit a TV image back to its controller as he steered it. During the war about 450 AZON radio-steered bombs were used by the AAF in Burma to knock down twenty-seven Japanese-held bridges that had resisted conventional bombing. A follow-on weapon, the RAZON (which could be steered in both range and azimuth, unlike the azimuth-only AZON), was employed in Korea to take out heavily defended bridges with similar success.

Although these early weapons were plagued with glitches, and a huge 12,000-pound version of RAZON (dubbed "TARZON") never worked right when it was tried in Korea, none of these problems involved fundamental engineering science. They were merely bugs such as badly fitting fuses, insufficient shielding against radio interference, or aerodynamic defects in the bomb casings. It would not have taken much time or money to fix them.

Yet after the Korean War, interest flagged and work on precision guidance languished. Besides the general deemphasis of nonnuclear missions, there was a bureaucratic tangle that had left conventional-weapons research an orphan within the Air Force. As part of the Defense Department's effort to avoid duplication between the services after unification in 1947, the Air Force did not even directly control most research on its own conventional weapons. The job of developing and testing explosive bombs was left where it had always been, with the Army Ordnance Department; incendiary bombs were the job of the Army's Chemical Service; armor-piercing bombs belonged to the Navy's Bureau of Ordnance. Nuclear weapons had no need for precision that could be measured in feet or even dozens of feet, and what little funding was available for conventional-weapons research and development in the 1950s went almost completely to a prosaic if much-needed effort to redesign the casings of the Air Force's and Navy's outmoded Second World War–era bombs, whose stubby shape had been determined chiefly by the desire to pack as many into a B-17 bomb bay as possible. More-aerodynamic, low-drag shapes were now needed that could be carried under the wings of jet fighter-bombers.

The stunning success of laser-guided bombs in Vietnam had vanquished official doubts about the effectiveness and reliability of precision weapons, but still did little to crack the budgetary stranglehold. The possibility of using lasers to guide weapons had first come up at a summer think-tank session of a defense science committee in 1958, two years before the first workable laser had been demonstrated. But the principle of the laser had been known for decades, and the concept of generating a coherent beam of electromagnetic energy of a single wavelength had been demonstrated with microwaves in 1954. One important property of coherent radiation is that it can be focused into a very narrow beam that remains narrow even over long distances. A guided bomb dropped from high altitude miles away could thus in theory steer itself to a direct impact using a simple optical seeker that detected laser light reflecting off the target as it was illuminated by a laser gun—with the laser likewise aimed from a safe spot miles away.

By 1964 there were stirrings of reawakened interest in "surgical strike," spurred by both the Cuban missile crisis and the growing American commitment in Vietnam. An Air Staff study in June 1964 criticized the neglect of conventional tactical ordnance and recommended the development of simple but high-accuracy conventional weapons. The Air Force had meanwhile regained responsibility for bombs from the Army and built up a small weapons-research lab at Eglin Air Force Base in Florida. It was there that a small staff known by the unglamorous name of Detachment 5 was searching for technologies that could have an immediate payoff in Vietnam.

Army scientists had been working on laser guidance since 1962, but when the Army lost interest they passed on their ideas to the Air Force. Colonel Joseph Davis, Detachment 5's commander, was interested at once. The first question was whether a man in a jet fighter could hold a laser gun steady on a target. Davis had himself flown over Eglin with a movie camera that he kept trained on a dam as the plane circled. When the film was developed, he saw that the crosshairs had remained right on the spot. Within a few months, Texas Instruments had a $99,000 contract to build twelve prototype "Paveway" laser-guided bombs. (Davis had told the TI engineers their proposal had to be less than $100,000.) By the end of the Vietnam War, American forces had dropped 28,000 Paveways. At first a handheld laser gun was used, but later a targeting pod that bore-sighted a laser with a TV or infrared camera was installed so the weapons officer could simply line up the target in the crosshairs of a video screen in the cockpit. The bombs from the start were able to achieve an average error of twenty feet from target; in time this was reduced to six feet. The technology was simple enough that the Paveways cost only about $8,000 each, yet each was as effective as twenty-five unguided bombs of equivalent weight. TV- and infrared-guided bombs were also deployed by the Air Force and Navy in Vietnam.

The most famous success scored by Paveways in Vietnam was the May 1972 attack on the Thanh Hoa bridge, which carried the only railroad and the principal highway across the Song Ma river about fifteen miles south of Hanoi. The North Vietnamese had spent seven years building the bridge after themselves blowing up its predecessor during the rebellion against the French. (The Communist fighters had run two locomotives packed with explosives head-on into each other on the bridge.) The new bridge was built to last. The U.S. Navy and Air Force had flown 871 sorties, losing 11 planes, in an effort to put the bridge permanently out of commission, but none of the attacks had done lasting damage. On May 13, fourteen fighter-bombers carrying 2,000- and 3,000-pound Paveways struck again, knocking one of the main spans off its concrete abutments.

Laser guidance had its limitations, notably the fact that lasers would not shine through dust, cloud, or haze. But the overall success rate of the Paveways had been extraordinary. Besides approaching a one shot–one kill ratio, which increased the capacity to destroy hard targets such as bridges and reduced the number of sorties required for any mission, the accuracy of laser-guided bombs had a profound political implication. Many targets that had been off-limits during Rolling Thunder to prevent civilian casualties in built-up areas were released for attack during the Linebacker strikes, precisely because of the assurance of accuracy that laser guidance now offered. Precision guidance at last promised to undercut what had from the earliest days of strategic bombing proved to be one of its profoundest limitations: the propaganda backlash that even unintended civilian casualties caused.

Yet Vietnam by 1972 was the wrong war to use to make this case, so sordid had the whole business become to much of the American public. The revolutionary implications of precision guidance demonstrated during the Linebacker strikes were lost amid the much greater attention that the use of B-52s had gained in the Christmas bombings that brought American involvement in the war to an end.

Worse, in the aftermath of Vietnam there was an alignment of politics—domestic, foreign, and interservice—which guaranteed that advanced conventional weapons would always be the first things to go when budgets shrank and the last things to be added when they grew. Personnel levels, force structure, and claims to future budget allocations always depended on the number of airplanes an air force had and the number of ships a navy had, and so those were the things military leaders wanted to buy first. They were also the things that sent the clearest political signal, both to domestic supporters of a strong defense and to foreign foes.

Since the Korean War, the three United States military services had been spending roughly 80 to 90 percent of their acquisition budgets on airplanes, ships, and vehicles, and only 10 to 20 percent on everything else, including weapons and electronics systems. Even the huge defense buildup that fulfilled President Ronald Reagan's 1980 election promise to strengthen national defense and stand up to Soviet aggression around the world did not change the picture; while defense budgets had doubled by the time he left office eight years later, Reagan's emphasis on making a high-visibility statement of resolve found its easiest expression in a commitment to new ICBMs and supersonic nuclear bombers and a 600-ship Navy, rather than in obscure technologies whose significance was little understood by the general public. Although total Air Force spending on conventional munitions increased by a factor of four between 1981 and 1987, it remained a tiny fraction of overall weapons budgets. In the mid-1980s, as the Air Force, Navy, and Marines were in the midst of spending $167 billion on new fighters, the entire program to acquire new laser-guided bombs came to $1.8 billion.

The few who argued the potential of advanced conventional weapons pointed out that, even though they were undeniably costly compared to ordinary bombs (the more sophisticated varieties, such as the Maverick air-to-ground antitank missile, could cost $100,000 each), this was still a lot cheaper than having a $20 million jet shot down. Precision guidance not only meant that far fewer passes over a heavily defended battlefield would be needed for each tank killed or bridge knocked down; it also meant that weapons could be released from much safer distances and altitudes, keeping planes out harm's way altogether. Measured in total cost per enemy target killed, precision guidance was clearly a bargain. In the early 1980s, defense analyst Seymour Deitchman calculated that 100 fighter-bombers carrying precision weapons could kill as many as 800 tanks per day, more than ten times the number that a force of 2,500 Second World War fighter-bombers could.

Moreover, if the Air Force was serious about cooperating with the Army to achieve battlefield effectiveness, as opposed to staging muscle-flexing shows of force, the entire focus on aircraft rather than munitions was backward. Lieutenant General Glenn Kent, a legendary operations analyst in the Air Force whose brilliance, tenacity, and sheer cussedness were apparent in the fact that he had managed to rise to three-star rank without ever becoming a rated pilot, made a point of always calling airplanes "platforms," a word deliberately chosen to put the emphasis on the munition that actually landed on the target, which after all was the thing that mattered. "Platform" sounded like a flatbed truck or the tripod that held up a machine gun, a none too subtle suggestion that the airplane ought to be considered of distinctly secondary importance. Because of pilots' traditional fixation with the glamour and performance of the airplane itself, Kent complained, "We have Space Age fighters dropping World War II iron bombs." The budgets that had put munitions programs "on a starvation diet," he said, ought to be "classified Top Secret to stop people from knowing how stupid we are."

But from the late 1970s on, several new currents in military thinking coalesced to save advanced conventional weapons from being swept aside in the usual tide of Pentagon budget politics. Most visible was the increasingly urgent need to find a new way of countering the Soviet conventional threat in Western Europe. Not only did Soviet ground forces outnumber NATO's by two to one in modern tanks, artillery pieces, and total number of divisions, but NATO's long-standing threat to answer a Soviet conventional attack with battlefield nuclear weapons was becoming increasingly unbelievable in the face of a like Soviet capability. NATO's commander, General Bernard Rogers, warned in repeated public statements that it would be a matter of only days before he would be forced to seek release of nuclear weapons in the event of a Soviet attack. What had been an extreme but rationally defensible posture at a time when NATO could threaten to use battlefield nuclear weapons without fear of nuclear retaliation was now reduced to a sort of God-knows-what-might-happen strategy that presented the Soviets with the possibility that, rational or not, a situation might get out of control in an "incalculable chain of nuclear escalation," with so many small nukes scattered all over the battlefield. It was not a comforting policy.

Rogers pressed for greatly improved NATO conventional capability as the only way out of this dilemma. In particular, he proposed that a combination of real-time intelligence, long-range weapons, and precision guidance could defeat the Warsaw Pact "follow-on forces" that would be fed from the Soviet Union through Eastern Europe to exploit an initial armored breakthrough. The "Rogers Plan," or FOFA—for "follow-on forces attack"—quickly became new buzzwords in NATO and raised the profile of smart-weapons technology among strategists and defense analysts. So did a new U.S. Army doctrine that paralleled this development. AirLand Battle had been conceived as an answer to criticisms that the Army had fallen into a static, Maginot Line strategy for defending Europe, and like FOFA it emphasized deep battlefield interdiction strikes to dis-

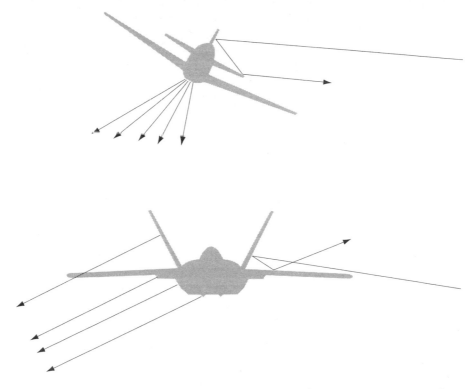

Two basic principles of stealthy design: avoid right angles (which always bounce a radar beam directly back to the source) and align surfaces so that radar reflections are concentrated in a limited range of angles.

rupt enemy movements. AirLand Battle also explicitly incorporated tactical air power into Army battlefield doctrine in a way that had never been done before.

President Jimmy Carter's Defense Secretary, Harold Brown, and his Undersecretary, William J. Perry, both Ph.D. scientists, had meanwhile come into office in 1977 with a fundamental conviction that electronics and information technology provided the only advantage the United States could hope to exploit in staying ahead of Soviet military strength across the board. There was no way that the West could match the Soviet Bloc in quantity, and the Soviets had shown little difficulty in matching the firepower of individual American weapons. But computers and electronics were another matter; the combination was not only an obvious American strength but it had the revolutionary potential to be a force multiplier on the battlefield, greatly increasing the effective firepower of an outnumbered and outgunned NATO.

One of the riskiest but most dramatic successes of the Brown-Perry philosophy was a program that remained shrouded in secrecy until 1988. The F-117 stealth fighter would, in retrospect, prove to be "one of the most pivotal contributions of the 1980s to the revolution in lethality and effectiveness of American air power," in the words of a RAND analyst, Benjamin S. Lambeth. Using a combination of radar-absorbing materials and geometric shapes that were a novelty for anything that was intended to fly through the air, the F-117's designers were able to reduce its radar reflection by 10 to 100 times from that of any existing fighter. With a payload of two 2,000-pound laser-guided bombs, and equipped with its own night-vision sensors and laser designator, the F-117's combination of precision, real-time targeting, and stealth embodied the new American approach of substituting technology for massive force in bringing air power to bear on a heavily defended battlefield.

Two other advanced-technology development programs launched around the same time underscored the all-important connection between precision strike and precision intelligence and targeting. In 1979 TAC began a project to equip the F-16 (and eventually also a two-seat ground-attack version of the F-15, the F-15E) with a night navigation and targeting system. Known as LANTIRN, it consisted of two externally carried pods, one of which superimposed an infrared image of the outside world on the pilot's forward view out the cockpit via a wide-screen head-up display; the other gave a higher-magnification infrared view of the ground, coupled to a laser designator for guiding Paveway bombs.

F–117
fighter-bomber
engines: 10,600–lb. thrust
    turbojet (2)
top speed: high subsonic
armament: two 2,000–lb. laser-
    guided bombs

At around the same time, a demonstration program to test a more ambitious concept for locating real-time targets across the battlefield was yielding promising results. Called Pave Mover, it consisted of an airborne radar that could spot moving targets on the ground and instantly relay their location to strike aircraft. The idea was to do for ground attack what AWACS had done for air-to-air combat. By 1985 work had begun on a full prototype of the system, known as JSTARS, which would fly on a modified Boeing 707.

Though the trends were in the right direction, it remained a constant fight throughout the 1980s to keep funding alive for LANTIRN, JSTARS, precision-guided bombs, and other members of this supporting cast. All faced repeated attempts to kill them off, and none had come along as quickly as their supporters had hoped, and certainly not in the quantities they had hoped. By 1990 the Air Force had an inventory of several thousand Paveways but only a handful of LANTIRN targeting pods. A fifth of the fifty-nine F-117s were unable to fly at any given time due to a shortage of spare parts. The only other Air Force planes with any capability of deploying laser-guided bombs were some sixty venerable F-111Fs equipped with Vietnam-vintage laser targeting systems. Only two

JSTARS, both experimental prototypes, were flying. They were about to be sent to Europe for a six-week series of tests when, on August 2, 1990, the world awoke to the news that a force of 120,000 Iraqi troops backed by 300 tanks had invaded Kuwait in the early morning hours.

Within five days, what would become a force of hundreds of thousands of American, British, and coalition troops and thousands of aircraft was on its way to the Persian Gulf. The immediate task was to defend Saudi Arabia. Iraqi forces in Kuwait had already sent three patrols to probe the Saudi border, and intelligence estimates warned that within days the Iraqi army could regroup to continue its offensive and take Saudi Arabia's oil fields.

The other urgent problem that confronted the American commander responsible for the region, Army General H. Norman Schwarzkopf, was what he would do if the Iraqis did something heinous, such as killing the Western hostages who had been rounded up in Baghdad, or launching a preemptive attack on United States or Saudi forces with chemical or biological weapons. Schwarzkopf's Central Command was little more than a planning headquarters in Tampa, Florida, and it had a limited planning staff at that. CENTCOM's air commander, Lieutenant General Charles A. Horner, was already in Riyadh overseeing the deployment of the first United States aircraft to Saudi bases. On August 8, the day after the deployment had begun, Schwarzkopf called the Air Staff in Washington to ask if someone there could help put together a plan for retaliatory air strikes against Baghdad. Schwarzkopf said he "needed it fast," and what he envisioned was something that could hit Iraqi President Saddam Hussein "at his heart," destroying strategic targets that "have value to him as a leader."

Schwarzkopf was clearly talking about only a retaliatory contingency option, not a plan for fighting a whole air war, but the Air Staff planning group that was handed the job had bigger ideas. As head of a small strategy and planning unit housed in the basement of the Pentagon, Colonel John A. Warden III had developed a theory that, in its essence, attempted to resuscitate the old Air Force strategic-bombing doctrine with an injection of new terminology and new technology. Warden's ideas flew in the face of the increasing tactical emphasis of the fighter generals, but they also had some undeniable political selling points that had strongly appealed to Secretary of the Air Force Donald B. Rice and Chief of Staff General Michael J. Dugan. Rice, a former head of the RAND Corporation and an astute politician, thought the Air Force needed a "strategic vision" that would emphasize its vital "role in national security," particularly on Capitol Hill, and particularly in an era when the Soviet Union was fading as the chief threat. More than that, it needed "a slogan—a catch phrase to essentially capture the essence of the Air Force."

Warden's "Checkmate Division" had been eager to oblige with both. The slogan was "Global Reach—Global Power," and soon it was plastered everywhere:

on briefing charts, on publications, on the walls of hangars, and as the title of a 1990 Air Force white paper by Rice's staff that advanced the idea that air power, because of its instant availability, precision, and lethality, might be the dominant factor at least in some future conflicts.

Warden's strategic vision was distilled in what he called his "five-ring model." In Warden's view, the major significance of the new precision and lethality of the air weapon was that for the first time they allowed the strategic heart of the enemy to be struck directly with non-nuclear weapons. The center ring, the bull's-eye in his diagram of enemy systems, was its leadership. The outer rings—which, in descending order of importance, were key production (such as oil and electric power), infrastructure (roads and railways), population, and lastly fielded forces—existed only to protect the leadership. The only reason there had ever been for attacking these outer rings was to get at the enemy leadership, Warden insisted. Only the inner ring could be coerced, only the inner ring held the key to military success that strategists going back to Clausewitz had identified as the whole point of war, which was to impose one's political will on an enemy. Once the inner ring was neutralized, the effect would be just like severing the brain from a human body: the enemy military force would be left as nothing but a "useless appendage."

Warden's theory was in one way an astonishing throwback. With only tiny variations, it might have been written by the Air Corps Tactical School in 1935. Warden spoke of "centers of gravity" rather than "vital centers," and he was able to point to real technologies such as stealth and precision guidance to carry it out, but the theory behind it was identical to all of the classic theories of strategic bombardment that had come before. Echoing Billy Mitchell, Warden insisted that targeting enemy armies would only weaken the real mission of air power, which was to strike directly at the enemy will to resist. To underscore the return to roots, Warden put up a sign on the door of the Checkmate Division just before the air war over Baghdad would begin a few months later: "USAF Air War Plans Division, 1941–1991," it declared. And to underscore the renunciation of what he insisted was a betrayal of those roots in the gradualism of Vietnam, Warden announced that the name of the plan he now proposed to develop in response to Schwarzkopf's request would be "Instant Thunder." Warden assembled his staff and exultantly declared what he had in mind. "This is not your Rolling Thunder. This is real war, and one of the things we want to emphasize right from the beginning is that this is not Vietnam! This is doing it right! This is using air power!"

Warden saw it as his great chance, and he was off and running. Two days later, he was in Tampa presenting his plan to Schwarzkopf, and it was now nothing less than a plan for winning the entire war in six days of strategic bombing. The air strikes would concentrate on air defenses and the inner three rings, eighty-four targets in all. About a thousand sorties would be flown each day, bombing presidential palaces, telephone exchanges, government ministries, internal security organs such as secret-police headquarters, and electric power, oil, bridges, and railways. The direct attacks on leadership would isolate and paralyze the Iraqi government; the attacks on industry and infrastructure would foment a coup or revolt by convincing "the Iraqi populace that a bright economic and political future would result from the replacement of the Saddam Hussein regime." Warden put up on the screen a slide that confidently listed the anticipated results, none more confident than this item near the top: "National Leadership: Incapacitated."

No attacks on Iraqi forces in Kuwait or along the Saudi border would be necessary at all. Bombing Saddam's seat of power in Baghdad would be a "left hook," delivering "a quick knockout blow," Warden told Schwarzkopf. It would be an opportunity for the greatest feat of generalship since MacArthur's landing at Inchon during the Korean War, which had outmaneuvered the North Korean Army. The CENTCOM commander seemed to like that.

The next day Warden repeated his presentation for Joint Chiefs Chairman Colin Powell, an Army four-star general. This time Warden was even more brashly confident. When Powell asked what he planned to do about all of the Iraqi tanks, Warden insisted they would not be a problem once the strategic campaign had "isolated" or even overthrown Saddam Hussein. "Now, General," Warden testily explained, "one of the things we really need to be careful about is that if there's some action on the ground you can't reroll the strategic air campaign. You've got to press on with the strategic air campaign. We made that mistake in World War II, and we don't want to do that again."

Powell was skeptical of "flyboy promises," and he pointed out that even if Warden was right in his flamboyant claim that the Iraqi Army would collapse and "walk home" once the strategic air campaign had done its work, this would still leave Iraq with a formidable military force that would continue to threaten the region with its 5,000 tanks. "I don't want them to go home," Powell said. "I want to leave smoking tanks as kilometer posts all the way to Baghdad."

Powell's skepticism was nothing compared to what Warden was about to receive from his fellow Air Force officers. It was almost a perfect role reversal from the past. Schwarzkopf, an Army infantry officer, seemed dazzled by the prospect of being able to win a war without risking any of his ground troops. The Air Force fighter generals, by contrast, couldn't contain their contempt for Warden's five-ring model (an "academic bunch of crap," in the view of TAC headquarters) and for his stirring renditions of the old party line about strategic bombing ("World War II bomber mentality alive and well," Horner's top planner, Brigadier General Buster C. Glosson, jotted down after receiving the Instant Thunder briefing).

Sent by Schwarzkopf to brief Horner himself in late August, Warden was met with a cold fury by the Central Command air commander. A week before, a carefully selected emissary from the Pentagon was dispatched to Riyadh to give Horner an advance look. Horner had all of a fighter general's contempt for the "air power airheads," as he called them, but he was also furious that Washington seemed to be trying to tell him how to plan his own air war. It seemed like the meddling of Vietnam all over again. When the hapless lieutenant colonel from the Pentagon arrived, Horner literally grabbed him by the tie and started berating him. When he tried to pull out his briefing slides Horner grabbed them and threw them against the wall.

It was a small preview of what awaited Warden a week later. To break the ice, Warden had brought a large bag filled with items he had heard were in short supply in the theater—razors, sun screen, lip balm—and presented it to Horner before beginning his briefing. "What is this shit?" Horner growled, and shoved the bag away. It was all downhill from there. The low point came when Horner, like Powell, asked what would happen if Iraqi tanks started pouring south.

"You're being overly pessimistic about those tanks," Warden replied. "Ground forces aren't important to the campaign."

Even Warden's deputies winced. Everyone else in the room froze. Horner said nothing for a minute, then turned around to his staff.

"I'm being very, very patient, aren't I?" he sarcastically asked no one in particular. "Yes, sir," someone answered.

"I'm really being nice not to make the kind of response that you all would expect me to make, aren't I?" "Oh, yes, sir," the self-appointed straight man again agreed. Horner didn't explode, but from that moment on he basically treated Warden as if he had ceased to exist. At the end of the session, Horner asked Warden's deputies one by one if they could stay and work with Glosson in drafting the air plan. Warden was on a plane home that same afternoon.

But the tantalizing possibility of an Iraqi strategic collapse from just a few days of strategic bombing still had Schwarzkopf mesmerized. And back in Washington Warden still had the ear of friends in high places, especially Secretary Rice. Glosson's planning staff in Riyadh was called the Black Hole, so tight was security surrounding the plans. But there was no shortage of speculation in Washington over how air power might substitute for a costly ground war in which, as think-tank experts such as Edward Luttwak were warning in newspaper articles nearly every day, thousands or even tens of thousands of American troops were sure to be killed even if everything went flawlessly. Powell sought to counter all the talk about easy victory through air power in a statement he gave before Congress in early December. Strategies that depended solely on "surgical air strikes," he told the lawmakers, "are designed to hope to win. They are not designed to win."

This was war, but it was still politics, too, and Rice was alarmed by what seemed a direct challenge to the Air Force's new profile and mission as laid out in the "Global Reach—Global Power" white paper. Rice had already been uneasy at the number of plum assignments in CENTCOM that had gone to Army generals. In Rice's view Air Force prestige was on the line, and he now started a counteroffensive. The first step was to have Warden brief top government officials on the latest Checkmate Division analyses showing how the air campaign could cause Iraqi opposition to disintegrate without a fight. American ground troops would be merely a "cocked fist" that would be "held in reserve," Warden explained in a briefing to Defense Secretary Dick Cheney in mid-December. Regardless of what Horner thought about the "air power airheads," they still had the backing of some important people in Washington.

The net result of these contrary tugs was a curiously hybrid air plan. In substance it had, by December, shifted almost completely under Glosson's direction to a massive, weeks-long campaign against Iraqi ground forces. The goal was no less than the outright destruction of 50 percent of tanks, artillery pieces, and armored personnel carriers *before* any ground offensive would even commence. Horner banished the name Instant Thunder, which he said he detested. And in the end the overwhelming weight of the entire air campaign, fully 67 percent of all coalition air strikes, would indeed be delivered against Iraqi ground forces.

Yet the outlines of Warden's strategic air campaign plan strangely survived through it all, and its rhetorical flourishes did, too. As the Black Hole refined the plan through the fall, Warden's original list of 84 strategic targets expanded to more than 200, then more than 400. His basic categories of strategic targets—leadership, command and control, electricity, oil, railroads—remained intact, as did the fundamental notion of making the strategic attack the opening shot of the air plan. The Black Hole's plan envisioned bringing the full force of United States air power to bear on Iraqi "leadership" during a Phase I "strategic air campaign" of six days: that was right out of Warden's Instant Thunder playbook. Only then would the air effort shift to securing air superiority over the area where Iraqi ground forces were actually dug in in the Kuwaiti theater of operations (Phase II) and attacking the fielded forces directly (Phase III).

Warden's supporters, meanwhile, were zealous in continuing to speak about "centers of gravity" and "incapacitating" the Iraqi leadership, and the virtues of bringing the war to "downtown Baghdad" on day one, and even as the plan was broadened and expanded they managed to keep some of Warden's dramatic rhetoric and sweeping air-power claims a part of it as well. "When taken in total, the result of Phase I will be the progressive and systematic collapse of Saddam Hussein's entire war machine and regime," declared one CENTCOM planning document. A former Checkmate Division planner who had stayed on to work in the Black Hole faithfully pushed the line that strategic attacks on sites of electric power and communications would also hasten the collapse by showing the Iraqi public that their leaders could not defend them; the message, he explained, was "Hey, your lights will come back on as soon as you get rid of Saddam." Glosson himself at times seemed to buy into it, too, suggesting that even he held out hopes that strategic strikes would lead to the overthrow of Saddam Hussein's regime by an angry populace.

Yet some of the effort to preserve Warden's "strategic air campaign" ended up being almost a semantic game, preserving it in name only as the expanded Phase I target list grew to encompass many targets that were strategic in name only. When Schwarzkopf began to insist that Iraq's elite Republican Guard armored units, which had spearheaded the invasion of Kuwait, be hit right at the start of the air campaign, the Republican Guard henceforth became a "strategic" target. (In all, 30 percent of the sorties in Phase I would be against the Republican Guard.) Attacks on airfields, air defense units, and nuclear, chemical, and biological weapons facilities during Phase I also became defined as "strategic." In the end only 2 percent of the Phase I sorties would be directed against actual "leadership" targets in downtown Baghdad. But that 2 percent would strongly shape public impressions of what the air campaign was, and what the new technologies of precision and stealth had done to overturn common assumptions of the past about the nature of air warfare.

From the start the attention was on Baghdad. At about 2:45 a.m. Iraqi time on January 17—it was 6:45 p.m. the evening before in Washington—CNN suddenly switched to an audio feed from its reporters in the Iraqi capital. Antiaircraft fire had erupted from positions all around the city, they were reporting. Air-raid warning sirens had just sounded. It was almost precisely 7:00 p.m. when the telephone link to Baghdad suddenly went dead.

At Glosson's "Black Hole" headquarters in Riyadh a cheer went up. CNN had just provided instant confirmation that the first bombs of the air campaign, a pair of 2,000-pound laser-guided bombs released by an F-117 stealth fighter over downtown Baghdad, had found their mark: the government telecommunications building.

Within minutes laser-guided bombs from seven more F-117s hit other communications centers, air defense headquarters, and command bunkers in Baghdad. Six minutes later a wave of fifty-four Tomahawk cruise missiles launched from ships hundreds of miles away in the Red Sea and the Persian Gulf arrived over the capital and set off a series of deafening booms as they crashed into electric power stations, presidential palaces, the Ministry of Defense building, and the headquarters of Saddam Hussein's ruling Ba'ath Party.

Over the coming days, television viewers would see over and again the image that more than any would become emblematic of the Desert Storm air war: a precision-guided weapon plunging down the air shaft of a government building in downtown Baghdad, hitting precisely in the crosshairs of the infrared targeting system that captured the video clip. "This is my counterpart's headquarters in Baghdad," deadpanned General Horner the next day as he showed reporters a strike video of the Iraqi Air Force headquarters being demolished by a laser-guided bomb in just such a pinpoint strike.

It was certainly the most vivid proof possible that the new technology of precision really worked. Reporters who were in Baghdad during the bombing and afterward found the precision almost eerie; not only were adjacent houses and businesses untouched by the air strikes, with life and business continuing as

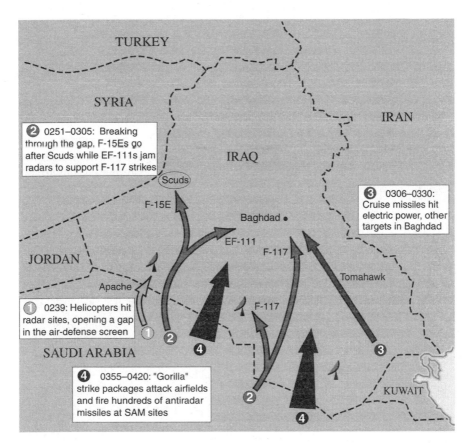

The attack on Iraq's air defense network that opened the Gulf War in the early morning hours of January 17, 1991, drew heavily on the Israeli experience in the Bekaa Valley.

usual in them, but in many cases the targeted buildings themselves, which had housed government ministries or telecommunications facilities, were still standing and at first glance seemingly intact. "Only from close up," reported Paul Lewis in the *New York Times,* "is it apparent that they are gutted shells, their innards either collapsed or trying to burst through the windows."

This was all true enough, yet the real air war was taking place largely off the world's television screens. The overwhelming weight of the first night's attacks had not been against these high-profile symbols of Saddam Hussein's regime and power in Baghdad but rather against the sprawling Iraqi air defense system. This was a "strategic" target in one sense, but in another sense it was simply a prerequisite for conducting any sustained air operations at all in the entire theater of operations. Following Israel's 1981 raid on its nuclear reactor at Osirak, Iraq had spent billions of dollars on radars, missiles, guns, data links, and computers to assemble one of the most formidable air-defense systems in the world. Baghdad itself, with sixty SAM batteries and some 2,000 antiaircraft guns, was the most heavily defended airspace on earth save Moscow. Beyond the Iraqi capital, the country was protected by belts of interlocking SAMs that were commanded by a centralized early-warning network. At the heart of the system was a computer and communications net that had been built by the French defense contractor Thomson-CSF. Seventy-three radar sites and four hundred observation posts fed data to seventeen regional command posts, located in hardened concrete shelters throughout the country. They in turn could automatically relay tracking data directly to SAM and AAA batteries. The data also was passed to four larger regional Sector Operations Centers, and then to Air Defense Headquarters in Baghdad, that oversaw the entire national air-defense operation. Microwave relays and buried fiber-optic cables handled the flow of data.

Buster Glosson's deputy for electronic warfare, Brigadier General Larry Henry, had studied the Bekaa Valley operation during an assignment at the National War College and was convinced that the Israeli method of using decoy drones followed by attacks with antiradar missiles could break the back of even the far larger Iraqi system. One of the more tangible results of Warden's brash persistence had been a secret project he had pushed, code-named Scathe Mean, which had quickly arranged to have some old Navy target-practice drones refitted as radar decoys. The first step was to take out the critical regional command centers in the south of the country so that the SAM batteries that threatened coalition aircraft would be cut off from the early-warning radar network and be forced to rely solely on their own tracking radars. When the decoys then provoked the SAM sites into turning those radars on, they would be hit with massive barrages of radar-seeking missiles. Finally, once the radars were destroyed—or their operators sufficiently intimidated to keep them off the air most of the time for fear of attracting a radar-seeking missile—the SAM sites could then be pounded methodically with precision-guided weapons and cluster bombs.

The official H-hour for the start of the war on Iraq was the 3 a.m. F-117 strikes on Baghdad. But the war had actually begun twenty-two minutes earlier when nine Apache attack helicopters, led by three U.S. Air Force special forces Pave Low choppers with special night-vision capability, approached two early-warning radar sites about forty miles apart along Iraq's southern border. Helicopters were chosen for the mission because they could fly below radar cover and make repeated attacks to ensure the targets were destroyed. At about four miles from their targets the Apaches fired twenty-seven laser-guided missiles, then moved closer at slow speeds pouring in unguided rockets and cannon fire. Minutes later, twenty-two aircraft poured through the hole that had been punched in the Iraqi radar screen; these headed north to jam Baghdad's radars in support of the F-117 strikes on the capital and to hit Iraq's fixed Scud missile launchers that threatened Israel from the western desert. At H–9 minutes, two F-117s took out the regional air-defense center at Nukhayb, 160 miles southwest of Baghdad, that the two demolished radar sites reported to, then continued on to strike the Western Sector Operations Center twenty minutes later.

The next wave was the roundhouse punch to these opening jabs. Two huge "gorilla" forces of aircraft now assembled and appeared to head straight for Baghdad: a large force of Navy strike aircraft coming from the west from carriers in the Red Sea, and twelve F-4G Wild Weasels and supporting aircraft from the south. They were preceded by a swarm of several dozen of the jet-powered Scathe Mean target drones that had been tricked up with radar reflectors and which arrived over the Iraqi capital at about 3:45 a.m. The Navy planes also carried air-dropped glider decoys with radio transmitters that imitated the electronic emissions of fighter-bombers. The effect was both to overload the Iraqi radar screens and provoke the SAM radars into action. The Iraqis promptly took the bait, and promptly discovered that the strike forces' true purpose was not Baghdad at all as they let fly with a hail of antiradar missiles. At one point in the battle, more than two hundred radar-homing missiles were in the air simultaneously. A quarter of the two thousand antiradar missiles used in the war were fired on the first day, half in the first week. After the first night's radar suppression strikes, A-10s came in and proceeded over the next three days to blast some seventy border radar posts that now didn't dare turn their radars on.

As a result of the attacks, radar emissions from SAM radars dropped off by more than 90 percent for the rest of the war. The threat never vanished completely, and postwar analysis found that the links of the air-defense network were harder to break than had been thought and the Iraqis continued to regenerate portions of the system throughout the war. And they still had infrared and optically guided SAMs; and, like the North Vietnamese, they could still try turning on SAM radars at just the last second before firing, especially if the SAMs were cued by other tracking radars. That that threat was very much alive was confirmed on day three when the coalition air forces sent a "gorilla" of ninety aircraft to hit Baghdad in daylight and two F-16s were shot down. It would be the one and only attempt to send nonstealthy aircraft over Baghdad during the entire war.

But the net effect of the attacks on the air-defense system was to clear a vast sanctuary above 10,000 feet for coalition aircraft to operate in. Overall the number of coalition aircraft shot down was astonishingly few. The attrition rate for the war came to 0.05 percent per combat sortie, about a tenth what it was in Linebacker II in Vietnam and a hundredth what it was for many of the Eighth Air Force's raids against Germany in the Second World War. During six weeks of bombing, coalition forces lost thirty-eight aircraft, half of them during the first week. In a typical six-week period in the Second World War, the United States and Britain lost one thousand aircraft.

Most of the aircraft lost over Iraq were shot down because they persisted in trying the low-level tactics that they had trained for or that their weapons systems required. Low-level bombing runs had always represented a somewhat desperate, last-ditch, even daredevil response to the SAM and AAA threat: it was like trying to get past an armed sentry by running as close to him as possible and hoping that he wouldn't have time to get off a shot in the surprise. RAF Tornado fighter-bombers suffered the most. They carried a special airfield-attack weapon that had been designed to operate from 200 feet in standard NATO fashion; it scattered a barrage of mines and cratering charges over runways and taxiways, and these submunitions would disperse too far if released from higher altitude. Four Tornados went down as they hit walls of AAA over heavily defended military airfields.

Although equipped with eight hundred aircraft, including advanced Soviet and French fighters, the Iraqi Air Force never marshaled a serious challenge to American and British air superiority. Partly that was because of a huge disparity in training and initiative (Horner remarked that "any of my captains could have run his air force and caused much more trouble than he did"), but it also was due to the enormous leverage that the dozens of Navy and Air Force AWACS planes gave the coalition forces in ensuring the interception and swift destruction of any Iraqi planes that did try to fight. A Navy F/A-18 pilot who shot down an Iraqi MiG on the first day of the war recalled that the entire process, from the AWACS radio call warning of the approaching enemy to the moment his Sidewinder missile slammed into it, took less than 40 seconds. More than 40 percent of air-to-air kills were beyond-visual-range shots in which AWACS played the indispensable role of electronically verifying the target, ensuring it was not a friendly aircraft. On one occasion eight Iraqi fighters were observed on an airfield ready for takeoff; the first two had barely gotten airborne when they erupted in fireballs, having been hit by missiles launched from unseen American fighters. The other six promptly taxied back to their hardened shelters.

Such a lopsided advantage actually disappointed many coalition fighter pilots who found themselves deprived of the chance to put into practice the dogfighting skills they had honed in years of exercises. A joke quickly began circulating:

Question: What are the three most feared words to an Iraqi fighter pilot?
Answer: "Cleared for takeoff."

Iraqi fighters scored only a single possible air-to-air kill, an F/A-18 that may have been hit by an infrared missile fired by a MiG-25 at one moment when it happened to be in a blind spot on the AWACS radars.

With the Iraqi Air Force pinned to the ground, coalition fighter-bombers began systematically picking off aircraft parked in concrete shelters, using special laser-guided bombs designed to penetrate bunkers and other hardened targets. About two-thirds of Iraq's six hundred aircraft shelters at forty-four airfields were destroyed in the strikes. By day eleven of the war the Iraqi Air Force had basically ceased flight operations—except to flee to Iran.

By the end of the first week the weight of the air campaign was already shifting to the Iraqi ground forces and their lines of supply. At a press conference in Washington a week into the war Colin Powell explained what was going to happen: "Our strategy to go after this army is very, very simple. First we're going to cut it off, and then we're going to kill it." That was not exactly a new strategy, but the use of air power for this task, on such a scale as it now set out to do, was something without precedent. As the official *Gulf War Air Power Survey* published after the war noted, "Never has an air force found itself in the position of 'preparing the battlefield' to the extent that ground commanders counted on air power being able to achieve a 50 percent level of destruction of the enemy's equipment."

By January 29, the end of the second week, the Black Hole planners had posted a sign on their wall: WE ARE NOT "*PREPARING* THE BATTLEFIELD," WE ARE *DESTROYING* IT.

It was not an idle boast. For one thing, precision guidance made it possible to cut off supplies and isolate a fielded army as never before. With only 450 bombing sorties, forty-one of the fifty-four road and rail bridges between Baghdad and Kuwait were destroyed, along with thirty-two pontoon bridges hastily erected by Iraqi engineers to take their place. The capacity of highways from Baghdad to the Kuwait theater of operations was cut to a tenth of what it had been. Although the Iraqis had stockpiled considerable amounts of food, water, and ammunition in the Kuwaiti theater, coalition air attacks on trucks and roads crippled their ability to distribute even what they had on hand. More than half of the trucks in the forward area were demolished in air strikes; to avoid drawing fire the remain-

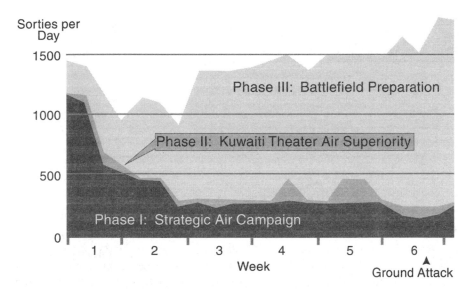

The phases of the Gulf War air campaign, January–February 1991.

ing trucks ceased moving in convoys and traveled only in small groups. Many Iraqi prisoners and deserters reported a lack of food and water in their units once the bombing began.

The interdiction campaign was nothing, however, compared to the direct strikes on the Iraqi Army that now began in earnest, too. The shift from low to medium altitude had suddenly made precision delivery essential even for many ground-attack missions that had been envisioned for fighter-bombers like the F-16 and the RAF Tornado equipped with unguided bombs. The F-16's computer "bombsight" continuously calculated a bomb impact point and displayed it with a small reticle on the head-up display, and using the standard NATO-style low-level approach followed by a pop-up to about four thousand feet and dive just in front of the target, F-16 pilots during exercises achieved remarkable accuracies even with "dumb" bombs, an average of about thirty feet. But now, trying to attack from medium altitude, those accuracies abruptly dropped to two hundred feet. None of the F-16s or Tornados had laser designators and so could not use precision weapons. As of January 23 only a dozen LANTIRN targeting pods were available, and F-15Es assigned to hunt mobile Scud missile launchers had first claim to them. The A-10, built specifically as a specialized close-support tank killer, had a fearsome 30mm Gatling gun whose depleted uranium shells could pierce a tank hull, and it could fire the Maverick antitank guided missile, but those were short-range weapons, too.

That left only the F-111Fs, which were the last airplanes anyone would have chosen for attacking enemy armor on the ground. But in late January F-111 pilots returning from strategic missions happened to notice that even well dug-in Iraqi tanks and other vehicles stood out distinctly at night on their infrared imaging displays. Exercises in December in Saudi Arabia had noted a similar phenomenon: tanks absorbed heat during the day and then lost it at a different rate than the surrounding sand at night. On February 5 an experiment was run; F-111s dropped eight 500-pound laser-guided bombs and killed four tanks and one artillery piece. Horner immediately ordered that all F-111Fs be shifted from the strategic campaign and be put to work taking out Iraqi vehicles.

One of the F-111 pilots who had flown the February 5 test mission remarked later, "If I had stood up at Staff College a year ago and proposed using the F-111F for this type of attack, I would probably have been laughed out of the room." The A-10 was really the plane that had been built for this job. Its designers had gone to enormous lengths to make this specialized attack jet survivable at low level over the battlefield: the pilot was enclosed in a titanium "bathtub," its engines were placed above the tail surface to shield their heat signature from infrared SAMs, and its rugged design and redundant control surfaces allowed it to keep flying even when large chunks of the plane had literally been blown away. Yet none of these measures proved nearly as good as the protection that being able to fly at 15,000 feet with a laser-guided bomb afforded the F-111. A-10 operations were sharply scaled back after two were shot down in one day on February 15.

With the shift of the F-111s to antitank operations, the number of armor and artillery kills piled up at a dizzying pace. F-111s carried out hundreds of strikes a night, destroying close to a thousand Iraqi armored vehicles over the next three weeks. (In a typical bit of boastful fighter-pilot understatement, the F-111 crews called the job "tank-plinking," which irritated Schwarzkopf: he wanted to call it "tank busting." Horner pulled the old trick of obeying orders to the letter: "I told the troops, 'General Schwarzkopf does not want you to call it tank-plinking,' " Horner later recalled, "and that way I ensured that it will forever be known as tank-plinking.")

Meanwhile JSTARS was demonstrating what would happen to Iraqi tanks that tried to move out of their dug-in positions. Two weeks into the war, in an effort to provoke coalition forces into ground fighting and to probe their dispositions and plans, the Iraqis launched an abortive attack with two heavy divisions from southeast Kuwait into Saudi Arabia and briefly occupied the coastal town of Al Khafji. The U.S. Air Force had originally turned down a request from CENTCOM to deploy JSTARS to the theater, insisting it was still experimental. But a U.S. Army corps commander, Lieutenant General Frederick M. Franks, Jr., had seen it work during the recent exercises in Europe and sang its praises to Schwarzkopf, and finally just a few days before the air war began the planes arrived in Saudi Arabia. The planes were indeed experimental; the commander of one recalled that it took "four very highly paid Ph.D.'s from the contractor" just to keep its software running. But it proved so effective in spotting the Iraqi armor movements during the battle of Al Khafji that entire Iraqi units were wiped out before they could even reach the fighting. A thousand attack sorties were launched over three days; 90 percent were able to locate their targets on the first pass thanks to real-time cueing from JSTARS. Most ran out of ammunition before they ran out of fuel. In one convoy, fifty-eight of seventy-one vehicles were destroyed. A prisoner from the Iraqi 5th Mechanized Division said later that his brigade had undergone more punishment in thirty minutes than in eight years of fighting in the Iran-Iraq War. By the time it was over, some 300 tanks, 150 armored personnel carriers, and 100 mobile artillery pieces had been destroyed.

A final blow against Iraqi ground forces from the air was provided in the form of psychologically devastating attacks by carpet-bombing B-52s and several huge, 15,000-pound BLU-82 bombs dropped from C-130 transports. The BLU-82 was known as the "daisy cutter" in Vietnam, where it was used to clear landing zones in the jungle. One was dropped in the desert near an Iraqi frontline position, followed by leaflets warning the Iraqis that they had "just experienced the most powerful conventional bomb dropped in the war" and promising that the next one would land directly on them. Investigations after the war concluded that about 9,000 Iraqi troops were killed by air attack before the start of the ground war, but 100,000 to 150,000 had deserted. Another 85,000 surrendered almost immediately once the ground fighting began. Iraqi prisoners spoke of the constant stress of knowing that coalition aircraft were always overhead, and not knowing if and when they might strike them, as *the* major factor that induced them to flee or give up.

Iraq had begun the war with the world's fourth most powerful military force: 60 divisions, 5,700 tanks, 5,000 armored fighting vehicles, 3,700 artillery pieces. Not just the Washington pundits but the United States military had been preparing for tens of thousands of casualties in a ground war. As it happened it was almost an anticlimax, over in four days. Although the 50 percent attrition goal had not quite been reached before the ground war began, air attacks had nonetheless eviscerated Saddam Hussein's army: 1,400 tanks and 2,000 other armored vehicles and artillery had been wiped out from the air before the first coalition tanks rolled across the Saudi-Iraqi border on February 24. The U.S. Army had shipped 220,000 rounds of tank ammunition to the theater, but in the actual event American tank crews fired less than 2 percent of it. Total American casualties for the entire operation were 148 dead and 458 wounded.

Nothing like this had ever happened in a war before.

Amid the exultation over such a victory, no one wanted to call attention to the fact that the strategic air campaign had been a bust, at least when measured against the promises that Warden and other enthusiasts had made for it. It certainly hadn't come anywhere near to inducing the promised "collapse of Saddam Hussein's entire war machine and regime." Warden's original claims for Instant Thunder had in any case been obscured in the subsequent planning and fudging of definitions; after the war, the U.S. Air Force tended to speak about what had actually been accomplished in the strategic attacks on Baghdad—"degrading" communications between the Iraqi leadership and its military forces in the field, forcing government facilities to use backup generators when electric power was lost, and causing the regime other "untold inconveniences"—as if such modest achievements had been the aim all along. Even while acknowledging that the attack on Iraq's fielded forces had constituted the overwhelming bulk of the air effort, the official Air Force line was to insist that nothing had been a wasted effort; the strategic strikes *together* with all of the rest of the air campaign had been "decisive."

In one sense it was amazing that in the wake of such a brilliant success, in which air power had almost singlehandedly defeated the mightiest army the United States had faced since the Second World War, the Air Force would go out of its way to chain itself to the historical dead weight of strategic bombing theory. As an analyst at the Institute for Defense Analyses, Caroline Ziemke, observed:

> By making strategic bombing synonymous with air power, strategic bombing theory effectively excluded some of the most potentially

decisive aspects of air power from its own scale of effectiveness. . . .
In the process, those who sought to prove that strategic bombardment
alone was a war-winning capability did a grave disservice to the
interests of broader air-power concepts.

But in a way it was even harder to admit mistakes in victory than in defeat.
Nobody was pointing fingers of blame or demanding explanations or searching
for scapegoats, and in the glow of triumph it was easier simply not to reopen all
of the old arguments that had reverberated between the Checkmate planners and
Horner's staff in the fall—especially after the images of precision strikes on targets in Baghdad had done so much to sell the public on the amazing new technical capabilities of air power.

The concern that any admission of error would tarnish the glowing reputation
air power had now secured for itself was apparent in the effort by some in the Air
Force even to block public release of the *Gulf War Air Power Survey,* which had
been prepared by an independent panel commissioned by Secretary Rice: as one
internal memo warned, the report's censure of the strategic air campaign's poor
results would cause the "actual accomplishments of airpower in the Gulf War" to
be "distorted or misinterpreted," and "even the positive statements" of the report
would be "lost" amid the criticisms. To his enduring credit, Rice released the
report anyway.

The *Survey* certainly did not mince words in dismissing the grandiose claims
Warden had made for the strategic air campaign in the first place, or in dissecting
its failures of execution. In particular, the idea of fomenting a revolt or coup
through strategic bombing was "dangerously optimistic" to the point of utter
naïveté:

> Six days of strategic bombardment in anything other than a nuclear
> context—which of course was completely unthinkable—had little
> chance of persuading Iraq to do the Coalition's bidding. The political
> strength of Saddam's regime was such that only a campaign aimed at
> breaking Iraq and probably involving tens of thousands of casualties
> could have toppled the dictator.

Despite all of the vague terminology about targeting "leadership," not a single top Iraqi leader was killed in the air strikes on presidential palaces and government buildings; they proved far too elusive a target. A lack of detailed
intelligence about inherently elusive targets was also to blame for the failure to
find and destroy a single mobile Scud launcher despite 2,500 Scud-hunting sorties, and for the emergence of the Iraqi nuclear weapons infrastructure virtually
intact at the end of the war.

Other strategic target sets were hit far more successfully but with equally little consequence. Despite the destruction of many critical command-and-control

"nodes" in Baghdad, Saddam was still able to communicate with his commanders in Kuwait throughout the war: the attacks did not come close to "decapitating" or "isolating" the Iraqi leadership as hoped for and Baghdad was even able to successfully order a redeployment of forces at the end of the war aimed at covering the retreat from Kuwait. Strategic strikes on key production were technically even more brilliant successes: electric generating capacity was cut by 88 percent, petroleum refining by 93 percent. Yet the *Survey* could find no evidence for any "cross-category" payoffs in fomenting public resistance or even degrading military effectiveness.

The reasons were all the familiar ones from history: the great difficulty of acquiring intelligence about a secretive regime; the ability of the enemy to hide, adapt, substitute, stockpile, and make use of redundant capacity. The experience of the Second World War had suggested that strategic interdiction of petroleum can be effective in a long war, but during the Gulf War Iraqi forces did not come close to running out of their considerable stockpiles of fuel.

For decades, the debate over the effectiveness of strategic bombing had turned on a debate over the technical ability to strike targets massively, quickly, and precisely. The *Survey* concluded that, far from having at last breathed life into the venerable theories of strategic attack, the awesome proof of that technical ability in the Gulf War had at last revealed that the patient had been dead for years: "If limitations to strategic air attack with strong antecedents in prior conflicts also manifested themselves in the near-ideal circumstances of Desert Storm, those limits should probably be construed as inherent features of strategic campaigns, not as aberrations or shortcomings that improved weaponry or other technical advances will overcome."

Following Desert Storm, Air Force politics and official doctrine continued to keep alive some Wardenesque rhetoric about targeting enemy "leadership" and other "strategic centers of gravity," but an ever stronger current continued to run the other way—notably among yet another generation of junior officers who, like the Vietnam generation before them, had seen tactical air power at work. Theses by majors and lieutenant colonels at the Air University in the late 1990s all but ridiculed Warden's "five ring" model and argued that what actually coerces an enemy into capitulating is when his army is wiped out. Other prominent voices in Air Force thinking agreed. Eliot Cohen, the director of the Gulf War Air Power Survey, commented that mechanistic theories like Warden's suffer from the same fault they had from time immemorial, in that they are not true military strategies at all: they fail to take into account the enemy's plans, intentions, strategies, psychology, culture, and unique vulnerabilities, simply reducing him to a "passive collection of targets." ("Social scientists make bad generals," Cohen sarcastically concluded.) And a RAND study for the Air Force pointed out what from any other source might have seemed simply the obvious but which coming from the Air Force was practically a revolutionary admission:

In every major conflict from World War II on, enemies have capitulated or acceded to peace terms demanded by the United States only
after their deployed forces have suffered serious battlefield defeats.
In future conflicts enemy leaders are likely to prove equally reluctant
to make concessions or terminate conflicts as long as they see a
chance to prevail on the battlefield.

Many who derided the Air Force's vestigial allegiance to Warden's theories
still acknowledged that an ability to hit certain "strategic" or "leadership" targets
could prove valuable, even when the major effects one was seeking to achieve
were on the battlefield. Destroying targets such as television transmitters or air-
defense command posts or chemical weapons storage sites could provide a tactical or psychological-warfare payoff, especially in a fast-paced war. An
extremely lucky shot at a "leadership" target might even kill an enemy dictator
outright. But short of such a coup, which all experience argued was implausible,
such strategic strikes were much more akin to special forces operations. No one
ever claimed that special forces operations would be decisive in a war.

There was, to be sure, one past limitation to strategic air attack that new technology had overcome: with the combination of stealth and precision it was for
the first time possible to strike at the heart of an enemy city with only minimal
risk to civilians, and minimal risks to one's own aircrews, too. But in a paradoxical way, even that astonishing revolution in the very nature of warfare had generated its own equal and opposite new limitation. The very promise of surgical
precision brought scrutiny of a kind to target selection that was unimaginable in
an earlier age when striking any target in an urban area inevitably meant heavy
civilian casualties. The technology itself invited the sort of political and legal
scrutiny that airmen had so railed against in Vietnam. And it also amplified the
political consequences of the mistakes when they did happen, both because the
weapons were so lethal and because the aura of perfection they had created
seemed to imply deliberate intent. If anything, it made strategic air power even
more vulnerable to efforts by totalitarian regimes to use civilians deliberately as
shields. After mistaken intelligence led to an F-117 strike with a bunker-
penetrating laser-guided bomb on what turned out to be a shelter filled with hundreds of civilians in Baghdad during the Gulf War, Powell ordered that no more
targets would be struck in the capital without his approval.

Many military men also were worried that the "Nintendo image" of precision
strikes and the extremely low American casualty rate might ironically have created an expectation of "antiseptic" warfare that would make it easier for an
American president to commit to future military action while also leaving the
public more intolerant than ever of seeing the fight through if American casualties did occur. Before the Gulf War, Saddam Hussein had boasted in an interview
on German television that once 5,000 American troops were killed, the Ameri-

can public would demand an end to the war. In one sense the victory against Saddam had vanquished the Vietnam syndrome; in another sense it had arguably only made it worse by creating an illusory, impossible standard of perfection. No one of course regretted the miraculously low American losses, but some Air Force officers worried that they had made it look too easy.

Both of these paradoxical limitations would work their strange effect in the brief but very peculiar air war over Serbia in the spring of 1999. During a NATO operation to pressure Serbian dictator Slobodan Milosevic to halt assaults on ethnic Albanians in the Serb province of Kosovo, it became almost a Kabuki-theater version of a strategic air campaign as night after night empty government buildings in downtown Belgrade were picked off with exquisite precision. Every target had to be approved individually by NATO political authorities. Information on the weapons to be used, their exact impact points, the expected casualties, and the expected physical damage was provided for each target.

At the same time NATO's commander decreed that the "first requirement" was to avoid any loss of aircraft, lest NATO casualties weaken European and American political support for the operation. An on-the-ground survey conducted after the war by an independent military expert, William Arkin of the organization Human Rights Watch, confirmed that "enormous efforts, largely successful" had indeed been made to avoid harm to civilians. Some 28,000 weapons had been dropped, 29 percent of them precision guided (versus about 8 percent of the 230,000 weapons dropped in Desert Storm), and the accuracy of the strikes had been "extraordinary." Yet many of the targets chosen seemed to be more symbolic than tied directly to any military objectives. Over the course of two and a half months NATO air strikes delivered fourteen million pounds of bombs, destroying or considerably damaging fifty bridges; ten military airfields; dozens of factories, oil refineries, electric power stations, and weapons depots; and fourteen official buildings, including the Belgrade headquarters of the ruling Socialist Party, the Interior Ministry, and Milosevic's presidential residence.

The war also saw the combat debut of the $1-billion-apiece B-2 stealth bomber. The eerie, bat-winged planes, originally built as nuclear bombers to strike the Soviet Union, flew their missions nonstop from their base in Knob Noster, Missouri, delivering precision bombs against a limited number of "high value" targets in Serbia. There were some strategic targets in heavily defended Belgrade that only the B-2 could strike. But the planes' employment spoke more about finding a new role for a hugely expensive Cold War–era investment than it did about the new strategic logic of air warfare. (The Air Force also made much of the B-2's demonstration of air power's "global reach" in flying directly from the continental United States to strike targets halfway around the world—glossing over the fact that the only reason the B-2 had flown such long missions was that it had to: unlike more-conventional aircraft, the B-2 could not be deployed to forward bases because of the specialized hangars and maintenance equipment needed to keep its delicate radar-absorbing coating in repair.)

Of course the larger question was whether using *any* aircraft to strike such a grab-bag of strategic targets had much of a point. As General John P. Jumper, commander of U.S. Air Forces Europe, dryly commented a month into the war, "Milosevic was killing people while we were killing things." Or, in the memorable phrase with which Arkin summed up the NATO strategy: "Smart bombs, dumb targeting."

In the end it did work, in a sense. Milosevic capitulated; civilian casualties from NATO bombing probably were fewer than 500; NATO combat fatalities were zero. Two NATO aircraft were shot down, an attrition rate of 0.005 percent, a tenth the Gulf War's rate. Some American air commanders insisted that had the attacks been directed with full force against "strategic centers of gravity" right from the start that the end would have been much swifter. Others pointed out that only when Russia, Serbia's chief ally, began to pressure Milosevic to back down and air strikes began taking a direct toll on Serb forces in Kosovo—and in particular when an offensive by the ethnic Albanians forced the Serbs to mass and at last become vulnerable to attack by B-52s in the final week of the war—did Milosevic accept NATO's terms. But virtually all American air commanders said that they never wanted to fight another air war like this one.

B-2
bomber
engines: 17,300-lb. thrust turbojet (4)
top speed: high subsonic
armament: 40,000-lb. bomb load

Like the attacks on strategic targets in Baghdad during Desert Storm, the air war over Serbia had, if nothing else, been a remarkable technical demonstration of new capabilities. The most significant change to have taken place since the Gulf War was the advent of a new family of guided weapons that could be delivered by just about any airplane under just about any conditions: night, bad weather, high altitude, the target in view or not. The JDAM (Joint Direct Attack Munition) steered itself to a digital map coordinate that could be entered electronically into the bomb's guidance system up to moments before being dropped. A Global Positioning System receiver in the bomb's control package compared its current location to the target location and sent steering corrections to its fins on the way down. The concept was simple and inexpensive: a $20,000 kit strapped onto an existing dumb bomb. It also permitted the attacking aircraft to "launch and leave," as there was no need to loiter in the vicinity holding a laser designator on the target until impact. With a thirty-foot accuracy, the JDAM did not quite have the precision of laser or TV guidance, but its all-weather capability more than offset that.

The JDAM also swept away the last remaining distinction between strategic and tactical aircraft, a line that had been blurring since Vietnam. A B-52 at 30,000 feet was now able to strike pinpoint targets as adeptly as a fighter-bomber. Only 10 percent of American combat aircraft were capable of delivering precision weapons during the Gulf War; by the time of the Kosovo operation 90 percent were.

It would be virtually 100 percent two years later when, in response to the September 11, 2001, terrorist attacks on New York and Washington, United States forces launched a swift campaign to oust the Taliban government of Afghanistan that had sheltered and supported the Al Qaeda terrorist organization. About 6,000 JDAMs and 6,000 laser-guided bombs were dropped in the campaign, accounting for more than 60 percent of the total bomb tonnage.

During the Gulf War the normal targeting cycle was seventy-two hours. Planning a strike required a complex process of generating what was called an Air Tasking Order to assemble a strike package. Because of a lack of digital communications links, the ATO had had to be delivered to Navy carriers on a disk carried by a messenger plane. The use of GPS coordinates for locating targets and aiming bombs now changed everything. Commandos on the ground in Afghanistan, equipped with handheld GPS receivers and laser range finders, could quickly fix the digital coordinates of an enemy position and call in a witheringly accurate air strike. Visual and radar imagery captured by drones, U-2 spy planes, and satellites—all relayed in real time to the coalition air operations headquarters in Saudi Arabia—could instantly be compared to a database of stored map images to refine the coordinates of targets picked up by these sensors or ground spotters. The turbojet-powered Global Hawk drone, with its synthetic aperture radar and moving-target indicator radar to locate enemy vehicles, was able to act as a sort of mini-JSTARS in itself. With a two-day endurance, built-in SAM countermeasures, and a satellite radio link for sending data and receiving steering instructions, Global Hawk was representative of a new generation of drones whose capabilities for the first time rivaled those of manned aircraft. The smaller, propeller-driven Predator drone was used to feed real-time TV images directly to commanders and gunship crews, and could even fire its own Hellfire laser-guided missiles by remote control.

One of the most dramatic displays of the speed, flexibility, and lethality that this combination of information technology and precision strike made possible came a month into the Afghan war. A commander of anti-Taliban forces advancing on the city of Kunduz urgently asked for an air strike within the next twenty-four hours against a Taliban position. An American special operations ground spotter relayed the request to the air operations center, which ordered a B-52 already in the air to contact the spotter; the spotter radioed the GPS coordinates directly to the B-52; and nineteen minutes later sixteen cluster bombs wiped out the enemy position. A particularly lethal weapon proved to be JDAM bombs fuzed to airburst over Taliban trenches. As the New York Times reported, this "new kind of American air power" allowed a "ragtag opposition" to rout the Taliban army. Or, as one Air Force colonel explained the success of the anti-Taliban Northern Alliance forces: "They had horses and air power."

A leaflet dropped during the Afghanistan air campaign. Parallel captions in Pushto and Arabic read, "Taliban, do you think that you are safe . . . in your tomb?" *(Library of Congress)*

Visiting Afghanistan after the war, the writer Michael Ignatieff described the almost surreal scene of a single American special forces soldier sitting in the corner of a warlord's compound in Mazar-i-Sharif during delicate negotiations between two rival local factions and a United Nations mediator. It was immediately clear that despite the considerable conventional show of force each of the battling militias put on throughout the city (tanks, machine-gun-equipped pickup trucks, tough-looking men brandishing grenade launchers), the "heavyset American is the one who matters," Ignatieff explained:

> He comes with a team that includes a forward air controller, who can call in airstrikes from the big planes doing Daytona 500 loops high in

the sky. No one knows how many CIA agents and special forces troops there are in the country. The number is small—perhaps as few as 350—but with up-links to air power and precision weapons, who needs regiments of ground troops? When you ask the carpet sellers in Mazar why there has been peace in the city, they point up into the air. Only America, the carpet sellers say, puts its peacekeepers in the sky.

In the Afghan war precision air power was no longer being reserved for a few "high value" strategic targets; precision had become ubiquitous. Following the war, the Air Force and Navy contracted Boeing to build an immense stockpile of JDAMs, some 230,000. Precision in these quantities was something more than just precision: it constituted an entire concept of air operations that departed wholesale from past practice and constraints. In Afghanistan many coalition air-craft took off without even knowing their targets. Many enemy ground forces were hit by bombs falling on them from airplanes they never saw, or even heard.

Making full use of the third dimension of space, and the fourth dimension of information warfare, this "new kind of American air power" could indeed oper-ate against an enemy force with impunity. Had H. G. Wells been thinking of sol-diers rather than the men, women, and children of the great metropolises in his forecast of a century earlier ("everybody, everywhere will be perpetually and constantly looking up, with a sense of loss and insecurity, with a vague distress of painful anticipations"), his prophecy could indeed have been said to have come to pass.

How much had changed since the Gulf War, not merely in the technological details of aerial weaponry but in the broad military implications of this new kind of air power, was evident in the lightning victory achieved by American and British forces in the spring of 2003 when they returned to Iraq to finish the job left undone twelve years earlier. Saddam Hussein's commanders, clearly antici-pating a replay of 1991 when coalition air forces conducted thirty-nine days of preliminary bombing before the ground offensive began, were continually caught off guard by the speed of the combined air and ground juggernaut that ran over them. In sixteen days U.S. mechanized infantry and marines were in Bagh-dad; in four more days the Iraqi regime fell; and less than a week after that all major combat operations had ended. It had taken a force of 550,000 troops in Gulf War I to accomplish the far more modest mission of ousting Iraqi forces from Kuwait; Gulf War II began with fewer than half that number of American and British troops, but with an air force that commanders estimated to be ten times more powerful in its destructive effect than the last time.

Air Force planners pressed from the start to have air operations incorporated as an integral component of the entire maneuver plan of coalition ground forces. In the past, Army and Air Force commanders alike had generally viewed battle-field air support as a matter of the Air Force "servicing" lists of targets that had been "nominated" by ground commanders; the Air Force command role was largely limited to affirming or vetoing close-air-support requests forwarded from the Army as the fight unfolded, and managing the purely technical details of logistics and coordinating air traffic over the battlefield. This made sense to the extent that Air Force officers could not be expected to be masters of the military art of planning and carrying out a land campaign; it reassured the Air Force it would be able to reject requests that threatened to disperse air power excessively (the old "penny packet" problem) while reassuring the Army that the flyboys would not go off and fight their own war, leaving the grunts in the lurch.

But with air power playing a role that now went well beyond mere support, the limitations of such a command structure were becoming painfully apparent. As good as air and ground coordination had generally been in Afghanistan, there were a number of incidents there in which Army commanders had drawn up battle plans without even consulting their Air Force counterparts. On March 1, 2002, literally less than twenty-four hours before the start of Operation Anaconda, an offensive designed to trap and eliminate a pocket of several hundred Al Qaeda troops in the mountains of eastern Afghanistan, Army Major General Franklin Hagenback called his counterpart in U.S. Central Command's Combined Air Operations Center to mention that he might need some air support. The air commander, Lieutenant General T. Michael Moseley, was flabbergasted; while on the phone he did a quick calculation and found that there was not even enough aviation fuel on hand at Bagram, the Air Force's main operating base in the country, to support more than one day of air strikes for an operation of this scope. The attack went ahead anyway, and almost immediately ran into trouble. Al Qaeda resistance was much fiercer than expected, Afghan troops who were supposed to lead the attack with American support performed poorly, and a helicopter attempting to insert U.S. special forces on a mountain ridge came under fire and crash-landed. In the end air power turned the tide: coalition aircraft dropped more than 2,500 bombs during the two-week offensive (Hagenback's original plan had predicted it would take seventy-two hours), with the most effective attack coming on March 10 when two A-10s were credited with killing 200 Al Qaeda and Taliban troops during a single mission.

A series of post-Vietnam defense reforms enacted to quell interservice squabbling and encourage joint operations had managed to prevent a repeat of the turf battles that had proved so debilitating in Vietnam, where the Air Force and Navy had each pursued their own, often incompatible, air strategies. The Department of Defense Reorganization Act of 1986 (generally known as the Goldwater–Nichols Act, after its two chief legislative sponsors) sought in particular to undercut the power of the separate services to interfere in wartime command decisions by establishing regional warfighting commands with a single unified commander. Under him was a single air commander who would coordinate the air operations of all four services (Air Force, Navy, Marines, and Army) in each theater. In practice this system had worked reasonably well in the first Gulf War, Kosovo, and Afghanistan. But developing and executing a truly joint air strategy still depended on the goodwill and cooperation of all the services; integrating air and ground planning even more so. And as Operation Anaconda and other such incidents showed, that goal remained a long way off.

During the planning for Gulf War II, several forces now came together to change that, however. For one thing, the sheer dominance of the air weapon put pressure on the other three services to take seriously their participation in the interservice institutions established by Goldwater–Nichols: it was a choice of either having an active presence at the joint Air Operations Center or losing any real say over the conduct of the air campaign. For another, the Air Force's own seriousness about making the battlefield the focus of its application of precision air power had produced a rising cadre of officers vastly more knowledgeable about land-warfare strategy and joint operations than their narrowly air-minded predecessors. In both the Afghanistan campaign and Gulf War II, Central Command's Air Operations Center brought in as the chief of its strategy division a graduate of the School of Advanced Airpower Studies. SAAS was located at Maxwell Air Force Base, the old home of the Air Corps Tactical School, but the resemblance between the two institutions ended there; founded in 1988 as a highly selective graduate school to educate Air Force strategists, SAAS had quickly developed a reputation for intellectual rigor, a comprehensive view of military history, and a scholarly independence from Air Force "party-line" thinking. The SAAS graduates who now took the lead in strategic planning at Central Command were officers who had the knowledge and background to develop their own target lists, and overall air strategy, for the land campaign.

The plan and execution of the Gulf War II air campaign spoke volumes about the transformation now taking place. Two early versions of the Gulf War II war plan that CENTCOM drew up in January and March 2002 envisioned as much as several weeks of intense preparatory bombing; the decision on when to send in the ground troops would be made only after assessing the effects of the air strikes. But by July a new plan, dubbed Hybrid, was in the works, and it proposed a dramatically compressed timetable, no more than three or four days between "A-Day" and "G-Day." There would be no time to sit back and assess effects before the tanks rolled. The planners acknowledged the risks, but Tommy Franks, the Army general who commanded CENTCOM, strongly endorsed the approach. He had already laid down the law that the campaign had to be "fast and final." The Hybrid plan would be it.

Moseley already had secured Franks's approval to begin taking apart the Iraqi air-defense network under the guise of enforcing the southern "no-fly" zone that had been in place over Iraq since the end of the first Gulf War; starting in mid-2002, whenever the Iraqis fired on coalition aircraft, American and British planes would reply by striking targets on a carefully planned list of some 350 fiber-optic cable relay stations and other critical communications nodes in the air-defense system. That was part of the reason the commanders were confident they would not need a long preliminary air campaign before starting the ground offensive this time. But a much larger reason was the confidence that the vastly increased effectiveness of air-to-ground munitions would be able to deliver a blow of unprecedented speed and magnitude against the Iraqi ground forces. The actual start of the war when it happened on March 20, 2003, compressed the schedule even more; so much so that major air and ground operations commenced almost simultaneously. Increasingly accurate Iraqi artillery fire aimed at the staging areas for coalition ground forces in Kuwait, and mounting fears that Saddam Hussein's forces were planning to sabotage Iraq's southern oilfields, forced Franks's hand. So did last-minute intelligence on Saddam Hussein's possible whereabouts. In the early morning of March 20, F-117s and Tomahawk cruise missiles struck a compound in Baghdad in an avowedly long-shot attempt to "decapitate" the regime. Hours later British and American troops were on the move into Iraq.

Once again, the media's attention was distracted by the telegenic images of the precision cruise-missile and bomb attacks on Baghdad that opened the air campaign in earnest the following night. But despite the impression left by spectacular fireballs erupting from Saddam's palaces and ministries in the darkened capital (and the incessant repetition by commentators of the catch phrase "shock and awe"), these opening salvos were extremely circumscribed. They were also extremely well focused on achieving specific effects that directly supported the campaign's overall strategic objectives. In sharp contrast to the sweeping promises of the past, the avowed purpose of the strategic strike was, as Moseley would later describe it, "setting the conditions for the regime to go away." And in sharp contrast to the grab-bag targeting of the past, the strategic strikes were narrowed to three specific categories: 59 "leadership" targets, including VIP residences, headquarters, and command centers; 112 communications targets; and 104 offices and facilities of the ruling Baath Party and Saddam's security services. The aim was explicitly to spare the country's economic infrastructure while disrupting the ability of top commanders to issue orders; to force their communications onto back-up radio systems, which could be intercepted, rather than secure fiber-optic communications systems, which could not; and to weaken the hold of the regime on the populace through its organs of local repression and terror. The highly visible attacks on the Baghdad leadership and its security services were indeed just one part of a much larger psychological-warfare operation that included the dropping of 42 million leaflets and carefully orchestrated statements by American officials, from the President on down, intended to persuade Iraqi generals and troops that they faced certain destruction if they put up a fight.

That was exactly what proceeded to happen. Several Republican Guard divisions did try to put up a fight, and their destruction followed at a murderous pace. By the third day of air operations nearly all of the roughly one thousand strike sorties a day were being flown against Iraqi fielded forces. Over the course

of the short war coalition air forces delivered 29,200 bombs and missiles, 68 percent of those precision guided; 79 percent of targets attacked were Iraqi troops
and vehicles. Those who saw later the results on the ground described a hellish
scene of destruction, images far too grisly to make it onto the front pages of
newspapers or television: columns of blasted Iraqi tanks and vehicles with
charred corpses of soldiers scattered about them. The most intense phase of air
attacks on the Iraqi armored units began during a pause in the ground advance
five days into the war, as the rapidly moving 3rd Division waited for supplies to
catch up and for a sandstorm to clear. With infrared sensors and radars that
allowed aircraft to see through the swirling sand, there was no "pause" in the air
war; on the contrary, as one official commented later, coalition air forces knew
the location of the Iraqi Republican Guard units better than their own commanders did.

"We're killing the Republican Guard," General Moseley announced at a
morning staff meeting, "but I want you to kill them faster." Tanker aircraft,
despite their serious vulnerability to ground fire, were sent to fly within Iraqi air
space so strike planes could remain on station longer. On April 3, Moseley himself flew on a tanker mission that came within sixty miles of Baghdad to make
the point that he viewed the risk as one worth taking. An airfield outside
Nasariyah was turned into a forward base for A-10s, giving them an additional
hour over target areas. The seizure of the airfield had been well planned and
studied in advance: Within twenty-four hours of the capture of the base by Army
troops an advance air team was on the spot; a day later C-130 transports were
landing with forklifts, other light equipment, and additional ground crews while
refueling trucks and fire equipment came in over land; two days after that A-10s
were flying from the base. The forward reach provided by the Nasariyah base
and the tanker flights allowed for a significant surge in strike sorties. At one
point in the fighting in and around Baghdad more than a hundred strike aircraft
were stacked up twenty-four hours a day waiting to go after targets that emerged.

By April 4, coalition intelligence officials estimated that the air attacks had
reduced two key Republican Guard divisions guarding the approaches to Baghdad to 18 and 44 percent of their effective combat strength. Apparently unable to
realize how quickly their armored forces were being chewed up by the air
attacks, Iraqi commanders kept feeding reinforcements from two more Republican Guard divisions north of Baghdad into the two now-doomed divisions to the
south. At one point the Air Operations Center issued a standing order that any
aircraft that still had bombs on board when returning from other missions would
call in for a GPS coordinate for a target in the Medina Republican Guard division to hit on their way home. In one attack carried out during the sandstorm, a
single strike with four JDAMs destroyed thirty Iraqi T-72 tanks that had been
massed in one spot.

Especially striking was that it was ground forces which had in many cases supported air forces in this fight—a complete reversal of a hundred years of combat experience, though one that a number of air-power thinkers in the late 1990s had predicted. A rapid, light ground advance had compelled the Iraqi ground commanders to move and concentrate their armored units south of Baghdad; when they did so they were detected and annihilated by air power. What opposition the swiftly advancing American troops encountered had been reduced by the earlier air attacks to dismounted infantry: the expected tank-on-tank battles against Saddam's elite armored forces never materialized. Rather than waiting for the Army to call in air strikes on enemy troop positions that they encountered, the Air Force was frequently using its own eyes in the sky, and in space, to find and identify enemy ground forces before they could even cause trouble. The Global Hawk drone proved an exceptionally powerful tool once again. Its data was fed directly by satellite to the Air Force's analysis center in Reno, Nevada; there analysts went through its imagery looking for anything of interest and adding digital coordinate reference points. A target picked up by Global Hawk could be in the hands of the Air Operations Center at Prince Sultan Air Base in Saudi Arabia in as little as twenty minutes. A special "time-sensitive targeting cell" at the center could have the target coordinates on a data link to a strike aircraft in as little as five or ten minutes more. The lesson of what this intelligence capability meant for air power, prestige, and influence was not lost on air commanders; any lingering unease that unmanned aerial vehicles might steal jobs from pilots paled by comparison. ("I love UAVs!" Moseley declared while briefing reporters during the war.)

As a number of Air Force thinkers began to point out, even conventional terms like *close air support* and *interdiction*—even *battlefield* itself—no longer made sense in this new kind of warfare in which air forces could locate and engage enemy ground forces across an entire theater of operations. Special-operations forces sent to western Iraq to hunt for Scud missiles were actually placed under the operational command of Moseley, the air component commander. Repeatedly they were challenged by Iraqi "anti-SOF" parties that outnumbered their small units five to one or more; repeatedly the Iraqis were taken out by JDAMs dropped on them from aircraft that flew constant patrols. In the more conventional fighting in the south, the traditional air-ground coordination system that had been so delicately negotiated over the years between the Army and Air Force quickly gave way to the new reality as well. By the letter of the law, the Air Force was not supposed to drop anything within an Army-designated Fire Support Coordination Line except under the direct control of the ground commander. The coordination line extended a set distance ahead of where friendly ground troops were, or would be expected to be in the near future; in a traditional ground battle with a well-defined front line that system made sense. But in the fast-paced—and now highly three-dimensional—ground fight in southern Iraq it did not. Moseley and the land commanders quickly agreed to use a "kill box" system instead, in which ten-by-ten mile grid squares were laid out over the whole region and the air and ground chiefs would continuously consult on which squares were open to air attack at any given time.

Mitchell and Douhet had foreseen air power eradicating the battlefront. Their reasoning was wrong, but their prophecy turned out to be right.

The signing of the peace that ended the Second World War had taken place on the deck of the battleship *Missouri,* lying at anchor in Tokyo Bay. Rumors of a plot by Japanese diehards to sink the ship in one last spectacular kamikaze attack had kept the American carriers on alert, just out of sight over the horizon. As General MacArthur's last words died away ("It is my earnest hope—indeed the hope of all mankind—that from this solemn occasion a better world shall emerge"), the clouds that had been hanging over the bay broke and sunlight glistened on the battleship's deck. And then overhead came the drone of an overawing display of American military might. It was a staggering sight, 462 B-29s making a long, sweeping turn over the fleet, then flying into the sunlight and vanishing at last from view over the snowcapped peak of Mount Fuji, the ancient symbol of the now defeated nation that floated like a cloud in the distance.

The B-29 was the airplane that had dropped the atomic bombs, the act that in retrospect would be the apogee of air-power absolutism. The American bombers with their shining metallic skins, smooth lines, soaring, three-story-tall tail fins were, like the whirring dynamos Henry Adams had stared at in the Paris Exposition of 1900, a perfect reflection of the idea they represented. One Japanese writer recalled how "the sight of a glistening B-29 trailing white vapor high in the sky" had held a sort of terrible fascination for Japanese civilians, generating hatred and fear but also an irresistible attraction with "its beauty and technological perfection." The planes, he recalled, "came to symbolize the superior strength and higher civilization of the United States."

The difference between the B-29 with its unmistakable message of air-power absolutism and the new American way of war that would crystallize a half century later was the difference between the frank display of technological brawn in Henry Adams's dynamos and the effortless power of the computer chip. Airmen had been drawn to the lure of strategic attack not just by the false promise of easy victory but by the real difficulties of intervening in the battlefield from the air. It was, they had always argued, too wasteful a use of scarce and expensive air resources to send them hunting down targets that the army was supposed to take care of anyway. Finding valid targets on the battlefield from the air, destroying them with any degree of certainty, and not getting killed in the process by antiaircraft fire was just too inefficient a process.

The new technologies of precision and information had now laid the last of those objections to rest. As Afghanistan and the two Gulf Wars had shown, it wasn't even necessary to get close to do close air support any longer. The ability to suppress air defenses and operate at medium and high altitudes with a near one bomb–one kill efficiency had swept aside all of the old ways of thinking about air power on the battlefield. With radars and sensors saturating the battlefield, enemy forces were increasingly being targeted before they ever came into contact with friendly ground forces at all. Strictly speaking, "close air support" meant missions within the range of artillery, supporting troops engaged in direct combat. In fact even in Desert Storm almost all CAS missions hit Iraqi troops and armor before they ever got that close.

And despite the howls of protest heard from Army partisans, who claimed that Defense Secretary Donald Rumsfeld had taken inordinate risks by launching Gulf War II with only two heavy armored divisions and no heavy armor reserve, the effect of air power on Iraqi tanks, surface-to-surface missiles, artillery, and troops left little doubt that Rumsfeld was right: far smaller, lighter, and more agile ground forces could now do jobs that once required huge armies. Many commentators were quick to interpret the "Rumsfeld doctrine" as a departure from the "Powell doctrine," former Joint Chiefs Chairman Colin Powell's principle that when the United States resorts to war it must employ overwhelming force to crush the enemy with minimal American casualties and swift results. On the contrary, Rumsfeld had concluded that air power had become so potent that achieving overwhelming force no longer depended upon the massive ground firepower once required. In this new way of war, ground forces would be needed primarily to induce the enemy to move and to occupy relatively undefended terrain.

The great historical joke on airmen was that having struggled for a century to escape the battlefield in their quest for equal status and independence—having fought so many bitter battles to free themselves from the indignity of providing "mere support" to ground forces—it was on the battlefield where air power finally achieved not merely equality, but its claim to ascendancy.

NOTES

Abbreviations

AAF	Craven and Cate, eds., *The Army Air Forces in World War II*
BA-MA	Bundesarchiv-Militärarchiv, Freiburg
GWAPS	Cohen, *Gulf War Air Power Survey*
IWM	Imperial War Museum, London
LOCMD	Library of Congress Manuscript Division, Washington, D.C.
NACP	National Archives at College Park, College Park, Maryland
OWW	McFarland, ed., *Papers of Orville and Wilbur Wright*
PRO	Public Record Office, Kew, United Kingdom
RAFM	Royal Air Force Museum, London
USAFHRA	US Air Force Historical Research Agency, Maxwell AFB, Alabama

1. Visions

page

3   *New York Times . . . Washington Post:* Crichton, *America 1900,* 4–5.

3   **"as old . . . as Shakespeare":** Sullivan, *Our Times,* 86.

4   **"American boy of 1854":** Adams, *Education of Henry Adams,* 53.

4   **"historical neck broken":** Ibid., 380–83.

4   **"everything was possible":** Jones, *Age of Energy,* 11.

4   **Inventions:** Schlereth, *Victorian America,* 22, 30, 115–16, 189; Sullivan, *Our Times,* 55.

5   **way news now traveled:** Furnas, *Americans,* 858; Schlereth, *Victorian America,* 177, 182; Sullivan, *Our Times,* 53.

5   *San Francisco Examiner:* Crichton, *America 1900,* 4–5.

5   **once unimaginable luxuries:** Schlereth, *Victorian America,* 24, 154–55, 180, 198–99.

5   **saw a doctor:** Schlereth, *Victorian America,* 284–85.

6   **one hundred million visitors:** Rydell, *World's a Fair,* 2.

6   **Columbian Exposition:** Burg, *Chicago's White City,* 81–84, 95, 132, 202.

6   *Ladies' Home Journal:* Crichton, *America 1900,* 259.

6   **"his own iceman":** *Golden Interlude,* 35.

6   **Newell Dwight Hillis:** Sullivan, *Our Times,* 60–61.

8   **"painful anticipations":** Wells, "Anticipations," 405–7.

8   *War in the Air:* Wells, *War in the Air,* 210–11, 393.

8   **"vital importance to the British Empire":** Gollin, *No Longer an Island,* 454.

8   **"before any practical flying":** Wells, *Experiment in Autobiography,* 569.

9   **steady stream of articles:** Chambers, "Debate over Modern War," 244.

9   **"more terrible than . . . kings":** Gilbert, *First World War,* 3.

9   **"cruel and squalid":** Churchill, *My Early Life,* 65.

9   **"armed rabbles":** Steinisch, "Militarism and Imperialism," 37.

9–10 **innovations in the science of weaponry:** Chambers, "Debate over Modern War," 258–59.

10   **"war as terrible as possible":** Maxim's introduction in Hearne, *Airships in Peace & War,* xxx.

10   **Jean de Bloch:** Chambers, "Debate over Modern War," 259–60.

10   **assembled at The Hague:** Parkinson, "Aeronautics at Hague Conference."

11   **"extravagant and impossible things":** Kennett, *History of Strategic Bombing,* 8.

12   *bluffeurs:* Crouch, *Bishop's Boys,* 317.

12   **"fliers or liars":** Ibid., 315.

12   **fitted out the back room:** Ibid., 112–14; Jakab, *Flying Machine,* 9.

12   **obtained from the Weather Bureau:** *OWW,* 12 n. 7.

12   **place with sustained winds:** Ibid., 23 n. 6.

13   **100-degree heat:** Jakab, *Flying Machine,* 87.

13   **"sails were rotten":** Fragmentary Memorandum by Wilbur Wright circa September 13, 1900, *OWW,* 23–25.

13   **sandstorms . . . "hen's eggs":** Letter, Orville Wright to Katharine Wright, October 14, 1900, *OWW,* 31–32; Letter, Orville Wright to Katharine Wright, July 28, 1901, *OWW,* 73.

14   **on matters of principle:** Crouch, *Bishop's Boys,* 60–62.

14   **oust Bishop Wright:** *OWW,* 238–39 n. 6, 493 n. 5.

14   **cranks and crackpots:** Crouch, *Bishop's Boys,* 137–38.

14   **"brute force" approach:** Anderson, *Aircraft Performance,* 9–11.

14   **a decade ahead:** Ibid., 13.

15   **inherent stability:** Jakab, *Flying Machine,* 48–49; Abzug and Larrabee, *Airplane Stability and Control,* 1–2.

15   **Lilienthal:** Anderson, *History of Aerodynamics,* 158–59, 163.

15   **"grand old man":** Jakab, *Flying Machine,* 83.

15   **Chanute . . . a national figure:** Anderson, *History of Aerodynamics,* 194–95; Jakab, *Flying Machine,* 43–44.

15   **missed the entire point:** Jakab, *Flying Machine,* 83–85.

16   **"skill rather than machinery":** Letter, Wilbur Wright to Octave Chanute, May 13, 1900. *OWW,* 15–19.

16   **"*first* instead of the *last*":** Wilbur Wright on "Some Aeronautical Experiments," September 18, 1901, *OWW,* 100.

16   **Orville Wright later recounted:** Orville Wright on the Wright Experiments of 1899, *OWW,* 8–9.

17   **"dramatic beyond its importance":** *OWW,* 9–10 n. 4.

17   **grocer's pull scale:** Jakab, *Flying Machine,* 100; Anderson, *History of Aerodynamics,* 458.

17   **clever calculation:** Letter, Wilbur Wright to Octave Chanute, December 3, 1900, *OWW,* 49–50.

17   **six-horsepower:** Wilbur Wright on "Some Aeronautical Experiments," September 18, 1901, *OWW,* 114.

18   **Chanute had reprinted:** *OWW,* 42 n. 8.

18   **For their 1901 glider:** Wilbur Wright on "Some Aeronautical Experiments," September 18, 1901, *OWW,* 107.

18   **bicycle wheel:** Letter, Wilbur Wright to Octave Chanute, October 6, 1901, *OWW,* 123–24.

19   **Lilienthal's tables . . . remarkably accurate:** Anderson, *History of Aerodynamics,* 210–15, 229–34.

19   **"most valuable technical data":** Ibid., 226.

19   **wind tunnel:** Jakab, *Flying Machine,* 127; Anderson, *History of Aerodynamics,* 223–25.

20   **1902 glider:** Jakab, *Flying Machine,* 153.

20   **Fokker E.III:** Anderson, *Aircraft Performance,* 16–17.

20   **"comparatively unoriginal aspect":** Jakab, *Flying Machine,* 184.

21   **"ludicrous position":** *OWW,* 600.

21   **efficiency of 70 percent:** Anderson, *Aircraft Performance,* 16–17.

21   **"some mixture of error":** Letter, Wilbur Wright to George Spratt, April 27, 1903, *OWW,* 307 n. 1.

22   **5 blanket nights:** Letter, Wilbur Wright to Bishop Milton Wright and Katharine Wright, November 23, 1903, *OWW,* 383 n. 8.

22   **"flying machine market":** Post Card, Orville Wright to Charles E. Taylor, October 20, 1903, *OWW,* 369–70.

22   **read a newspaper account:** *OWW,* 390 n. 5.

22   **cabled home:** Telegram, Wilbur Wright to Bishop Milton Wright, December 15, 1903, *OWW,* 393.

23   **"most famous photographs":** Jakab, *Flying Machine,* 211.

23   **distorted and embroidered accounts:** *OWW,* 412 n. 1.

24   **"from *any* individual or institution":** Letter, Octave Chanute to Wilbur Wright, January 14, 1904; and Wilbur Wright to Octave Chanute, January 18, 1904, *OWW,* 414–15.

24   **"great Government":** Gollin, *No Longer an Island,* 196.

24   **wary of being swindled:** Report from Lt. Col. Gleichen, August 17, 1906, Correspondence with the Wright brothers, AIR 1/728/176/3/33, PRO.

24   **"Governments often appropriate inventions":** Gollin, *No Longer an Island,* 96–97.

24   **"mountebank business":** Letter, Wilbur Wright to Octave Chanute, January 29, 1910, *OWW,* 983.

25   **attempt at blackmail:** *OWW,* 413 n. 2.

25   **Cabot brothers:** Gollin, *No Longer an Island,* 51–52; Kelly, ed., *Miracle at Kitty Hawk,* 122–23.

25   **"official insanity":** Gibbs-Smith, *Rebirth of European Aviation,* 191.

25   **"without expense to the United States":** Gollin, *No Longer an Island,* 94–95; Kelly, ed., *Miracle at Kitty Hawk,* 136–37.

25   **"flat turndown":** Letter, Wilbur Wright to Octave Chanute, June 1, 1905, *OWW,* 495.

25   **first-ever circular flight:** Wilbur Wright's Diary E, September 20, 1904, *OWW,* 456.

25–26   **Capper . . . had come:** Gollin, *No Longer an Island,* 63–65.

26   **filed a lengthy report:** Aeronautics, Visit to the United States of America in Sept. to Novr. 1904 by Brevet Lieutenant-Colonel J. E. Capper, AIR 1/1608/204/85/36, PRO.

26   **"give Great Britain the first chance":** Gollin, *No Longer an Island,* 69.

26   **wrote to the War Office:** Ibid., 102–5.

27   **"see your machine fly":** Ibid., 123.

27   **For $100,000:** Letter, Wright Cycle Co. to British Military Attaché, July 31, 1906, Correspondence with the Wright brothers, AIR 1/728/176/3/33, PRO.

27   **"I cannot advise any further action":** Memorandum to D.F.W. from Colonel, Balloon Factory, September 6, 1906, Correspondence with the Wright brothers, AIR 1/728/176/3/33, PRO.

28   **"where you were in 1904":** Letter, Octave Chanute to Wilbur Wright, November 1, 1906, *OWW,* 733.

28   **"almost as bad as reality":** Gollin, *No Longer an Island,* 182–83.

28   **son of a famous general:** Ibid., 168–71.

28   **at Blair Atholl:** Ibid., 271–72.

29   **"Le bluff continue":** Kelly, *Wright Brothers,* 189.

29   **"We are as children" . . . "not worth knowing":** Gibbs-Smith, *Rebirth of European Aviation,* 286–87.

29   **"Veelbur Reet" . . . Louvre:** Crouch, *Bishop's Boys,* 383, 387.

29   **"balancing planes":** Gibbs-Smith, *Rebirth of European Aviation,* 135.

30   **control system . . . propeller design:** Anderson, *Aircraft Performance,* 18.

30   **"solved the problem of flight":** Gollin, *No Longer an Island,* 362.

30   **"Specifications for a Military Flying Machine":** Ibid., 310–13.

30   **"jump over the moon!":** Letter, Milton Wright to Wilbur Wright, September 6, 1908. Kelly, ed., *Miracle at Kitty Hawk,* 305–6.

30   **bids . . . mere formality:** Squier, "Military Aeronautics," Appendix No. 1, 304; Letter, Wilbur Wright to Octave Chanute, January 27, 1908, *OWW,* 855.

31   **Herring's skullduggery:** Letter, Orville Wright to Lieutenant Colonel C. DeForest Chandler, December 15, 1921. Kelly, ed., *Miracle at Kitty Hawk,* 407–8.

31   **"technical delivery" . . . "innovation trunk":** Squier, "Military Aeronautics," 252; Crouch, *Bishop's Boys,* 380.

31   **"best customer":** Letter, Orville Wright to Wilbur Wright, November 14, 1908, *OWW,* 938.

31   **acceptance tests:** Kelly, ed., *Miracle at Kitty Hawk,* 342; *OWW,* 960–61.

2. Bogeymen

page

32   **"further wars practically impossible":** Letter, Orville Wright to C. M. Hitchcock, June 21, 1917, *OWW,* 1104–5.

32   **Aerial Navigation Subcommittee:** Jones, *Origins of Strategic Bombing,* 26.

32   **Nicholson spoke for the Army:** Gollin, *No Longer an Island,* 414, 420.

33   **"certain death":** Grey, "General Staff and Aviation," 583.

33   **chief object of the Corps:** Squier, "Military Aeronautics," 304.

34   **employ of the Balloon Factory:** Aeronautical Work on the Existing Site of the R.A.E., South Farnborough, AIR 1/729/176/4, PRO; Turnill and Reed, *Farnborough,* 25–27.

34   **grown up in Davenport . . . sued by the real Buffalo Bill:** I am grateful to Jean Roberts for supplying this information, and for copies of relevant documents. For the lawsuit, see J 4/8743, PRO; for the true details of Cody's birth, see TS 17/1253, PRO.

35   **"real scientific Department of the State":** Gollin, *Impact of Air Power,* 3, 22.

35   **Wind-tunnel research:** Anderson, *History of Aerodynamics,* 268, 292.

36   **adoption of ailerons:** Anderson, *Aircraft Performance,* 20–21; Gibbs-Smith, *Rebirth of European Aviation,* 135.

36   **"avaricious monopolists":** Crouch, *Bishop's Boys,* 418–19.

36   **cross-licensing agreement:** Ibid., 466–67.

36   **drain . . . Wrights' own energy:** Ibid., 447; Abzug and Larrabee, *Airplane Stability and Control,* 3.

37   **"to seek their solution":** Anderson, *History of Aerodynamics,* 383.

37   **"civilian and an engineer":** Gollin, *Impact of Air Power,* 107.

37   **"impossible operation of war":** Gollin, *No Longer an Island,* 426–27.

38   **Reims:** Gollin, *Impact of Air Power,* 89–90.

38   **six million marks:** Kennett, *First Air War,* 11.

38   **bring Wilbur Wright to Rome:** Kennett, *First Air War,* 8; Crouch, *Bishop's Boys,* 387–88.

38   **Aero Club of Padua . . . and Switzerland:** Kennett, *First Air War,* 14–15; Kennett, *History of Strategic Bombing,* 16–17.

39   **"no longer an island":** Gollin, *No Longer an Island,* 193.

39   **"controls the fate of nations":** Ibid., 381.

39   **"invasion novels":** Clarke, *Voices Prophesying War,* 112–13, 153.

40   **P. G. Wodehouse:** Wodehouse, *Swoop,* 6, 18.

40   **Aerial League:** Gollin, *No Longer an Island,* 454–55.

41   **"nerve centers":** Ibid., 457.

41   "inflict enormous damage": Gollin, "Phantom Airship Scare," 53.

41   "What other people": Kennett, *First Air War,* 14.

41   Illustrated postcards: Jones, *Origins of Strategic Bombing,* 34.

41   hundred thousand troops: Gollin, "Phantom Airship Scare," 49.

41   sure were German Zeppelins: Ibid., 53–56; Gollin, *Impact of Air Power,* 53.

42   "well-known nervousness": Robinson, *Zeppelin in Combat,* 50–54.

42   Blériot . . . H. G. Wells: Gollin, *Impact of Air Power,* 68–72.

42   French Army . . . had bought: Morrow, *Great War,* 13.

43   Montagu complained: Gollin, *Impact of Air Power,* 35.

43   "migratory birds": Ibid., 79.

43   "We require an air service": Grey, "Lord Haldane," 652.

43   Riley E. Scott: Scott, "Panama Canal."

44   haul a Lewis machine gun: Chandler, "Lewis Aeroplane Gun."

44   Michelin prize: "Foreign Notes"; "Scott's Bomb-Dropper."

44   "destroy any large fortress": "Aero Drops a Shell," *Washington Post,* October 10, 1911.

45   "a ghost of a show": Scott, "Dropping Bombs," 388.

45   TERRORIZED TURKS SCATTER: Kennett, *History of Strategic Bombing,* 13.

45   had hit a hospital: Ibid., 59.

46   Douhet . . . declared: Kennett, *First Air War,* 18.

46   "dampened the enemy's ardor": Ibid., 14.

46   fléchettes: Kennett, *History of Strategic Bombing,* 15. A French Army captain, Sazerac de Forge, was reported to have conducted experiments with anti-Zeppelin fléchettes in 1910, dropping them from the Eiffel Tower: Aeronautical Reports 1910, General Staff, War Office, AIR 1/729/176/5, PRO.

46   "Absolutely worthless": Kennett, *History of Strategic Bombing,* 15.

46   reporters . . . genuinely surprised: Kennett, *First Air War,* 19.

47   "wasting his time": Gollin, *Impact of Air Power,* 199.

47   "c'est zéro": Morrow, *Great War,* 35.

47   French Army's fall maneuvers: Ibid., 15; Kennett, *First Air War,* 17.

47   U.S. Army's Field Exercises: Foulois, "Early Flying Experiences," 61–62.

47   Frederick Sykes: Gollin, *Impact of Air Power,* 198–99.

48   book-length manuals: Kennett, *First Air War,* 30.

48   Sykes, in his 1911 report: Gollin, *Impact of Air Power,* 200.

48   "public mind has been disturbed": Jackson, "Defence of Localities," 701–2.

48   Other experts suggested: Scott, "Dropping Bombs"; Poutrin, "Aéroplanes dans la guerre," 383; Hicks, "Command of the Air," 353.

48   fourth arm . . . third service: Hicks, "Command of the Air," 348; Kennett, *History of Strategic Bombing,* 41–44.

49   "permanent . . . autonomous": Poutrin, "Aéroplanes dans la guerre," 385.

49   idle and premature speculation: Morrow, *Great War,* 14.

49   Many of the French builders: Ibid., 12.

49–50   Gnome . . . speed records: "80 H.P. Gnome Engine"; Morrow, *Great War,* 27, 33; Christienne and Lissarrague, *French Military Aviation,* 40.

50   "weakened their prestige": Notes on the Duty and Performance Expected of Aircraft on or about July 1914, AIR 1/728/176/2/1, PRO, p. 1.

51   "mere ostentation": Ibid., pp. 7–8.

51   The BE2 airplane: Comparison of German and R.A.F. Machines, AIR 1/731/176/6/1, PRO; Morrow, *Great War,* 54.

51   without touching the stick: Macmillan, "More about the Spin," 26.

51   no engine above 100 horsepower: Joint Air War Committee, 1 April 1916, AIR 1/2319/223/24, PRO.

51   only engines available: Notes on the Duty and Performance Expected of Aircraft on or about July 1914, AIR 1/728/176/2/1, PRO, pp. 17–20.

51   **Kaiser sponsored a prize:** Morrow, *Great War,* 37.
51   **"catching an enemy":** Notes on the Duty and Performance Expected of Aircraft on or about July 1914, AIR 1/728/176/2/1, PRO, p. 8.
51   **"more of Jules Verne":** Christienne and Lissarrague, *French Military Aviation,* 51.
51   **1,000 operational front-line aircraft:** Kennett, *First Air War,* 21.
51   **appropriations in 1913:** Holley, *Ideas and Weapons,* 28–29.
52   **spanned 100 feet . . . a ton of gasoline:** Cochrane, Hardesty, and Lee, *Aviation Careers of Sikorsky,* 192; Finne, *Igor Sikorsky,* 47.
52   **did not go well:** Finne, *Igor Sikorsky,* 47–50.
53   **"almost outrageous":** Sikorsky, *Story of Winged-S,* 17.
53   **"reliable information . . . non-existent":** Ibid., 49.
54   **narrowly avoided killing himself:** Ibid., 58–59.
54   **"Start the construction immediately":** Ibid., 73.
54   **through congratulating them:** Ibid., 115.

3. Realities

page
55   **"Why four great powers":** John Burns, quoted in Gilbert, *First World War,* 23.
56   **flow of spare parts:** Morrow, *Great War,* 62.
56   **"one of the rare generals":** Christienne and Lissarague, *French Military Aviation,* 78.
56   **two million tons of fodder:** Winter and Baggett, *Great War,* 85.
57   **"complete and accurate information":** Raleigh and Jones, *War in the Air,* I:329.
58   **"no Tannenberg":** Kennett, *First Air War,* 31.
58   **sixty-five escadrilles:** Christienne and Lissarague, *French Military Aviation,* 82.
58   **"in blind panic":** Kennett, *First Air War,* 30.
58   **every role and mission:** Boyne, "World War One Aviation."
58   **"German Army is at the gates":** "Bombards Paris from Air," *New York Times,* August 31, 1914.
58   **nothing but pistols:** Duty and Performance Expected of Aircraft on or about July 1914, AIR 1/728/176/2/1, PRO.
59   **"armed with rifles":** Raleigh and Jones, *War in the Air,* I:328.
59   **outré improvisations:** Ibid., II:138.
59   **popular aviation press:** See, for example, Grey, "Aircraft in the War," 191.
59   **narrow V-shaped gap:** Duty and Performance Expected of Aircraft on or about July 1914, AIR 1/728/176/2/1, PRO, pp. 12–15.
59   **captured a German order:** Kennett, *First Air War,* 66; Christienne and Lissarague, *French Military Aviation,* 80–81.
59   **Joseph Frantz:** Christienne and Lissarague, *French Military Aviation,* 81.
60   **colored signal lights:** Offence versus Defence in the Air, General Staff, October, 1917, AIR 10/398, PRO.
60   **"clock code":** Raleigh and Jones, *War in the Air,* II:86.
60   **white panels:** Communications between Ground and Airplanes, Bulletin No. 66, Air Service Bulletins, Gorrell History L-6, RG 120, NACP.
60   **shelled their own infantry:** Christienne and Lissarague, *French Military Aviation,* 91.
60   **"not regarded as of much interest":** Offence versus Defence in the Air, General Staff, October, 1917, AIR 10/398, PRO.
60   **400,000 aerial pictures:** Great Britain, *Synopsis of Air Effort,* 18.
61   **"powers of a major general":** Kennett, *First Air War,* 39–40.
61   **Neuve Chapelle:** Boyne, "World War One Aviation."
62   **whenever enemy aircraft appeared:** Duty and Performance Expected of Aircraft on or about July 1914, AIR 1/728/176/2/1, PRO.

62   **"fighting would be necessary":** Offence versus Defence in the Air, General Staff, October, 1917, AIR 10/398, PRO.

62   **heavy, two-seat fighter:** A report from a British inspector who visited the front in August 1915 urged that both small fast scouts and a "heavy fighting type of aeroplane" be developed; the latter "must, therefore, have such a range of direction for its gun fire as will enable it to neutralise its deficiency in speed and climb when molested by an enemy scout." Memorandum of Visit to France, August 4 to 11, 1915, Item 33, Directorate of Aircraft Equipment Files, AIR 1/2429/305/30/1, PRO.

63   **Roland Garros:** Christienne and Lissarague, *French Military Aviation,* 92; Morrow, *Great War,* 91–92.

63   **aerial engagements per month:** Offence versus Defence in the Air, General Staff, October, 1917, AIR 10/398, PRO.

63   **Fokker E.I:** Morrow, *Great War,* 104–5.

64   **halt all flights of monoplanes:** Letter to Secretary of State for War J. B. Seely, September 14, 1912, Royal Aircraft Factory, AIR 1/729/5/72, PRO.

64   **FE2:** The origin of the letter designations for the Royal Aircraft Factory's models was more than a bit eccentric. Pusher-type aircraft were originally designated FE for "Farman experimental," the Farman biplane being an early exemplar of that type; tractor biplanes were similarly termed BE for "Blériot experimental"; and those with a forward canard were SE for "Santos-Dumont experimental." The designations in time were conveniently reinterpreted to stand for "Fighter experimental," "British experimental," and "Scouting experimental." RE was added for "Reconnaissance experimental." See Turnill and Reed, *Farnborough,* 36; *Jane's Aircraft of World War I,* 35.

64   **match for the Fokker:** Boyne, "World War One Aviation."

64   **"as good as dead":** Lewis, *Sagittarius Rising,* 53.

64   **by mistake in a fog:** Christienne and Lissarague, *French Military Aviation,* 92.

64   **"stunting":** Duty and Performance Expected of Aircraft on or about July 1914, AIR 1/728/176/2/1, PRO.

65   **"hard and fast" edict:** Raleigh and Jones, *War in the Air,* II:156–57; Christienne and Lissarague, *French Military Aviation,* 97.

66   **in all of the Boer War:** Gilbert, *First World War,* 132.

66   **"The world is mad":** Ibid., 257.

66   **7,000 horses:** Ibid., 235.

66   **"and destroy him":** Kennett, *First Air War,* 71.

67   **fifty yards above and behind:** Notes on Recent Air Service Operations, Gorrell History C-15, RG 120, NACP, p. 85.

67   **"true defence lies in attack":** Offence versus Defence in the Air, General Staff, October, 1917, AIR 10/398, PRO.

67   **"relentless and incessant":** Ibid.

68   **first ten weeks of the battle:** Morrow, *Great War,* 168.

68   **"further such patrols penetrate":** Fighting in the Air, AIR 10/324, PRO.

68   **499 aircrew killed:** Morrow, *Great War,* 173–74.

68   **"lasting . . . three weeks":** Lewis, *Sagittarius Rising,* 154.

69   **"does away with their efficacy":** Aerial Fighting—Notes on Formations etc. Used in Battle of the Somme, Report by Commandant du Peuty, September 24, 1916, AIR 2/127, PRO.

69   **thirty-three *Jagdstaffeln*:** Morrow, *Great War,* 152; Boyne, "World War One Aviation."

69   **Luftstreitkräfte:** Morrow, *Great War,* 159.

69   **"BRING DOWN YOUR HUN!":** Notes on Aerial Fighting, Air Council, September 1918, AIR 10/63, PRO.

69   **"into the store":** Kennett, *First Air War,* 77.

69   **Boelcke alone shot down:** Boyne, "World War One Aviation."

70   **"very agile in the air":** Memorandum of Visit to France, August 4 to 11, 1915, Item 33, Directorate of Aircraft Equipment Files, AIR 1/2429/305/30/1, PRO.

70   **"Power is essential":** Report of Visit to the Expeditionary Force by Captain R. Bagnall-Wild, December 13, 1915, AIR 1/2430/305/30/15, PRO.

70   **"our home-designed machines":** Report on I.A.'s Visit Overseas, December 1915, AIR 1/2429/305/30/5, PRO.

70   **turning back at the gates:** Turnill and Reed, *Farnborough,* 41.

70   **poured out its bile:** See, for example, the November 18, 1914; December 16, 1914; and August 11, 1915, issues of *The Aeroplane.*

70   **"born an Irish gentleman":** Grey, "More R.A.F. Morals."

70   **"murdering our pilots":** Turnill and Reed, *Farnborough,* 41.

71   **Handley Page firm:** AIR 1/729/176/5/70, PRO.

71   **five thousand scientists:** Royal Aircraft Factory Establishment as of 26.4.17, AIR 1/731/176/6/18, PRO.

71   **had totaled only 77:** Report of Committee on Royal Aircraft Factory, May 12, 1916, Burbridge Committee, AIR 1/731/176/6/3, PRO, para. 7.

71   **file full of letters:** Report by the President of the Air Board to the War Committee, July 19, 1916, Burbridge Committee, AIR 1/731/176/6/3, PRO.

72   **"tone of certain letters":** Reminiscences of Sir Henry Fowler, Sept. 1916–1918, AIR 1/732/176/6/146, PRO.

72   **"white feathers":** Turnill and Reed, *Farnborough,* 41.

72   **fifty thousand parts:** Kennett, *First Air War,* 101.

72   **"necessities of war":** Reminiscences of Sir Henry Fowler, Sept. 1916–1918, AIR 1/732/176/6/146, PRO.

72   **"Christmas trees":** Duty and Performance Expected of Aircraft on or about July 1914, AIR 1/728/176/2/1, PRO, p. 9.

72   **£900 per plane:** Joint Air War Committee, April 1, 1916, AIR 1/2319/223/24, PRO.

72   **Air Board:** Morrow, *Great War,* 180–81.

73   **"have a dictatorship":** Letter, Brancker to Trenchard, April 13, 1916, Trenchard Papers 76/1/5, RAFM.

73   **safety factor of four:** D. S. MacInnes, Lieut-Colonel, Assistant Director of Military Aeronautics, AIR 1/2430/305/30/17, PRO.

73   **cylinders . . . flying off:** Kennett, *First Air War,* 103–4.

73   **"windage":** Whitford, *Fighter Design,* 56.

73   *révision générale:* Kennett, *First Air War,* 106.

73   **one pound per horsepower:** Memorandum of Visit to France, August 4 to 11, 1915, Item 33, Directorate of Aircraft Equipment Files, AIR 1/2429/305/30/1, PRO.

73   **Unlike automobile engines:** Coatalen, "Aircraft and Motor Car."

73   **specialty steels and aluminum:** Morrow, *Great War,* 93.

74   **120 miles per hour:** Boyne, "World War One Aviation."

74   **220-horsepower Hispano-Suiza:** Christienne and Lissarrague, *French Military Aviation,* 149.

74   **oil pipe . . . would burst:** Morrow, *Great War,* 290–91.

74   **captured Hispano-Suiza:** Ibid., 164–65, 232–33.

74   **V-8 and V-12 engines:** Ibid., 345, 369.

75   **teach himself to fly:** Foulois, "Early Flying Experiences," 27–31.

75   **first simulator:** Arnold, *Global Mission,* 18–19.

76   **"only acrobats":** Gibbs-Smith, *Rebirth of European Aviation,* 301–3.

76   **unintuitive control system:** Abzug and Larrabee, *Airplane Stability and Control,* 5.

76   **killed in training accidents:** Frisbee, ed., *Makers of the Air Force,* 16–17.

76   *rouleurs:* Kennett, *First Air War,* 124–25; Bingham, *Explorer in Air Service,* 132–34.

76   **shut down their schools:** Christienne and Lissarrague, *French Military Aviation,* 152.

76   **Central Flying School:** Great Britain, *Synopsis of Air Effort,* 4.

77   **"odious Huns":** Morrow, *Great War,* 175; Kennett, *First Air War,* 125.

77   **close-order drill:** Kennett, *First Air War,* 123.

77   **Trenchard insisted . . . "strong argument for joining":** Morrow, *Great War,* 174.

77   **adding specialized courses:** Christienne and Lissarrague, *French Military Aviation,* 152–54.

77   **five hours in the model:** Great Britain, *Synopsis of Air Effort,* 4.

77   **"camera guns":** Memoir, Flying Officer D. R. Goudie, RFC Collection 80/26/1, IWM; Bingham, *Explorer in Air Service,* 165–66.

77   **formation flying . . . patrol work:** Bingham, *Explorer in Air Service,* 152.

78   **"in some cases the test ride":** Hudson, *Hostile Skies,* 111.

78   **$25,000 . . . $4,000:** Bingham, *Explorer in Air Service,* 47; Correspondence Relating to the Liquidation of Stock of Aircraft and Engines, 1919–1924, Entry 127, RG 18, NACP.

78   **scores of specialized facilities:** Great Britain, *Synopsis of Air Effort,* 4; Christienne and Lissarrague, *French Military Aviation,* 154.

78   **"51 were obsolete":** Hudson, *Hostile Skies,* 2–3.

78   **five thousand pilots:** Air Services Lesson Learned During the Present War, Gorrell History A-1, RG 120, NACP, p. 83.

78   **"useful people":** Bingham, *Explorer in Air Service,* 28–29.

78   **May 21, 1917, the first students:** Ibid., 21.

79   **"'high degree of disorder'":** Ibid., xii.

79   **only 767:** Cameron, *Training to Fly,* 196.

79   **"worse to come":** Bingham, *Explorer in Air Service,* 80–81.

79   **"A damn sight better":** Cameron, *Training to Fly,* 153.

79   **"land without cracking up":** Ibid., 126.

79   **course in navigation:** Bingham, *Explorer in Air Service,* 139–40.

80   **"torque . . . would turn them over":** Springs, *Nocturne Militaire,* 16.

80   **"Only one was killed":** Grider, *War Birds,* 86–87.

80   **"Over and over":** Bingham, *Explorer in Air Service,* 50.

80   **"my fainting has always":** Ibid., 69.

81   **first psychological study:** Rippon and Manuel, "Essential Characteristics of Aviators"; "Attributes of the Successful Flying Officer."

81   **"Homer in the original Greek":** Kennett, *First Air War,* 136.

81   **Rickenbacker:** Hudson, *Hostile Skies,* 72.

81   **Lufbery:** Flammer, "Lufbery," 14–15.

82   **"white as snow":** Diary, October 19, 1917, Captain O. L. Beater, RFC Collection 86/65/1, IWM.

82   **"absolutely ripping time" . . . "colossally thrilled":** Letters, July 24, 1917; July 2, 1918; August 1, 1918; September 26, 1918, Lieutenant Y. E. S. Kirkpatrick, RFC Collection 73/235/1, IWM.

82   **Sixteen percent of French pilots:** Morrow, *Great War,* 366–67.

82   **"what we all feared most":** Memoir, "An Encounter with Richthofen's Circus," Captain H. Brokensha, RFC Collection 73/185/1, IWM.

82   **Parachutes:** Training Bulletin No. 4, November 6, 1918, History of the Training Section, Gorrell History J-1, RG 120, NACP ("For some time we have been hearing reports of successful jumps from planes by German aviators equipped with parachutes. . . . Everything considered, parachutes should be adopted as they will result in the avoidance of a great many casualties").

82   **"explosive bullet hit":** Buckley, *Squadron 95,* 133–35.

83   **grease as protection:** Protection Against Cold and Frostbite, Air Service Bulletin No. 21, Gorrell History L-5, RG 120, NACP.

83   **Oxygen masks:** Inhalation of Oxygen While Flying at High Altitudes, Report of Technical Section of Military Aeronautics, July 1918, AIR 1/2429/30/4, PRO.

83   **"splitting wood"** . . . **"stunt":** Kennett, *First Air War,* 127; Letter, November 8, 1917, 2nd Lieutenant H. G. Downing, RFC Collection 88/7/1, IWM.

83–84 **"we do see life"** . . . **"five minutes later":** Letters, July 24, 1917; October 20, 1917, 2nd Lieutenant H. G. Downing, RFC Collection 88/7/1, IWM.

84   **"Dead or Crippled for Life":** Cameron, *Training to Fly,* 169.

84   **"Thank God!"** . . . **poems and songs:** Quoted in Air Service Poems and Cartoons, Gorrell History M-11, RG 120, NACP; Brewers, *Riders of the Sky,* 111–12.

85   **one American flag** . . . **"went off with his wife":** Springs, *Nocturne Militaire,* 41–42, 102, 105, 114–15.

85   **blood of their victims:** Kennett, *First Air War,* 168–69.

85   **"Assassin!":** Ibid., 173.

86   **"God, it was great!":** Ibid., 216.

86   **"every atom of personality":** Ibid., 168.

86   **"indoctrinating":** Foulois, "Early Flying Experiences," 45–48.

86   **Carl Spaatz:** Quoted in Kennett, *First Air War,* 225.

86   **"more or less of a mob":** U.S. Air Force, *Development of Air Doctrine,* 24.

86   **"slovenly, unmilitary"** . . . **"tin majors":** Bingham, *Explorer in Air Service,* 23, 43; Grider, *War Birds,* 109.

86   **"no more rough landings":** Cameron, *Training to Fly,* 126; Bingham, *Explorer in Air Service,* 103.

88   **"Death was anonymous":** Kennett, *First Air War,* 154–60.

88   **"THE BOMBER":** Air Service Poems and Cartoons, Gorrell History M-11, RG 120, NACP.

88   **"enjoy his little holiday":** *Sunday Herald,* August 4, 1918.

88   **"flood of bunk":** Grider, *War Birds,* 253.

88   **dropped notes:** An example of such a note, addressed to "das Royal Flying Corps" in early February 1917, is in AIR 1/435/15/273/11, PRO.

88   **donated their furs:** Kennett, *First Air War,* 153.

89   **"visitors turn and stare":** Letter, October 22, 1916, Lt. William Leefe Robinson VC, RFC Collection 90/3/1, IWM.

89   **Faulkner . . . Lafayette Escadrille:** Kennett, *First Air War,* 228.

4. Grand Plans

page

90   **"two bright stars":** Air Raids 1915, AIR 1/2319/223/30/1, PRO.

90   **killing one man and one woman:** Air Raids and Bombardments, AIR 1/720/46/1, PRO.

90   **read an English newspaper:** Robinson, "Zeppelin Bomber," 126.

90   **"five o'clock Taube":** Kennett, *History of Strategic Bombing,* 21, 23.

91   **"soil of Old England!":** Robinson, "Zeppelin Bomber," 136.

91   **Tirpitz:** Murray, *Luftwaffe,* 5; Robinson, *Zeppelin in Combat,* 50–54.

91   **"panic in the population":** Robinson, "Zeppelin Bomber," 133.

91   **"commandeer it for its own purposes":** Ibid., 134.

91   **"above all on royal palaces":** Ibid., 137.

92   **"attitude of prayer":** Castle, *Fire over England,* 60, 66–67.

92   **enemy spies "signaling":** Intelligence Circular No. 5, April 1916, GHQ Home Forces, AIR 1/720/36/1, PRO.

92   **"protection against Air Raids":** Tab 48, Air Defence of Great Britain, AIR 9/69, PRO.

92   **"My jaw dropped":** Gollin, *Impact of Air Power,* 214–15.

93   **"naval harbors, oil tanks":** Ibid., 202–3.

93   **"pissed on Churchill's plant":** Ibid., 181.

93   **"pom-poms":** Ibid., 210–11.

93   **eight thousand shells:** Kennett, *History of Strategic Bombing,* 46.

93   **shed at Düsseldorf:** Raleigh and Jones, *War in the Air,* I:389–90.

94   **Friedrichshafen:** Ibid., 395–401.

94   **demanded that the Navy get out:** Jones, *Origins of Strategic Bombing,* 97–98, 103.

94   **"air raids have not hitherto succeeded":** Intelligence Circular No. 1, January 1916, GHQ Home Forces, AIR 1/720/36/1, PRO.

95   **using radio fixes:** Robinson, "Zeppelin Bomber," 137, 145.

95   **"showered us with confetti":** Hearne, *Zeppelins and Super-Zeppelins,* 2.

95   **Blackouts of cities:** Tab 49, Air Defence of Great Britain, AIR 9/69, PRO.

95   **be awarded a knighthood:** "R.F.C. in the War."

95   **network of sound detectors:** Ferris, "Fighter Defence," 856.

95   **dangerous night missions:** Notes and Correspondence on Night Flying (Home Defence) June and November and December 1916, AIR 1/2430/305/30/12, PRO; Notes on Recent Air Service Operations, Gorrell History C-15, RG 120, NACP, p. 107.

95   **turned on end and slowly fell:** Letter, October 22, 1916, Lt William Leefe Robinson VC, RFC Collection 90/3/1, IWM; Robinson, *Zeppelin in Combat,* 172–73.

96   **"Cheers thundered all around us":** Letter, September 4, 1916, Mrs M Dayrell-Browning, IWM.

96   **jammed the lanes:** Robinson, *Zeppelin in Combat,* 178.

96   **"How disturbing these attacks are":** Robinson, "Zeppelin Bomber," 144.

96   **"morale of the English people":** Kennett, *History of Strategic Bombing,* 24–25.

96   **system . . . quickly broke down:** Jones, *Origins of Strategic Bombing,* 130–34.

96   **162 people were killed:** Air Raids and Bombardments, AIR 1/720/46/1, PRO.

97   **French soldiers had mutinied:** Gilbert, *First World War,* 333–34.

97   **"large proportion of whom were aliens":** Tabs 14, 22, Air Defence of Great Britain, AIR 9/69, PRO.

97   **"bravery and prudence":** "Air-Raid Psychology."

97   *Flight* **called . . . "bomb Germany with compound interest":** "Air Raids and Some Conclusions"; Kennett, *History of Strategic Bombing,* 26–27, 34.

97   **"cost the British each day":** Kennett, *History of Strategic Bombing,* 25.

98   **"independent means of war operations":** Smuts's report can be found in Raleigh and Jones, *War in the Air,* Appendix II; also in War Cabinet, Committee on Air Organisation and Home Defence Against Air Raids, Trenchard Papers 76/1/2, RAFM.

98   **"strangle the infant at birth":** Beaverbrook, *Men and Power,* 220.

98–99   **entrance examination . . . did it in a week:** Raleigh and Jones, *War in the Air,* I:418–19.

99   **"a more gigantic waste":** Morrow, *Great War,* 320.

99   **insubordinate . . . "dull, unimaginative mind":** Jones, *Beginnings of Strategic Air Power,* xix; Morrow, *Great War,* 319.

99   **Trenchard was offered his choice:** Letter, William Weir, May 6, 1918, Trenchard Papers 76/1/20, RAFM.

100   **"bloody paralyser" . . . "seriously worry Germany":** Morrow, *Great War,* 122, 320.

101   **"public opinion get carried away":** Kennett, *First Air War,* 56.

101   **"Actual experience" . . . "just as demoralising":** Long Distance Bombing, November 26, 1917; The Scientific and Methodical Attack of Vital Industries, May, 1918, Trenchard Papers 76/1/67, RAFM.

101   *On ne bombarde pas:* Bomb Targets, French Bombing Policy, AIR 1/1976/204/273/40, PRO.

101   **"growing dread":** Intelligence, Air Raids on Germany Effect on Morale etc, 20th April–4th October 1918, AIR 1/460/15/312/99, PRO.

103   **"the German Jew":** Bombing Objectives, Wing Commander Randall, December 4, 1917, AIR 1/460/15/312/97, PRO.

103   **"some magic in the air":** Kennett, *First Air War,* 42.

103   **official communiqués:** Ibid., 51.

103   **"results ... have been negligible":** Memorandum on Bomb Dropping in the Western Theatre of War from 1st March to the 20th June 1915, AIR 1/921/204/5/889, PRO.

103   **GHQ ordered:** Order, Chief of the General Staff, July 24, 1915, AIR 1/921/204/5/889, PRO.

104   **upwind or downwind:** Cherry, "Aeroplane as Long Range Gun."

104   **"Experimental Flight" had been formed:** Notes on History of Orfordness, HTT 5/151, Tizard Papers, IWM.

104   **"He turned to his shelves":** Clark, *Tizard,* 11.

106   **improvised bombsights:** Short Notes on the Evolution and Theory of Bomb Sights, AIR 1/674/21/6/77, PRO; Introduction to Bombsights and Methods of Bombing, HTT 5/150, Tizard Papers, IWM.

106   **system for measuring bomb ballistics:** Lectures on Bomb-Sighting, AIR 1/2103/207/31. PRO.

106   **performance of aircraft in flight:** Tizard, "Methods of Measuring."

107   **earned his wings ... "never heard of you":** Clark, *Tizard,* 31–33.

108   **Bradshaw's railway guide:** Jones, *Origins of Strategic Bombing,* 44.

108   **"bombed 4 different villages":** Lieut Commdr Lord Tiverton RNVR, Scheme for Bombing German Industrial Centres—September 1917, AIR 1/462/15/312/121, PRO.

108   **"drop the bombs when he thought best":** High Altitude Bomb Sights, Equipment Booklets, Gorrell History I-4, RG 120, NACP.

108   **"course-setting" bombsight:** Nomenclature of Navigational Instruments Now in Use (September 1918), AIR 1/1084/204/5/1410, PRO; R.N.A.S. Course-Setting Bomb Sight—Mark I, AIR 1/2103/207/30/34, PRO.

108   **cloud flying:** Points of Interest in Cloud Flying, HTT 5/150, Tizard Papers, IWM.

109   **"I have a shrewd suspicion":** Memorandum to Strategic Council, April 24, 1918, AIR 1/450/15/3/312/4, PRO.

109   **German raids on London:** Jones, *Beginnings of Strategic Air Power,* 19.

109   **full-scale outlines of German target factories:** Memorandum, Tiverton to D.F.O., June 10, 1918, AIR 1/460/15/312/97, PRO.

109   **"unenlightened populace quiet":** Memorandum, Tiverton to C.A.S., May 22, 1918, Bombing Industrial Objectives in Germany, AIR 1/460/15/312/101, PRO.

110   **"nothing has been done":** Jones, *Origins of Strategic Bombing,* 188.

110   **crop blight:** Proposed Destruction of German Harvest, AIR 1/2319/223/29/3, PRO.

110   **delayed-action bombs:** Bombing Industrial Objectives in Germany, Nov 1917–Nov 1918, AIR 1/460/15/312/101, PRO.

110   **only 5 percent ... against factories:** Statistical Analysis of Aerial Bombardments, Statistics Branch, War Department, November 7, 1918, Gorrell History B-7, RG 120, NACP, p. 85.

110   **repaired ... in an hour and a half:** Ibid., p. 99.

110   **month of May alone:** Ibid., p. 92.

110   **85 percent of its bombing missions:** Results of Raids Carried Out on Germany by the 8th Brigade and the Independent Force, R.A.F., AIR 10/1214, PRO.

110   **"I particularly dislike":** Memorandum, Scientific and Methodical Attack of Vital Industries, May 26, 1918, Trenchard Papers 76/1/67, RAFM.

110   **"Neither side has been able to win":** Letter, Orville Wright to C. M. Hitchcock, June 21, 1917, *OWW,* 1104–5.

111   **"Yankee punch in the war":** Hudson, *Hostile Skies,* 12; Morrow, *Great War,* 267.

111   **"splendid scale" ... "dealing with a miracle":** Morrow, *Great War,* 267–68.

111   **"ramrod straight" ... convert to the cause:** Hudson, *Hostile Skies,* 46–47.

111   **train to Paris:** Mitchell, *Memoirs of World War I,* 14–17.

112   **proposal for a huge American contribution:** Air Service Lessons Learned during the Present War, p. 83, Gorrell History A-1, RG 120, NACP.

112   **Mitchell's role . . . minor at best:** Nalty, ed., *Winged Shield,* I:42.
112   **General Staff thought it was preposterous:** Holley, *Ideas and Weapons,* 45.
112   **"GREATEST OF AERIAL FLEETS":** *New York Herald,* June 18, 1917.
112   **largest single appropriation:** Hudson, *Hostile Skies,* 6.
112   **total of eighty-seven:** Casari, *Aviation Serial Numbers,* 57.
112   **"FIFTY THOUSAND OPEN ROADS":** Arnold, *Global Mission,* 58
112   **Spruce Production Regiments:** Nalty, ed., *Winged Shield,* I:47.
112   **Liberty was the work:** Knappen, *Wings of War,* 79–83, 89.
112   **standardize . . . European models:** Bolling, "Report of Aeronautical Commission"; Holley, *Ideas and Weapons,* 61.
113   **wasted tens of millions:** American Aircraft Production, AIR 1/2430/305/31/1, PRO.
113   **only American-built planes:** Division of Military Aeronautics, Summary of Activities, April to November 1918, Entry 11, RG 18, NACP.
113   **80 percent . . . supplied by French:** Morrow, *Great War,* 338.
113   **"No amount of money":** Ibid., 266.
113   **"carpetbaggers":** Mitchell, *Memoirs of World War I,* 165–66.
114   **relieved from duty:** Memorandum, B. D. Foulois to Commander-in-Chief American Expeditionary Forces, June 4, 1918, para. 112a, Air Services Lessons Learned During the Present War, Gorrell History A-1, RG 120, NACP.
114   **"more than half way":** Letter, J. W. McAndrews to Brigadier General B. D. Foulois, June 8, 1918, para. 112b, Air Services Lessons Learned During the Present War, Gorrell History A-1, RG 120, NACP.
114   **Foulois . . . stepped aside:** paras. 114–15, Air Services Lessons Learned During the Present War, Gorrell History A-1, RG 120, NACP.
114   **Battle of Cambrai:** Greenhous, "Counter Anti-Tank Role."
115   **forces of small, fast airplanes:** History of Tank and Aeroplane Cooperation, AIR 1/725/97/10, PRO; Hallion, *Strike from the Sky,* 20–23.
115   **"condition akin to panic":** Hallion, *Strike from the Sky,* 19.
116   **strong southwest wind:** Hopper, "American Day Bombardment," 92.
116   **"Horses fell":** Rickenbacker, *Fighting the Flying Circus,* 233–34.
116   **Meuse-Argonne:** Morrow, *Great War,* 337.
116   **barrage of leaflets:** The AEF Air Service and Propaganda against the Enemy, Gorrell History M-8, RG 120, NACP.
117   **offensive on the Meuse:** Gilbert, *First World War,* 490, 494.
117   **two to three thousand aircraft:** Christienne and Lissarrague, *French Military Aviation,* 158; Morrow, *Great War,* 345; Great Britain, *Synopsis of Air Effort,* 17.
117   **"practically held court":** Arnold, *Global Mission,* 85.
118   **"Vive nôtre Général":** Mitchell, *Memoirs of World War I,* 293.
118   **"man with a goatee":** Von Kármán, *Wind and Beyond,* 35–41.
119   **boundary-layer . . . lifting-line:** Anderson, *History of Aerodynamics,* 251–55, 282–87.
120   **aspect ratio of 4.1:** Anderson, *History of Aerodynamics,* 262.
121   **outclimb the Camel and Spad:** Whitford, *Fighter Design,* 23–24.
121   **Fokker triplane:** Boyne, "World War One Aviation."
121   **fled to Holland:** Morrow, *Great War,* 353.

5. Lessons Learned and Mislearned

page
125   **"fade into oblivion":** Mitchell, *Memoirs of World War I,* 277, 301.
125   **twenty times . . . two hundred:** Nalty, ed., *Winged Shield,* I:78.
125   **cancelled $100 million . . . went for $250:** Cunningham, *Aircraft Industry,* 38; Morrow, *Great War,* 357, 361; Correspondence Relating to the Liquidation of Stock of Aircraft and Engines, 1919–1924, Entry 127, RG 18, NACP.

126   **"Not a dollar":** Nalty, ed., *Winged Shield,* I:78.

126   **"no right or title":** Frisbee, ed., *Makers of the Air Force,* 20–21.

126   **"entirely incapable":** Memorandum, December 30, 1919, Item 16, Air Service, Army & Navy Recommendations File, Box 32, Mitchell Papers, LOCMD.

127   **stage the sort of stunts:** Nalty, ed., *Winged Shield,* I:86–91.

127   **raced the pigeons:** Arnold, *Global Mission,* 100–101.

127   **"a spoiled brat":** Nalty, ed., *Winged Shield,* I:93.

127   **number of tickets sold:** McElvaine, *Great Depression,* 208.

130   **Badische works . . . "impossible undertaking":** Results of Air Raids on Germany Carried out by the 8th Brigade and the Independent Force, R.A.F., January 1st–November 11th, 1918, AIR 10/1214, PRO, pp. 2–3, 63.

130   **Thionville station:** Kennett, *History of Strategic Bombing,* 49.

131   **"proportion of 20 to 1":** Trenchard, "Despatch," 135.

132   **"not productive":** Results of Air Service Efforts as Determined by Investigation of Damage Done in Occupied Territory, Gorrell History R-1, RG 120, NACP, pp. 6–7.

132   **"objective of air attack":** Smith, *British Air Strategy,* 62.

132   **"defeat the enemy nation":** Minutes of Conference Held in CAS's Room, Air Ministry, July 19, 1923, AIR 19/92, PRO.

132   **"rifleman or the sailor is protected":** Memorandum by the Chief of the Air Staff, May 1928, Webster and Frankland, *Strategic Air Offensive,* Appendix 2, IV:71–76.

133   **"our civilisation will fall":** Smith, *British Air Strategy,* 46–47.

133   **"morons volunteering":** Arthur Harris, quoted in Smith, *British Air Strategy,* 64.

133   **Churchill was named:** Omissi, *Colonial Control,* 8.

133   **esprit de corps:** Cabinet Memorandum, The Separate Existence of the Royal Air Force and the Air Ministry Scheme of Expansion for Home Defence, July 5, 1923, Trenchard Papers 76/1/48, RAFM.

133   **Battle of Cambrai:** Hallion, *Strike from the Sky,* 21.

134   **21 aircraft and 9 airships:** Notes on Miscellaneous Points Affecting Home Defence, AIR 19/91, PRO.

134   **commercial airline company:** Arguments Put Forth for Having Fighters, Tab 56, Air Defence of Great Britain, AIR 9/69, PRO.

134   **defense was viewed as "insidious":** Lecture by D.C.A.S. to Naval Staff College, March 10, 1924, Tab 54, Air Defence of Great Britain, AIR 9/69, PRO.

134   **calculated that a French attack:** Air Staff Notes on Enemy Air Attack on Defended Zones in Great Britain, AIR 19/91, PRO; Expansion of the Royal Air Force for Home Defence, AIR 19/92, PRO.

134   **twenty new home squadrons:** Memorandum from CAS, June 1922, AIR 19/90, PRO.

134   **subcommittee . . . under Lord Salisbury:** Role of the R.A.F. in War and the Strategical Use of Air Power, Part One: Policy 1917–1923, Air Ministry, December 1942, Trenchard Papers 76/1/357, RAFM, p. 3.

135   **"forty-eight extra bombers":** Minutes of Conference Held in CAS's Room, Air Ministry, July 19, 1923, AIR 19/92, PRO.

135   **"deterrent":** Note to C.I.D., 107-A, May 30, 1922, AIR 19/91, PRO.

135   **"planning seemed unnecessary":** Smith, *British Air Strategy,* 47–48.

135   **"squeal before we did":** Ibid., 61.

135   **38 out of 1,346:** Jones, *Beginnings of Strategic Air Power,* 63.

135   **"will be no repair":** Dye, "Logistics Doctrine."

135   **walking sticks and razors:** Jones, *Beginnings of Strategic Air Power,* 56–57.

136   **"only appropriate use for airpower":** Robertson, "Royal Air Force Doctrine," 48.

136   **"one and indivisible":** Smith, *British Air Strategy,* 58–59.

136   **never even read . . . Douhet:** Buckley, *Air Power,* 76.

137   **never learned to fly:** Kennett, *History of Strategic Bombing,* 55.

137   "national totality" . . . "mobilize at all!": Douhet, *Command of the Air,* 5, 9–10, 58, 185.

137–38   "inhuman and atrocious" . . . "patricide": Ibid., 181, 189.

138   "worthless, superfluous, harmful": Ibid., 101.

138   "excellent exposition": U.S. Air Force, *Development of Air Doctrine,* 50. The U.S. Air Service had a typescript translation of *The Command of the Air* in its files as early as 1923: Futrell, *Ideas, Concepts,* I:39.

138   "key weapon of war": Groves, "For France to Answer," 145–46.

139   "nuclear warfare today": Bialer, *Shadow of the Bomber,* 158.

139   "knock out blow to England": Ibid., 135–36.

139   "go to Berchtesgaden": Buckley, *Air Power,* 114.

139   "revolting and un-English": Kennett, *History of Strategic Bombing,* 70.

139   "baby-killers": McFarland, *Pursuit of Precision,* 81.

139   *Infantry Journal:* "War à la Douhet."

140   declared a jihad . . . delusional: Jardine, "Mad Mullah."

140   Colonial Office . . . approached Hugh Trenchard: Towle, *Pilots and Rebels,* 12.

140   "no conceivable circumstances": Omissi, *Colonial Control,* 14.

141   clothes of the great man: Jardine, "Mad Mullah."

141   Z Unit: The Role of the R.A.F. in War and the Strategical Use of Air Power, Part Three. Operations 1919–1939. Somaliland and N.W. Frontier, Trenchard Papers 76/1/357, RAFM, p. 2.

141   six million pounds: Notes on the History of the Employment of Air Power, Air Staff Memorandum No. 48, AIR 10/1367, PRO.

141   "mere 'mopper up' ": Omissi, *Colonial Control,* 14–15.

141   "somewhat of a hoax": Ibid., 16.

141   only made the Army's job harder: Liddell Hart, *Britain Goes to War,* 128–29.

141   "the decisive factor": Notes on the History of the Employment of Air Power, Air Staff Memorandum No. 48, AIR 10/1367, PRO.

142   "caused a panic": Towle, *Pilots and Rebels,* 37.

142   "achieved more than 60,000": Liddell Hart, *Britain Goes to War,* 127.

142   18 million pounds: Churchill, Cabinet Paper on Mesopotamian Expenditure, May 1, 1920, and Memorandum, Churchill to C.A.S., February 29, 1920, Trenchard Papers 76/1/35, RAFM.

142   Trenchard . . . "sooner the Air Force crashes": Omissi, *Colonial Control,* 22–24.

143   RAF would command: Corum, "Air Control," 19.

143   "once properly learnt": Chamier, "Use of the Air Force," 210.

143   "God knows where": Smith, *British Air Strategy,* 29.

143   RAF was assiduous: Cochrane, "Royal Air Force at Aden"; Bottomley, "Royal Air Force on North-West Frontier"; Memorandum of Air Action Carried out during the Past 18 Months, Air Staff, June 1929, Trenchard Papers 76/1/35, RAFM; Short Summary of Akhwan Raids in Iraq, February 29, 1928, Trenchard Papers 76/1/35, RAFM.

143   Palestine and Transjordan: Cabinet Paper, The Fuller Employment of Air Power in Imperial Defence, Air Ministry, November 1928, Trenchard Papers 76/1/42, RAFM.

143   "I cannot emphasize too much": Omissi, *Colonial Control,* 34.

144   "heavy hand" . . . "Humanity was the same": Corum, "Air Control," 28–29; Omissi, *Colonial Control,* 110–11.

144   "brought him to heel": Edmonds, "Air Strategy," 195.

144   "three-stage" theory: Ibid., 194; Omissi, *Colonial Control,* 110–11.

144   only an experienced air officer: Notes on the Regulation of Air Control in Undeveloped Countries, Air Staff, November 21, 1928, Trenchard Papers 76/1/35, RAFM.

145   "Director of R.A.F. Training": Towle, *Pilots and Rebels,* 17.

145   lost only fourteen pilots: Ibid., 19.

145   **Navigation was taken care of:** Omissi, *Colonial Control,* 146.

145   **had no bombsight:** Corum, "Air Control," 27.

145   **Wapiti . . . Hinaidi:** Jarrett, ed., *Biplane to Monoplane,* 58.

145   **even half of their bombs:** Corum, "Air Control," 27.

145   **"used energetically":** Omissi, *Colonial Control,* 138.

145   **Churchill was aghast:** Ibid., 41.

146   **"industrial disturbances":** Ibid., 41. An RAF manual issued in 1931 subsequently instructed that in the event of "industrial unrest," the air force might be called upon to "maintain communications," and air force personnel might be deployed to guard government property on the ground, but emphasized: "NO OFFENSIVE ACTION WILL BE TAKEN FROM THE AIR" and "No aeroplanes are to be armed or to carry bombs under any circumstances." See Orders for the Royal Air Force (Home) in the Event of Industrial Unrest, AIR 10/1483, PRO.

146   **took very little destruction:** Gray, "Myths of Air Control," 43–44.

146   **"the fair sex":** Peck, "Aircraft in Small Wars," 549.

146   **Hendon Pageant:** Omissi, *Colonial Control,* 171–73.

147   **watching the Navy squirm:** Nalty, ed., *Winged Shield,* I:93.

148   **Two months later . . . *New York Times:*** Zimmerman, "Sinking of *Ostfriesland.*"

148   **"warships to attack":** Hone, Friedman, and Mandeles, *Aircraft Carrier Development,* 27; Futrell, *Ideas, Concepts,* I:36.

148   **stand bareheaded:** Nalty, ed., *Winged Shield,* I:94.

148   **new bombsight:** McFarland, *Pursuit of Precision,* 28–29.

149   **outraged radio message:** Zimmerman, "Sinking of *Ostfriesland.*"

149   **at 12:33 p.m.:** Mitchell, "Bombing of Battleships," 61.

149   **"still the backbone" . . . Borah:** Ransom, "Battleship Meets Airplane," 22.

149   **"solved, and is finished":** Mitchell, "Bombing of Battleships," 52.

149   **"unsinkable" . . . "wept aloud":** Layman, "Day the Admirals Wept," 74, 76.

150   **"No seacraft" . . . "blown up in the air":** Comments of General Mitchell on Conclusion Drawn from 1923 Bombing Tests, Bombing Maneuvers off Cape Hateras Report File, Box 35, Mitchell Papers, LOCMD.

150   **"third-class foe" . . . "incapable":** Scrapbook 1923–1925, Box 63, Mitchell Papers, LOCMD.

150   **"convert his enemies" . . . "render him unfit":** Hurley, *Billy Mitchell,* 97–98.

150   **Newspaper cartoonists:** Scrapbook 1925, Box 62, Mitchell Papers, LOCMD.

150   **Rickenbacker sent a cable:** Scrapbook 1923–1925, Box 63, Mitchell Papers, LOCMD.

150   **"almost treasonable":** Charge Sheet Against General Mitchell, Court-Martial 1925 File, Box 38, Mitchell Papers, LOCMD.

151   **"delusions of grandeur" . . . MacArthur:** Grant and Katz, *Great Trials,* 267, 270.

151   **"bow and arrow men":** Statement of William Mitchell—February 1st, 1926, Press Releases on Trial and Court Martial File, Box 38, Mitchell Papers, LOCMD.

151–52   **greet him at Union Station . . . "below average":** Huston, ed., *Arnold's Diaries,* I:18–20.

152   **Air Corps Act:** U.S. Air Force, *Development of Air Doctrine,* 28–29.

6. The Quest for Precision

page
153   **"years ahead of his time":** Arnold, *Global Mission,* 158.

154   **stressed-skin . . . Albatross:** Jakab, "Wood to Metal," 915–16.

155   **early as 1920 . . . "branch of pure science":** Hanle, *Aerodynamics to America,* 85, 88.

155   **Munk . . . was hired:** Anderson, *History of Aerodynamics,* 290.

156   **Variable Density Tunnel:** Munk and Miller, *Variable Density Tunnel.*

156   **"Dr. Munk . . . is sure":** Hansen, *Engineer in Charge,* 85.

156   resigned en masse: Anderson, *History of Aerodynamics*, 290, 303.
156   highly mathematical nature: Hanle, *Aerodynamics to America*, 91.
156   "undisputed leader": Anderson, *History of Aerodynamics*, 303.
156   "designer's bible": Munk and Miller, *Tests at Full Reynolds Number;* Anderson, *History of Aerodynamics*, 343.
157   "otherwise be unattainable": Jones, "Streamline Aeroplane," 358.
158   157 to 177 miles per hour: Gray, *Frontiers of Flight*, 113; Weick, *Drag and Cooling.*
158   thirty-by-sixty-foot: DeFrance, *Full-Scale Tunnel.*
158   "drag clean-up": Coe, *Drag Cleanup Tests.*
158   ever higher horsepower: Schlaifer and Heron, *Aircraft Engines and Fuels*, 7, 29, 31.
159   heavy government subsidies: Fairey, "Expenditure on Aviation," 414–15.
159   factor of five: Breguet, "Aerodynamical Efficiency."
159   three hundred airplanes . . . 1,186: Hanle, *Aerodynamics to America*, 1; Futrell, *Ideas, Concepts,* I:62.
159   "dark horse": Shirer, *Twentieth Century Journey*, 324.
159   "stay awake thirty-six hours": Ibid., 330.
160   "torrents of words" . . . "get the money now": Ibid., 339–41.
161   leapt to 48,000: Futrell, *Ideas, Concepts,* I:61–62.
161   engine wing-mountings: Wood, *Nacelle-Propeller Combinations.*
161   orders for fifty-nine: Jarrett, ed., *Biplane to Monoplane*, 19–20.
161   acquired the "von": Von Kármán, *Wind and Beyond*, 17.
162   men from Mars: Rhodes, *Making of Atomic Bomb*, 106–7.
162   "WHAT IS THE FIRST BOAT": Von Kármán, *Wind and Beyond*, 120.
162–63 "goal of my life" . . . "German science": Ibid., 141, 146.
163   "definitely better beer": Hanle, *Aerodynamics to America*, 133.
163   "wad of putty": Von Kármán, *Wind and Beyond*, 170.
164   DC-3: Oleson, "Douglas Aircraft"; Anderson, *History of Aerodynamics*, 358; Jarrett, ed., *Biplane to Monoplane*, 44–45.
164   lift-to-drag ratio . . . wing loading: Anderson, *History of Aerodynamics*, 360.
164   ten thousand DC-3s: Jarrett, ed., *Biplane to Monoplane*, 45.
164   variable-pitch prop: Martin, "Boeing Designs"; Schlaifer and Heron, *Aircraft Engines and Fuels*, 57.
165   high-lift devices: Lee, *Fighter Facts*, 15.
166   Eduard Lindeman: Hanle, *Aerodynamics to America*, 8.
166   fortune in mining . . . Guggenheim Fund: Hallion, *Legacy of Flight*, 21–24, 45, 86, 111.
166   "a good pilot" . . . radio beacons: Hallion, *Test Pilots*, 98–100.
167–68 Doolittle . . . Elmer A. Sperry: Hallion, *Legacy of Flight*, 111, 117.
168   "greatest single step in safety": Hallion, *Test Pilots*, 102–3.
169   "magnificent" job: Frisbee, ed., *Makers of the Air Force*, 29–31.
169   sixty-six crashes . . . "prisoner of ease": Nalty, ed., *Winged Shield*, I:123–25.
170   "reverence for marksmanship": AAF, I:597.
170   "shank of the drill": Gorrell, "American Proposal," 104–5.
171   "powerful aids to recruitment": Quoted in Futrell, *Ideas, Concepts,* I:39.
171   800 feet . . . Army board: McFarland, *Pursuit of Precision*, 27, 41.
171   To calibrate the sight: "Drift" Bomb Sights, Instructions for Use, History of Strategical Section, Gorrell History B-6, RG 120, NACP.
172   tug on strings: McFarland, *Pursuit of Precision*, 35.
172   Sperry bombsight: Ibid., 28–29.
172–73 Pee Dee River bridge . . . "question of cost": Ibid., 40–42.
173   invented gravity . . . "Old Man Dynamite": Ibid., 48–53.
174   "ingenious and inhuman": Gill, "Young Man," 28.
174   "only against area targets": McFarland, *Pursuit of Precision*, 81.

175   USS *Pittsburgh:* Ibid., 72.
175   **Navy had thousands:** Ibid., 141, 146–48.
175   **"Bombardier's Oath":** Ibid., 155, 276 n. 18.
175   **pickle barrel:** Ibid., 5, 242 n. 11.
176   **"utterly absurd":** U.S. Air Force, *Development of Air Doctrine,* 90.
176   **coveted spots . . . 320 generals:** Cameron, *Training to Fly,* 275; Finney, *History of Tactical School,* 25.
176   **"heavily striking vital points":** U.S. Air Force, *Development of Air Doctrine,* 41.
176   **"basic arm" . . . "strategical objectives":** Finney, *History of Tactical School,* 63–64; U.S. Air Force, *Development of Air Doctrine,* 56.
176   **one-day course:** Hughes, *Overlord,* 57.
177   **Herbert Dargue:** McFarland, *Pursuit of Precision,* 90–91.
177   **"industrial web":** Wilson, "Origin of a Theory."
178   **"fell into our laps":** U.S. Air Force, *Development of Air Doctrine,* 81.
178   **"war-making capacity":** Clodfelter, "Pinpointing Devastation," 84–85.
178–79 **"will of the people" . . . "be evacuated":** Pape, *Bombing to Win,* 63–64; Finney, *History of Tactical School,* 65; U.S. Air Force, *Development of Air Doctrine,* 115.
179   **war-winning strategy:** Finney, *History of Tactical School,* 73.
180   **Kuter now calculated:** McFarland, *Pursuit of Precision,* 94–98.
180   **GHQ Air Force:** *AAF,* I:31–32; Wolk, *Air Force Independence,* 15.
180   **the B-9:** Nalty, ed., *Winged Shield,* I:138; Jarrett, ed., *Biplane to Monoplane,* 19, 62.
181   **B-10:** Jarrett, ed., *Biplane to Monoplane,* 62–63.
181   **five-year expansion plan:** Futrell, *Ideas, Concepts,* I:67.
181   **transferring $7.5 million:** Nalty, ed., *Winged Shield,* I:136.
181   **circular to manufacturers:** Futrell, *Ideas, Concepts,* I:69–70.
181   **Egtvedt flew directly:** Martin, "Boeing Designs."
181   **model 299:** U.S. Air Force, *Development of Air Doctrine,* 46–47.
182   **"put your hand on":** Arnold, *Global Mission,* 154.
182   **hundred miles of the coast:** Nalty, ed., *Winged Shield,* I:147–48.
182   **B-17A:** Jarrett, ed., *Biplane to Monoplane,* 70.
183   **impact . . . Air Corps Tactical School:** Finney, *History of Tactical School,* 68.

7. The Fight for the Fighter

page

184   **eighty cigarettes a day:** Perret, *Winged Victory,* 27.
184   **"impossible for fighters to intercept":** Chennault, *Way of a Fighter,* 22.
184   **"figment of the imagination" . . . "mild annoyance":** U.S. Air Force, *Development of Air Doctrine,* 56; Hughes, *Overlord,* 57.
185   **"fantastic and arbitrary restrictions":** Chennault, *Way of a Fighter,* 26.
185   **"Next Great War" . . . exercise at Fort Knox:** The Role of Defensive Pursuit, File number 248.282-4, USAFHRA, pp. 12–13, 24.
185   **pursuit course counted:** Hughes, *Overlord,* 57.
185   **"no 'normal' role":** U.S. Air Force, *Development of Air Doctrine,* 62.
186   **virtual outcast:** Chennault, *Way of a Fighter,* 26–28.
186   **"foundation for future wars":** The Expansion of the Royal Air Force 1934–1939, AIR 41/8, PRO, p. 12.
187   **"timely redress of inequality":** Churchill, *Second World War,* I:73.
187   **"sincerity of Germany" . . . "office seeker":** Manchester, *Last Lion,* II:92, 99–103.
187   **"must have a convention" . . . "always get through":** Bialer, *Shadow of the Bomber,* 14, 20–21.
187   **"terrorizing the civilian population":** "General Report of Jurists," 249–50.
188   **"no adequate protection":** Spaight, *Air Power and War Rights,* 18–19.

188   **"except for police purposes":** The Expansion of the Royal Air Force 1934–1939, AIR 41/8, PRO, pp. 28–29.

188   **Germany's claim . . . "irrefutable":** Shirer, *Rise and Fall,* 210.

188   **"disarm to the level of Germany":** Churchill, *Second World War,* I:111.

189   **"putting our defences in order":** Dunbabin. "British Rearmament," 591.

189   **fifty-two . . . twenty-three to follow:** RAF Narrative: The Air Defence of Great Britain, Volume I, The Growth of Fighter Command July 1936–June 1940, AIR 41/14, PRO, pp. 15, 28.

189   **Churchill took to the floor:** Churchill, *Second World War.* I:116–17.

189   **the men were opposites:** Jones, *Wizard War,* 14.

189   **"defeatist attitude":** Clark, *Tizard,* 107–8.

190   **"death of Versailles":** Shirer, *Rise and Fall,* 284.

190   **1932 exercise:** Report on Air Exercises 1932 Held by Air Defence of Great Britain Command from 18th to 21st July, 1932, AIR 10/1523, PRO; Ferris, "Fighter Defence," 875–77.

191   **operations-room system:** Hough and Richards, *Battle of Britain,* 15–16; The Growth and Progress of Operations Rooms, AIR 16/195, PRO.

191   **eighty seconds from warning:** Ferris, "Fighter Defence," 864.

191   **"fighters as few as . . . will permit":** Ibid., 852.

192   **twenty-eight squadrons:** RAF Narrative: The Expansion of the Royal Air Force 1934–1939, AIR 41/8, PRO, p. 71.

192   **"likely to lose the next war":** Rowe, *Story of Radar,* 4–5.

192   **"No avenue, however fantastic":** Clark, *Tizard,* 110–11.

193   **phoned Robert Watson-Watt:** Ibid., 113.

193   **"metallic objects miles away":** Marconi, "Radio Telegraphy," 237.

194   **fluttered whenever an airplane passed:** Clark, *Tizard,* 114–15.

194   **"We now have in embryo":** Hough and Richards, *Battle of Britain,* 50.

194   **bombarded Tizard:** Correspondence with W. S. Churchill, M.P., Air Defence Research Committee, AIR 19/25–26, PRO; Air Defense Research Committee, HTT 99, Tizard Papers, IWM; C.S.S.A.D., HTT 111, Tizard Papers, IWM.

195   **flat in St. James's:** Clark, *Tizard,* 130–31.

195   **"slow-motion picture":** Letter, Churchill to Kingsley Wood, June 9, 1938, Correspondence with W. S. Churchill, M.P., Air Defence Research Committee, AIR 19/25, PRO.

195   **Germans were already far ahead:** Price, *Instruments of Darkness,* 60–61.

196   **"quite hopeless":** Third Progress Report, 3 October 1936, Special Interception Experiments at Biggin Hill, Part I, AIR 16/179, PRO.

196   **three miles . . . 90 percent of the time:** Third Progress Report, 3 October 1936, and Sixth Report, 18 January 1937, Special Interception Experiments at Biggin Hill, Part I, AIR 16/179, PRO.

197   **contributed £100,000 . . . 408 miles per hour:** Quill, *Birth of a Legend,* 45–47.

197   **enclosing the radiator:** Jarrett, ed., *Biplane to Monoplane,* 146: Smith, "Development of Spitfire," 343.

198   **Ethylene glycol:** Schlaifer and Heron, *Aircraft Engines and Fuels,* 238.

198   **school was not possible . . . "existing financial conditions":** Memorandum from Air Ministry, October 28, 1928, Study of Air Fighting Tactics at Northolt, Vol. I, AIR 16/305, PRO; Memorandum, Air Council, December 5, 1932, Study of Air Fighting Tactics at Northolt, Vol. I, AIR 16/305, PRO.

198   **list of questions:** Notes on Design and Tactics of ADGB Fighters, 23 April 1934, Study of Air Fighting Tactics at Northolt, Vol. I, AIR 16/305, PRO; Letter from Hugh Dowding to Tizard, September 17, 1938, Correspondence with Dowding, HTT 39, Tizard Papers, IWM.

199   **"Fighter Attack No. 1":** Fighter Tactics 1938, AIR 16/42, PRO.

199   **with a mallet:** Jarrett, ed., *Biplane to Monoplane,* 62.

200   **eight guns would thus be necessary:** RAF Narrative: The Expansion of the Royal Air Force, 1934–1939, AIR 41/8, PRO, pp. 84–85.

200   **"Union Travel Society":** Drum, *German Air Force,* 14, 52; Proctor, *Luftwaffe,* 20–21.

201   **thirty-two different parties:** Thomas and Witts, *Guernica,* 16.

201   **chance to discomfit France:** Corum, *Luftwaffe,* 188.

201   **"Hispano-Moroccan Transport":** Howson, *Aircraft of Spanish Civil War,* 207; Drum, *German Air Force,* 13.

202   **direct hit on a government battleship:** Proctor, *Luftwaffe,* 28–29.

202   **thirty-one I-16s . . . SB-2 bombers:** Howson, *Aircraft of Spanish Civil War,* 15, 193.

203   **"Max Winklet" . . . combat pay:** Ries and Ring, *Legion Condor,* 44–45; Drum, *German Air Force,* 47.

203   **Bf 109:** Corum, *Luftwaffe,* 192.

203   **regulations governing officer selection:** Ibid., 66–68.

204   **Wilberg's Jewish ancestry:** Ibid., 57.

204   **130 experienced air officers:** Ibid., 59–63.

204   **air base at Lipetsk:** Ibid., 76, 117–18; Heyman, "NEP and Industrialization," 43–44.

204   **up-to-date fighter doctrine:** Corum, *Luftwaffe,* 130.

205   **"Conduct of the Aerial War":** Ibid., 134–44; Cooling, ed., *Case Studies in Air Superiority,* 31.

205   **"whoever falls, falls!":** Howson, *Aircraft of Spanish Civil War,* 209.

205   **Battle of Guadalajara:** Poulain, "Aircraft and Mechanized Warfare," 365–66; Hallion, *Strike from the Sky,* 97–101.

205   **"most dramatic examples":** Corum, *Luftwaffe,* 193.

205   **"not Abyssinia":** Larrazabel, *Air War over Spain,* 128.

206   **At Ochandiano:** Entries for April 3 and 4, 1937, Richthofen Diaries, N 671/2, BA-MA.

206   **radius of turns:** For a good explanation of the relationship between wing loading and fighter performance, see Lee, *Fighter Facts,* 6–14.

206   **kicked out of the Navy:** Smith, "Rebel of '33."

206   **"entire enemy air force":** Tinker, *Some Still Live,* 278.

206   **American military attaché:** Richardson, "Airpower Concepts," 16; Hallion, *Strike from the Sky,* 91.

206   **attack on Bilbao:** Corum, *Luftwaffe,* 196.

206   **"Even the best bombers":** Poulain, "Role of Aircraft," 583, 586.

207   **"population stands up very well":** Ibid., 585.

207   **"charming it with their guitar":** Wyden, *Passionate War,* 446–47.

208   **"no stampede":** Kennett, *History of Strategic Bombing,* 98–99.

208   **"become automatically united":** Strauss, "Psychological Effects," 277–78.

208   **Luftwaffe study:** Corum, *Luftwaffe,* 211, 222.

209   **doctorate in engineering . . . self-control:** Thomas and Witts, *Guernica,* 28–29.

209   **"golden rule":** Ibid., 63.

209   **"wonder we hit anything":** Ibid., 151.

210   **"usual mix":** Ibid., 205–6.

210   **"technical success":** Entry for April 30, 1937, Richthofen Diaries, N671/2, BA-MA.

210–11   **"unparalleled" . . . "radio general":** Wyden, *Passionate War,* 357–60.

211   **three hundred civilians were killed:** Ries and Ring, *Legion Condor,* 63–64.

211   **"Anybody who draws any lessons":** Arnold, "Air Lessons," 17.

212   **"Air Power has not been tested":** Kenney, "Airplane in Modern Warfare."

212   **RAF's reaction . . . simply ignored:** Corum, "Spanish Civil War," 315–18.

212   **"unwise to base conclusions":** Third Interim Report of Bombing Committee, p. 8, Bombing and Air Fighting Committees Interim Reports Jan 1934–Dec 1938, AIR 5/1143, PRO.

212   **"gentlemen-pilots club":** Corum, "Spanish Civil War," 331.

212   **Air Staff argued—"logically":** Bialer, *Shadow of the Bomber,* 57–58.
212   **British Air Intelligence:** Watt, "British Intelligence," 258, 267–68.
213   **mastery of instrument flying:** Drum, *German Air Force,* 174–78.
213   *Flivos:* Ibid., 198–200.
213   **neutralizing enemy flak:** Proctor, *Luftwaffe,* 149.
213   **Russian I-16 pilots:** Ries and Ring, *Legion Condor,* 53.
213   **Werner Mölders:** Proctor, *Luftwaffe,* 256.
214   **twenty-two Republican fighters:** Ibid., 216.
214   **ten thousand bombs a day:** Hallion, *Strike from the Sky,* 107.
214   **"man who could be relied upon":** Shirer, *Rise and Fall,* 387.
214   **"normal trade" . . . pay the indemnity:** Dunbabin, "British Rearmament," 598–99.
215   **equipped with Hurricanes:** RAF Narrative: The Air Defence of Great Britain, Volume I, The Growth of Fighter Command July 1936–June 1940, AIR 41/14, PRO, pp. 66–68.
215   **"yokel in a field":** Letter, Dowding to Tizard, June 12, 1939, Interception Experiments and Exercises, AIR 16/251, PRO.
215   **telephone lines . . . failed:** Minor Home Defence Air Exercise No. 7, May 9, 1939, AIR 16/118, PRO.
216   **twenty-nine of the thirty-five:** Hough and Richards, *Battle of Britain,* 64.
216   **F. Sidney Cotton:** Photographic Reconnaissance, Vol. I to April 1941, AIR 41/6, PRO, pp. 37–42.
216   **screaming headlines:** Shirer, *Rise and Fall,* 564.
216   **810 bombers:** Stokesbury, *World War II,* 69–71.

8. Finest Hour

page

219   **"vilest of the vile!":** Churchill, *Second World War,* I:323.
219   **war machine might bog down:** Peszke, "Forgotten Campaign," 65.
219   **"old woman":** Hooton, *Phoenix Triumphant,* 175.
219   **obsolete aircraft:** Corum, *Luftwaffe,* 271.
220   **"Beppo" Schmid:** Boog, "German Air Intelligence," 122.
220   **lacked a coherent operational doctrine:** Cynk, "Polish Air Force," 181.
220   **reconnaissance flights:** Hooton, *Phoenix Triumphant,* 176.
220   **striking at the Polish rail system:** Corum, *Luftwaffe,* 272.
220   **Cracow's airfields:** Hooton, *Phoenix Triumphant,* 179–80.
220   **50 percent . . . first two days:** Orwovski, "Polish Air Force," pt. 1, 381.
220   **Poles abandoned:** Hooton, *Phoenix Triumphant,* 181; Murray, "Luftwaffe against Poland," 77.
221   **locate gasoline tank cars:** Orwovski, "Polish Air Force," pt. 2, 398.
221   **Ju 52 transports:** Corum, *Luftwaffe,* 273.
221   *Flivos:* Hallion, *Strike from the Sky,* 132.
221   **deafening roar . . . four thousand incendiaries:** Hallion, *Strike from the Sky,* 133.
222   **"shambles" . . . coal scuttles:** Hooton, *Phoenix Triumphant,* 185–88; Murray, *Luftwaffe,* 32; Bekker, *Luftwaffe War Diaries,* 58.
222   **Luftwaffe lost . . . 7 percent:** Murray, "Luftwaffe against Poland," 77; Bekker, *Luftwaffe War Diaries,* 364–65.
223   **"decision on the battlefield":** Corum, *Luftwaffe,* 274.
223   **"led to expect by our Military Advisers":** Chamberlain to Churchill, September 16, 1939, Gilbert, ed., *Churchill War Papers,* I:101.
223   **"English and French friends":** Churchill, *Second World War,* I:589.
223   **"attracted to the political aspects":** Overy, "Hitler and Air Strategy," 411.
223   **"Prague would be in ruins":** Murray, *Luftwaffe,* 28.

224   **During the Munich crisis:** Ray, *Night Blitz,* 48–49.
224   **"leaflets . . . replaced by bombs":** *The Aeroplane,* September 14, 1939, quoted in Ray, *Night Blitz,* 67.
224   **evacuated to the countryside . . . auxiliary firemen:** Ray, *Night Blitz,* 48–49.
224   **Gallup Poll:** Manchester, *Last Lion,* II:608–9.
224   **By the spring only 1 percent:** Overy, *Battle of Britain,* 4.
224   **September 4 . . . December 18:** Webster and Frankland, *Strategic Air Offensive,* I:192–97.
225   **AA guns . . . 88mm flak gun:** Westermann, *Flak,* 58–59, 75; Hallion, *Strike from the Sky,* 134.
225   **"horror of the experts":** Entry for May 1, 1937, Richthofen Diaries, N 671/2, MA-BA.
225   **700 88mm guns:** Harvey, "French Armée de l'Air," 449.
225   **seventy-five minutes:** Corum, *Luftwaffe,* 277.
225   **caught on the ground:** Griffin, "Battle of France," 147; Murray, "Luftwaffe against Poland," 81; Hooton, *Phoenix Triumphant,* 240.
226   **Air-driven sirens:** Hallion, *Strike from the Sky,* 139–41.
226   **"shatter their nerves":** Horne, *To Lose a Battle,* 291.
226   **to fly low . . . 66 to 100 percent:** Hallion, *Strike from the Sky,* 142–43.
227   **practice dive-bombing:** Overy, *Battle of Britain,* 8.
227   **"contrived ambiguity":** Young, "Strategic Dream," 63.
227   **strategic bombers in disguise:** Christienne and Lissarrague, *French Military Aviation,* 259.
227   **hundred Curtiss P-36s:** Haight, "France's Search for Military Aircraft," 149.
227   **Cot . . . issued regulations:** Corum, "Spanish Civil War," 322–24.
228   **no idea about how to use aviation:** Kirkland, "French Air Force."
228   **"haven't any work for them":** Harvey, "French Armée de l'Air," 457.
228   **postmortem on the action:** Sir Robert Brooke-Popham's Committee on RAF War Experiences in France, AIR 2/5251, PRO, p. 3.
228   **Sortie rates:** Kirkland, "French Air Force."
229   **British base at Gibraltar:** Harvey, "French Armée de l'Air," 451.
229   **"We have been defeated":** Churchill, *Second World War,* II:42–47.
229   **agreed to send ten more:** Ibid., 49–51.
229   **400 Hurricanes:** Ray, *Battle of Britain,* 29.
229   **40 a week:** United Kingdom, Weekly Output of Aircraft: Fighters, C.S.B. Statistics Vol. I, AIR 22/293, PRO.
229   **Curtiss and Morane fighters:** Kirkland, "French Air Force."
229   **"pathetic inefficiency":** Ray, *Battle of Britain,* 26–27; Hooton, *Phoenix Triumphant,* 204.
230   **taking off every fifty minutes:** Murray, "Luftwaffe against Poland," 85; Murray, *Luftwaffe,* 42.
230   **Luftwaffe had lost . . . 1,400:** Griffin, "Battle of France," 152; Murray, *Luftwaffe,* 44; Ray, *Battle of Britain,* 29.
230   **"most dreadful rot":** Overy, *Battle of Britain,* 10–11.
230   **"shall never surrender":** Churchill, *Second World War,* II:117–18.
231   **Pope Pius XII . . . "appeal to reason":** Shirer, *Rise and Fall,* 747, 753–55.
231   **"on speaking terms":** Ray, *Battle of Britain,* 45.
231   **Hitler showed little interest:** Ibid., 41–42.
231   **Göring issued orders:** Murray, *Luftwaffe,* 47.
233   **Germans . . . within striking distance:** Ibid., 50; Ray, *Battle of Britain,* 53.
233   **148 . . . versus 286:** Hough and Richards, *Battle of Britain,* 357–59.
233   **abandoned . . . numbered attacks:** Ibid., 95, 313.
233   **250 yards . . . gun sight that reflected:** Whitford, *Fighter Design,* 181–82.
234   **deferred his retirement:** Letter, Dowding to Newall, July 7, 1940; Letter, Newall to

Dowding, August 12, 1940, Employment and Retirement of Air Chief Marshal Sir Hugh Dowding, AIR 19/572, PRO.

234   **"dismal Jimmy":** Haslam, "Lord Dowding," 176; Ray, *Battle of Britain,* 19.

234   **"boxes with coils":** Price, *Instruments of Darkness,* 48.

234   **"second Wagner":** Overy, "Hitler and Air Strategy," 417; Overy, *Battle of Britain,* 30.

235   **Führer Order No. 17:** Hough and Richards, *Battle of Britain,* 137.

235   OPERATION EAGLE: Ibid., 154.

235   **Spitfire was significantly faster:** Overy, *Battle of Britain,* 39, 56–57.

236   **Schmid's errors:** Cox, "RAF and Luftwaffe Intelligence," 436–37.

236   **Britain as a possible target:** Corum, *Luftwaffe,* 283.

236   **16 Me 110s . . . 100 Ju 88 bombers:** Hough and Richards, *Battle of Britain,* 142–45.

236   **Kenley:** Ibid., 205–7.

237   **bomber formations head-on:** Author's interview with Harold Bird-Wilson, June 1990.

237   **thirteen minutes . . . flying inland:** Ray, *Battle of Britain,* 84.

237   **"someone else may be shooting":** Author's interview with C. S. Bamberger, June 1990.

237   **quarter of its pilots:** Air Defence of Great Britain: Appendices and Maps, AIR 41/16, PRO, p. 45.

237   **twice as fast as factories:** "United Kingdom, Weekly Output of Aircraft: Fighters," C.S.B. Statistics Vol. I, AIR 22/293, PRO; Overy, *Battle of Britain,* 161.

237   **training course . . . half its length:** Murray, *Luftwaffe,* 53.

238   **"next time I fired":** Author's interview with C. S. Bamberger, June 1990.

238   **Göring fell for their ploy:** Ray, *Battle of Britain,* 69–70.

238   **"doubled the efficiency":** Ibid., 57.

238–39   **serviceable fighters . . . mere 100:** Overy, *Battle of Britain,* 80.

239   **"call me Meier!":** Shirer, *Rise and Fall,* 517 n.

239   COWARDLY BRITISH ATTACK . . . *"raze* their cities":** Ibid., 779–80.

239   **"a heavy strain":** Lecture by Hptm. Otto Bechtle at Berlin-Gatow, Battle of Britain: German Account and Other Papers, AIR 40/2444, PRO, p. 4.

239   **During . . . August each side:** Overy, *Battle of Britain,* 80.

239   **"taken personal command":** Hough and Richards, *Battle of Britain,* 254.

240   **Luftwaffe lost 298 aircraft:** Overy, *Battle of Britain,* 95.

240   **boyish exuberance:** See, for example, letters of Squadron Leader J. M. V. Carpenter, Fighter Command Collection, IWM.

240   **"confidence and calmness":** Overy, *Battle of Britain,* 100.

242   **"scruffy looking chap":** Hough and Richards, *Battle of Britain,* 211.

242   **Eighteen thousand tons:** Ray, *Night Blitz,* 260, 264.

242   **In June the British War Cabinet:** Overy, *Battle of Britain,* 89.

242   **"cruel, wanton":** Ray, *Night Blitz,* 106–7.

242   **forty thousand . . . millions more:** Ibid., 187, 206, 260.

243   **"exterminating attack":** Jones, *Wizard War,* 183.

243   **Thames docks . . . "give it 'em back!":** Ray, *Night Blitz,* 18–19, 104–5.

243   **"psychiatrists were surprised":** Mackintosh, *War and Mental Health,* 27.

243   **Home Intelligence . . . "improper behavior":** Overy, *Battle of Britain,* 101.

243   **poorly organized rest shelters:** Ray, *Night Blitz,* 143.

243   **Savoy Hotel:** Ibid., 142.

244   **"other shadowy men":** Monsarrat, *Breaking In,* 288.

244   **"social fault line":** Overy, *Battle of Britain,* 108.

244   **"Onward Christian Soldiers!":** *Flugblatt-Propaganda,* 314.

244   **"breaking down the class structure":** Ray, *Night Blitz,* 145.

244   **naval mines:** Churchill, *Second World War,* II:363–64.

245   **Tizard argued . . . night attacks:** Clark, *Tizard,* 158–59.

245   **first airborne test:** Letter, A. P. Rowe to Dowding, June 15, 1939, Interception Experiments and Exercises, AIR 16/251, PRO.

245   burst into flames ... "philosopher and a plumber": Zimmerman, "British Radar Organization," 90, 95–96, 98.

246   single Beaufighter in the air: Haslam, "Lord Dowding," 181.

246   11,000 night sorties: Ray, *Battle of Britain,* 159.

246   eleven gun-aiming radar units: Ray, *Night Blitz,* 111.

246   Hurricanes ... blue crayon: Ray, *Battle of Britain,* 140, 147.

246   Night Air Defence Committee: Ray, *Night Blitz,* 118.

246   "nearly broke my heart": Haslam, "Lord Dowding," 183.

246   "Divine Will": Ray, *Battle of Britain,* 189–90.

247   Knickebein: Jones, *Wizard War,* 85, 92–95.

247   "I'll take it!": Ibid., 3.

247   Lorenz "blind-landing" receiver: Price, *Instruments of Darkness,* 20–21.

247   "now that you mention it": Jones, *Wizard War,* 94.

248   "Well yes it would!": Ibid., 101.

248   "Sherlock Holmes or Monsieur Lecoq": Churchill, *Second World War,* II:384–85.

248   technical meeting ... "Aspirins": Jones, *Wizard War,* 103–4, 127–28.

249   X-Gerät: Ray, *Night Blitz,* 93–94; Jones, *Wizard War,* 146–52.

250   Coventry ... a canard: DeWeerd, "Churchill, Coventry, and Ultra"; Gilbert, *Finest Hour,* 912–14.

250   traveled by landline: Bennett, *Behind the Battle,* 139.

250–51 Luftwaffe orders ... *coventrate:* Ray, *Night Blitz,* 153–58.

251   "starting so many fires": Harris, *Bomber Offensive,* 83.

251   most important he had written: Churchill, *Second World War,* II:558.

251   like Guam: Perret, *Winged Victory,* 36; Huston, ed., *Arnold's Diaries,* I:99.

251   "foundation of victory": Churchill, *Second World War,* II:558–67.

251   "the whole program": Burns, *Roosevelt: Soldier,* 25.

252   British ... all the new planes: Perret, *Winged Victory,* 45; Arnold, *Global Mission,* 233.

252   German aircraft industry: Murray, *Luftwaffe,* 92–102.

252   "wonder weapons": Overy, "Hitler and Air Strategy," 415–16.

252   aerodynamic insanity: Corum, *Luftwaffe,* 268.

253   "Jews" in the Air Ministry: Overy, *Battle of Britain,* 56.

253   400 percent ... "cars and refrigerators": Murray, *Luftwaffe,* 103–4.

253   Ford ... Knudsen: *AAF,* VI:320.

253   "expect blacksmiths": Perret, *Winged Victory,* 43.

253   factory floor space: *AAF,* VI:318.

254   output ... would triple: Ibid., 333.

254   "modification centers": Ibid., 336–37.

254   300,000 ... 435 airplanes: Ibid., 287, 331.

254   25 percent ... ninety Luftwaffe bombers: Ray, *Night Blitz,* 239, 241.

9. Air Versus Sea

page
255   was almost worthless: Budiansky, *Battle of Wits,* 41–42.

255   Japanese Navy intended them to see: Coox, "Japanese Air Forces," 75.

255   "rapidly as American personnel" ... "daring but incompetent": Kahn, "Germany and Japan," 476–77.

255   buying frenzy: Coox, "Japanese Air Forces," 80–81.

256   Lake Kasumigaura: Ohmae and Pineau, "Japanese Naval Aviation," 70.

256   Herbert Smith: Coox, "Japanese Air Forces," 80–81.

256   Gloster and Blackburn: Jarrett, ed., *Biplane to Monoplane,* 120.

256   In 1911 the Royal Navy: Aircraft Carriers, Part I: 1914–1918; Part IV: Aeroplanes Carried in Fighting Ships, AIR 1/2103/207/31, PRO.

257   **3,500 aircraft:** Hone and Mandeles, "Interwar Innovation," 64.

257   **job of aircraft at sea:** Till, "Airpower and the Admiralty," 343.

257   **Mahan . . . Japanese home waters:** Miller, *War at Sea,* 10; Spector, *Eagle against Sun,* 43–4; Goldstein and Dillon, eds., *Pearl Harbor Papers,* 5–6; Matsuo, *How Japan Plans to Win,* 284–85.

257   **exception to the British view:** Till, "Airpower and the Admiralty," 343; Hone and Mandeles, "Interwar Innovation," 64.

258   **"mental soundness was doubted":** Goldstein and Dillon, eds., *Pearl Harbor Papers,* 6–7.

259   **thousand antiaircraft guns:** Lowry and Wellham, *Attack on Taranto,* 63.

259   **torpedo missed the destroyer:** Miller, *War at Sea,* 122.

260   **"almost like heroes":** Lowry and Wellham, *Attack on Taranto,* 73–74

260   **"Light Brigade to do it once":** Ibid., 83.

261   **Japanese delegation arrived:** Ibid., 92–93.

261   **"deal a crushing blow":** Goldstein and Dillon, eds., *Pearl Harbor Papers,* 13–14.

261   **Yamamoto . . . threaten to resign:** Prange, *Verdict of History,* 507–8.

262   **"risky and illogical":** Goldstein and Dillon, eds., *Pearl Harbor Papers,* 119.

262   **Chief of the Technical Division:** Hone and Mandeles, "Interwar Innovation," 70.

262   **"one step beyond":** Horikoshi, *Eagles of Mitsubishi,* 150.

263   **dive-bombing . . . circular error:** McFarland, *Pursuit of Precision,* 66, 106–10, 194; Wildenberg, *Destined for Glory,* 220.

264   **Washout rates:** Coox, "Japanese Air Forces," 79–80.

264   **impressive improvements . . . "stunt-like" flights:** Goldstein and Dillon, eds., *Pearl Harbor Papers,* 6, 11, 145.

264   **scale models:** Ibid., 27.

264   **wooden fins:** Lowry and Wellham, *Attack on Taranto,* 91–92.

264   **"no serious menace":** Melhorn, *Two Block Fox,* 64–65.

264   **"capital ship of the future":** Reynolds, *Fast Carriers,* 1.

264   **"essential arm of the fleet":** Turnbull and Lord, *United States Naval Aviation,* 161.

265   **Johnny Come Latelys:** Wildenberg, *Destined for Glory,* 3.

265   **King . . . Halsey:** Miller, *War at Sea,* 190; Wildenberg, *Destined for Glory,* 163.

265   **"master of publicity" . . . superb administrator:** Hone, "Navy Air Leadership," 109; Hone and Mandeles, "Interwar Innovation," 73–75.

265   **Aircraft . . . had to be pushed forward:** Stern, *Lexington Class Carriers,* 119.

266   **exercises in 1934:** Hone, Friedman, and Mandeles, *Aircraft Carrier Development,* 53–54.

266   DEAL A FATAL BLOW: Goldstein and Dillon, eds., *Pearl Harbor Papers,* 96.

266–67   **stunned bandsmen . . . "Too bad":** Miller, *War at Sea,* 201, 206.

267   **Japanese study . . . "soul of the Emperor":** Goldstein and Dillon, eds., *Pearl Harbor Papers,* 121, 287–91.

267   **"I have to report" . . . "weak and naked":** Churchill, *Second World War,* III:620.

267   **"clinched the conviction":** Prange, *Verdict of History,* 542.

268   **ordered a complete reorganization:** Miller, *War at Sea,* 232–33.

268   **"Shangri-La":** Burns, *Roosevelt: Soldier,* 224.

268   **come from Midway:** Prange, *Miracle at Midway,* 21–25.

269   **"lacks the will to fight":** Ibid., 181.

269   **"never exchanged a shot":** Churchill, *Second World War,* IV:247.

269   **JN-25 traffic:** Budiansky, *Battle of Wits,* 12–13.

270   **Army B-17s:** Wildenberg, *Destined for Glory,* 212.

270   **believed . . . plenty of time:** Isom, "Battle of Midway."

272   **"Shimatta":** Prange, *Miracle at Midway,* 265.

272   **platform for antiaircraft guns:** Buckley, *Air Power,* 190.

272   *Essex* **class carriers:** Reynolds, *Fast Carriers,* 38–39, 53–54.

273   **effective radar system:** Wooldridge, ed., *Carrier Warfare,* 111.
273   **needed fast tankers:** Hone, Friedman, and Mandeles, *Aircraft Carrier Development,* 53–54, 70.
273   **logistics, deeming it "boring":** Coox, "Japanese Air Forces," 84–85.
273   **failures of American torpedoes:** Miller, *War at Sea,* 484–85.
273   **2,000-horsepower American engines:** Jarrett, ed., *Biplane to Monoplane,* 107–8.
274   **U-570:** van der Vat, *Atlantic Campaign,* 297–98.
274   **damaging attacks . . . deterrent effect:** *AAF,* I:515, 531, 535.
274–75 **seventeen knots . . . eight knots:** Miller, *War at Sea,* 34.
275   **Hitler . . . finally agreed:** van der Vat, *Atlantic Campaign,* 256.
275   **139 ships . . . 700 ships:** Churchill, *Second World War,* III:782.
275   **31 million tons:** Miller, *War at Sea,* 177.
275   **alarming to Churchill:** Churchill, *Second World War,* III:147.
275   **"belongs to me!":** van der Vat, *Atlantic Campaign,* 256; Miller, *War at Sea,* 171–72.
275   **breaking of the naval Enigma:** Erskine, "Naval Enigma," 499.
276   **"situation is so serious":** Churchill, *Second World War,* IV:119.
276   **600 thousand tons:** Ibid., IV:126.
276   **"bloody incompetence":** The Americans, the Navy Department, and U-boat Tracking, ADM 223/286, PRO.
276   **"Battle of the Atlantic is being lost":** Cohen and Gooch, *Military Misfortunes,* 62.
276   **control of the Admiralty:** van der Vat. *Atlantic Campaign,* 226.
276   **I Bomber Command:** *AAF,* I:541.
277   **"purely defensive":** van der Vat, *Atlantic Campaign,* 458–59; Overy, *Air War,* 71.
277   **800 minimum . . . took away half of them:** van der Vat, *Atlantic Campaign,* 297, 385.
277   **"absolute priority" . . . subsequent report:** Ibid., 386–87.
278   **"diversion to the Navy":** *AAF,* I:539.
278   **"basic requirements" . . . "needle in a haystack":** Konvitz, "Bombs, Cities, and Submarines," 27.
278   **reinforced concrete:** *AAF,* II:245.
278   **eight hundred tons of bombs:** Konvitz, "Bombs, Cities, Submarines," 29.
278–79 **precision was illusory . . . 250 sorties:** *AAF,* II:249–53.
279   **Lorient . . . Saint-Nazaire was next:** Konvitz, "Bombs, Cities, Submarines," 38, 42–43.
279   **"No dog or cat":** *AAF,* II:315–16.
279   **political intervention:** Churchill, *Second World War,* IV:130; *AAF,* II:386–88; Overy, *Air War,* 71–72.
279   **destroyed 41 . . . more than 300:** Slessor, *Central Blue,* 470–75.
280   **infrared-masking paint:** McCue, *U-boats in Biscay,* 28.
280   **lock it in the boat's safe:** Budiansky, *Battle of Wits,* 294.
280   **single German aircraft carrier:** van der Vat, *Atlantic Campaign,* 388–89.
280   **"milk cow":** McCue, *U-boats in Biscay,* 154–56.

10. The Temporary Triumph of Tactical Aviation

page
281   **"we and the enemy were ignorant":** British Bombing Survey Unit, The Strategic Air War Against Germany, AIR 10/3866, PRO, p. 1.
281   **suppressed the report:** Zuckerman, *Apes to Warlords,* 337.
282   **150 or so:** Ray, *Night Blitz,* 246.
282   **"Western Air Plans":** Smith, *British Air Strategy,* 339; British Bombing Survey Unit, The Strategic Air War Against Germany, AIR 10/3866, PRO, pp. 1–4.
282   **22 percent . . . within five miles:** Werrell, "Strategic Bombing," 704; Noble and Frankland, *Strategic Air Offensive,* IV:205–13.
282   **"war is not going to be won":** Clark, *Tizard,* 308–9.

283   "Like you . . . the Germans": Ibid., 313.
283   "known to have been achieved": Bennett, *Behind the Battle,* 149, 156.
283   "seen in any photograph": Hinsley, et al., *British Intelligence,* III(1):297.
283   "almost irrelevant": Bennett, *Behind the Battle,* 154.
283   "halfway stage": Harris, *Bomber Offensive,* 77–78.
283   politically inescapable: Bennett, *Behind the Battle,* 148.
283   "focussed on the morale": Webster and Frankland, *Strategic Air Offensive,* IV:144.
283   Harris chose to ignore this: Bennett, *Behind the Battle,* 152.
284   "evidence of panic" . . . "break the spirit": Zuckerman, *Apes to Warlords,* 142; British
         Bombing Survey Unit, The Strategic Air War Against Germany, AIR 10/3866, PRO, p. 7.
284   "dehousing" . . . "persevere": Kennett, *History of Strategic Bombing,* 129; Bennett,
         Behind the Battle, 148, 152.
284   "seek no mercy": Bennett, *Behind the Battle,* 152.
284   an agnostic . . . "panacea mongering": Buckley, *Air Power,* 157–58, 163.
284–85 "pull the plug out of Hitler's bath": Perret, *Winged Victory,* 242.
285   ultralogical toughness: Astor, *Mighty Eighth,* 15–16.
285   "fall guy": Buckley, *Air Power,* 164.
285   mass raids: Pape, *Bombing to Win,* 269–70.
285   raid on Cologne: Kennett, *History of Strategic Bombing,* 133–34.
285   impervious . . . "Victory, speedy and complete": Bennett, *Behind the Battle,* 148, 163.
286   ABC-1: *AAF,* I:137–38.
286   in the mad rush: Nalty, ed., *Winged Shield,* I:187.
287   old Munitions Building: Gaston, *Planning the Air War,* 22.
287   "only land armies can finally win": U.S. Air Force, *Development of Air Doctrine,*
         124–25.
287   "heavy drain" . . . "means of livelihood": Gaston, *Planning the Air War,* 32–33; Clod-
         felter, "Pinpointing Devastation," 91.
287   124 targets: *AAF,* I:599; Clodfelter, "Pinpointing Devastation," 91.
287   "American propensity": Barry Watts, quoted in Clodfelter, "Pinpointing Devastation,"
         97.
287   three hundred fully operational: *AAF,* II:718.
287–88 British press . . . afraid to fly: Perret, *Winged Victory,* 245, 253; *AAF,* I:662–63,
         II:299–302.
288   "central road" . . . "decisive air offensive": Arnold, *Global Mission,* 265, 322–23.
288   fighters . . . light bombers and dive-bombers: U.S. Air Force, *Development of Air
         Doctrine,* 125–26; *AAF,* I:148–49.
288   "if it becomes necessary": Clodfelter, "Pinpointing Devastation," 90.
288   "immature": Hughes, *Overlord,* 77.
288   "event of a Russian collapse": Pape, *Bombing to Win,* 259 n. 10.
289   con man: Orange, *Coningham,* 7–9.
289   accepted into the RFC: Ibid., 14–15.
289   covered in blood: Ibid., 28.
289   "convincing English gentleman": Ibid., 50.
290   "What a queer fellow": Ibid., 91, 110.
290   at Dunkirk: Syrett, "Tunisian Campaign," 158.
290   "Royal Absent Force": Hallion, *Strike from the Sky,* 151.
290   Northern Ireland: Gooderson, *Air Power at the Battlefront,* 24.
290   more staff officers than aircraft: Syrett, "Tunisian Campaign," 158.
291   the Army's demands: Gooderson, *Air Power at the Battlefront,* 35.
291   eighty miles apart . . . radio links: Hallion, *Strike from the Sky,* 154.
291   "such a mischievous practice": Ibid., 153; Orange, *Coningham,* 79.
292   "eating out of his hand": Orange, *Coningham,* 79.
292   Coningham stressed practice: Ibid., 93.

292   **Air Support Control center:** Gooderson, *Air Power at the Battlefront,* 26; Hallion, *Strike from the Sky,* 154–56.

292   **three hours to thirty-five minutes:** Hallion, *Strike from the Sky,* 160.

292   **"corps commander really wants":** Orange, *Coningham,* 112.

293   **Bletchley Park code breakers:** Budiansky, *Battle of Wits,* 270–71.

293   **"The convoy sets sail":** Hinsley, ed., *British Intelligence,* II:424–25.

293   **"fights like a savage":** Hallion, *Strike from the Sky,* 161.

293   **organic . . . observation squadrons:** Maycock, "Tactical Air Doctrine," 187.

294   **"damned fool decisions" . . . "morale would be bolstered":** Syrett, "Tunisian Campaign," 167–69.

294   **"worst performance of U.S. Army":** Bradley and Blair, *A General's Life,* 128.

294   **cabled all air commands:** *AAF,* II:157.

294   **sent every ranking officer:** Syrett, "Tunisian Campaign," 174.

294   **FM 100-20:** Huston, "Tactical Use of Air Power," 167–68.

295   **"Total lack of air cover" . . . "mail them each a medal":** Orange, *Coningham,* 146–48; Syrett, "Tunisian Campaign," 177–78.

296   **"war, for God's sake!":** Hughes, *Overlord,* 152; Hallion, *Strike from the Sky,* 176.

296   **Me 109s . . . "Rhubarbs":** Gooderson, *Air Power at the Battlefront,* 57–58; Hallion, *Strike from the Sky,* 156–57.

297   **three missions a day:** Gooderson, *Air Power at the Battlefront,* 66.

298   **75 percent . . . executed:** Whiting, "Soviet Aviation under Stalin," 62–63.

298   **Ground attacks invariably:** Wagner, ed., *Soviet Air Force,* 202–3.

298   **"very effective and unpleasant":** Hallion, *Strike from the Sky,* 240–41.

299   **"specious appearance":** Bennett, *Behind the Battle,* 157.

299   **"existence of the other two services":** Ibid., 154.

299   **"highly dubious operation":** Mark, *Aerial Interdiction,* 230.

299   **confused the lines of authority:** Ibid., 220–21.

300   **ensured of proper "credit":** Huston, ed., *Arnold's Diaries,* II:108–9.

300   **lightweight . . . threatening to resign:** Hughes, *Overlord,* 120–21.

300   **Eisenhower . . . forced a decision:** Mark, *Aerial Interdiction,* 230.

300   **"invasion can't succeed":** Ibid., 230.

300   **"get off scot-free":** Zuckerman, *Apes to Warlords,* 286.

301   **"had a new entry":** Ibid., 258.

301   **eighty rail centers . . . a failure:** *AAF,* III:150–54; Mark, *Aerial Interdiction,* 227–28, 233, 238.

301–2   **Leigh-Mallory ruled . . . 4,400 tons:** *AAF,* III:156–59; Hughes, *Overlord,* 130; Mark, *Aerial Interdiction,* 256–57.

302   **"air power was decisive":** U.S. Strategic Bombing Survey, *Summary (European),* 15–16.

303   **"not . . . any German Air Force":** Hughes, *Overlord,* 136.

303   **ten thousand planes:** Mark, *Aerial Interdiction,* 243.

303   **"whiskery unkempt ruffians":** Memoir, Flight Lieutenant George Millington, Fighter Command Collection 92/29/1, IWM.

303   **"kitchen sink" . . . "Bradley liked me":** Hughes, *Overlord,* 157; Hallion, *Strike from the Sky,* 199.

303   **put up standing patrols:** Jacobs, "Tactical Air Doctrine," 43; Hallion, *Strike from the Sky,* 201.

304   **"Get the hell out of here!":** Bradley, *Soldier's Story,* 338; Hughes, *Overlord,* 183–84.

304   **throughout the First Army . . . "no answer":** Hallion, *Strike from the Sky,* 199, 205–6.

304   **Kampfgruppe Heintz . . . 2nd SS:** Mark, *Aerial Interdiction,* 248–49; Hallion, *Strike from the Sky,* 204.

305   **100 bridges, 4,000 locomotives:** Pape, *Bombing to Win,* 279.

305   **half to flak:** Hallion, *Strike from the Sky,* 225.
305   **Savio River:** Appendix B, School of Air Support R.A.F. Old Sarum Special Senior Course 29th April–2nd May 1945, Box 4, Quesada Papers, LOCMD.
305   **3,500 bombs or 800 rockets:** Appendix D, ibid. Senior Course 29th April–2nd May 1945, Box 4, Quesada Papers, LOCMD.
305   **301 tanks:** Gooderson, *Air Power at the Battlefront,* 119.
305   **Monte Cassino:** Huston, "Tactical Use of Air Power," 175.
305   **emplacements almost unscathed:** *AAF,* III:170.
306   **killing 101:** U.S. Air Force, *Bombers in Tactical Role,* 84–85.
306   **"8th and 9th Luftwaffe":** Memorandum, W. B. Smith, Lt. Gen. USA, to Spaatz, August 16, 1944, Personal Diary August 1944, Box 15, Spaatz Papers, LOCMD.
306   **Cheers were heard:** Gooderson, "Heavy and Medium Bombers," 388.
306   **Sikorsky's VS-300:** Cochrane, Hardesty, and Lee, *Aviation Careers of Sikorsky,* 120–32.
307   **Fritz X:** Bogart, "German Remotely Piloted Bombs"; Price, *Sky Warriors,* 62.
307   **foil-backed materials:** Radar Camouflage, AVIA 7/2320, PRO.

11. The Allied Bomber Offensive

page
308   **"war's last beauty parlor":** Fussell, *Wartime,* 132.
308   **besieged by applicants . . . Cadet Branch:** *AAF,* VII:498–99, 562; Cameron, *Training to Fly,* 385.
309   **193,000 American pilots:** *AAF,* VII:577, 584, 589; Goldman, *Morale in AAF,* 12.
309   **"ill-fated combination" . . . "heat and cold":** Goldman, *Morale in AAF,* 8, 11.
309   **"inch deep in mud":** Letter, George James Hull to Joan Kirby, September 29, 1943, Bomber Command Collection, IWM.
309   **"tea and coca urn":** Manuscript memoir, Flight Lieutenant D. G. Hornsey, Bomber Command Collection, IWM, p. 69.
310   **purgatory as privates:** Goldman, *Morale in AAF,* 14.
310   **"saddest, poorest":** Stiles, *Serenade,* 141–42.
310   **Morse code:** *AAF,* VII:560.
310   **fast and got the wrong answers:** Cameron, *Training to Fly,* 386.
310   **risking frostbite:** Nalty, ed., *Winged Shield,* I:284.
311   **"I hate the R.A.F.":** Letter, George James Hull to Joan Kirby, November 4, 1943, Bomber Command Collection, IWM.
311   **"Queen of the May":** Account of Raid by No. 218 Squadron, Bomber Command, on 7 November 1941, J. P. Dobson, Bomber Command Collection 92/2/1, IWM.
311   **checklist of fifty-six items:** Astor, *Mighty Eighth,* 372–73.
311   **"combat box":** *AAF,* II:332; Nalty, ed., *Winged Shield,* I:283; Perret, *Winged Victory,* 254–55.
312   **"nothing shall deter him":** Lead Crew Manual 3rd Bombardment Division, Manuals (doctrines and guides), Box B4, LeMay Papers, LOCMD.
312   **statistical control unit . . . postwar study:** Nalty, ed., *Winged Shield,* I:265; Worden, *Fighter Generals,* 11.
313   **1918 RFC lecture:** Notes for Lecture to Officers as to How to Escape if Taken Prisoner, AIR 1/1976/204/273/48, PRO, p. 6.
313   **harrowing account:** Squadron Leader S. A. Booker, Bomber Command Collection 97/9/1, IWM.
313   **simply lynched:** Astor, *Mighty Eighth,* 407; Kennett, *History of Strategic Bombing,* 162.
313   **"impressed with my survival":** Astor, *Mighty Eighth,* 420–21.
313   **only 1.5 percent:** Nalty, ed., *Winged Shield,* I:264.
313   **"discover" mechanical troubles:** Goldman, *Morale in AAF,* 42.

314   **"fringe merchants":** Fussell, *Wartime,* 255.

314   **crew weight loss:** McFarland and Newton, *To Command the Sky,* 110–11.

314   **steel-plate jockstrap:** Perret, *Winged Victory,* 248.

314   *flieger sheit:* Astor, *Mighty Eighth,* 394.

314   **"Nobody expected to live":** McFarland, *Pursuit of Precision,* 187.

314   **Flight surgeons urged:** Nalty, ed., *Winged Shield,* I:264.

315   **morale plummeted:** Werrell, *Blankets of Fire,* 204.

315   **medals:** Perret, *Winged Victory,* 415; Goldman, *Morale in AAF,* 50–52.

315   *Ops Room twats . . . Ten thousand dollars:* Fussell, *Wartime,* 263–65.

316   **"doubts" . . . "killing sheep":** Survey of Combat Crews in Heavy Bombardment Groups in ETO, Research Branch, Special Service Division, HQ ETO, June 1944, Official Diary June 1944, Box 18, Spaatz Papers, LOCMD.

316   **"assault on German agriculture":** Fussell, *Wartime,* 16.

316   **abandon daylight bombing:** *AAF,* I:602.

316   **"devil shall get no rest":** Eaker Oral History, May 22, 1962, File number K239.0512-627, USAFHRA, p. 6; Perret, *Winged Victory,* 243–44.

316   **"Your primary objective":** Churchill, *Second World War,* V:519–20.

317   **"firestorm":** Buckley, *Air Power,* 158–59; Kennett, *History of Strategic Bombing,* 147–48.

317   **Harris's running tally:** Pape, *Bombing to Win,* 271–72.

317   **"cost Germany the war":** Webster and Frankland, *Strategic Air Offensive,* II:190.

317   **"The maintenance of morale":** Combined Bomber Offensive Progress Report, February 4, 1943–November 1, 1943, Box 67, Spaatz Papers, LOCMD, pp. 5–6.

317   **"including the Russian":** British Bombing Survey Unit, The Strategic Air War Against Germany, AIR 10/3866, PRO, p. 11.

317   **Berlin was a more modern city:** Kennett, *History of Strategic Bombing,* 154–55.

317   **"not . . . an overwhelming success":** Harris, *Bomber Offensive,* 187.

317   **Absenteeism rates:** Buckley, *Air Power,* 166.

318   **left homeless . . . "discouraged workers":** Pape, *Bombing to Win,* 271–73.

318   **Within days of the Hamburg raid:** Konvitz, "Why Cities Don't Die."

318   **49 percent of the German GNP:** Gaston, *Planning the Air War,* 33.

318   **twice as many machine tools:** British Bombing Survey Unit, The Strategic Air War Against Germany, AIR 10/3866, PRO, p. 83.

318   **no more than 1 percent:** Pape, *Bombing to Win,* 273 n. 49.

318   **"However dissatisfied":** U.S. Strategic Bombing Survey, *Summary (European),* 4, 16.

319   **"Dropping-on-the-leader":** McFarland, *Pursuit of Precision,* 171.

319   **Gee, Oboe . . . H2X:** British Bombing Survey Unit, The Strategic Air War Against Germany, AIR 10/3866, PRO, pp. 44–45.

319   **tried to banish the term:** Huston, ed., *Arnold's Diaries,* II:142.

320   **1.1 miles . . . 2.5 miles:** Memorandum, Conference on Bombing Accuracy, 22–23 March 1945, Bombing Accuracy File, Box 76, Spaatz Papers, LOCMD.

320   **"drenching":** *AAF,* III:723.

320   **"poor for morale":** Memorandum to Commanding General, Eighth Air Force, June 3, 1944, Subject: ORS Reports on Bombardment Groups' Bombing Accuracy, Official Diary June 1944, Box 18, Spaatz Papers, LOCMD.

320   **near-total failures:** *AAF,* III:667.

320   **objective differences:** Crane, "Evolution of U.S. Bombing," 38.

320   **Gallup Poll:** Quester, "Impact of Strategic Warfare," 182.

320   **Three days after Pearl Harbor:** Crane, "Evolution of U.S. Bombing," 20.

320   *Time . . . Harper's:* Hopkins, "Bombing and Conscience," 463.

320   **religious shrines:** Crane, "Evolution of U.S. Bombing," 20 n. 24.

321   **"Christian people" . . . "God has given us the weapons":** Hopkins, "Bombing and Conscience," 467–71.

321   "'pull our punches'": Memorandum, T. J. Hanley, Jr., to Assistant Chiefs of Air Staff, April 30, 1943, Reel 114, Arnold Papers, LOCMD.
321   "Are we beasts?": Bennett, *Behind the Battle,* 170.
321   "tarred with the morale bombing": Letter, Spaatz to Arnold, August 27, 1944, Personal Diary August 1944, Box 15, Spaatz Papers, LOCMD.
321   "man in the street": Schaffer, "American Military Ethics," 328.
321   final morale-breaking attack: Pape, *Bombing to Win,* 262.
322   "proof positive of Allied air power": Schaffer, "American Military Ethics," 327–28.
322   "What do we say?" . . . "no change": Ibid., 331–32.
322   Committee of Operations Analysts: *AAF,* II:349, 353–54.
323   half of all production . . . stockpiling: *AAF,* II:357; Report of Committee of Operations Analysts, March 8, 1943, Box 146, Spaatz Papers, LOCMD.
323   "got Schweinfurt": *AAF,* II:704.
323   damage to machine tools: U.S. Strategic Bombing Survey, *Summary (European),* 5; U.S. Strategic Bombing Survey, *Machine Tools,* 1–2.
323   "brief shortage" . . . sleeve bearings: "Defeat," HQ Army Air Forces, Mediterranean Theater of Operations, Intelligence Section, Speeches and Printed Matter File, Box 7, Quesada Papers, LOCMD, p. 54; U.S. Army Air Force, *War against Luftwaffe,* 205.
323   "not a tank": Speer, *Inside the Third Reich,* 286.
323   36 B-17s . . . 138 had been damaged: U.S. Strategic Bombing Survey, *Summary (European),* 5.
323   burning haystacks: Nalty, ed., *Winged Shield,* I:285.
324   108 RAF Spitfires: Perret, *Winged Victory,* 245–46.
324   "does not diminish" . . . 102 enemy planes: Ibid., 246, 249.
324   "really dumb": Huston, ed., *Arnold's Diaries,* I:2, 60.
324   "hell of a time" . . . "lazy, cowardly": Ibid., II:33, 51.
324   flying parallel: McFarland and Newton, *To Command the Sky,* 96.
324–25   By April 1943 . . . "very large flash": U.S. Army Air Force, *War against Luftwaffe,* 64–68.
325   YB-40: Ibid., 71–72.
325   "premature" . . . number-four priority: McFarland and Newton, *To Command the Sky,* 106, 112, 145.
325   bouncing responsibility: U.S. Army Air Force, *War against Luftwaffe,* 73–75.
326   "What's the idea": Speer, *Inside the Third Reich,* 290.
326   Göring later acknowledged: Interrogations: Goering, Hermann, May 10, 1945, Box 134, Spaatz Papers, LOCMD.
326   "The first duty": McFarland and Newton, *To Command the Sky,* 160.
326   wiped out 30 percent . . . flying hours: Murray, "Attrition and the Luftwaffe"; U.S. Strategic Bombing Survey, *Over-all Report (European),* 21.
326–27   bombing airframe factories: U.S. Army Air Force, *War against Luftwaffe,* 203–4; U.S. Strategic Bombing Survey, *Summary (European),* 6–7; British Bombing Survey Unit, The Strategic Air War Against Germany, AIR 10/3866, PRO, pp. 57–58; U.S. Strategic Bombing Survey, *Over-all Report (European),* 18; "Defeat," HQ Army Air Forces, Mediterranean Theater of Operations, Intelligence Section, Speeches and Printed Matter File, Box 7, Quesada Papers, LOCMD, p. 53.
327–28   Oil . . . heavily on imports: Hinsley, et al., *British Intelligence,* III(2):917.
328   70 percent . . . within 2,000 feet: British Bombing Survey Unit, The Strategic Air War Against Germany, AIR 10/3866, PRO, p. 47.
328   Oshima . . . aviation gasoline: U.S. Strategic Bombing Survey, *Effects on German Economy,* 79; Hinsley, et al., *British Intelligence,* III(2):507, 925; British Bombing Survey Unit, The Strategic Air War Against Germany, AIR 10/3866, PRO, p. 117.
328–29   Battle of the Bulge . . . Vistula: U.S. Strategic Bombing Survey, *Summary (European),* 9.
329   "the magnetism": Webster and Frankland, *Strategic Air Offensive,* III:84.

329   **"I have always held":** British Bombing Survey Unit, The Strategic Air War Against Germany, AIR 10/3866, PRO, p. 27.

330   **"We won the war":** Hughes, *Overlord,* 310.

330   **two million soldiers . . . optical instruments:** Overy, *Air War,* 122.

330   **600,000 German civilians:** Werrell, "Strategic Bombing," 709.

330   **2,400,000 men:** "Statistical Portrait of USAAF," 30.

330   **bomber-crew casualties:** Werrell, "Strategic Bombing," 708.

330   **great ironies:** Murray, "Attrition and the Luftwaffe."

331   **"look into the atomic bomb":** Goudsmit, *Alsos,* 15.

331   **"execrable drawing":** Rhodes, *Making of Atomic Bomb,* 274–75.

331   **Frisch:** Ibid., 323–25.

332   **"troy ounce":** Ibid., 490.

333   **"both let out a yell":** Goudsmit, *Alsos,* 68–71.

333   **"you're all second-raters":** *Operation Epsilon,* 71, 77.

334   **meeting was a disaster:** Giovannitti and Freed, *Decision to Drop Bomb,* 64–66.

334   **Szilard told Byrnes:** Rhodes, *Making of Atomic Bomb,* 635–38.

334   **"Greek tragedy":** Giovannitti and Freed, *Decision to Drop Bomb,* 325.

335   **lowest VD rate:** Rhodes, *Making of Atomic Bomb,* 654.

335   **"all sons of bitches":** Ibid., 675.

335–36 **"imperial destiny" . . . astonished his military aide:** Bix, *Hirohito,* 481–83.

336   **invasion of Okinawa:** Miller, *War at Sea,* 371–72.

336   **Anami, insisted:** Pape, *Bombing to Win,* 124.

336   **"die-for-the-emperor campaign":** Bix, *Hirohito,* 494–95.

336   **"Little Tokyos":** Memorandum for the Chief of Air Staff, Subject: Test of Incendiaries, May 5, 1944, Reel 115, Arnold Papers, LOCMD.

337   **"three-billion-dollar gamble":** *AAF,* V:7.

337   **Originally justified . . . China:** Ibid., 11, 17.

337   **missions to be aborted:** Ibid., 567; Gilium, "The Beast," 232–33.

337   **Tailwinds:** Buckley, *Air Power,* 192.

337–38 **"tough guy" . . . "angry stimulation":** Worden, *Fighter Generals,* 56–57; LeMay and Kantor, *Mission,* 19, 37–39.

338   **grimly determined:** Perret, *Winged Victory,* 255.

338   **started pulling fuel out, too:** *AAF,* V:599.

338   **"hell of a lot of publicity":** Ibid., 608.

338   **334 B-29s . . . napalm:** Ibid., 614–16; U.S. Strategic Bombing Survey, *Effects of Incendiary Attacks,* 67, 90, 116–17.

338   **LeMay publicly insisted:** Werrell, *Blankets of Fire,* 157.

338   **"Had to be done":** Crane, "Evolution of U.S. Bombing," 36; Hurley and Ehrhart, eds., *Air Power and Warfare,* 200–201.

338–39 **4.7 square miles . . . killed an estimated 35,000:** U.S. Strategic Bombing Survey, *Effects of Atomic Bombs,* 33; *AAF,* V:722–25.

339   **"more pressing business":** Pape, *Bombing to Win,* 124.

339   **"Hirohito sought to justify":** Bix, *Hirohito,* 529.

339   **"without losing face":** Giovannitti and Freed, *Decision to Drop Bomb,* 322–23.

340   **Mine-laying:** Werrell, *Blankets of Fire,* 170; *AAF,* V:672–74; Hallion, "World War II as Turning Point."

340   **Marshall sent them a curt cable:** Personal Eyes Only for Spaatz from Marshall, August 10, 1945, Personal Diaries August 1945, Box 21, Spaatz Papers, LOCMD.

340   **"vindicated and mature":** Spaatz to Norstad, August 14, 1945, Personal Diaries August 1945, Box 21, Spaatz Papers, LOCMD.

340   **"shattered once and for all":** Letter, Lovett to Spaatz, August 19, 1945, Personal Diaries August 1945, Box 21, Spaatz Papers, LOCMD.

341   **"doubting Thomases"**: Letter, Arnold to Spaatz, August 19, 1945, Personal Diaries August 1945, Box 21, Spaatz Papers, LOCMD.

341   **210 B-29s**: U.S. Strategic Bombing Survey, *Effects of Atomic Bombs,* 33.

341   **"wanton, indiscriminate bombing"**: Personal to Spaatz from Norstad, August 8, 1945, Personal Diaries August 1945, Box 21, Spaatz Papers, LOCMD.

12. Strategic Air Command

page
345   **not "victory"**: Goldman, *Crucial Decade,* 4.

345   **$100 billion . . . "breakdown of Army discipline"**: Manchester, *Glory and Dream,* 397, 408–9.

345   **"disintegration"**: McCullough, *Truman,* 474.

345   **"you had, like measles"**: Goldman, *Crucial Decade,* 28.

346   **"fat, overfed, lonely"**: Ibid., 33–34.

346   **"cry-baby"**: McCullough, *Truman,* 47.

346   **40 percent of the GNP**: Worden, *Fighter Generals,* 27.

346   **"get into mischief"**: McCullough, *Truman,* 476.

346   **"would be ironical"**: Goldman, *Crucial Decade,* 57.

347   **"cold war"**: Ibid., 60.

347   **"Global New Dealism"**: Ibid., 77.

347   **in the Red Army**: Manchester, *Glory and Dream,* 410.

348   **"an additional weapon"**: Board Report re Atomic Energy 10/23/45, file 384.3 (17 August 1945) Atomic Section 1, Air Force Plans Decimal File 1942–1954, Entry 335, RG 341, NACP.

348   **"requisite of national survival"**: Armitage and Mason, *Nuclear Age,* 16–17.

348   **"Air Power all-important"**: Greenwood, "Postwar Strategic Air Force," 220.

348   **"backbone of our Air Force"**: Futrell, "Air Power Concept," 256–57.

348   **"any part of the world"**: Moody, *Strategic Air Force,* 65.

348   **Bradley . . . to agree**: Futrell, *Ideas, Concepts,* I:255.

349   **"Atomic bombs will be used"**: Report by the Joint Strategic Plans Committee to the Joint Chiefs of Staff on General Guidance on Strategic Concepts, 15 March 1948, file OPD 381 (19 Feb 1946) Section 1 (Strategic Guidance for Planning), Air Force Plans Decimal File 1942–1954, Entry 335, RG 341, NACP, pp. 5–6.

349   **Douhet's principles . . . "To Kill a Nation"**: Futrell, *Ideas, Concepts,* I:239: Futrell, "Air Power Concept," 258.

349   **"If a shooting war" . . . "voice of doom"**: Moody, *Strategic Air Force,* 230–31.

349   **Manual 1-2**: Worden, *Fighter Generals,* 68.

349   **Manual 1-8**: Clodfelter, *Limits of Air Power,* 27–28.

350   **"ax-grinders"**: Futrell, *Ideas, Concepts,* I:238–39.

350   **"economical . . . basis"**: Eaker Oral History, File number K239.0512-627, USAFHRA, p. 6.

350   **"one real defense"**: Greenwood, "Postwar Strategic Air Force," 218–19.

350   **Marshall and Eisenhower**: Wolk, *Air Force Independence,* vii, 38.

351   **"Mimeograph Machines" . . . "war of annihilation"**: "The B-36 Controversy in Retrospect" by Dr. Murray Free, SAFAS 1962, Congressional Hearings and Reports, Box 102, LeMay Papers, LOCMD, pp. 20–27.

351   **"Must the Italian Douhet"**: Futrell, *Ideas, Concepts,* I:251–52.

352   **"end of the runway"**: Ibid., I:237.

352   **Chief of Naval Operations was fired**: Nalty, ed., *Winged Shield,* I:424.

352   **only thirteen atomic bombs**: Cochran et al., *U.S. Warhead Production,* 14–15.

352   **SAC had no fighters**: Borowski, *Hollow Threat,* 81, 104, 166–67.

353   **"Air power romanticism":** Rostow, *United States in the World,* 224.
353   **"Our fellas":** McCullough, *Truman,* 649–50.
353   **George Kennan:** Futrell, *Ideas, Concepts,* I:238.
353   **"increase their will to fight":** Armitage and Mason, *Nuclear Age,* 183.
354   **German aircraft industry . . . Tu-4:** Ibid., 142–45.
354   **"only one thing worse":** Goldman, *Crucial Decade,* 100.
354   **Johnson . . . Truman:** Nalty, ed., *Winged Shield,* I:420.
355   **Heavy Bombardment Committee:** Moody, *Strategic Air Force,* 181.
355   **improvements . . . innovations:** Anderson, *Aircraft Performance,* 33–34.
355   **dismissed the jet:** Holley, "Jet Lag," 124–26; Schlaifer and Heron, *Aircraft Engines and Fuels,* 334–35.
355   **supersonic wind tunnel:** Von Kármán, *Wind and Beyond,* 223–24.
356   **wrote a term paper:** Early History of the Whittle Jet Propulsion Gas Turbine, AIR 62/4, PRO.
356   **daredevil pilot:** "Sir Frank Whittle," Obituary, *The Daily Telegraph,* August 10, 1996.
356   **Whittle's great innovation:** The Development of Jet Propulsion and Gas Turbine Engines in the United Kingdom, AIR 62/2, PRO, para. 3.
356–57   **"ram" effect . . . capital of £2,000:** Early History of the Whittle Jet Propulsion Gas Turbine, AIR 62/4, PRO.
357   **worst disadvantages:** The Development of Jet Propulsion and Gas Turbine Engines in the United Kingdom, AIR 62/2, PRO, para. 56.
357   **intervention of Henry Tizard:** Clark, *Tizard,* 96–98.
357   **only for "special purposes":** Report on Whittle Jet Propulsion System by A. A. Griffith, D. Eng., AIR 62/6, PRO, p. 12.
357   **£25 per hour:** The Development of Jet Propulsion and Gas Turbine Engines in the United Kingdom, AIR 62/2, PRO, para. 71.
358   **"war winning" . . . "Action This Day":** Clark, *Tizard,* 302–4.
358   **£100,000 ex gratia:** "Sir Frank Whittle," Obituary, *The Daily Telegraph,* August 10, 1996.
358   **Me 262's debut:** Price, "Messerschmitt 262," 55–57. The poor reliability of the Me 262's engines proved a more significant obstacle to getting the plane into operation than did Hitler's oft-cited order to have it reconfigured as a fighter-bomber.
358   **discourage risk-taking:** Holley, "Jet Lag," 135–36.
359   **Pratt & Whitney and Wright:** Whitford, *Fighter Design,* 73.
359   **"burble":** Stack, Lindsey, and Littell, *Compressibility Burble;* Anderson, *History of Aerodynamics,* 406.
359   **"suddenly the plane was gone":** Hallion, *Test Pilots,* 186.
360   **Jones had dropped out:** Anderson, *History of Aerodynamics,* 426.
360   **"hocus pocus":** Ibid., 427–28.
361   **Jones's findings:** Jones, *Pointed Wings.*
361   **Busemann:** Anderson, *History of Aerodynamics,* 425.
361   **proposed B-52:** Moody, *Strategic Air Force,* 238.
361   **Ulysses S. Grant:** Worden, *Fighter Generals,* 56.
362   **"this afternoon":** Ibid., 59.
362   **two miles. "Not one airplane":** Borowski, *Hollow Threat,* 166–67; Moody, *Strategic Air Force,* 233.
363   **"incompetent . . . unfortunate":** Worden, *Fighter Generals,* 62.
363   **manuals and checklists:** LeMay and Kantor, *Mission,* 439–40.
363   **drawing of their target:** Borowski, *Hollow Threat,* 169.
363   **with the perks to match:** Ibid., 178; Worden, *Fighter Generals,* 63; LeMay and Kantor, *Mission,* 437–39; Statement by General Curtis E. LeMay before the Banking and Currency Committee of the House of Representatives, 24 May 1956, Congressional Hearings and Reports, Box 102, LeMay Papers, LOCMD.

363   **bombing accuracy . . . SAC had tripled:** Strategic Air Command Progress Analysis, 1 November 1948 to 31 October 1953, SAC Progress Analyses, Box B98, LeMay Papers, LOCMD.

363   **San Francisco . . . "bombed":** LeMay and Kantor, *Mission*, 436.

364   **"not the slightest chance" . . . "everybody is going to lose":** Futrell, *Ideas, Concepts*, I:443–44.

365   **"Research and No Development":** Kaplan, *Wizards of Armageddon*, 57–59.

365   **Bernard Brodie:** Ibid., 45–47.

366   **"New hardware" . . . 44 percent:** Worden, *Fighter Generals*, 71, 94 n. 67, 112.

366   **"behave your damn selves":** LeMay and Kantor, *Mission*, 481.

366   **"kill the bastards!":** Kaplan, *Wizards of Armageddon*, 245–46.

366   **"designated ground zeroes":** The Operational Side of Air Offense, Remarks by General Curtis E. LeMay to the USAF Scientific Advisory Board at Patrick Air Force Base, 21 May 1957, Item B-60725, Top Secret Files, Box B206, LeMay Papers, LOCMD, p. 3.

366–67   **budget . . . "curse of bigness":** Worden, *Fighter Generals*, 66–67, 109–10.

367   **use a slide rule:** Holley, "Jet Lag," 138.

367   **panel under von Kármán:** Arnold, *Global Mission*, 532–33.

367   **"hogwash":** Tierney, "Take the A-Plane."

367   **LeMay pushed it:** The Operational Side of Air Offense, Remarks by General Curtis E. LeMay to the USAF Scientific Advisory Board at Patrick Air Force Base, 21 May 1957, Item B-60725, Top Secret Files, Box B206, LeMay Papers, LOCMD, p. 9.

367   **"if Russia beats us":** Tierney, "Take the A-Plane."

367   **"fantastic" talk:** Futrell, *Ideas, Concepts*, I:219–20.

368   **"Maginot Line" . . . number of ICBMs:** Worden, *Fighter Generals*, 119, 124.

368   **26,500 warheads:** Clodfelter, *Limits of Air Power*, 27.

368   **performed "marvelously" . . . "ugly mistake":** Hughes, *Overlord*, 311–12.

369   **staff of sixty-six:** Futrell, "Air Power Concept," 259; Worden, *Fighter Generals*, 39–40.

369   **"outnumbered and outvoted":** Worden. *Fighter Generals*, 93 n. 53.

369   **come directly from SAC:** Clodfelter, *Limits of Air Power*, 29.

369   **"afford the luxury":** Worden, *Fighter Generals*, 81; Futrell, *Ideas, Concepts*, I:441.

369   **Mustangs . . . "be relearned":** Armitage and Mason, *Nuclear Age*, 22–24

369–70   **army of oxcarts . . . caves:** Clodfelter, *Limits of Air Power*, 21–22; Armitage and Mason, *Air Power*, 43–44; Pape, *Bombing to Win*, 150.

370   **"fizzle":** Mark, *Aerial Interdiction*, 323.

370   **"aberration," a "special case":** Armitage and Mason, *Nuclear Age*, 44; Hamilton, "Doctrine since Vietnam," vi.

370   **"fatal to the United States" . . . a "fad":** Worden, *Fighter Generals*, 52 n. 72, 64–65.

370   **"obsolete explosives":** A Decade of Security Thru Airpower, Official Documents, Aircraft Operations, Box B96, LeMay Papers, LOCMD.

370   **"intelligence and resolution":** Futrell, *Ideas, Concepts*, I:451.

370   **"not American":** Clodfelter, *Limits of Air Power*, 25.

370   **deranged pronouncements:** Bendersky, *The "Jewish Threat,"* 404–9.

371   **"war fit a weapon":** Futrell, "Air Power Concept," 269.

371   **21,000 feet . . . Yalu:** Mark, *Aerial Interdiction*, 298, 323–24; Armitage and Mason, *Nuclear Age*, 29–31, 43.

371   **thirteen-to-one . . . "trains":** Mark, *Aerial Interdiction*, 291–93, 313–14.

372   **"strategic umbrella":** Worden, *Fighter Generals*, 61.

372   **"best preparation":** Clodfelter, *Limits of Air Power*, 30–31.

372   **"got the atom bomb":** Liebling, *Earl of Louisiana*, 121.

372   **"token" surface forces:** The Operational Side of Air Offense, Remarks by General Curtis E. LeMay to the USAF Scientific Advisory Board at Patrick Air Force Base, Florida, 21 May 1957, Item B-60725, Top Secret Files, Box B206, LeMay Papers, LOCMD, pp. 13–14.

372   **high-explosive ordnance:** Futrell, "Air Power Concept," 266.

372   **surface-to-surface missiles:** Futrell, *Ideas, Concepts,* I:441–42.

373   **"accumulate the initial effect":** Tabs D and E, Tactical Air Operations, Presented to the Review Board June 21, 1949, by Tactical Air Command, Speeches and Printed Matter File, Box 7, Quesada Papers, LOCMD.

373   **By 1957 two-thirds:** Futrell, *Ideas, Concepts,* I:532.

374   **"We had better get at it":** Letter, Vannevar Bush to General Omar N. Bradley, April 13, 1950, Item B-4560, Top Secret File, Box B195, LeMay Papers, LOCMD.

374   **"more powerful the weapons"** . . . **"all-out war":** Kissinger, *Nuclear Weapons,* 1, 7–8, 29.

374–75   **"certain to increase"** . . . **mobility and flexibility:** Ibid., 11, 29, 125–31.

375   **RAND study . . . keep it to themselves:** Clodfelter, *Limits of Air Power,* 32–34.

375   **"inevitable nuclear confrontation":** Worden, *Fighter Generals,* 149.

13. Hard Knocks

page

376   **"raggedy-ass":** Manchester, *Glory and Dream,* 1040.

376   **seen . . . as colonialists:** Sheehan, *Bright Shining Lie,* 144.

376   **"sense of inevitable victory":** Halberstam, *Best and Brightest,* 123.

377   **"Whiz Kids":** Ibid., 226–29.

377   **"Every quantitative measurement":** Sheehan, *Bright Shining Lie,* 289–90.

377   **masquerading as . . . objectivity:** Halberstam, *Best and Brightest,* 234–35.

377   **forbidden any offensive actions:** Sheehan, *Bright Shining Lie,* 122–24.

377–78   **million refugees . . . Diem's South Vietnamese Army:** Ibid., 136–37; Halberstam, *Best and Brightest,* 149–50.

378   **"mood":** Clodfelter, *Limits of Air Power,* 52.

378   **"clarify and simplify":** Mrozek, *Air Power in Vietnam,* 62.

378   **"test case":** Clodfelter, *Limits of Air Power,* 41.

378   **"*future* destruction":** Pape, *Bombing to Win,* 179.

379   **"signal U.S. determination"** . . . **"worse to come":** Drew, *Rolling Thunder,* 32–33.

379   **"general frustration":** Clodfelter, *Limits of Air Power,* 82–83.

379   **"Stone Age":** LeMay and Kantor, *Mission,* 565.

379   **"any two-week period":** Clodfelter, *Limits of Air Power,* 206.

379   **Johnson's mistrust:** Drew, *Rolling Thunder,* 48.

379   **"hornswoggled":** Thompson, *Hanoi and Back,* 137.

379   **"outdoing Hitler":** Entries for June 1, 1945, and June 6, 1945, Stimson, *Diaries,* reel 9.

379   **"Bomb, bomb, bomb":** Halberstam, *Best and Brightest,* 564.

380   **"bomb an outhouse":** Karnow, *Vietnam,* 415.

380   **were off-limits:** Thompson, *Hanoi and Back,* 75–76; Clodfelter, *Limits of Air Power,* 85–86.

381   **twentieth parallel . . . "next logical targets":** Pape, *Bombing to Win,* 186; Clodfelter, *Limits of Air Power,* 80.

381   **allotment of sorties:** Clodfelter, *Limits of Air Power,* 86, 130.

381   **"pussy-footing":** Ibid., 145.

381   **242 targets:** Pape, *Bombing to Win,* 188.

381   **"all worthwhile fixed targets":** Thompson, *Hanoi and Back,* 68.

381   ***never* have worked . . . "doctrinal void":** Clodfelter, *Limits of Air Power,* 145; Drew, *Rolling Thunder,* 47–48.

382   **already in the Stone Age:** Halberstam, *Best and Brightest,* 462.

382   **12 percent of . . . GNP:** Drew, *Rolling Thunder,* 36.

382   ***more* vulnerable:** Clodfelter, *Limits of Air Power,* 125–26.

382   **"bicycle brigades":** Salisbury, *Behind the Lines,* 89.

382–83   **oil-storage . . . 7,200 tons a day:** Clodfelter, *Limits of Air Power,* 111–12, 135–36.

383   **750 feet:** Drew, *Rolling Thunder,* 37–38.

383   **"enormous weight" . . . "ocarina":** Thompson, *Hanoi and Back,* 44–45.

383   **"accidental gore":** Hoeffding, *Bombing North Vietnam,* 17.

383   **eight million tons:** Thompson, *Hanoi and Back,* 301; Clodfelter, *Limits of Air Power,* 129.

384   **"McNamara's Wall":** Nalty, ed., *Winged Shield,* II:291–92; Rosenau, *Special Forces,* 11–13.

384   **never fell below . . . minimum:** Gilster, *Southeast Asia,* 20–21, 24–27.

385   **gunships:** Mrozek, *Air Power in Vietnam,* 126–27; Werrell, "USAF Technology in Vietnam," 91–93.

386   **"organic" . . . loophole:** Wolf, ed., *Roles and Missions,* 241–45; Futrell, *Ideas, Concepts,* II:516–17.

387   **"Air Cavalry":** Weinert, *Army Aviation,* 181.

387   **concentrated . . . counterattack:** Everett-Heath, *Helicopters in Combat,* 23.

387   **antitank weapons . . . air-mobile units:** Weinert, *Army Aviation,* 159–63.

387   **astonished:** Mrozek, *Air Power in Vietnam,* 31.

387–88   **entire battalion . . . 20mm cannons:** Everett-Heath, *Helicopters in Combat,* 50–52, 60–61.

388   **UH-1 . . . ACH-47:** Flintham, *Air Wars and Aircraft,* 294–95.

388   **"overenthusiasm":** Sbrega, "Southeast Asia," 455.

388   **"direct aerial fire support":** Costello, *Close Air Support,* 21; Futrell, *Ideas, Concepts,* II:518.

388   **Ap Bac:** Everett-Heath, *Helicopters in Combat,* 72.

388   **Lam Son 719:** Armitage and Mason, *Nuclear Age,* 91–92.

389   **2,400 helicopters:** Everett-Heath, *Helicopters in Combat,* 111.

390   **fuel tanks . . . 400 had been lost:** Werrell, "USAF Technology in Vietnam," 89–91.

391   **smoke trail:** Fallows, *National Defense,* 96.

391   **all three missiles:** Thompson, *Hanoi and Back,* 91; Hallion, *Storm over Iraq,* 29–30; Whitford, *Fighter Design,* 194–96.

392   **ran out of bombs:** Hallion, *Storm over Iraq,* 20.

392   **"agonizing relearning":** Futrell, *Ideas, Concepts,* II:288.

393   **"political ploy":** Ibid., II:289.

393   **flak trap:** Thompson, *Hanoi and Back,* 35–36.

393   **"MiG in there":** Ibid., 98–100.

393   **Wild Weasels:** Ibid., 105, 205.

394   **"get out of hand":** Clodfelter, *Limits of Air Power,* 154.

394   **fifteen B-52s:** Nordeen, *Missile Age,* 60; Thompson, *Hanoi and Back,* 264–65.

394   **lethality of . . . AAA:** Nordeen, *Missile Age,* 27–28; Thompson, *Hanoi and Back,* 311.

395   **"war that is finished":** Karnow, *Vietnam,* 667.

395   **"timorous military amateurs":** Clodfelter, *Limits of Air Power,* 207.

395   **"additional evidence":** Mrozek, *Air Power in Vietnam,* 22.

396   **hundred or more trucks:** Nordeen, *Missile Age,* 43.

396   **"Neither the war":** Clodfelter, *Limits of Air Power,* 146.

396   **doctrine manuals:** Hamilton, "Doctrine since Vietnam," xxii–xxiii.

396   **"ruining SAC" . . . B-1:** Worden, *Fighter Generals,* 195, 219.

396   **"comic book":** Drew, "Air Power Wilderness."

397   **"bound to be pyrrhic":** Muravchik, "Vietnam Paradigm," 17.

397   **every four-star:** Worden, *Fighter Generals,* xi n. 3, 226.

398   **Red Flag:** Hallion, *Storm over Iraq,* 32–33.

398   **"unprecedented cooperative effort":** Lambeth, *Transformation,* 83–85.

398   **Project Warrior:** Hamilton, "Doctrine since Vietnam," xiii, xxvi, xxx.

398   **McPeak observed:** Lambeth, *Transformation,* 90.

398   **"24 karat pain":** Hammond, *Mind of War,* 2–6.

398   **thirteen-hour briefing:** Ibid., 120–21; Fallows, *National Defense,* 28.

399   **"AVOID GROUND":** Thompson, *Hanoi and Back,* 245–46.

399   **super F-111:** Hallion, *Storm over Iraq,* 37.

400   **"do better than this":** Hammond, *Mind of War,* 77–79.

401   **"Not a pound":** Hallion, *Storm over Iraq,* 291.

401   **"procure the F-16":** Hammond, *Mind of War,* 95–96.

401   **forty hours of maintenance . . . every twelve minutes:** Whitford, *Fighter Design,* 20; Fallows, *National Defense,* 41.

401   **simplistic nostalgia:** Hallion, *Storm over Iraq,* 52.

402   **half as much maintenance:** Lambeth, *Transformation,* 81; Whitford, *Fighter Design,* 20.

402   **Soviet field army . . . half-life:** Deitchman, "Future of Tactical Air," 37–38.

403   **two Syrian helicopters:** Nordeen, *Fighters over Israel,* 166.

403–4 **204 aircraft . . . "thought I'd be court-martialed":** Oren, *Six Days of War,* 174–75.

404   **Christmas Day 1969:** Nordeen, *Fighters over Israel,* 102–3.

404   **massive buildup:** Nordeen, *Missile Age,* 123–24, 127–29.

404   **14 percent . . . "have to find an answer":** Armitage and Mason, *Nuclear Age,* 127; Nordeen, *Missile Age,* 143; Grant, "Bekaa Valley," 59.

404   **meticulously rehearsed:** Tice, "Unmanned Aerial Vehicles"; Grant, "Bekaa Valley"; Nordeen, *Missile Age,* 161.

404   **"Syrians were waiting":** Nordeen, *Fighters over Israel,* 170–71.

405   **eighty-five Syrian jets:** Lambeth, *Lebanon Air War,* 9–10.

405   **"customers wear wings":** Clark, "Uninhabited Aerial Vehicles," 31.

14. Precision, at Last

page

406   **"magic weapon":** Fallows, *National Defense,* 56–57.

406   **two dozen . . . projects:** National Defense Research Committee, *Guided Missiles;* Summary of Projects on Jet Propulsion and Guided Missiles, 16 October 1944, Bulky Files, Oct. 1942–May 1944, 452.1 Airplanes, Entry 294B, RG 18, NACP.

406   **AZON:** Mets, *Surgical Strike,* 16–20.

407   **Army Ordnance Department . . . low-drag shapes:** Ibid., 34–39.

407   **summer think-tank:** Ibid., 50–51.

407   **Air Staff study . . . "Detachment 5":** Ibid., 47–56.

408   **Colonel Joseph Davis:** Loeb, "Bursts of Brilliance."

408   **28,000 Paveways:** U.S. Air Force, *Statistical Digest,* Table 37.

408   **twenty feet . . . $8,000:** Mets, *Surgical Strike,* 67, 71, 83.

408   **run two locomotives:** Thompson, *Hanoi and Back,* 234–35.

408   **871 sorties . . . targets that had been off-limits:** Mets, *Surgical Strike,* 85–90.

409   **80 to 90 percent . . . $167 billion:** Deitchman, "Revolution in Conventional Weapons," 36; Lambeth, *Transformation,* 76; Deitchman, "Weapons, Platforms," 89.

409   **a bargain . . . 800 tanks:** Deitchman, "Future of Tactical Air," 43; Deitchman, *Military Power and Technology,* 46; Deitchman, "Weapons, Platforms," 93.

410   **"how stupid we are":** Author's interview with Glenn Kent, February 1987.

410   **"incalculable chain":** Rogers, "Follow-on Forces Attack."

412   **"revolution in lethality":** Lambeth, *Transformation,* 74–75.

412   **JSTARS:** Hallion, *Storm over Iraq,* 310–11.

412   **LANTIRN . . . Vietnam-vintage:** Winnefeld, Niblack, and Johnson, *League of Airmen,* 73, 271.

413   **sent three patrols:** Gordon and Trainor, *Generals' War,* 51.

413   **retaliatory air strikes:** Reynolds, *Heart of Storm,* 22–24.

413   "strategic vision" . . . "slogan": Faulkenberry, *Global Reach*, 16, 21 n. 19.
414   "five-ring model": Warden, "Enemy as System."
414   throwback: West, *Warden and Tactical School*, 33.
414   "1941–1991": Clodfelter, "Pinpointing Devastation," 101.
414   "using air power!": Reynolds, *Heart of Storm*, 28–29.
415   "the Iraqi populace": Pape, *Bombing to Win*, 223.
415   "Incapacitated" . . . Inchon: Gordon and Trainor, *Generals' War*, 82, 89; Pape, *Bombing to Win*, 221.
415   "Now, General": Reynolds, *Heart of Storm*, 72.
415   "flyboy promises": Atkinson, *Crusade*, 60.
415   "smoking tanks": Reynolds, *Heart of Storm*, 72–73.
415   "academic bunch of crap": Ibid., 49.
415   "bomber mentality": Winnefeld, Niblack, and Johnson, *League of Airmen*, 68.
415   "air power airheads": Atkinson, *Crusade*, 61.
416   briefing slides: Reynolds, *Heart of Storm*, 91–92.
416   winced . . . "very, very patient": Atkinson, *Crusade*, 62; Reynolds, *Heart of Storm*, 128.
416   Edward Luttwak: Muravchik, "Vietnam Paradigm," 18–19.
416   "hope to win": Gordon and Trainor, *Generals' War*, 179.
417   "cocked fist": Ibid., 189.
417   fully 67 percent: *GWAPS*, II(II):148.
417   more than 200: *GWAPS*, I(I):146.
417   "systematic collapse": Lambeth, *Transformation*, 147.
417   by an angry populace: *GWAPS*, I(I):93.
418   Republican Guard: Winnefeld, Niblack, and Johnson, *League of Airmen*, 130.
418   cheer went up: Gordon and Trainor, *Generals' War*, 216.
418   "my counterpart's headquarters": Hallion, *Storm over Iraq*, 174.
419   "Only from close up": Ibid., 200.
419   spent billions of dollars: *GWAPS*, II(II):130–34; Gordon and Trainor, *Generals' War*, 102–8.
420   Larry Henry: Gordon and Trainor, *Generals' War*, 111–12.
420   Helicopters were chosen: Everett-Heath, *Helicopters in Combat*, 37.
420–21   Nukhayb . . . "gorilla" forces: *GWAPS*, II(I):122–33.
421   A-10s: Gordon and Trainor, *Generals' War*, 121.
421   radar emissions . . . 90 percent: *GWAPS*, II(I):156
421   attrition rate: *GWAPS*, II(II):114–16, 142; Overy, *Air War*, 77; Winnefeld, Niblack, and Johnson, *League of Airmen*, 312.
422   "any of my captains": Mackenzie, "Chuck Horner," 60.
422   40 seconds: Nordeen, *Missile Age*, 215.
422   AWACS . . . "'Cleared for takeoff'": Lambeth, *Transformation*, 114–15.
422   aircraft shelters: *GWAPS*, II(II):149–57.
422   ceased flight operations: Ibid., II(II):120.
423   "cut it off": Ibid., II(II):159.
423   "Never has an air force": Ibid., II(I):249.
423   DESTROYING IT: Hallion, *Storm over Iraq*, 209.
423   road and rail bridges: Davis, *Decisive Force*, 42; Hallion, *Storm over Iraq*, 193.
423–24   a tenth . . . lack of food: *GWAPS*, II(II):187, 197–200.
424   thirty feet . . . two hundred feet: Hallion, *Storm over Iraq*, 212.
424   dozen LANTIRN: Winnefeld, Niblack, and Johnson, *League of Airmen*, 250 n. 16.
424   On February 5: *GWAPS*, II(I):204–5.
424   "laughed out of the room": Lambeth, *Transformation*, 125.
424   A-10 operations: *GWAPS*, II(I):279–80; Hallion, *Storm over Iraq*, 210–11.
425   dizzying pace: *GWAPS*, II(II):373–74, 391–92.

425   **"tank-plinking":** Hallion, *Storm over Iraq,* 217.
425   **Frederick M. Franks, Jr.:** Grier, "Joint STARS."
425   **"Ph.D.'s"** . . . **In one convoy:** Hallion, *Storm over Iraq,* 220–21.
425   **A prisoner . . . 300 tanks:** *GWAPS,* II(II):234–42.
425   **leaflets . . . 150,000 had deserted:** Hosmer, *Psychological Effects,* 147, 154, 163–64; *GWAPS,* II(II):220–21.
426   **1,400 tanks:** *GWAPS,* II(II):260–61.
426   **tank ammunition:** Atkinson, *Crusade,* 342.
426   **"untold inconveniences":** Davis, *Decisive Force,* 54.
426   **"synonymous with air power":** Lambeth, *Transformation,* 266.
427   **"distorted or misinterpreted":** Gordon and Trainor, *Generals' War,* 520 n. 4.
427   **"dangerously optimistic":** *GWAPS,* II(I):330–31.
427   **single top Iraqi leader:** Pape, *Bombing to Win,* 230.
427–28   **Scud-hunting . . . able to communicate:** U.S. General Accounting Office, *Desert Storm,* 151–55; *GWAPS,* II(II):276; Keaney and Cohen, *Revolution in Warfare,* 60.
428   **strikes on key production:** *GWAPS,* II(II):301, 304, 309.
428   **substitute . . . redundant capacity:** Gilster, "Desert Storm."
428   **"inherent features of strategic campaigns":** *GWAPS,* II(II):363–64.
428   **Wardenesque rhetoric:** See, for example, U.S. Air Force, *Aerospace Intelligence,* 37–38; this official Air Force publication, dated June 2001, offers Warden's five-ring model as a guide for identifying "high value targets" and asserts that attacking fielded forces represents a "high-cost and low-payoff strategy."
428   **ridiculed Warden's:** Hinman, *Politics of Coercion,* 20–21; Fadok, *Strategic Paralysis,* 27–29; West, *Warden and Tactical School,* 35.
429   **"In every major conflict":** Hosmer, *Psychological Effects,* xxi–xxii.
429   **in Baghdad . . . Powell ordered:** Gordon and Trainor, *Generals' War,* 326–27.
429   **"antiseptic":** Mackenzie, "Chuck Horner"; Keaney and Cohen, *Revolution in Warfare,* 214.
429   **interview on German television:** *GWAPS,* I(I):64.
430   **NATO political authorities. . . . "first requirement":** U.S. Air Force, *Air War over Serbia,* 10, 23.
430   **8 percent . . . in Desert Storm:** Keaney and Cohen, *Revolution in Warfare,* 191.
430   **accuracy . . . "extraordinary":** Arkin, "Smart Bombs."
430   **military objectives:** Beagle, *Effects-Based Targeting,* 72–79.
430   **bridges . . . presidential residence:** U.S. Department of Defense, *Kosovo Report,* 82.
431   **"killing things":** U.S. Air Force, *Air War over Serbia,* 25.
431   **Others pointed out:** "NATO's Most Lethal Airstrike Ended a Battle, Perhaps a War," *Washington Post,* June 26, 1999; Hinman, *Politics of Coercion,* 51.
432   **10 percent of . . . combat aircraft:** U.S. Department of Defense, *Kosovo Report,* 88.
432   **ATO . . . messenger plane:** Loeb, "Bursts of Brilliance," 26.
432   **nineteen minutes . . . New York Times reported:** "Use of Pinpoint Air Power Comes of Age in New War," *New York Times,* December 24, 2001.
433   **"heavyset American":** Ignatieff, "Nation-Building Lite."
434   **ten times more powerful:** "Intense, Coordinated Air War Backs Baghdad Campaign," *Washington Post,* April 6, 2003.
435   **Operation Anaconda:** Grant, "Airpower of Anaconda"; author's interview with Col. Tom Ehrhard, May 2003.
436   **Hybrid . . . "fast and final":** Author's interview with Col. Mace Carpenter, July 2003.
437   **59 "leadership" . . . 104 offices:** Ibid.
438   **29,200 bombs . . . 79 percent:** U.S. Air Force, *IRAQI FREEDOM,* 5, 11.
438   **"kill them faster" . . . Iraqi T-72 tanks:** "An Air War of Might, Coordination and Risks," *Washington Post,* April 27, 2003.
438   **seizure of the airfield:** Author's interview with Maj. Gen. Nick Williams, July 2003.

438   **18 and 44 percent:** "An Air War of Might, Coordination and Risks," *Washington Post,* April 27, 2003.

438   **still had bombs on board:** Author's interview with Brig. Gen. Dan Darnell, July 2003.

439   **air-power thinkers . . . predicted:** Bingham, "Theater Warfare"; Bingham, "Second Airpower Revolution"; Costello, *Close Air Support,* 62; Hinman, *Politics of Coercion,* 36.

439   **Global Hawk . . . "kill box":** Author's interview with Brig. Gen. Dan Darnell, July 2003.

439   **"I love UAVs!":** Moseley briefing, Prince Sultan Air Base, Saudi Arabia, April 5, 2003.

439   **terms . . . no longer made sense:** Meilinger, "New Airpower Lexicon."

440   **long, sweeping turn:** Miller, *War at Sea,* 532; *AAF,* V:734.

440   **"glistening B-29":** Kosaka, *100 Million Japanese,* 27.

441   **ever got that close:** Costello, *Close Air Support,* 31–32.

BIBLIOGRAPHY

Unpublished Sources

BUNDESARCHIV-MILITÄRARCHIV, FREIBURG, GERMANY (BA-MA)
N671 Personal war diary of Wolfram von Richthofen

IMPERIAL WAR MUSEUM, LONDON (IWM)
HTT Henry T. Tizard Papers
Personal memoirs, diaries, letters: Royal Flying Corps, Fighter Command, Bomber Command

LIBRARY OF CONGRESS MANUSCRIPT DIVISION, WASHINGTON, D.C. (LOCMD)
Henry Harley Arnold Papers
Curtis E. LeMay Papers
William Mitchell Papers
Elwood R. "Pete" Quesada Papers
Carl Spaatz Papers

NATIONAL ARCHIVES AT COLLEGE PARK, COLLEGE PARK, MARYLAND (NACP)
Records of the Army Air Forces, Record Group 18
Gorrell's History of the American Expeditionary Forces Air Service, 1917–1919. Records of the
    American Expeditionary Forces (World War I), Record Group 120
Records of Headquarters, United States Air Force, Record Group 341

PUBLIC RECORD OFFICE, KEW, UNITED KINGDOM (PRO)
ADM 223 Admiralty: Intelligence Reports and Papers
AIR 1 Air Historical Branch: Papers (Series I)
AIR 2 Air Ministry: Registered Files
AIR 5 Air Historical Branch: Papers (Series II)
AIR 9 Air Ministry: Directorate of Operations and Intelligence and Directorate of Plans: Registered
    Files
AIR 10 Air Publications and Reports
AIR 16 Air Ministry: Fighter Command: Registered Files
AIR 19 Air Ministry: Private Office Papers
AIR 22 Air Ministry: Periodical Returns, Intelligence Summaries, and Bulletins
AIR 40 Air Ministry, Directorate of Intelligence: Intelligence Reports and Papers
AIR 41 Air Historical Branch: Narratives and Monographs

AIR 62 Air Commodore Sir Frank Whittle: Papers
AVIA 7 Royal Radar Establishment and Predecessors: Registered Files

ROYAL AIR FORCE MUSEUM, LONDON (RAFM)
Hugh Trenchard Papers

U.S. AIR FORCE HISTORICAL RESEARCH AGENCY, MAXWELL AIR FORCE BASE, ALABAMA (USAFHRA)
USAFHRA Document Collection

Published Primary Sources (Official Reports, Contemporaneous Views, Memoirs, Reprints of Original Documents)

Adams, Henry. *The Education of Henry Adams.* 1918. Reprint. New York: Modern Library, 1931.
"Air-Raid Psychology and Air-Raid Perils." *The Lancet* 193 (1917): 540–41.
"Air Raids and Some Conclusions." *Flight* 9 (1917): 973–74.
Arnold, H. H. "Air Lessons from Current Wars." *U.S. Air Services,* May 1938, 17.
———. *Global Mission.* New York: Harper, 1949.
"The Attributes of the Successful Flying Officer." *The Lancet* 195 (1918): 425–26.
Beaverbrook, Max Aitken, Baron. *Men and Power, 1917–1918.* New York: Duell, Sloan, and Pearce, 1957.
Bingham, Hiram. *An Explorer in the Air Service.* New Haven, Conn.: Yale University Press, 1920.
Bolling, R. C. "Report of Aeronautical Commission." *Air Power Historian* 7 (1960): 222–32.
Bottomley, N. H. "The Work of the Royal Air Force on the North-West Frontier." *Journal of the Royal United Service Institution* 84 (1939): 769–80.
Bradley, Omar N. *A Soldier's Story.* New York: Henry Holt, 1951.
Bradley, Omar N., and Clay Blair. *A General's Life: An Autobiography.* New York: Simon & Schuster, 1983.
Breguet, Louis. "Aerodynamical Efficiency and the Reduction of Air Transport Costs." *Aeronautical Journal* 26 (1922): 307–13.
Brewers, Leighton. *Riders of the Sky.* Boston: Houghton Mifflin, 1934.
Buckley, Harold. *Squadron 95.* Paris: Obelisk, 1933.
Chamier, J. A. "The Use of the Air Force for Replacing Military Garrisons." *Journal of the Royal United Service Institution* 66 (1921): 205–16.
Chandler, C. De F. "The Lewis Aeroplane Gun." *Journal of the United States Artillery* 38 (1912): 222–23.
Chennault, Claire. *Way of a Fighter.* New York: Putnam, 1949.
Cherry, R. G. "The Aeroplane as a Long Range Gun." *Journal of the Royal Artillery,* June 1919, 129–36.
Churchill, Winston S. *My Early Life: A Roving Commission.* New York: Charles Scribner's Sons, 1930.
———. *The Second World War.* 6 vols. New York: Houghton Mifflin, 1948–1953.
Coatalen, Louis. "Aircraft and Motor Car Engine Design." *The Aeroplane* 12 (May 16, 1917): 1254–64.
Cochrane, R. A. "The Work of the Royal Air Force at Aden." *Journal of the Royal United Service Institution* 76 (1931): 88–103.
Cohen, Eliot A. *Gulf War Air Power Survey.* 5 vols. Washington, D.C.: GPO, 1993.
DeFrance, Smith J. *The N.A.C.A. Full-Scale Wind Tunnel.* Report No. 459. National Advisory Committee for Aeronautics, 1934.
Douhet, Giulio. *The Command of the Air.* Translated by Dino Ferrari. New York: Coward-McCann, 1942. Reprint. Washington, D.C.: Office of Air Force History, 1983.
Drum, Karl. *The German Air Force in the Spanish Civil War.* Air Force Historical Studies No. 150. Maxwell Air Force Base, Ala., 1957.

Edmonds, C. H. K. "Air Strategy." *Journal of the Royal United Service Institution* 64 (1924): 191–208.

"The 80 H.P. British-Built Gnome Engine." *Flight* 6 (1914): 1159–61.

Finne, K. N. *Igor Sikorsky: The Russian Years.* 1930. Translated and adapted by Von Hardesty. Washington, D.C.: Smithsonian Institution Press, 1987.

Flugblatt-Propaganda im 2. weltkrieg. Vol. 2: *Aus Deutschland 1939/40.* Erlangen, Germany: Verlag D + C, 1982.

"Foreign Notes." *Aero and Hydro,* September 28, 1912, 562.

Foulois, Benjamin D. "Early Flying Experiences." *Air Power Historian* 2 (1955): 17–35, 45–65.

"General Report of the Commission of Jurists at The Hague." *American Journal of International Law* 17 Supplement (1923): 249–51.

Gilbert, Martin, ed. *The Churchill War Papers.* New York: Norton, 1993.

Gilium, E. M. " 'The Beast': Living with the 3350: A History of the How and Why of the Marriage of the B-29 and the R-3350." *Journal of the American Aviation Historical Society,* Fall 2000, 231–37.

Gill, Brendan. "Profiles: Young Man Behind Plexiglass." *New Yorker,* August 12, 1944, 26–37.

Goldstein, Donald M., and Katherine V. Dillon, eds. *The Pearl Harbor Papers: Inside the Japanese Plans.* Washington, D.C.: Brassey's, 1993.

Gorrell, Edgar S. "An American Proposal for Strategic Bombing in World War I." *Air Power Historian* 5 (1958): 102–17.

Goudsmit, Samuel A. *Alsos.* New York: Henry Schuman, 1947.

Great Britain. Air Ministry. *Synopsis of British Air Effort during the Great War.* London· HMSO, 1919.

[Grey, C. G.] "The General Staff of the War Office and Aviation." *The Aeroplane* 1 (November 23, 1911): 583–84.

———. "Lord Haldane and Military Aviation." *The Aeroplane* 1 (December 14, 1911): 651–52.

———. "Aircraft in the War." *The Aeroplane* 7 (August 26, 1914): 191–94.

———. "More R.A.F. Morals." *The Aeroplane* 10 (May 10, 1916): 742–43.

[Grider, John MacGavock.] *War Birds: Diary of an Unknown Aviator.* New York: George H. Doran, 1926.

Groves, P. R. C. "For France to Answer." *The Atlantic Monthly.* February 1924, 145–53.

Harris, Arthur. *Bomber Offensive.* London: Collins, 1947.

Hearne, R. P. *Airships in Peace and War.* London: John Lane, 1910.

———. *Zeppelins and Super-Zeppelins.* London: John Lane, 1916.

Hicks, W. Joyson. "The Command of the Air." *National Review* (London) 59 (1912): 346–58.

Hoeffding, Oleg. *Bombing North Vietnam: An Appraisal of Economic and Political Effects.* RAND Memorandum RM-5213, December 1966.

Hopper, Bruce C. "American Day Bombardment in World War I." *Air Power Historian* 4 (1957): 87–97.

Horikoshi, Jiro. *Eagles of Mitsubishi: The Story of the Zero Fighter.* Translated by Shojiro Shindo and Harold N. Wantiez. Seattle: University of Washington Press, 1981.

Huston, John W., ed. *American Airpower Comes of Age: General Henry H. "Hap" Arnold's World War II Diaries.* 2 vols. Maxwell Air Force Base, Ala.: Air University Press, 2002.

Jackson, Louis. "The Defence of Localities Against Aerial Attack." *Journal of the Royal United Service Institution* 58 (1914): 701–26.

Jane's Fighting Aircraft of World War I. Reprint. London: Random House, 2001.

Jardine, Douglas J. "The Mad Mullah of Somaliland." *African Society Journal* 208 (July 1920): 109–21.

Jones, B. Melvill. "The Streamline Aeroplane." *Journal of the Royal Aeronautical Society* 33 (1929): 358–71.

Jones, Robert T. *Properties of Low-Aspect-Ratio Pointed Wings at Speeds below and above the Speed of Sound.* Report No. 835, National Advisory Committee for Aeronautics. Langley Field, Va., 1946.

Jones, R. V. *The Wizard War: British Scientific Intelligence 1939–1945*. New York: Coward, McCann & Geoghegan, 1978.

Kelly, Fred C., ed. *Miracle at Kitty Hawk: The Letters of Wilbur and Orville Wright*. New York: Farrar, Straus and Young, 1951.

Kenney, George. "The Airplane in Modern Warfare." *U.S. Air Services,* July 1938, 17.

Kissinger, Henry. *Nuclear Weapons and Foreign Policy*. New York: Harper, 1957.

LeMay, Curtis E., and McKinlay Kantor. *Mission with LeMay*. Garden City, N.Y.: Doubleday, 1965.

Lewis, Cecil. *Sagittarius Rising*. New York: Harcourt, Brace, 1936.

Liddell Hart, B. H. *Paris: Or the Future of War*. New York: Dutton, 1925.

———. *When Britain Goes to War: Adaptability and Mobility*. London: Faber and Faber, 1935.

McFarland, Marvin W., ed. *The Papers of Wilbur and Orville Wright*. 2 vols. New York: McGraw-Hill, 1953. Reprint. North Stratford, N.H.: Ayer, 1998.

Mackintosh, James M. *The War and Mental Health in England*. New York: The Commonwealth Fund, 1944.

Marconi, Guglielmo. "Radio Telegraphy." *Proceedings of the Institute of Radio Engineers* 10 (1922): 215–38.

Matsuo, Kinoaki. *How Japan Plans to Win*. Translated by Kilsoo K. Haan. Boston: Little, Brown, 1942.

Mitchell, William. "The Bombing of the Battleships." *Air Power Historian* 4 (1957): 51–65.

———. *Memoirs of World War I*. New York: Random House, 1960.

Monsarrat. Nicholas. *Breaking In, Breaking Out*. New York: Morrow, 1971.

Munk, Max M., and Elton W. Miller. *Model Tests with a Systematic Series of 27 Wing Sections at Full Reynolds Number*. Report No. 221, National Advisory Committee for Aeronautics. Langley Field, Va., 1926.

———. *The Variable Density Wind Tunnel of the National Advisory Committee for Aeronautics*. Report No. 227, National Advisory Committee for Aeronautics. Langley Field, Va., 1926.

National Defense Research Committee. *Guided Missiles and Techniques*. Summary Technical Report of Division 5, NDRC, Vol. 1. Washington, D.C., 1946.

Operation Epsilon: The Farm Hall Transcripts. Berkeley, Calif.: University of California Press, 1993.

Peck, R. H. "Aircraft in Small Wars." *Journal of the Royal United Service Institution* 73 (1928): 535–50.

Poulain, Didier. "Aircraft and Mechanized Land Warfare: The Battle of Guadalajara, 1937." *Journal of the Royal United Service Institution* 83 (1938): 362–67.

———. "The Role of Aircraft in the Spanish Civil War." *Journal of the Royal United Service Institution* 83 (1938): 581–86.

Poutrin, Lieutenant breveté. "Les aéroplanes dans la guerre future." *Revue générale de l'aeronautique militaire* 1 (1911): 382–85.

"The R.F.C. in the War." *Flight* 6 (1914): 951–52.

Rickenbacker, Edward V. *Fighting the Flying Circus*. New York: Frederick A. Stokes, 1919.

Rippon, T. S., and E. G. Manuel. "The Essential Characteristics of Successful and Unsuccessful Aviators." *The Lancet* 195 (1918): 411–15.

Robida, Albert. *La guerre au vingtième siècle*. Paris, 1887.

Rogers, Bernard. "Follow-On Forces Attack (FOFA): Myths and Realities." *NATO Review,* December 1984, 1–9.

Rowe, A. P. *One Story of Radar*. Cambridge: Cambridge University Press, 1948.

Salisbury, Harrison E. *Behind the Lines—Hanoi*. New York: Harper & Row, 1967.

Scott, Riley E. "Dropping Bombs from Flying Machines." *Scientific American,* October 28, 1911, 388–89.

———. "Can the Panama Canal Be Destroyed from the Air?" *Sunset* 32 (1914): 775–84.

"Scott's Bomb-Dropper Tried in France." *Aero,* March 9, 1912, 459.

Shirer, William L. *Twentieth Century Journey*. New York: Simon & Schuster, 1976.

Sikorsky, Igor I. *The Story of the Winged-S*. Revised edition. New York: Dodd, Mead, 1958.

Slessor, John. *The Central Blue: The Autobiography of Sir John Slessor, Marshal of the R.A.F.* New York: Praeger, 1957.

Smith, J. "The Development of the Spitfire and Seafire." *Journal of the Royal Aeronautical Society* 51 (1947): 339–83.

Spaight, J. M. *Air Power and War Rights.* London: Longmans, 1924.

Speer, Albert. *Inside the Third Reich.* Translated by Richard and Clara Winston. New York: Macmillan, 1970.

Springs, Elliott White. *Nocturne Militaire.* New York: George H. Doran, 1927.

Squier, George O. "The Present Status of Military Aeronautics." *Flight* 1 (1909): 121.

Stack, John, W. F. Lindsey, and Robert E. Littell. *The Compressibility Burble and the Effect of Compressibility on Pressures and Forces Acting on an Airfoil.* Report No. 646, National Advisory Committee for Aeronautics. Langley Field, Va., 1939.

Stiles, Bert. *Serenade to the Big Bird.* New York: Norton, 1947.

Stimson, Henry L. *The Henry Lewis Stimson Diaries in the Yale University Library.* New Haven, Conn.: Manuscripts and Archives, Yale University Library, 1973. Microfilm.

Strauss, E. B. "The Psychological Effects of Bombing." *Journal of the Royal United Service Institution* 84 (1939): 267–82.

Tinker, F. G., Jr. *Some Still Live.* New York: Funk & Wagnalls, 1938.

Tizard, H. T. "Methods of Measuring Aircraft Performance." *Aeronautical Journal* 21(2): 2–16 (April–June 1917).

Trenchard, H. M. "Despatch from Major-General Sir H. M. Trenchard, K.C.B., D.S.O., Commanding the Independent Force, Royal Air Force." *London Gazette,* Tenth Supplement, December 31, 1918, 133–38.

U.S. Air Force. *USAF Statistical Digest 1974.* Washington, D.C., 1975.

———. *Aerospace Intelligence Preparation of the Battlespace.* Air Force Pamphlet 14–118. Washington, D.C., 2001.

———. *Operation IRAQI FREEDOM—By the Numbers.* 2003.

U.S. Department of Defense. *Report to Congress: Kosovo/Operation Allied Force After-Action Report.* Washington, D.C., 2000.

U.S. Strategic Bombing Survey. *Summary Report (European War).* Report No. 1. Washington, D.C., 1945.

———. *Over-all Report (European War).* Report No. 2. Washington, D.C., 1945.

———. *The Effects of Strategic Bombing on the German War Economy.* Report No. 3. Washington, D.C., 1945.

———. *Machine Tools and Machinery as Capital Equipment.* Report No. 54. Washington, D.C., 1947.

U.S. Strategic Bombing Survey [Pacific War]. *The Effects of the Atomic Bombs on Hiroshima and Nagasaki.* Report No. 3. Washington, D.C., 1946.

———. *The Effects of Incendiary Bomb Attacks on Japan.* Report No. 90. Washington, D.C., 1947.

Von Kármán, Theodore. *The Wind and Beyond.* Boston: Little, Brown, 1967.

"War à la Douhet." *Infantry Journal* 44 (1937): 432–33.

Warden, John. "The Enemy as a System." *Airpower Journal* 9(1): 40–55 (Spring 1995).

Weick, Fred E. *Drag and Cooling with Various Forms of Cowling for a "Whirlwind" Radial Air-Cooled Engine—1.* Report No. 313, National Advisory Committee for Aeronautics, 1930.

Wells, H. G. "Anticipations: An Experiment in Prophesy: War." *North American Review,* September 1901, 401–12.

———. *War in the Air.* London, 1908.

———. *Experiment in Autobiography.* New York: Macmillan, 1934.

Wilson, Donald. "Origin of a Theory for Air Strategy." *Aerospace Historian* 18 (1971): 19–25.

Wodehouse, P. G. *The Swoop! Or How Clarence Saved England: A Tale of the Great Invasion.* London: Alston Rivers, 1909.

Wolf, Richard I., ed. *United States Air Force Basic Documents on Roles and Missions.* Washington, D.C.: Office of Air Force History, 1987.

Wood, Donald H. *Tests of Nacelle-Propeller Combinations in Various Positions with Reference to Wings.* Report No. 415. National Advisory Committee for Aeronautics, 1933.

Wooldridge, E. T., ed. *Carrier Warfare in the Pacific: An Oral History Collection.* Washington, D.C.: Smithsonian Institution Press, 1993.

Zuckerman, Solly. *From Apes to Warlords.* New York: Harper & Row, 1978.

Secondary Sources

Abzug, Malcolm J., and E. Eugene Larrabee. *Airplane Stability and Control: A History of the Technologies That Made Aviation Possible.* Cambridge: Cambridge University Press, 1997.

Anderson, John David. *A History of Aerodynamics and Its Impact on Flying Machines.* Cambridge: Cambridge University Press, 1997.

———. *Aircraft Performance and Design.* Boston: WCB/McGraw-Hill, 1999.

Arkin, William. "Smart Bombs, Dumb Targetting." *Bulletin of the Atomic Scientists,* May–June 2000, 46–53.

Armitage, M. J., and R. A. Mason. *Air Power in the Nuclear Age.* Urbana, Ill.: University of Illinois Press, 1983.

Astor, Bruce. *The Mighty Eighth: The Air War in Europe as Told by the Men Who Fought It.* New York: Dell, 1998.

Atkinson, Rick. *Crusade: The Untold Story of the Persian Gulf War.* Boston: Houghton Mifflin, 1993.

Beagle, T. W., Jr. *Effects-Based Targeting: Another Empty Promise?* Maxwell Air Force Base, Ala.: Air University Press, 2001.

Bekker, Cajus [pseud. for Hans Dieter Berenbrok]. *The Luftwaffe War Diaries.* Translated by Frank Ziegler. London: Macdonald, 1967.

Bendersky, Joseph W. *The "Jewish Threat": Anti-Semitic Politics of the U.S. Army.* New York: Basic Books, 2000.

Bennett, Ralph. *Behind the Battle: Intelligence in the War with Germany 1939–1945.* Revised edition. London: Pimlico, 1999.

Bialer, Uri. *The Shadow of the Bomber: The Fear of Air Attack and British Politics, 1932–1939.* London: Royal Historical Society, 1980.

Bingham, Price T. "Theater Warfare, Movement, and Airpower." *Airpower Journal* 12(2): 15–26 (Summer 1998).

———. "The Second Airpower Revolution." *Marine Corps Gazette* 84(1): 23–26 (January 2000).

Bix, Herbert P. *Hirohito and the Making of Modern Japan.* New York: HarperCollins, 2000.

Bogart, Charles H. "German Remotely Piloted Bombs." *Proceedings of the United States Naval Institute,* November 1976, 62–68.

Boog, Horst. "German Air Intelligence in World War II." *Aerospace Historian* 33 (1986): 121–29.

Borowski, Harry. *A Hollow Threat: Strategic Air Power and Containment Before Korea.* Westport, Conn.: Greenwood Press, 1982.

Boyne, Walter J. "Technological Progress in World War One Aviation." American Institute of Aeronautics and Astronautics, Paper No. 78-3002.

Buckley, John. *Air Power in the Age of Total War.* Bloomington, Ind.: Indiana University Press, 1999.

Budiansky, Stephen. *Battle of Wits: The Complete Story of Codebreaking in World War II.* New York: Free Press, 2000.

Burg, David F. *Chicago's White City of 1893.* Lexington, Ky.: University Press of Kentucky, 1976.

Burns, James McGregor. *Roosevelt: The Soldier of Freedom.* New York: Harcourt Brace Jovanovich, 1970.

Cameron, Rebecca Hancock. *Training to Fly: Military Flight Training, 1907–1945.* Washington, D.C.: Air Force History and Museums Program, 1999.

Casari, Robert B. *U.S. Army Aviation Serial Numbers and Orders 1908–1923 Reconstructed.* Apollo, Pa.: Closson Press, 1995.

Castle, H. G. *Fire over England: The German Air Raids of World War I.* London: Leo Cooper, 1982.

Chambers, John Whitelay, II. "The American Debate over Modern War, 1871–1914." In *Anticipating Total War: The German and American Experiences, 1871–1914,* edited by Manfred F. Boemeke, Roger Chickering, and Stig Förster. Cambridge: Cambridge University Press, 1999.

Christienne, Charles, and Pierre Lissarrague. *A History of French Military Aviation.* Trans. by Frances Kianka. Washington, D.C.: Smithsonian Institution Press, 1986.

Clark, Richard M. *Uninhabited Combat Aerial Vehicles.* CADRE Paper No. 8. Maxwell Air Force Base, Ala.: Air University Press, 2000.

Clark, Ronald W. *Tizard.* Cambridge, Mass.: MIT Press, 1965.

Clarke, I. F. *Voices Prophesying War, 1763–1984.* New York: Oxford University Press, 1966.

Clodfelter, Mark. *The Limits of Air Power: The American Bombing of North Vietnam.* New York: Free Press, 1989.

———. "Pinpointing Devastation: American Air Campaign Planning before Pearl Harbor." *Journal of Military History* 58 (1994): 75–101.

Cochran, Thomas B., et al. *U.S. Nuclear Warhead Production.* Nuclear Weapons Databook, vol. II. New York: Ballinger, 1987.

Cochrane, Dorothy, Von Hardesty, and Russell Lee. *The Aviation Careers of Igor Sikorsky.* Seattle: University of Washington Press, 1989.

Coe, Paul L. *Review of Drag Cleanup Tests in Langley Full-Scale Tunnel (from 1935 to 1945) Applicable to Current General Aviation Airplanes.* NASA Technical Note D-8206. Washington, D.C., 1976.

Cohen, Eliot A., and John Gooch. *Military Misfortunes: The Anatomy of Failure in War.* New York: Free Press, 1990.

Cooling, Benjamin Franklin, ed. *Case Studies in the Achievement of Air Superiority.* Washington, D.C.: Center for Air Force History, 1994.

Coox, Alvin D. "The Rise and Fall of the Imperial Japanese Air Forces." *Aerospace Historian* 27 (1980): 74–86.

Corum, James S. *The Luftwaffe: Creating the Operational Air War, 1918–1940.* Lawrence, Kans.: University Press of Kansas, 1997.

———. "The Spanish Civil War: Lessons Learned and Not Learned by the Great Powers." *Journal of Military History* 62 (1998): 313–34.

———. "Air Control: Reassessing the History." *Royal Air Force Air Power Review,* Summer 2001, 15–35.

Costello, Peter A. *A Matter of Trust: Close Air Support Apportionment and Allocation for Operational Level Effects.* Maxwell Air Force Base, Ala.: Air University Press, 1997.

Cox, Sebastian. "A Comparative Analysis of RAF and Luftwaffe Intelligence in the Battle of Britain, 1940." *Intelligence and National Security* 5(2): 425–43 (April 1990).

Crane, Conrad C. "Evolution of U.S. Strategic Bombing of Urban Areas." *The Historian* 50 (1987): 14–39.

Craven, Wesley Frank, and James Lea Cate, eds. *The Army Air Forces in World War II.* 6 vols. Chicago: University of Chicago Press, 1948–1955.

Crichton, Judy. *America 1900: The Turning Point.* New York: Henry Holt, 1998.

Crouch, Tom. *The Bishop's Boys: A Life of Wilbur and Orville Wright.* New York: Norton, 1989.

Cunningham, William Glenn. *The Aircraft Industry: A Study in Industrial Location.* Los Angeles: L. L. Morrison, 1951.

Cynk, Jerzy B. "The Truth About the Operational Doctrine of the Polish Air Force—A Rebuttal." *Aerospace Historian* 25 (1978): 176–82.

Davis, Richard G. *Decisive Force: Strategic Bombing in the Gulf War.* Washington, D.C.: Air Force History and Museums Program, 1996.

Deitchman, Seymour J. "The Future of Tactical Air Power in Land Warfare." *Astronautics and Aeronautics,* July–August 1980, 34–53.

———. *Military Power and the Advance of Technology.* Boulder, Colo.: Westview Press, 1983.

——. "Weapons, Platforms, and the New Armed Services." *Issues in Science and Technology,* Spring 1985, 83–99.

——. "Exploiting the Revolution in Conventional Weapons." *Aerospace America,* June 1987, 34–37.

DeWeerd, Harvey A. "Churchill, Coventry, and Ultra." *Aerospace Historian* 27 (1980): 227–29.

Drew, Dennis M. *Rolling Thunder 1965: Anatomy of a Failure.* Airpower Research Institute Report No. AU-ARI-CP-86-3. Maxwell Air Force Base, Ala.: Air University Press, 1986.

——. "Two Decades in the Air Power Wilderness: Do We Know Where We Are?" *Air University Review* 37(6): 2–13 (September–October 1986).

Dunbabin, J. P. D. "British Rearmament in the 1930s: A Chronology and Review." *Historical Journal* 18 (1975): 587–609.

Dye, Peter. "Logistics Doctrine and the Impact of War." Paper presented at symposium, Turning Points in Air Power History. Royal Air Force Museum, Hendon, U.K., July 10–11, 2001.

Erskine, Ralph. "Naval Enigma: A Missing Link." *International Journal of Intelligence and Counterintelligence* 3 (1989): 493–508.

Everett-Heath, John. *Helicopters in Combat: The First Fifty Years.* London: Arms and Armour, 1992.

Fadok, David S. *John Boyd and John Warden: Air Power's Quest for Strategic Paralysis.* Maxwell Air Force Base, Ala.: Air University Press, 1995.

Fairey, Richard. "Some Aspects of Expenditure on Aviation." *Journal of the Royal Aeronautical Society* 54 (1950): 405–32.

Fallows, James. *National Defense.* New York: Vintage, 1981.

Faulkenberry, Barbara J. *Global Reach—Global Power: Air Force Strategic Vision, Past and Future.* Maxwell Air Force Base, Ala.: Air University Press, 1996.

Ferris, John. "Fighter Defence before Fighter Command: The Rise of Strategic Air Defence in Great Britain, 1917–1934." *Journal of Military History* 63 (1999): 845–84.

Finney, Robert T. *History of the Air Corps Tactical School, 1920–1940.* U.S. Air Force Historical Studies No. 100. Maxwell Air Force Base, Ala.: Air University, 1955. Reprint. Washington, D.C., Center for Air Force History, 1992.

Flammer, Philip M. "Lufbery: Ace of the Lafayette Escadrille." *Air Power Historian* 8 (1961): 13–22.

Flintham, Victor. *Air Wars and Aircraft.* New York: Facts on File, 1990.

Frisbee, John L., ed. *Makers of the United States Air Force.* Washington, D.C.: Office of Air Force History, 1987.

Furnas, J. C. *The Americans: A Social History of the United States, 1857–1914.* New York: Putnam, 1969.

Fussell, Paul. *Wartime: Understanding and Behavior in the Second World War.* New York: Oxford University Press, 1989.

Futrell, Robert Frank. "The Influence of the Air Power Concept on Air Force Planning, 1945–1962." In *Military Planning in the Twentieth Century,* edited by Harry R. Borowski. Proceedings of the Eleventh Military History Symposium, U.S. Air Force Academy, October 10–12, 1984. Washington, D.C.: Office of Air Force History, 1986.

——. *Ideas, Concepts, Doctrine: Basic Thinking in the United States Air Force.* 2 vols. Maxwell Air Force Base, Ala.: Air University Press, 1989.

Gaston, James. *Planning the American Air War: Four Men and Nine Days in 1941,* Washington, D.C.: National Defense University Press, 1982.

Gibbs-Smith, Charles Harvard. *The Rebirth of European Aviation, 1902–1908: A Study of the Wright Brothers' Influence.* London: HMSO, 1974.

Gilbert, Martin. *The First World War.* New York: Henry Holt, 1994.

——. *Winston S. Churchill: Finest Hour, 1939–1941.* London: Heinemann, 1983.

Gilster, Herman L. *The Air War in Southeast Asia: Case Studies of Selected Campaigns.* Maxwell Air Force Base, Ala.: Air University Press, 1993.

——. "Desert Storm: War, Time, and Substitution Revisited." *Airpower Journal* 10(1): 1–10 (Spring 1996).

Giovannitti, Len, and Fred Freed. *The Decision to Drop the Bomb.* New York: Coward-McCann, 1965.

The Golden Interlude, 1900–1910. Alexandria, Va.: Time-Life Books, 1992.

Goldman, Eric F. *The Crucial Decade—and After: America, 1945–1960.* New York: Vintage, 1960.

Goldman, Martin R. R. *Morale in the AAF in World War II.* USAF Historical Studies No. 78. Maxwell AFB, Ala.: USAF Historical Division, Air University, 1953.

Gollin, Alfred M. "England Is No Longer an Island: The Phantom Airship Scare of 1909." *Albion* 13 (1981): 43–57.

———. *No Longer an Island: Britain and the Wright Brothers, 1902–1909.* Stanford, Calif.: Stanford University Press, 1984.

———. *The Impact of Air Power on the British People and Their Government, 1909 14.* Stanford, Calif.: Stanford University Press, 1989.

Gooderson, Ian. "Heavy and Medium Bombers: How Successful Were They in Tactical Close Air Support During World War II?" *Journal of Strategic Studies* 15 (1992): 367–99.

———. *Air Power at the Battlefront: Allied Close Air Support in Europe 1943–45.* London: Frank Cass, 1998.

Gordon, Michael R., and Bernard E. Trainor. *The Generals' War: The Inside Story of the Conflict in the Gulf.* Boston: Little, Brown, 1995.

Grant, Rebecca. "The Bekaa Valley War." *Air Force Magazine,* June 2002, 58–62.

———. "The Airpower of Anaconda." *Air Force Magazine,* September 2002, 60–68.

Grant, Robert, and Joseph Katz. *The Great Trials of the Twenties.* Rockville Centre, N.Y.: Sarpedon, 1998.

Gray, George W. *Frontiers of Flight: The Story of NACA Research.* New York: Knopf, 1948.

Gray, Peter W. "The Myths of Air Control and the Realities of Imperial Policing." *Royal Air Force Air Power Review* 4(2): 37–52 (Summer 2001).

Greenhous, Bereton. "Close Support Aircraft in World War I: The Counter Anti-Tank Role." *Aerospace Historian* 21 (1974): 87–93.

Greenwood, John T. "The Emergence of the Postwar Strategic Air Force, 1945–1953." In *Air Power and Warfare,* edited by Alfred F. Hurley and Robert C. Ehrhart. Proceedings of the Eighth Military History Symposium, United States Air Force Academy, 18–20 October 1978. Washington, D.C.: Office of Air Force History, 1979.

Grier, Peter. "Joint STARS Does Its Stuff." *Air Force Magazine,* June 1991, 38–42.

Griffin, David E. "The Role of the French Air Force: The Battle of France 1940." *Aerospace Historian* 21 (1974): 144–53.

Haight, John McVickar, Jr. "France's Search for American Military Aircraft: Before the Munich Crisis." *Aerospace Historian* 25 (1978): 141–52.

Halberstam, David. *The Best and the Brightest.* New York: Random House, 1972.

Hallion, Richard. *Legacy of Flight: The Guggenheim Contribution to American Aviation.* Seattle: University of Washington Press, 1977.

———. *Test Pilots: The Frontiersmen of Flight.* Revised ed. Washington, D.C.: Smithsonian Institution Press, 1988.

———. *Strike from the Sky: The History of Battlefield Air Attack, 1911–1945.* Washington, D.C.: Smithsonian Institution Press, 1989.

———. *Storm over Iraq: Air Power and the Gulf War.* Washington, D.C.: Smithsonian Institution Press, 1992.

———. "World War II as a Turning Point in Air Power Doctrine." Paper presented at symposium, Turning Points in Air Power History. Royal Air Force Museum, Hendon, U.K., July 10–11, 2001.

Hamilton, Robert J. "Green and Blue in the Wild Blue: An Examination of the Evolution of Army and Air Force Airpower Thinking and Doctrine since the Vietnam War." Thesis, School of Advanced Airpower Studies, Air University, Maxwell Air Force Base, Ala., 1993.

Hammond, Grant T. *The Mind of War: John Boyd and American Security.* Washington, D.C.: Smithsonian Institution Press, 2001.

Hanle, Paul A. *Bringing Aerodynamics to America.* Cambridge, Mass.: MIT Press, 1982.

Hansen, James R. *Engineer in Charge.* NASA History SP-4305. Washington, D.C., 1987.

Harvey, A. D. "The French Armée de l'Air in May–June 1940: A Failure of Conception." *Journal of Contemporary History* 25 (1990): 447–65.

Haslam, E. B. "How Lord Dowding Came to Leave Fighter Command." *Journal of Strategic Studies* 4 (1981): 175–86.

Heyman, Neil M. "NEP and Industrialization to 1928." In *Soviet Aviation and Air Power: A Historical View,* edited by Robin H. Higham and Jacob Kipp. Boulder, Colo.: Westview Press, 1978.

Hinman, Ellwood P. *The Politics of Coercion: Toward a Theory of Post–Cold War Conflict.* CADRE Paper No. 14. Maxwell Air Force Base, Ala.: Air University Press, 2002.

Hinsley, F. H., et al. *British Intelligence in the Second World War.* 5 vols. London: HMSO, 1979–90.

Holley, I. B. *Ideas and Weapons.* New Haven, Conn.: Yale University Press, 1953. Reprint. Washington, D.C.: GPO, 1983.

———. "Jet Lag in the Army Air Corps." In *Military Planning in the Twentieth Century,* edited by Harry R. Borowski. Proceedings of the Eleventh Military History Symposium, U.S. Air Force Academy, October 10–12, 1984. Washington, D.C.: Office of Air Force History, 1986.

Hone, Thomas C. "Navy Air Leadership: Rear Admiral William A. Moffett as Chief of the Bureau of Aeronautics." In *Air Leadership: Proceedings of a Conference at Bolling Air Force Base, April 13–14, 1984,* edited by Wayne Thompson. Washington, D.C.: Office of Air Force History, 1986.

Hone, Thomas C., and Mark D. Mandeles. "Interwar Innovation in Three Navies: U.S. Navy, Royal Navy, Imperial Japanese Navy." *Naval War College Review* (Spring 1987): 63–83.

Hone, Thomas C., Norman Friedman, and Mark D. Mandeles. *American and British Aircraft Carrier Development, 1919–1941.* Annapolis, Md.: Naval Institute Press, 1999.

Hooton, E. R. *Phoenix Triumphant.* London: Arms and Armour, 1994.

Hopkins, George E. "Bombing and the American Conscience During World War II." *The Historian* 28 (1966): 451–73.

Horne, Alistair. *To Lose a Battle; France 1940.* Boston: Little, Brown, 1969.

Hosmer, Stephen T. *Psychological Effects of U.S. Air Operations in Four Wars, 1941–1991: Lessons for U.S. Commanders.* Santa Monica, Calif.: RAND, 1996.

Hough, Richard, and Denis Richards. *The Battle of Britain.* New York: Norton, 1989.

Howson, Gerald. *Aircraft of the Spanish Civil War 1936–39.* Washington, D.C.: Smithsonian Institution Press, 1990.

Hudson, James J. *Hostile Skies: A Combat History of the American Air Service in World War I.* Syracuse, N.Y.: Syracuse University Press, 1968.

Hughes, Thomas Alexander. *Overlord: General Pete Quesada and the Triumph of Tactical Air Power in World War II.* New York: Free Press, 1995.

Hurley, Alfred F. *Billy Mitchell: Crusader for Air Power.* New York: Franklin Watts, 1964.

Hurley, Alfred F., and Robert C. Ehrhart, eds. *Air Power and Warfare.* Proceedings of the Eighth Military History Symposium, United States Air Force Academy, 18–20 October 1978. Washington, D.C.: Office of Air Force History, 1979.

Huston, James A. "Tactical Use of Air Power in World War II: The Army Experience." *Military Affairs* 14 (1950): 166–85.

Ignatieff, Michael. "Nation-Building Lite." *New York Times Magazine,* July 28, 2002.

Isom, Daniel Woodury. "The Battle of Midway: Why the Japanese Lost." *Naval War College Review,* Summer 2000, 60–100.

Jacobs, William A. "Tactical Air Doctrine and AAF Close Air Support in the European Theater, 1944–1945." *Aerospace Historian* 27 (1980): 35–49.

Jakab, Peter L. *Visions of a Flying Machine: The Wright Brothers and the Process of Invention.* Washington, D.C.: Smithsonian Institution Press, 1990.

———. "Wood to Metal: The Structural Origins of the Modern Airplane." *Journal of Aircraft* 36 (1999): 914–18.

Jarrett, Philip, ed. *Biplane to Monoplane: Aircraft Development 1919–39*. London: Putnam Aeronautical Books, 1997.

Jones, Howard Mumford. *The Age of Energy*. New York: Viking, 1971.

Jones, Neville. *The Origins of Strategic Bombing: A Study of the Development of British Air Strategic Thought and Practice up to 1918*. London: William Kimber, 1973.

———. *The Beginnings of Strategic Air Power: A History of the British Bomber Force, 1923–1939*. London: Frank Cass, 1987.

Kahn, David. "The United States Views Germany and Japan in 1941." In *Knowing One's Enemies: Intelligence Assessment Before the Two World Wars*, edited by Ernst R. May. Princeton, N.J.: Princeton University Press, 1984.

Kaplan, Fred. *The Wizards of Armageddon*. New York: Simon & Schuster, 1983.

Karnow, Stanley. *Vietnam: A History*. New York: Viking, 1983.

Keaney, Thomas A., and Eliot A. Cohen. *Revolution in Warfare? Air Power in the Persian Gulf*. Annapolis, Md.: Naval Institute Press, 1995.

Kelly, Fred C. *The Wright Brothers*. New York: Harcourt, Brace, 1943.

Kennett, Lee. *A History of Strategic Bombing*. New York: Charles Scribner's Sons, 1982.

———. *The First Air War, 1914–1918*. New York: Free Press, 1991.

Kirkland, Faris R. "The French Air Force in 1940: Was It Defeated by the Luftwaffe or Politics?" *Air University Review* 36(6): 101–18 (September–October 1985).

Knappen, Theodore M. *Wings of War: An Account of the Important Contribution of the United States to Aircraft Invention, Engineering, Development and Production during the World War*. New York: Putnam, 1920.

Konvitz, Josef W. "Why Cities Don't Die." *American Heritage of Invention and Technology*, Winter 1990, 58–63.

———. "Bombs, Cities, and Submarines: Allied Bombing of the French Ports, 1942–43." *International History Review* 14 (1992): 23–44.

Kosaka, Masataka. *100 Million Japanese: The Postwar Experience*. Tokyo: Kodansha International, 1972.

Lambeth, Benjamin S. *Moscow's Lessons from the 1982 Lebanon Air War*. Santa Monica, Calif.: RAND, 1984.

———. *The Transformation of American Air Power*. Ithaca, N. Y.: Cornell University Press, 2000.

Larrazábal, Jesús Salas. *Air War over Spain*. Translated by Margaret A. Kelley. London: Ian Allan, 1974.

Layman, R. D. "The Day the Admirals Wept: *Ostfriesland* and the Anatomy of a Myth." In *Warship 1995*, edited by John Roberts. London: Conway Maritime Press, 1995.

Lee, John G. *Fighter Facts and Fallacies*. New York: William Morrow, 1942.

Liebling, A. J. *The Earl of Louisiana*. New York: Simon & Schuster, 1961.

Loeb, Vernon. "Bursts of Brilliance." *Washington Post Magazine*, December 15, 2002.

Lowry, Thomas P., and John W. G. Wellham. *The Attack on Taranto: Blueprint for Pearl Harbor*. Mechanicsburg, Pa.: Stackpole Books, 1995.

McCue, Brian. *U-Boats in the Bay of Biscay: An Essay in Operations Analysis*. Washington, D.C.: National Defense University Press, 1990.

McCullough, David. *Truman*. New York: Simon & Schuster, 1992.

McElvaine, Robert S. *The Great Depression*. New York: Times Books, 1984.

McFarland, Stephen L. *America's Pursuit of Precision Bombing, 1910–1945*. Washington, D.C.: Smithsonian Institution Press, 1995.

McFarland, Stephen L., and Wesley Phillips Newton. *To Command the Sky: The Battle for Air Supremacy over Germany, 1942–1944*. Washington, D.C.: Smithsonian Institution Press, 1991.

Mackenzie, Richard. "A Conversation with Chuck Horner." *Air Force Magazine*, June 1991, 57–64.

Macmillan, Norman. "More about the Spin." *Aeronautics* 43(2): 26–28 (December 1960).

Manchester, William. *The Glory and the Dream*. New York: Bantam, 1975.

———. *The Last Lion: Winston Spencer Churchill*. 2 vols. Boston: Little, Brown, 1983–1988.

Mark, Eduard. *Aerial Interdiction in Three Wars.* Washington, D.C.: Center for Air Force History, 1994.

Martin, George C. "Boeing Aircraft Designs—1928–1953." American Institute for Aeronautics and Astronautics, Paper No. 78–3004.

Maycock, Thomas J. "Notes on the Development of AAF Tactical Air Doctrine." *Military Affairs* 14 (1950): 186–91.

Meilinger, Phillip S. "Towards a New Airpower Lexicon: Or, Interdiction: An Idea Whose Time Has Finally Gone?" *Airpower Journal* 7(2): 39–47 (Summer 1993).

Melhorn, Charles M. *Two Block Fox: The Rise of the American Aircraft Carrier, 1911–1929.* Annapolis, Md.: Naval Institute Press, 1974.

Mets, David R. *Nonnuclear Aircraft Armament. The Quest for a Surgical Strike: The United States Air Force and Laser Guided Bombs.* Eglin Air Force Base, Fla.: Office of History, Armament Division, Air Force Systems Command, 1987.

Miller, Nathan. *War at Sea: A Naval History of World War II.* New York: Oxford University Press, 1996.

Moody, Walton S. *Building a Strategic Air Force.* Washington, D.C.: Air Force History and Museums Program, 1996.

Morrow, John H., Jr. *The Great War in the Air: Military Aviation from 1909 to 1921.* Washington, D.C.: Smithsonian Institution Press, 1993.

Mrozek, Donald J. *Air Power and the Ground War in Vietnam.* Washington, D.C.: Pergamon-Brassey's, 1989.

Muravchik, Joshua. "End of the Vietnam Paradigm?" *Commentary,* May 1991, 17–23.

Murray, Williamson. "Attrition and the Luftwaffe." *Air University Review* 34(3): 66–77 (March–April 1983).

———. *Luftwaffe.* Baltimore, Md.: Nautical & Aviation Publishing, 1985.

———. "The Luftwaffe against Poland and the West." In *Case Studies in the Achievement of Air Superiority,* edited by Benjamin Franklin Cooling. Washington, D.C.: Center for Air Force History, 1994

Nalty, Bernard C., ed. *Winged Shield, Winged Sword: A History of the United States Air Force.* 2 vols. Washington, D.C.: Air Force History and Museums Program, 1997.

Nordeen, Lon. *Fighters over Israel.* New York: Orion Books, 1990.

———. *Air Warfare in the Missile Age.* 2nd edition. Washington, D.C.: Smithsonian Institution Press, 2002.

Ohmae, Toshikazu, and Roger Pineau. "Japanese Naval Aviation." *Proceedings of the United States Naval Institute,* December 1972, 68–77.

Omissi, David E. *Air Power and Colonial Control: The Royal Air Force 1919–1939.* Manchester, U.K.: Manchester University Press, 1990.

Orange, Vincent. *Coningham: A Biography of Air Marshal Sir Arthur Coningham.* Washington, D.C.: Center for Air Force History, 1992.

Oren, Michael B. *Six Days of War.* New York: Oxford University Press, 2002.

Orwoyski, J. S. "Polish Air Force versus Luftwaffe." Parts 1 and 2. *Air Pictorial.* October, November 1959.

Overy, R. J. "Hitler and Air Strategy." *Journal of Contemporary History* 15 (1980): 405–21.

———. *The Air War 1939–1945.* New York: Stein and Day, 1981.

———. *The Battle of Britain: The Myth and the Reality.* New York: Norton, 2001.

Pape, Robert. *Bombing to Win: Air Power and Coercion in War.* Ithaca, N.Y.: Cornell University Press, 1996.

Parkinson, Russell J. "Aeronautics at The Hague Conference of 1899." *Air Power Historian* 7 (1960): 106–11.

Perret, Geoffrey. *Winged Victory: The Army Air Forces in World War II.* New York: Random House, 1993.

Peszke, Michael Alfred. "The Forgotten Campaign: Poland's Military Aviation in September, 1939." *The Polish Review* [New York] 39 (1994): 51–72.

Prange, Gordon. *Miracle at Midway.* New York: McGraw Hill, 1981.

———. *Pearl Harbor: The Verdict of History.* New York: McGraw-Hill, 1986.

Price, Alfred. *Instruments of Darkness: The History of Electronic Warfare.* Reprint. Los Altos, Calif.: Peninsula, 1987.

———. *Sky Warriors.* 1994. Reprint. London: Cassell Military Classic, 1998.

———. "The Messerschmitt 262 Jet Fighter: Missed Opportunity or Impossible Dream?" *Royal Air Force Air Power Review,* Summer 2001, 52–66.

Proctor, Raymond L. *Hitler's Luftwaffe in the Spanish Civil War.* Westport, Conn.: Greenwood Press, 1983.

Quill, Jeffrey. *Birth of a Legend: The Spitfire.* Washington, D.C.: Smithsonian Institution Press, 1986.

Quester, George H. "The Impact of Strategic Warfare." *Armed Forces and Society* 4 (1978): 179–206.

Raleigh, Walter, and H. A. Jones. *The War in the Air.* 6 vols. Oxford: Clarendon Press, 1922–1927.

Ransom, Harry H. "The Battleship Meets the Airplane." *Military Affairs* 23 (Spring 1959): 21–27.

Ray, John. *The Battle of Britain: Dowding and the First Victory, 1940.* London: Cassell, 2000.

———. *The Night Blitz, 1940–1941.* London: Cassell, 2000.

Reynolds, Clark G. *The Fast Carriers: The Forging of an Air Navy.* New York: McGraw-Hill, 1968.

Reynolds, Richard T. *Heart of the Storm: The Genesis of the Air Campaign against Iraq.* Maxwell Air Force Base, Ala.: Air University Press, 1995.

Rhodes, Richard. *The Making of the Atomic Bomb.* New York: Touchstone, 1988.

Richardson, Dan. "The Development of Airpower Concepts and Air Combat Techniques in the Spanish Civil War." *Air Power History,* Spring 1993, 13–21.

Ries, Karl, and Hans Ring. *The Legion Condor: A History of the Luftwaffe in the Spanish Civil War, 1936–39.* Translated by David Johnston. West Chester, Pa.: Schiffer Military History, 1992.

Robertson, Scot. "The Development of Royal Air Force Strategic Bombing Doctrine between the Wars: A Revolution in Military Affairs?" *Airpower Journal* 12(1): 37–52 (Spring 1998).

Robinson, Douglas H. "The Zeppelin Bomber: High Policy Guided by Wishful Thinking." *Air Power Historian* 8 (1961): 130–47.

———. *The Zeppelin in Combat.* London: Foulis, 1962.

Rosenau, William. *Special Operations Forces and Elusive Enemy Ground Targets: Lessons from Vietnam and the Persian Gulf War.* Santa Monica, Calif.: RAND, 2001.

Rostow, Walt W. *The United States in the World Arena: An Essay in Recent History.* New York: Harper, 1960.

Rydell, Robert W. *All the World's a Fair: Visions of Empire at American International Expositions, 1876–1916.* Chicago: University of Chicago Press, 1984.

Sbrega, John J. "Southeast Asia." In *Case Studies in the Development of Close Air Support,* edited by Benjamin Franklin Cooling. Washington, D.C.: Center for Air Force History, 1990.

Schaffer, Ronald. "American Military Ethics in World War II: The Bombing of German Civilians." *Journal of American History* 67 (1980): 318–34.

Schlaifer, Robert, and S. D. Heron. *Development of Aircraft Engines and Fuels.* Boston: Division of Research, Graduate School of Business Administration, Harvard University, 1950.

Schlereth, Thomas J. *Victorian America: Transformations in Everyday Life, 1876–1915.* New York: HarperCollins, 1991.

Sheehan, Neil. *A Bright Shining Lie: John Paul Vann and America in Vietnam.* New York: Random House, 1988.

Shirer, William L. *The Rise and Fall of the Third Reich.* New York: Simon & Schuster, 1960.

Smith, Malcolm S. *British Air Strategy between the Wars.* Oxford: Clarendon Press, 1984.

Smith, Richard K. "Rebel of '33." *Shipmate,* March 1977, 31–34.

Spector, Ronald. *Eagle against the Sun: The American War with Japan.* New York: Free Press, 1985.

"A Statistical Portrait of USAAF in World War II." *Air Force Magazine,* June 1995, 30–36.

Steinisch, Irmgard. "A Comparative Study of Militarism and Imperialism in the United States and

Imperial Germany, 1871–1914." In *Anticipating Total War: The German and American Experiences, 1871–1914,* edited by Manfred F. Boemeke, Roger Chickering, and Stig Förster. Cambridge: Cambridge University Press, 1999.

Stern, Robert C. *The Lexington Class Carriers.* Annapolis, Md.: Naval Institute Press, 1993.

Stokesbury, James L. *A Short History of World War II.* New York: Morrow, 1980.

Sullivan, Mark. *Our Times: America at the Birth of the Twentieth Century.* Edited with new material by Dan Rather. New York: Scribner, 1996.

Syrett, David. "The Tunisian Campaign." In *Case Studies in the Development of Close Air Support,* edited by Benjamin Franklin Cooling. Washington, D.C.: Center for Air Force History, 1990.

Thomas, Gordon, and Max Morgan Witts. *Guernica, the Crucible of World War II.* New York: Stein and Day, 1975.

Thompson, Wayne. *To Hanoi and Back: The USAF and North Vietnam, 1966–1973.* Washington, D.C.: Air Force History and Museums Program, 2000.

Tice, Brian P. "Unmanned Aerial Vehicles: The Force Multiplier of the 1990s." *Airpower Journal* 5(1): 41–55 (Spring 1991).

Tierney, John. "Take the A-Plane: The $1,000,000,000 Nuclear Bird that Never Flew." *Science 82,* January–February 1982, 46–55.

Till, Geoffrey. "Airpower and the British Admiralty between the World Wars." In *Changing Interpretations and New Sources in Naval History,* edited by Robert William Love, Jr. Papers from the Third United States Naval Academy History Symposium. New York: Garland Publishing, 1980.

Towle, Philip Anthony. *Pilots and Rebels: The Use of Aircraft in Unconventional Warfare 1918–1988.* London: Brassey's, 1989.

Turnbull, Archibald D., and Clifford L. Lord. *History of United States Naval Aviation.* New Haven, Conn.: Yale University Press, 1949.

Turnill, Reginald, and Arthur Reed. *Farnborough: The Story of RAE.* London: Hale, 1980.

U.S. Air Force. Headquarters. *The Air War over Serbia: Aerospace Power in Operation Allied Force.* Washington, D.C., 2000.

———. Historical Division. *The Employment of Strategic Bombers in a Tactical Role.* USAF Historical Studies, No. 88. Maxwell Air Force Base, Ala., 1954.

———. *The Development of Air Doctrine in the Army Air Arm, 1917–1941.* USAF Historical Studies, No. 89. Maxwell Air Force Base, Ala., 1955.

U.S. Army Air Forces. *The War against the Luftwaffe: AAF Counter-Air Operations, April 1943–June 1944.* AAF Historical Study No. 110. 1945.

U.S. General Accounting Office. *Operation Desert Storm: Evaluation of the Air Campaign.* Washington, D.C., 1997.

Van der Vat, Dan. *The Atlantic Campaign.* London: Grafton, 1988.

Wagner, Ray, ed. *The Soviet Air Force in World War II.* Translated by Leland Fetzer. Garden City, N.Y.: Doubleday, 1973.

Watt, Donald Cameron. "British Intelligence and the Coming of the Second World War in Europe." In *Knowing One's Enemies: Intelligence Assessment Before the Two World Wars,* edited by Ernst R. May. Princeton, N.J.: Princeton University Press, 1984.

Webster, Charles, and Noble Frankland. *The Strategic Air Offensive against Germany, 1939–45.* 4 vols. London: HMSO, 1961.

Weinert, Richard P., Jr. *A History of Army Aviation, 1950–1962.* TRADOC Historical Monograph Series. Fort Monroe, Va.: U.S. Army Training and Doctrine Command, 1991.

Werrell, Kenneth P. "The Strategic Bombing of Germany in World War II: Costs and Accomplishments." *Journal of American History* 72 (1986): 702–13.

———. *Blankets of Fire: U.S. Bombers over Japan during World War II.* Washington, D.C.: Smithsonian Institution Press, 1996.

———. "Did USAF Technology Fail in Vietnam?" *Airpower Journal* 12(1): 187–99 (Spring 1998).

West, Scott D. *Warden and the Air Corps Tactical School: Déjà vu?* Maxwell Air Force Base, Ala.: Air University Press, 1999.

Westermann, Edward B. *Flak: German Anti-Aircraft Defenses, 1914–1945*. Lawrence, Kans.: University Press of Kansas, 2001.

Whitford, Ray. *Fundamentals of Fighter Design*. Shrewsbury, U.K.: Airlife Publishing, 2000.

Whiting, Kenneth R. "Soviet Aviation and Air Power under Stalin, 1928–1941." In *Soviet Aviation and Air Power: A Historical View*, edited by Robin H. Higham and Jacob Kipp. Boulder, Colo.: Westview Press, 1978.

Wildenberg, Thomas. *Destined for Glory: Dive Bombing, Midway, and the Evolution of Carrier Airpower*. Annapolis, Md.: Naval Institute Press, 1998.

Winnefeld, James A., Preston Niblack, and Dana J. Johnson. *A League of Airmen: U.S. Air Power in the Gulf War*. Santa Monica, Calif.: RAND, 1994.

Winter, Jay, and Blaine Baggett. *The Great War and the Shaping of the Twentieth Century*. New York: Penguin, 1996.

Wolk, Herman S. *The Struggle for Air Force Independence 1943 to 1947*. Washington, D.C.: Air Force History and Museums Program, 1997.

Worden, Mike. *Rise of the Fighter Generals: The Problem of Air Force Leadership 1945–1982*. Maxwell Air Force Base, Ala.: Air University Press, 1998.

Wyden, Peter. *The Passionate War: The Narrative History of the Spanish Civil War, 1936–39*. New York: Simon & Schuster, 1983.

Young, Robert J. "The Strategic Dream: French Air Doctrine in the Inter-War Period, 1919–39." *Journal of Contemporary History* 9(4): 57–76 (October 1974).

Zimmerman, David. "British Radar Organization and the Failure to Stop the Night-time Blitz." *Journal of Strategic Studies* 21(3): 86–106 (September 1998).

Zimmerman, Gene T. "More Fiction than Fact—The Sinking of the *Ostfriesland*." *Warship International* 12 (1975): 142–54.

IND£X